personal finance

Elizabeth B. Goldsmith
Florida State University

Australia • Canada • Mexico • Singapore • Spain • United Kingdom • United States

Publisher: Peter Marshall
Development Editor: Laura Graham
Editorial Assistant: Keynia Johnson
Marketing Manager: Becky Tollerson
Project Editor: Sandra Craig
Print Buyer: Barbara Britton
Permissions Editor: Joohee Lee
Production, Text and Cover Design,
Illustration, Composition: Delgado Design, Inc.
Art Editor: Ed Smith
Photo Researcher: Mark Bogorad
Copy Editor: Chuck Cox
Cover Illustrator: Craig Smallish
Cover Printer: Phoenix Color
Printer: Courier/Kendallville

Printed in the United States of America
1 2 3 4 5 6 7 04 03 02 01 00

For permission to use material from this text, contact us by
Web: http://www.thomsonrights.com
Fax: 1-800-730-2215
Phone: 1-800-730-2214

Library of Congress Cataloging-in-Publication Data

Goldsmith, Elizabeth B.
Personal Finance / Elizabeth B. Goldsmith
p. cm.
Includes bibliographical references and index.
ISBN: 0-534-54495-9 (alk. paper)
1. Finance, Personal. 2. Investments

HG179 .G663 2000
333.024—dc21 00-033364

Wadsworth/Thomson Learning
10 Davis Drive
Belmont, CA 94002-3098
USA

For more information about our products, contact us:
Thomson Learning Academic Resource Center
1-800-423-0563
http://www.wadsworth.com

International Headquarters
Thomson Learning
International Division
290 Harbor Drive, 2nd Floor
Stamford, CT 06902-7477
USA

UK/Europe/Middle East/South Africa
Thomson Learning
Berkshire House
168-173 High Holborn
London WC1V 7AA
United Kingdom

Asia
Thomson Learning
60 Albert Street, #15-01
Albert Complex
Singapore 189969

Canada
Nelson Thomson Learning
1120 Birchmount Road
Toronto, Ontario M1K 5G4
Canada

Brief Contents

Contents

CHAPTER 8 PROPERTY, LIABILITY, AND AUTOMOBILE INSURANCE 181

CHAPTER 9 HEALTH CARE, DISABILITY, AND LONG-TERM CARE 207

CHAPTER 10 LIFE INSURANCE 227

Preface

The decisions individuals make about how to manage their finances directly influence the degree of satisfaction and success they can ultimately achieve. *Personal Finance* is about how people choose to spend, save, invest, and protect their assets. Personal finance touches every aspect of our lives, including where we go to school, what kind of housing we choose and where we live, where we work and how we get there. The goal of this book is to help readers attain the knowledge and the skills they need to make sound decisions, and to successfully navigate the financial hurdles, pitfalls, and the windfalls they may encounter along the way, in order to make the best possible use of their resources.

Part I of the text looks at the basics of planning and managing finances. Chapters 1–5 cover career planing, record keeping, taxes, and the management of cash, savings, credit and loans. Part II covers major purchasing decisions, including the purchase of housing and transportation in Chapter 6, and the costs of caring for others, such as dependent children or aging parents in Chapter 7—coverage that is unique to this text. Chapter 8, 9, and 10 explore property, health, and life insurance purchasing decisions. Part III begins with the fundamentals of investing in Chapter 11, and goes on to explore stocks and bonds in Chapters 12 and 13. Chapter 14 investigates other investment options, such as mutual funds and real estate. The last section, Part IV, looks at planning for the future, including retirement and estate planning in Chapters 15 and 16.

Throughout the text, real-life examples covering many different lifestyles; family situations; and ethnic, religious, and socioeconomic groups are used to illustrate the concepts presented. In addition, each chapter includes relevant websites and helpful tips for using electronic resources.

This text is designed to be practical, inspiring, and, above all, user friendly. It looks not just at money, but at the role that finances play in creating a fulfilling life. The information presented here provides a solid foundation in the basics of personal financial management—a foundation on which students will continue to build throughout their lives.

Distinguishing Features

A number of chapter features appear throughout the text to enhance the learning environment for student:

- **Did You Know** features on the opening page of every chapter highlight an interesting fact from the chapter to generate interest in the topic to be covered.
- **Chapter Objectives** outline the key points to be discussed, and appear in the margins in the body of the chapter where the material related to the objectives can be found.
- **Chapter Overviews** at the beginning of every chapter provide an introduction to the subject and preview the organization of the material to follow.

In addition, the following features complement each chapter of the text:

* **Margin Definitions** appear in the places where key words and concepts are first introduced. The corresponding terms are boldfaced where they appear in the text. Terms are also compiled in a glossary at the end of the book, for easy reference.
* **Consumer Alert Boxes** within the chapters provide useful cautionary information about potential financial pitfalls or misunderstandings.
* **FAQs: Frequently Asked Questions** near the end of every chapter provide, an approachable question-and-answer format, tips and hints that students can apply to their own financial management strategies.
* **Electronic Resources** sections in every chapter provide addresses for and descriptions of key websites where related information may be found and additional topics may be explored.
* **Get Online Boxes** feature Internet activities designed to aid students in learning to use Internet tools discriminatingly.

Each chapter concludes with a number of pedagogical features designed to aid students in reviewing and reinforcing the concepts presented:

* **Chapter Summaries** recap the main points covered in the chapter.
* **Review Questions** provide a list of ten true/false and multiple -choice questions to test students' knowledge and comprehension of the chapter. Answers to each question are provided in the Appendix.
* **Discussion Questions** are rigorous critical-thinking questions designed to provide a more in-depth review of the content of the chapter.
* **Decision-Making Cases** present scenarios of specific individuals and families involved in making financial management decisions, and encourage students to apply the chapter concepts to real-life situations. These cases may be used for class discussions or written assignments.
* **InfoTrac Exercises** encourage students to make use of the InfoTrac College Edition, provided free with every new edition of this text, to explore topics related to the chapter content, and to stay abreast of current trends and research in the area of personal finance.

Supplements

The following supplementary teaching materials are available to accompany this text:

* **Instructor's Resource Manual**—This resource includes Lecture Outlines and Lecture Enrichment tips for each chapter, and an extensive Test Bank to enhance instructors' teaching of the material.
* **Examview Computerized Testing**—This electronic version of the print Test Bank, compatible with both Windows and Macintosh platforms, allows instructors to create, edit, store, and print exams from the questions provided.
* **PowerPoint Presentation**—Designed to enhance classroom lectures, this collection of slides contains key figures, tables, and illustrations from the text.
* **Spreadsheet Workbook**—This provides worksheet templates for students to use to practice the calculations presented in the text. The workbook is three-hole punched and perforated for the students' and instructors' convience.
* **Internet Trifold**—Free with every new copy of the text, this supplement provides

lists of important financial websites, organized by category, on a standing three-paneled card for easy reference.

* **InfoTrac College Edition**—Free with every new copy of the text, this supplement provides access to an online database of full-length articles from more than 900 scholarly and popular periodicals. Exclusive to Wadsworth and other ITP Higher Education companies.
* **The Wadsworth Human Ecology Resource Center Web Site**—This powerful online resource provides updated links to a number of Internet sites, including those listed in the text, as well as a number of quality assessment tools that students who buy this book can access for free. Also includes updates and features for both students and instructors at humanecology.wadsworth.com.

Acknowledgments

I would like to recognize the following members of the Wadsworth team for their contributions to this book: Peter Marshall, Publisher; Laura Graham, Associate Developmental Editor; Becky Tollerson, Marketing Manager; and Sandra Craig, Project Editor. Thanks also to Ed Smith of Delgado Design, Inc., for his contributions to the later stages of the manuscript.

Professionals in personal finance (bankers, stockbrokers, insurance specialists, accountants, career counselors, and credit union managers) reviewed early drafts of this book to ensure that the content was up-to-date and applicable to the real world. My particular thanks in this regard go to Marcus Beck, Dan Clark, Jeff Clark, Donna Huston, Ted Hunt, Kevin Eastman, and Janet Lenz. Jeanne M. Hogarth of the Federal Reserve Board is also acknowledged for her help with the time value of mony discussion in Chapter 1.

Thanks also are extended to my sons, David and Andrew, and my husband, Ron, for their patience and support.

The instructors who served as reviewers of this text were absolutely crucial to the successful completion of the project. Special thanks go to all of the following:

M. J. Alhabeeb, University of Massachusetts–Amherst

Anne Bailey, Miami University

Cheryl Buehler, University of Tennessee–Knoxville

Sugato Chakravarty, Purdue University

Sharon DeVaney, Purdue University

Jonathan Fox, Ohio State University

Vickie Hampton, University of Texas–Austin

Charles Hatcher, University of Georgia

Deborah Haynes, Montana State University

Sandra Houston, University of Missouri–Columbia

Wendall Hull, University of Alabama

Craig Israelson, University of Missouri–Columbia

Dan Klein, Bowling Green State University

Carole Makela, Colorado State University

Allen Martin, California State University–Northridge
Jerry Mason, Texas Tech University
Nancy Porter, Clemson University
Elizabeth Shields, California State Unversity–Fresno
Cathy Solheim, Auburn University
Barbara Stewart, University of Houston
Judy VanName, University of Delaware
David Yarborough, University of Southwestern Louisiana
Virginia Zuiker, University of Minnesota

Elizabeth B. Goldsmith

About the Author

Elizabeth B. Goldsmith is Professor of Resource Management and Consumer Economics at Florida State University, where she has received the University Teaching Award and the Teaching Incentive Award for classroom teaching excellence. She received her Ph.D. from Michigan State University, from which she was recently named Outstanding Alumna. After completing her doctorate, Dr. Goldsmith taught at New Mexico State University and the University of Alabama, before joining Florida State.

Dr. Goldsmith has published journal articles, encyclopedia chapters, and a textbook, Resource Management for Individuals and Families, now in its second edition, also published by Wadsworth. She served as associate editor of the Journal of Family and Consumer Sciences, and as an advisor to The Wall Street Journal, the University of the West Indies, the United Nations, the National Park Service, and the White House. Her finance research focuses on gender, money and e-commerce. She has presented papers in Australia, Finland, Germany, Great Britain, Malta, Mexico, Northern Ireland, and Sweden.

part one

planning and managing finances

chapter one

personal finance and career planning

DEVELOPING PLANS
SETTING GOALS
TAKING CHARGE

DID YOU KNOW?

– A college graduate can expect to earn twice as much as a high school graduate over a lifetime.

– Half of job seekers find jobs through personal referral.

OBJECTIVES

After reading Chapter 1 you should be able to do the following:

1. Describe the three steps in the financial planning process.
2. Identify your financial goals.
3. Identify the four key players in the economy.
4. Describe the economic cycle and how it affects personal finance.
5. Understand the time value of money.
6. Recognize the importance of career planning, salary, and benefits to overall financial well-being.

CHAPTER OVERVIEW

The purpose of this first chapter is to introduce the topic of personal finance and show how it fits into your life and career plans. **Personal finance** is how people spend, save, invest, and protect their financial resources. **Resources** (time, energy, and money) are what individuals use to get what they want and to reach goals. The primary goal of the study of personal finance is to make individuals more knowledgeable about their finances in order to increase their life satisfaction.

Personal finance
The study of how people spend, save, invest, and protect their financial resources.

Resources
What individuals use to get what they want and to reach goals.

The second half of the chapter provides information on job benefits and retirement plans so you can weigh choices when comparing offers. Since, as both a consumer and an investor you operate within the global marketplace, this chapter also discusses factors operating in the overall economy. It also covers the three steps in the financial management process, the time value of money, and the effect that education has on future earnings. The chapter ends with a section on Electronic Resources with websites leading to more information.

IMPORTANCE OF STUDYING PERSONAL FINANCE

Personal finance is important to study because:

- we live in an increasingly complex world with more options than ever before. Developing strategies to cope with these options is critical to one's sense of well-being and security.
- of the slow growth in personal income (wages and salaries). In recent years people have sought other ways (saving, investing) to make their money increase.
- of changes in the labor market and the stock market. Job security is not what it once was and a rapidly rising stock market in the 1990s caught the attention of potential investors. Now over half of American households own mutual funds.
- it is a fact of life. On the day that you arrive your parents apply for a Social Security number for you. Hospitals include a Social Security application as part of their birth registration procedures. Once you have a number, bank and investment accounts can be opened in your name, and your parents can claim you as a dependent on their taxes. So it can be said that personal finance starts on day one.

Objective 1:
Describe the three steps in the financial planning process.

Although personal finance is a fundamental part of life, it is not easy to master. Many people are less than successful at managing money as evidenced by the rising amount of debt and personal bankruptcies. Fewer people seem to be able to deal with the present, let alone plan for the future. Understanding the planning process is key to moving in the right direction.

Personal Financial Planning
The process of managing finances to reach goals and to provide satisfaction.

Personal financial planning is the *process* of managing your finances to reach goals and to increase personal satisfaction. The word "process" implies a series of steps or stages. The three steps in the personal financial planning process are setting financial

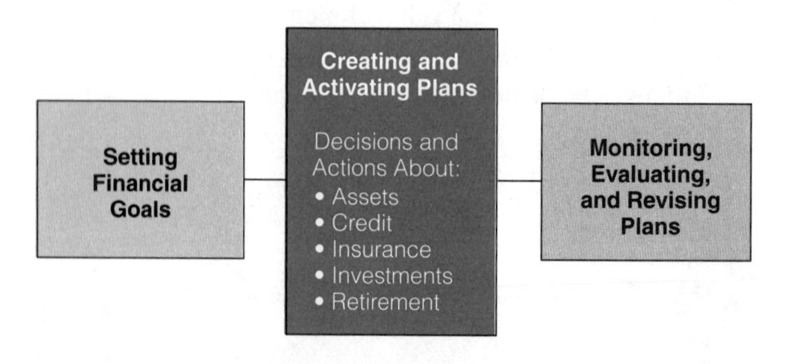

EXHIBIT 1.1 Personal Financial Planning Process

goals; creating and activating plans; and monitoring, evaluating, and revising plans, as shown in Exhibit 1.1.

Objective 2:
Identify your financial goals.

Values
Principles that guide behavior.

Attitudes
Likes and dislikes.

Goals
End results.

Needs
What you must have.

Wants
What you would like to have.

Step 1: Setting Financial Goals

In this first step you decide what you want to achieve. For example, are you saving for anything currently? If so, how much money will you need? Financial goals should be based on **values** (principles that guide behavior) and **attitudes** (likes and dislikes). **Goals** are end results, the things worth striving for. A basic goal in financial planning is to maximize income and wealth, but how people do this is affected by their own unique blend of values and attitudes. Individuals differ regarding how thrifty they are and what attitudes they hold about money.

To determine goals, questions to ask include: What kind of work do you want to do? What would be satisfying to you? Where do you want to live? Ultimately you should separate your *needs* (what you must have) from your *wants* (what you would like to have). You may want an Africa Safari for a dream vacation, but your savings suggest the "Animal Kingdom" at Disney World. Another consideration is: How long can you live on a tight budget? What is a tight budget? Becka, a student at an expensive private school, was told by her parents that she had to cut her telephone bill to $1,000 a month. Becka protested that this was too harsh. What do you think? Personal finance deals with more than money. It is about determining the lifestyle desired, the quality of one's daily life, relationships, and all that these encompass.

Once you have determined your general approach to money, needs, and spending, you can move on to setting specific goals. Besides values and attitudes, your current financial situation and potential financial situation should be taken into account. To be successful, goals should be

- *flexible*: change with times and conditions/events.
- *realistic*: based on your income and life situation.
- *specific and measurable*: for example, a goal of saving $200 in the next three months is more specific than saying "I'm going to try to save more." Or, you don't want just any car, you want a $30,000 car.
- *prioritized*: ranked according to importance. For example, you want a house more than a boat.
- *action-oriented*: the achievement of goals requires action.

As noted, to be successful goals need to be flexible. They should change over time. For example, when Eric graduated in merchandising from a midwestern university he went to New York to start his career. After working there for two years, he decided to return to his home state of Missouri and go into management at a retail store. After being hired, Eric made up a wish list and divided it by time into short-term (within a year) and long-range goals. His short-term goals included paying off the last $5,000 of his student loan and keeping up with his normal expenses, including his apartment rent. Paying off the loan was a high priority for Eric because he wanted to move on financially. Eric's long-range goals included setting up a monthly budget, joining a fitness club, and saving $150 a month. Eric thought he might eventually want his own store, so he needed to build his savings and establish a good credit record to be able to apply for a small business loan, probably in five to ten years.

Eric's goals could be further subdivided into monetary goals (pay off the loan and pay the rent) and nonmonetary goals (joining a fitness club). Exhibit 1.2 gives you the opportunity to write down your financial goals and the actions you plan to take to achieve them along with estimated time for completion, estimated cost, priority, and a final space to check off when the goal is reached.

EXHIBIT 1.2
Setting Financial Goals

Below write your goals and fill in the other blanks. Examples of Eric's goals, as described in the text, are given.

Short-Term Financial Goals (one year or less)

Goals	Actions to Take	Estimated Completion	Estimated Cost	Priority	Check When Completed
Eric's Goal: Payoff student loan	Send monthly $417 check straight to loan company before paying anything else	1 year	$5,000	High	

Your goals:
1. ______
2. ______
3. ______

Long-Range Financial Goals (more than one year)

Goals	Actions to Take	Estimated Completion	Estimated Cost	Priority	Check When Completed
Eric's Goal: Save $150/month	Payroll Deduction to savings account	5 years	$9,000+ interest	Medium	

Your goals:
1. ______
2. ______
3. ______

Opportunity costs
The cost of what is given up.

Opportunity Costs Most of the decisions that you make have a cost associated with them. In goal setting the concept of **opportunity costs** is critical. Opportunity costs refer to what a person gives up in order to do or have something else. For example, if you decide to go to law school after graduating from college, this puts off working full-time for three years. The initial tradeoffs in this decision involve both time and money. Some of the most difficult life decisions involve opportunity costs.

Risk
The possibility of experiencing harm, suffering, danger, or loss.

Risk Another concept that impacts on goal setting is risk. **Risk** is defined as the possibility of experiencing harm, suffering, danger, or loss. In personal finance the emphasis is on the potential loss of, or lack of, money. More than anything else, investment and insurance decisions come down to the ability to handle risk; therefore it is important to explore risk further.

Within money risk there are several subcategories, including

* *income risk*: Losing a job or other source of income is always a possibility.
* *inflation risk*: Rising prices affect how far your dollars stretch. If your investments are not earning more than the cost of living increases each year then you are losing purchasing power. **Inflation** is defined as a rise in price levels. *Over the last twenty years the inflation rate has averaged about three percent* per year.

Inflation
Rising in prices.

* *interest rate risk*: Changing interest rates can have a negative effect on your investments. The "real" value of your investments could erode due to the rising cost of living.
* *liquidity risk*: Some investments are more difficult than others to convert into cash. **Liquidity** refers to how readily something can be converted into cash. For example, a savings account where you can get money out in a few minutes is more liquid than real estate investments, coin collections, art, or business interests that are difficult to sell at short notice.

Liquidity
How readily something can be turned into cash.

* personal risk: Health, safety and other risks are involved in money decisions. Should you pay more to buy a new car with air bags or buy a used car without them? Can you risk going without health insurance for a few months when you are between jobs?
* *status risk*: Clothing, neighborhoods, brands, and cell phones convey images.
* *time risk*: How long can you afford to put off saving for a child's college tuition or your retirement? Should current needs be met before future savings are addressed?

A prudent financial plan takes into account all these types of risks, an individual's personality and earning capacity, and the usual performance of certain types of investments. To apply the concept of risk, did any of the goals or actions that you listed in Exhibit 1.2 involve risk?

Risk aversion
Avoidance of risk.

In economics the avoidance of risk is called **risk aversion.** It is assumed that individuals will try to avoid or reduce risks in order to minimize problems and maximize positive outcomes. On the other hand, there is a risk/return tradeoff which means that historically investments that produce higher returns generally are riskier. So that not taking some risk, may result in another kind of risk, the risk of earning less money than you should have by being too conservative. Exhibit 1.3 illustrates the hidden risk of playing it too safe.

What kind of investor or money manager are you? The key factors that determine type are:

* *risk tolerance* (the ability to accept risk)

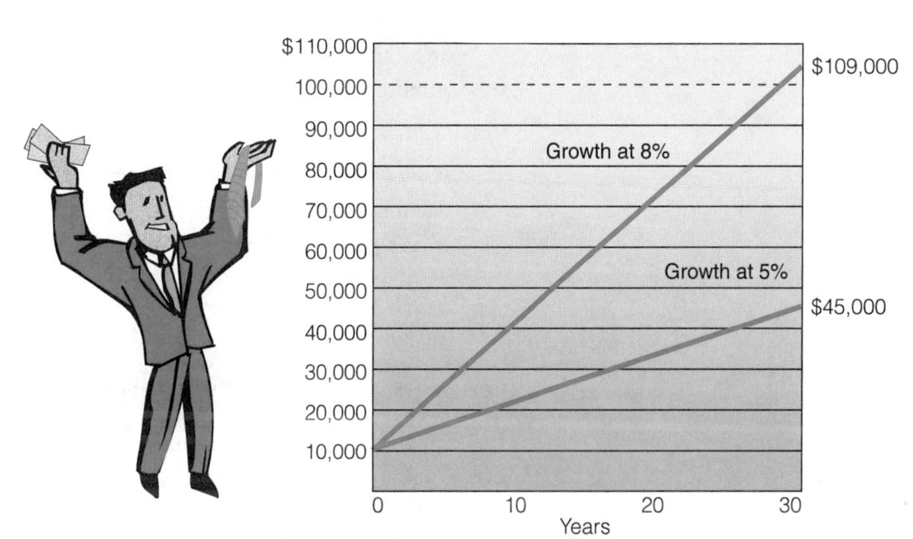

No one wants to take unnecessary risks, but risk is always present. If you do not take some risk, you may never reach your financial goal. This chart shows the difference in outcome when $10,000 is invested at a 5% growth rate versus 8% growth rate after 30 years.

EXHIBIT 1.3
The Potential Hidden Risk of Playing it Too Safe

* *goal time horizon* (the span of time between today and goal achievement). For example, young people have a long time horizon to retirement so they can afford to take more risks with their money than older people who have a short time horizon to retirement. The amount of time available affects the investment decisions that are made.

Regarding risk tolerance, an individual may have:

* an aggressive approach, wherein substantial volatility and short-term financial setbacks are accepted as part of maximizing growth.
* a conservative approach, meaning an inability to tolerate more than minimum volatility with the goal of preserving principal (the basic amount invested, exclusive of earnings).

The approach to risk tolerance and the realities of goal time horizon affect much of personal finance. More discussion of risk will be given in the insurance and investment chapters.

Step 2: Creating and Activating Action Plans

Once goals are set and you have determined how you feel about risk, then you are ready to design and activate a plan. An action plan begins with figuring out which steps to take first. What do you want to do? Where do you need to go? What forms do you have to fill out? As shown in Exhibit 1.1, this phase may involve decisions about managing assets, credit, insurance, investments, and retirement.

If at any time in this step you are unsure what to do next, *ask yourself: What is the best use of my time and my money right now?* The answer to this question will send you in the right direction.

In Exhibit 1.2 a space was provided for you to write down specific actions that you plan to take to accomplish the goals you have set. Possible actions may fall under the following categories:

* *Staying on course*: for example, holding onto the checking and savings accounts you already have. In Exhibit 1.2 Eric's plan to finish paying off his student loan would fall under this category.
* *Expanding*: adding new pieces such as increasing the amount of life insurance coverage on an existing policy or increasing your contribution to an already established retirement plan.
* *Cutting back*: Spending less, selling assets such as real estate or stocks.
* *Embarking on a new course*: trying something for the first time such as having a regular savings plan or purchasing stocks, bonds, or mutual funds. Eric's plan to begin a new savings program of $150 a month would fall under this category.

Step 3: Monitoring, Evaluating, and Revising Plans

This step is often the least popular, but it is the most critical from a learning point of view. Goals need to be reexamined and updated, investments have to be checked periodically to see how they are progressing. Mistakes have to be faced up to and remedied. Financial losses happen all the time, for example, when the stock market drops and investment and retirement plans decrease in value, or whenever a house is sold at a loss. The good news is that financial gains happen all the time as well.

To keep up-to-date on your financial situation, you should completely *evaluate your financial plan at least once a year.* Tax time in the spring is a natural time for an annual review since tax forms require an accounting of the previous year's earnings and losses. Another time for an annual review may be when an employer offers an open enrollment period for adding more insurance or changing retirement options. An annual review may not be enough, however, if there are a number of significant changes in your life such as entering graduate school or getting married.

Objective 3:
Identify the four key players in the economy.

THE GENERAL ECONOMY AND FOUR KEY PLAYERS

Economic theory assumes that individuals seek to maximize satisfaction and avoid loss from the choices made. It assumes that individuals are rational (capable of reason) and will make the best choices they can. In order to make rational decisions, people observe the movements of the marketplace and on that basis make decisions about whether it is time to spend, save, invest, or protect assets. No one should make financial plans without considering changes in the economy such as inflation and interest rates.

Economics
The study of the economy, how wealth is created and distributed.

Wealth
Total value of all items owned.

Financial assets
Intangibles or paper assets such as savings and securities.

Tangible assets
Physical assets such as homes or cars.

Economy
The economic system of a country or region.

Economics is the study of the economy, how wealth is created and distributed, and the forces of supply and demand. **Wealth** is the total value of all items owned. It includes both financial and tangible assets. **Financial assets** are intangible. This category includes paper assets, such as savings and securities, such as stocks and bonds. **Tangible assets** are physical assets, such as homes, computers, or cars. As the examples show, assets are anything with commercial or exchange value owned by an individual or an organization.

The word "economy" refers to the economic system of a country or region. The United States economy is influenced by the interaction of the four key players diagrammed in Exhibit 1.4: consumers, government, business, and media.

Consumers

Consumers participate in the economy whenever they spend, invest, save, or react to information about products and services. A constant dilemma consumers face is how much to spend on current needs and how much to save for future needs. The **average propensity to consume** refers to the percentage of each dollar of income that an individual spends, on average, for current consumption.

Average propensity to consume
Percentage of each dollar of income that an individual spends, on average, for current consumption.

Consumption
Using up of goods and services.

Level of living
Current state of living.

Standard of living
Quality of life one seeks, implies prosperity.

Consumption refers to the using up of goods and services. This includes both present consumption and future consumption. Individuals live their actual **level of living,** but save and invest to achieve a higher **standard of living** or quality of life. Standard of living implies a degree of prosperity, including the comforts, luxuries, and necessities one seeks.

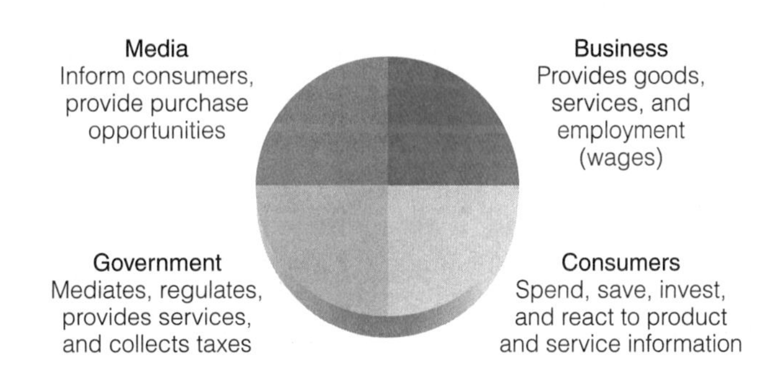

EXHIBIT 1.4
Four Key Players in the Economy

A sound financial plan makes provision for the future. Deciding what to buy and when to buy occupies a lot of consumers' time. Collectively, the decisions made by millions of consumers (referred to as mass consumption) have a tremendous impact on the general economy.

Obviously consumers vary widely in the amount of money they have. The marketplace tries to reach everyone, so that is why on the same street there are restaurants offering $5 and $50 meals, and the same store sells $29 jeans and $59 jeans. Since 1980 the wealthiest fifth of the population has seen its income grow by 21 percent, while wages for the bottom 60 percent have stagnated or even dipped, according to Census Bureau data.[1] Families in the top 5% of household earners have seen their incomes shoot up in the last 25 years, whereas those on the lower end have mostly been stagnating (see Exhibit 1.5).

Government

The government mediates or regulates the consumer–business–media exchange. For example, in 2000 the government was engaged in a monopoly case involving Microsoft. The fundamental issue was whether consumers were getting a fair shake. Through intervention on behalf of consumers, the government tries to ensure a fair playing field for all groups.

Besides its mediating/regulating function, the government provides services such as police and fire protection, highways, national defense, and welfare. *The federal government is the largest single employer in the United States* because it employs members of the armed forces, postal workers, and all other national-level government workers. It plays an important role in product testing, especially foods and medicines. In order to provide the services that it does, the government collects taxes. In 2000, income taxes reached an all-time high, going as high as 40 percent of income for those in the highest tax brackets.

Federal Reserve System
Regulates U.S. monetary system including maintaining an adequate money supply.

While various government agencies test and regulate products and provide services, the **Federal Reserve System** (nicknamed *The Fed*), the nation's central bank, has the responsibility of regulating the U.S. monetary and banking system, including maintaining an adequate money supply. The Fed does this by influencing borrowing, interest rates, and the buying and selling of government securities. Established by the Federal Reserve Act of 1913, The Fed looks out for the consumer interest while also providing a conducive environment for business expansion by keeping interest rates at

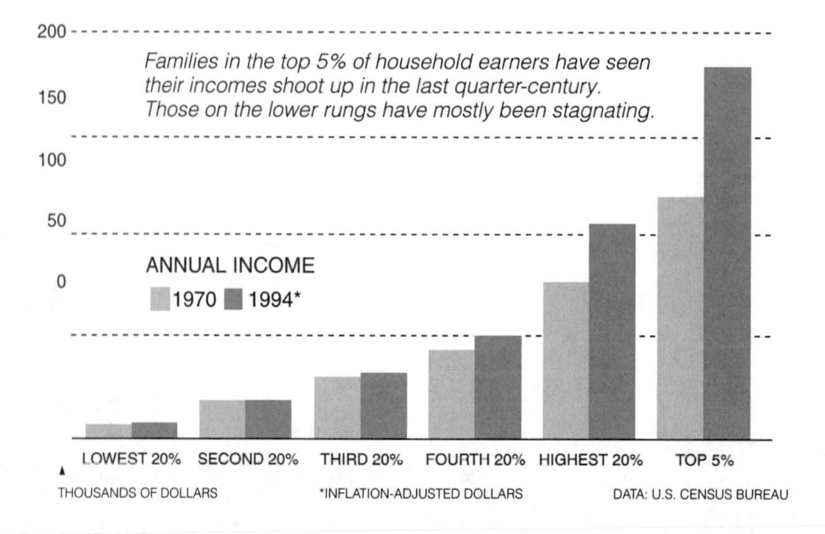

EXHIBIT 1.5
The Rich Get Richer

an appropriate level. Although the Federal Reserve System's governing board is appointed by the President and confirmed by the Senate, it functions as a nonpolitical independent entity.

Business

Business provides goods, services, and employment. The word "business" encompasses a wide range of commercial, industrial, and professional dealings. Business plays an important role in the circulation of money as part of the free enterprise system. The prices of goods and services are determined by the forces of supply and demand.

Free choice and competition form the basis of our economic system. Consumers affect the market by what they buy. Companies, if they are to succeed, respond to current consumer demands and anticipate future consumer needs. Businesses exist in a highly competitive market that is regulated by government and consumer demand.

Media

The two major media categories are *print* —which includes newspapers, magazines, direct mail, and outdoor (billboards, advertisements on buses)—and *broadcast*—which includes radio, television, cable, and the Internet. Media inform the public about currently happening and upcoming events, services, and products. Before 1930 newspapers and magazines carried the bulk of advertising. In the 1930s and 1940s radio had its heyday. Since the 1950s, television has captured the biggest share of advertising dollars. A newer form of media, computer online services, offers products and service information through the Internet.

The primary goal of advertising is to attract attention. Celebrity spokespersons, humor, fear, and sensory imaging are designed to attract attention. Once potential consumers are attracted, the next step is for them to comprehend what the product or service is. (Have you ever looked at an advertisement and wondered, "What are they selling?"). Finally, the advertiser wants the consumer to remember the product or service and buy it. City dwellers see or hear many more advertisements in a day than rural dwellers. The average person is exposed to 250–1,000 advertisements a day depending on his or her lifestyle.

Objective 4:
Describe the economic cycle and how it affects personal finance.

THE ECONOMIC CYCLE

Theoretically the economy follows a cycle with a wavelike motion (see Exhibit 1.6). The **economic cycle** (sometimes referred to as the business cycle) refers to periodic expansions and contractions in economic activity. It affects how you should handle your money. This would be easy if the cycle were totally predictable, but it is largely unpredictable especially in terms of how long each phase of the cycle may take. The challenge lies in anticipating the changes it might make and responding appropriately.

The economic cycle is made up of four stages:

Expansion
Growing economic activity, low unemployment rate.

Expansion: a preferred stage with prosperity, when unemployment is low, and retail sales and economic activity are growing. This is the time when consumers buy cars, homes, and other expensive items since interest rates and inflation are relatively low and employment is high. As this book went to press, *the United States had been in an expansion stage since 1990.* This represented the longest expansion stage during peacetime surpassing the 1961–1969 run during the Vietnam War era.

Recession
Temporary slowing of the economy.

Recession: a temporary slowing of the economy (or downturn in the economy), when unemployment is higher than desired, and economic activity is slow. The National

EXHIBIT 1.6
The Economic Cycle
The economy goes through various stages over time although depressions are rare. The stages tend to be cyclical and affect the levels of employment, production, and consumption.

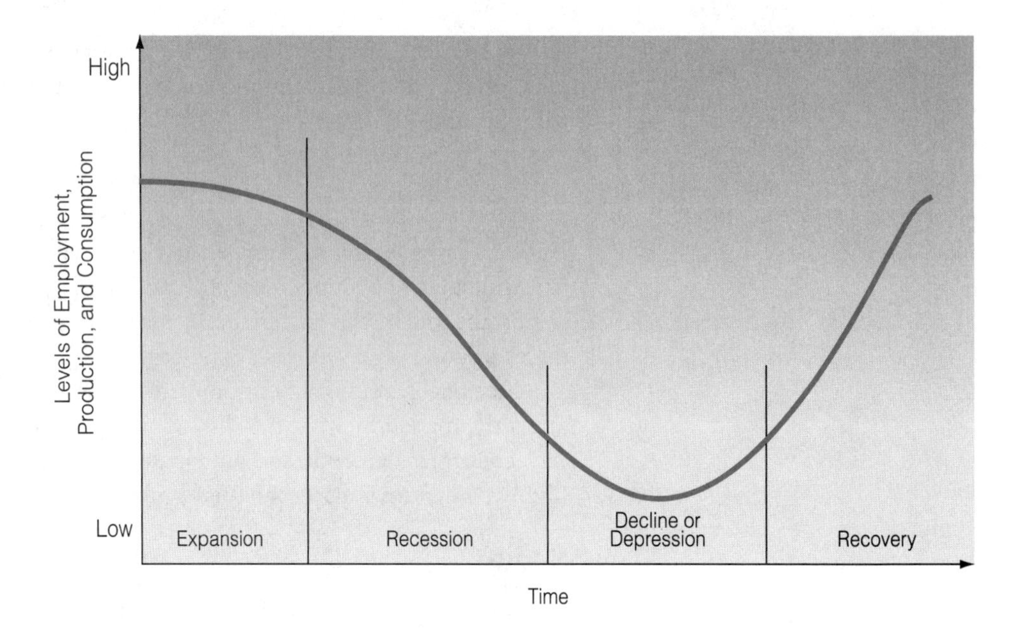

Bureau of Economic Research defines a recession as a recurring period of decline in total output, income, employment, and trade, usually lasting from six months to a year and marked by widespread contractions in many sectors of the economy. During a recession consumers are cautious about spending.

Depression
Downward trend, high unemployment, growth at standstill.

Decline or Depression: an undesirable, downward trend featuring high unemployment (with 10 percent or more of the working-age population unemployed) and economic growth at a standstill. Most notable is the Great Depression of the 1930s, which struck Europe and the United States hardest, but had worldwide effects. Consumers have reduced purchasing power and confidence during a depression or decline.

Recovery
Hopeful stage, unemployment lessening, economy moving upward.

Recovery: a hopeful stage when things start looking better, the nation is emerging from a recession or depression, the level of unemployment is lessening, retail sales improve, and the economy is moving upward. Consumers start buying more and have more confidence in the economy.

Indicators of the Direction of the Economy

Besides understanding that the economy runs in cycles it is important to know about the indicators that show the direction of the economy. This section will discuss the following indicators:

- inflation.
- the index of leading economic indicators.
- the consumer price index.
- interest rates.
- the Gross Domestic Product.

As mentioned earlier, *inflation refers to rising prices.* So as inflation increases, the buying power of the dollar decreases. In other words inflation devalues money, as shown in Exhibit 1.7. Inflation hurts people on fixed incomes especially some dis-

EXHIBIT 1.7
How Inflation Devalues Money

Value of $1,000 at various inflation rates after 5 years				
	Average Annual Rate of Inflation			
	2%	4%	6%	8%
5 years	$904	$815	$734	$659

abled and retired people because their incomes can not keep up with rising prices

The rate of inflation varies. Most recently inflation has been low compared to the 10 to 12 percent inflation in the late 1970s and early 1980s. As mentioned earlier, knowing the current inflation rate is one of the best ways to determine how well you are doing financially. For example, if the inflation rate is 2.5 percent, you are beating it if your mutual fund is bringing in over three percent. Conversely, if your savings account is paying only two percent interest you are not keeping up with inflation.

Index of Leading Economic Indicators
Composite index averaging 11 components of growth.

The **Index of Leading Economic Indicators (LEI)** released every month by the U.S. Commerce Department's Bureau of Economic Analysis, is a composite index averaging 11 components of growth from different segments of the economy. *A falling index over a three month period indicates a slowing down (a decline) in the economy whereas a climbing index is an indicator of prosperity, growth.* It should be pointed out that no index is foolproof. However, in recent years the LEI has been mostly reliable, signaling economic downturns eight to 18 months in advance.[2]

Consumer price index
Widely used measure of average changes in prices.

Another index is the **consumer price index (CPI)**, published each month by the Bureau of Labor Statistics in the Department of Labor. The CPI measures the average change in the prices consumers pay for a fixed basket of 400 goods and services, including clothing, food, transportation, health care, shelter, entertainment, and utilities. It moves up and down as the rate of inflation changes. The CPI provides a yardstick, a guideline as to the cost of goods and services, but because you do not buy all 400 of the goods and services each year, it is not a totally accurate picture of how your actual expenses may be running. In recent years several revisions of the CPI have been suggested to make it a more appropriate measure. The CPI is important because it is used as a cost-of-living benchmark to determine tax brackets and increases in Social Security payments and other government payments.

Purchasing power
Amount of goods and services an individual's money can buy.

Interest rates
The cost of borrowing money.

As prices rise, more income is needed to buy things. This refers to **purchasing power,** the amount of goods and services an individual's money can buy. **Interest rates** represent the cost of money—how much it is going to cost to borrow money—usually expressed as an annual rate such as 7%. People with poor credit ratings are often forced to pay much higher interest rates compared to people with good credit ratings. This is because they are perceived as a greater credit risk to the lender.

Interest rates are a key factor in deciding when to borrow money for major purchases such as cars or houses. Naturally you would rather pay 7.5 percent interest on a new car loan than 10 percent interest. When this book went to press, home mortgage loans were at 8 percent, up from six months earlier when they hovered around 7 percent. People were kicking themselves for not buying a home when interest rates were lower. Interest rates are affected by inflation. When inflation rises, the cost of borrowing money rises.

Gross domestic product
Market value of goods and services produced in one year.

Another statistic or measure to watch that provides a general idea of the condition or state of the economy is the **Gross Domestic Product (GDP)**. GDP is the market value of the goods and services produced in one year in a single country. The GDP includes consumer and government purchases, domestic investments, and net exports. Growth in the U. S. economy is measured by the inflation-adjusted GDP, or real GDP. The Commerce Department releases figures for GDP on a quarterly basis. During economic expansion, GDP rises steadily. GDP affects **personal income**, the amount of money individuals receive in the form of wages and salaries (when looked at nationally, the biggest part, averaging 57%), investment income, and transfer payments such as unemployment compensation and Social Security. When the GDP is up so is personal income. For example, from January 1998 to January 1999, GDP rose and Americans saw a robust increase in their personal incomes.

Objective 5:
Understand the time value of money.

THE TIME VALUE OF MONEY

As mentioned earlier, a basic problem is deciding whether to buy now (spend) or wait (save) and buy something in the future. Because the economy was in an expansion phase in the 1990s, the tendency during that decade was to spend (buy homes, cars etc.) rather than to save. But, if the economy during that decade had taken a downturn and inflation and interest rates had risen significantly more people may have hesitated before spending and waited for prices (relative to income) and interest rates to come down.

One of the most fundamental concepts in financial management is the time value of money. According to Jeanne M. Hogarth of Consumer and Community Affairs of the Federal Reserve Board, the **time value of money** is the value derived from the use of money over time to increase its total by investment and reinvestment. This may refer to either "present value" or "future value" calculations.

Time value of money
Theory that the value derived from the use of money over time increases its total by investment and reinvestment.

Present value
An asset's current value.

Future value
The value in the future of a current amount of money or asset.

Present value calculations state the value today of an amount that would exist in the future with a stated investment rate called the "discount rate." For example, with a 10 percent annual discount rate, the present value of $110 one year from now is $100. **Future value** calculations state the value in the future of a known amount today with a stated investment rate. For example, with a 10 percent annual investment rate, the future value in one year of $100 today is $110. In either case the interest rate used reflects the lost opportunities for return from alternative investments.

Another way to explain the time value of money is that a dollar today may be worth more than a dollar received in the future. Theoretically as long as you earn a positive return on your investments, you should prefer to receive equal amounts of money sooner rather than later. So if someone offered you $100 today or $100 a year from now, you would be better off taking it today because of inflation.

EXHIBIT 1.8
The Future Value of $1
To be used in determining time value of money

	Rate of Interest									
Number of Years	1%	2%	3%	4%	5%	6%	7%	8%	9%	10%
1/2	1.0050	1.0100	1.0149	1.0198	1.0247	1.0296	1.0344	1.0392	1.0440	1.0488
1	1.0100	1.0200	1.0300	1.0400	1.0500	1.0600	1.0700	1.0800	1.0900	1.1000
2	1.0201	1.0404	1.0609	1.0816	1.1025	1.1236	1.1449	1.1664	1.1881	1.2100
3	1.0303	1.0612	1.0927	1.1249	1.1576	1.1910	1.2250	1.2597	1.2950	1.3310
4	1.0406	1,0824	1.1255	1.1699	1.2155	1.2625	1.3108	1.3605	1.4116	1.4641
5	1.0510	1.1041	1.1593	1.2167	1.2763	1.3382	1.4026	1.4693	1.5386	1.6105
10	1.1046	1.2190	1.3439	1.4802	1.6289	1.7908	1.9672	2.1589	2.3674	2.5937
15	1.1610	1.3459	1.5580	1.8009	2.0789	2.3966	2.7590	3.1722	3.6425	4.1772
20	1.2202	1.4859	1.8061	2.1911	2.6533	3.2071	3.8697	4.6610	5.6044	6.7275
30	1.3478	1.8114	2.4273	3.2434	4.3219	5.7435	7.6123	10.0627	13.2677	17.4494

The time value of money should be considered when making major financial plans, determining when to buy a house, taking out a loan versus paying cash, estimating retirement needs, and analyzing investments.

How do you find out the time value of money specific to a purchase or investment decision? There are two ways:

1. You can ask a finance professional, such as a banker to do the calculations for you. For example, you could ask, "If I buy this $5,000 certificate of deposit at 5 1/2 percent interest and hold it for two years, what will it be worth at the end of two years?" The banker would provide the answer by pushing a button on a calculator.
2. You could perform the calculations yourself, using one of four methods that will be applied to the situation of a recent college graduate, Val Trent. Val is considering what to do with an extra $1000 he has. One of his friends wants to borrow it, and he says he will pay Val back $1100 in a lump sum after three years. The other main option Val is considering is to invest the $1000 in a treasury bond for three years, paying 6 percent annual interest. Here are the four methods he used to calculate his return. Each method supplies the same answer.

Basic Math Method: Assuming a rate of interest of 6 percent, the future value (FV) at the end of one year would be $1060 [$1000 + (0.06 X $1000]. The FV after 2 years would be $1123.60 [1060 + (0.06 X $1060)]. The FV after 3 years would be $1191.02 [1123.60 + (0.06 X 1123.60)]. Thus Val would earn more by not lending the money to his friend because the bond will give him $91.02 more ($1191.02 versus $1100).

Basic Calculation Method Using the following equation results in the same answer, where i represents the interest rate:

$$\begin{aligned} FV &= (\text{Sum of money})\ (i + 1.0)\ (i + 1.0)\ (i + 1.0)\ldots \\ &= (\$1000)\ (1.06)\ (1.06)\ (1.06) \\ &= \$1191.02 \end{aligned}$$

Calculator Method Using a calculator will provide the same answer, employing the following equation, where i represents the interest rate and n represents the number of years:

$$\begin{aligned} FV &= (\text{Sum of money})\ (1.0 + i)n \\ &= (\$1000)\ (1.06)^3 \\ &= (\$1000)\ (1.191016) \\ &= \$1191.02 \end{aligned}$$

Table Method Using Exhibit 1.8, go across the top row to the 6 percent column. Read down the 6 percent column and across the row for 3 years to locate the factor 1.191. Multiply by the present value of the cash asset ($1000) to arrive at a future value ($1191).

It is assumed in these calculations that the interest earned was not withdrawn but was left in to reinvest. The calculations given in Val's decision situation are examples of the concept of **compound interest** which is the calculation of interest on interest as well as the interest on the original investment. If the interest was not reinvested, but withdrawn then this would be called **simple interest**. For example, if you took $1,000 at a 6 percent rate for 30 years it would come out this way:

Compound interest
Calculation of interest on interest.

Simple interest
Interest on original sum of money.

$$\begin{aligned} \text{Simple interest} &= (\text{Principal})\ (\text{Rate})\ (\text{Time}) \\ &= (\$1000\quad (0.06)\ (30) \\ &= \$1800 \end{aligned}$$

This $1800 in simple interest when added to the $1000, would yield a total of $2800, which is $2946 less than the $5743 that compounding would yield. Compound interest is a benefit to consumers and an attractive feature in savings accounts and other forms of investing. Chapter 4 on Managing Cash and Savings will explain compound interest further.

Objective 6:
Recognize the importance of career planning, salary, and benefits to overall financial well-being.

THE CHANGING WORLD OF WORK AND THE IMPORTANCE OF EDUCATION

The first half of the chapter stressed the importance of making a financial plan and being aware of economic cycles, economic indicators, and the time value of money. Now the chapter turns to career concerns because of the many factors that influence your financial situation, *none is more important than your employment situation. The main reason people work is to earn money.*

One of life's realities is that an engineering graduate starting out with a $72,000 salary will have more to work with than a humanities graduate starting out at $23,000. This pay differential highlights supply and demand in the labor market. For example, labor predictions indicate there will be more white-collar workers and fewer blue-collar workers.[3] The demand for service, construction, and retail trade workers will increase, and it is predicted that more people will work at home, at least part-time. See Exhibit 1.9 for Bureau of Labor Statistics predictions of occupations with the largest job growth.

Education increases potential earning power. According to a nationwide poll of college students conducted by the University of California at Los Angeles (UCLA), recent college freshmen, when asked the main reasons why they came to college, said "to get a better job" and "to make more money." *A college graduate can expect to earn twice as much as a high school graduate over the course of a lifetime of employment.* More than half of today's high school graduates go on to some form of higher education, and about 20 percent of the adult population in the United States has a four-year college degree. To conclude, there is no question that investing in a college education is worth it from a financial standpoint.

Geographic Region and Community Size

Employment opportunities and pay vary by region as well as by field. Employers in urban areas tend to pay higher salaries than those in rural areas but these higher salaries are offset by higher living costs and property taxes. Chambers of Commerce can be contacted for statistics on community living costs or you can go to the Internet for this type of information.

Within all states there are pockets of high growth and the availability of jobs in particular specialties varies greatly by region. Growing states such as Nevada, tend to have more employment opportunities than slower growth states. According to the *U.S. Statistical Abstract*, the states with the largest populations are in descending order California, Texas, New York, Florida and Pennsylvania.

Marital Status, Households, and Earnings

Many households today have two earners. Dual-earner households have more to save, spend, and invest than single-earner households. They tend to be better educated, more mobile, and more likely to own their own home than one-earner families.[4]

[Accessibility Information]

Employment Projections

Occupations With the Largest Job Growth, 1996-2006

Occupations are ranked based on projected numerical changes in employment, 1996-2006; numerical employment changes are in thousands of jobs

Occupation	Employment		Change		Education and training category
	1996	2006	Number	Percent	
Cashiers	3,146	3,677	530	17	Short-term on-the-job training
Systems analysts	506	1,025	520	103	Bachelor's degree
General managers and top executives	3,210	3,677	467	15	Work experience plus bachelor's or higher degree
Registered nurses	1,971	2,382	411	21	Associate's degree
Salespersons, retail	4,072	4,481	408	10	Short-term on-the-job training
Truck drivers light and heavy	2,719	3,123	404	15	Short-term on-the-job training
Home health aides	495	873	378	76	Short-term on-the-job training
Teacher aides and educational assistants	981	1,352	370	38	Short-term on-the-job training
Nursing aides, orderlies, and attendants	1,312	1,645	333	25	Short-term on-the-job training
Receptionists and information clerks	1,074	1,392	318	30	Short-term on-the-job training
Teachers, secondary school	1,406	1,718	312	22	Bachelor's degree
Child care workers	830	1,129	299	36	Short-term on-the-job training
Clerical supervisors and managers	1,369	1,630	262	19	Work experience in a related occupation
Database administrators, computer support specialists, and all other computer scientists	212	461	249	118	Bachelor's degree
Marketing and sales worker supervisors	2,316	2,562	246	11	Work experience in a related occupation
Maintenance repairers, general utility	1,362	1,608	246	18	Short-term on-the-job training
Food counter, fountain, and related workers	1,720	1,963	243	14	Short-term on-the-job training
Teachers, special education	407	648	241	59	Bachelor's degree
Computer engineers	216	451	235	109	Bachelor's degree
Food preparation workers	1,253	1,487	234	19	Short-term on-the-job training
Hand packers and packagers	986	1,208	222	23	Short-term on-the-job training
Guards	955	1,175	221	23	Short-term on-the-job training
General office clerks	3,111	3,326	215	7	Short-term on-the-job training
Waiters and waitresses	1,957	2,163	206	11	Short-term on-the-job training
Social workers	585	772	188	32	Bachelor's degree
Adjustment clerks	401	584	183	46	Short-term on-the-job training
Cooks, short order and fast food	804	978	174	22	Short-term on-the-job training
Personal and home care aides	202	374	171	85	Short-term on-the-job training
Food service and lodging managers	589	757	168	28	Work experience in a related occupation
Medical assistants	225	391	166	74	Moderate-term on-the-job training

SOURCE: "Occupational employment projections to 2006,"*Monthly Labor Review*, November 1997, pp. 58-83.
Download PDF version (183K). (See How to view PDF files for help.)

Employment Projections Home Page

BLS Home Page

Occupational Outlook Program
Bureau of Labor Statistics
OOHInfo@bls.gov
Last modified: June 30, 1998
URL: http://stats.bls.gov/emptab2.htm

EXHIBIT 1.9 Occupations with the Largest Job Growth, 1996-2006

Family household
Two or more related persons, one of whom owns or rents the living quarters.

NonFamily household
Householders who live either alone or with others to whom they are not related.

The Census Bureau defines a **family household** as consisting of two or more related persons, one of whom owns or rents the living quarters. This category includes married couples with or without children under 18 living at home, single fathers, single mothers, and other types of families. A **nonfamily household** is defined as households made up of householders who live either alone or with others to whom they are not related.

The percentage of nonfamily households is growing. In 1995 about 30 percent of all households were classified as nonfamily households. It is predicted that this will grow to 33 percent by 2005. This trend affects consumption patterns. Housing, appliances, and food container sizes will diminish to meet reduced needs, but there may be a growth in upgraded household furnishings, vacations, luxury items, and sports and entertainment items.[5]

Gender Gap

The gender gap is the difference in earnings between employed men and women. Today women earn 76 cents to every dollar that men earn, but the gap is closing. In 1990 the gap was 72 cents for every dollar. According to the Census Bureau, there is far less of an earnings gap between younger men and women than between older men and women. Another trend is that one-third of U.S. wives outearn their husbands.

Age and Stage in the Lifecycle

*People between the ages of 45–54 have the highest median income*s, according to the Bureau of Labor Statistics. High incomes are associated with managerial or professional positions.

When people are in their twenties, career exploration is common. By the thirties more commitment to a profession is typical. In the forties many people reconsider their options, perhaps make a job switch or open their own business, or recommit to the career they started earlier. During the fifties many people hit their earnings peak and turn to mentoring younger employees and planning seriously for retirement. Some retire early and begin a second career or business.

Taking a Personal Inventory

The best career decisions start with a self-assessment which means a self-directed search of who you are and what you really want. Even if you are sure about the kind of work you want to do, it is still worthwhile to consider the following four things:

1. Your career goals. What kind of job you want when you graduate versus the kind of job you will want ten years from now.

2. An occupational cluster. This means to think of a job family rather than a specific job. For example, a graduate with a marketing degree could work in retailing, advertising, public relations, management, sales, or research.

3. Your job skills, volunteer experience, preferred leisure activities, talents, and strengths.

4. Education and internships. What courses did you like or dislike? Do you need more education or training to get ahead? More and more students are finding that interning is the best way to get on-the-job experience, to find out ahead of time jobs they like or dislike, and to obtain job offers.

SOURCES OF CAREER INFORMATION AND CONDUCTING A JOB SEARCH

Many sources of information are available at university career services centers. They also offer career fairs and interview services, matching you with employers who come to campus. These services are free. Outside employment agencies may charge you or employers a placement fee, so it is wise to use university services. Other sources of career information are libraries, newspapers, electronic bulletin boards and databases, the World Wide Web (for addresses, see the Electronic Resources section at the end of the chapter); community and professional organizations; contacts through relatives, friends, and co-workers, and direct solicitations of employers.

According to *Who's Hiring Who, How to Find That Job Fast*, nearly fifty percent of job seekers find jobs through personal referral. The next most common method is through direct contact with employers.[6] Only five percent of job seekers find jobs through want ads.[7]

Applying for Employment

Most job applications require a resumé with a cover letter, as well as a personal interview.

A *resumé* is a summary of education, training, and experience. A sample resume is given in Exhibit 1.10. The newest forms of job hunting involve posting your resume on the Internet and searching the Web for job openings (see the Electronic Resources section for sample sites). Your campus career services center can help with putting together a scannable resume.

A *cover letter* is the correspondence sent with the resumé that indicates your interest in a specific job and a request for an interview. Cover letters should be customized for each job. A sample cover letter is given in Exhibit 1.11.

An *interview* can take place over the Internet or in person. Regardless of the place of the interview, the first 30 seconds of the interview are crucial in establishing rapport. A second interview at the organization's headquarters may be required, especially if the first interview was on campus or was a virtual interview via the Internet.

Salary and Employee Benefits

A major factor in job selection is the salary and benefits package. Thinking in terms of an acceptable salary range will be more helpful than having a non-negotiable figure in mind. Talk to advisors and people in similar positions to get an idea of what an acceptable range is or check the Internet.

Typical fringe benefits include life, health, and disability insurance; retirement plans; and reimbursement for moving, travel, and educational expenses. *Health insurance is the most expensive benefit.*

Cafeteria-style benefits
Allow for employee choice of benefits.

Employers usually offer **cafeteria-style benefits** which are plans that allow employees to base their job benefits on a credit system and on personal needs. These flexible plans may also allow for a selection of suppliers (dentists, doctors, health maintenance organizations). One of the choices offered may be to choose minimum or maximum benefits or the option of converting benefits to more take-home pay or extra vacation time.

Since health insurance is the most costly benefit, take time in selecting health and dental care and medical insurance plans, consider the types of services and suppliers; the direct costs to you (insurance premiums) for example and other anticipated out-of-pocket costs such as deductibles.

487 Clark Drive
Tallahassee, FL 33071

Phone (850) 577-9922
E-mail kcm@garnet.acns.fsu.edu

Kelly C. Martin

Objective

Seeking full-time employment in fashion merchandising where I can use my knowledge and experience.

Education

Florida State University, Tallahassee, FL
Bachelor of Science: Merchandising, Fall 2002
GPA: 3.5

Valencia Community College: Orlando, FL
Associate of Arts: Spring 2000
GPA: 3.0

Work Experience

Burdines, Miami, FL
(Intern, Fall 2002)

- Performed a number of tasks in central buying office.
- Maintained inventories of shipments and transfers.
- Wrote productivity reports to higher management.

GAP, Tallahassee, FL
(Assistant manager, 9/00-8/02)

- Provided professional assistance to customers.
- Trained in human resources: recruited and trained sales people, set schedules, made daily priority lists and performance appraisals.
- Initiated merchandise analysis such as inventory, sales counts, and purchase orders.

Olive Garden Restaurant, Orlando, FL
(Hostess, summer 2000)

- Prepared schedules and seating charts.
- Greeted customers.
- Communicated ideas and trends to management.

Awards received

Fashion Inc. President, 2000-2001

Dean's List, Fall 2000-Spring 2002

Student Leadership Award, April 1999

EXHIBIT 1.10
Resume

487 Clark Drive
Tallahassee, FL 33071
September 28, 2002

Mr. Anthony Barton
Senior Buyer
Bebe
1 Ivy Cresent
Lynchburg, VA 24513

Dear Mr. Barton,

I am applying for a position of liason to the assistant buyer that is posted on Bebe's Web page. This position is exactly the one I am seeking based on my education, experience, and interest in fashion. I feel I could make a meaningful contribution to the Bebe team.

I will be receiving my degree in Merchandising with a minor in Business at the close of this semester. Throughout my years at Florida State University, I have been active in on-campus activities while working part-time as Assistant Manager at the GAP. My current internship at the Burdines' buying office in Miami has reinforced my interest in working in the buying field.

You may refer to my attached resume for a summary of my qualifications. I would welcome the opportunity to discuss them with you in person. I can be reached at (850) 577-9922 or at kcm.garnet.acns.fsu.edu.

Thank you for your attention and consideration.

Sincerely,

Kelly C. Martin

Enclosure

EXHIBIT 1.11
Cover Letter

Benefits Online

The newest twist is the offering of benefit information online so employees can check how their benefits are faring on a daily basis if so desired. Currently 80 percent to 90 percent of all companies are either investigating or implementing interactive computer technology that will allow workers to access benefit information, make changes, and conduct related business by using personal computers.[8]

Retirement

Another benefit worthy of special attention because of long-term effects is retirement plan options. In addition to Social Security benefits, many employers contribute to a pension plan and you may be able to add more of your own funds. Retirement will be discussed thoroughly in Chapter 15, but a few initial decisions confronting new employees involve tax implications and whether or not to be vested.

Tax implications should be considered before selecting a retirement plan or other type of employer-sponsored program. A tax-exempt benefit means you do not have to pay tax. An example of a tax-exempt benefit would be a life-insurance policy that is not taxed on the first $100,000. A tax-deferred plan means you will have to pay taxes sometime in the future usually at retirement or when you leave the organization. Depending on your age the amount may be reduced. For example, at age 59 1/2, you can withdraw as much as you want from Individual Retirement Accounts (IRAs) and other retirement plans without paying a 10 percent penalty to the government.

Vesting refers to the point at which retirement payments made by the employer on your behalf belong to you even if you leave the organization. Vesting schedules may be given in sliding scales such as being fully vested at ten years of employment and 50 percent vested at five years. Often school systems or government agencies have vesting programs. Human resource personnel can explain the benefit options and the retirement policies of your organization.

Summary of Benefits

Dual- career couples should check the plans offered to each spouse and choose the appropriate benefits. In selecting benefits, whether single or married, *your primary objective is to secure the best benefits to meet your needs and those of dependent family members*. As a minimum, you should enroll in employer-sponsored benefit plans including health and disability and 401(k) (retirement) plans.

A good benefits package can increase your total compensation 30 percent or more so it is important to choose wisely and to revisit the options annually. For example, you may not need a dental insurance plan when first employed, but you may need one later when you foresee extensive dental work ahead. Many places of employment offer open enrollment periods once or twice a year.

Besides the conventional benefits already discussed and the usual sick and vacation leave policies, there are many other types of benefits called **perquisites** (more commonly known as "perks"). These may include use of a company car, stock options, computers at home, membership in clubs and trade associations, travel, college tuition waivers, interest free loans, child-care vouchers, and so on. These add to the value of the employment package and should be used when comparing offers.

Perquisites
Job extras beyond usual benefits.

CAREER PATHS

Career development is a lifelong process. Traditionally labor statistics indicated that most workers changed jobs five to seven times in their lifetime. More recently, because of the strong job market, the number of changes is rising, especially for young workers. Another new phenomenon is termed "exploding" offers, which must be accepted in a week or less. Companies in competitive industries are offering these to students whom they do not want to shop around. The problem is that this kind of pressure may result in panicky decisions or can lead to the new hire reneging later. A factor to consider in accepting any job offer deals with the financial and personal costs and benefits involved as they relate to your needs and goals as well as those of the members of your family.

Another factor affecting career paths is the overall economy. Being aware of trends, such as downsizing or outsourcing, is crucial in today's job market. To keep ahead of things, the most secure employees are lifelong learners in charge of their own careers such people take advantage of training opportunities.

To summarize this section of the chapter, career and benefit plans are not static. They require constant renewal with a focus on the future. To get going financially, earning an income is the first step. The remaining steps involve knowing how much to save, spend, and invest, and how to protect your money. Future chapters will show you how to do all of these.

ELECTRONIC RESOURCES

There are thousands of websites on personal finance. At the end of each chapter in this book there will be websites pertinent to the contents of the chapter. Many people use the Internet for financial information because it is free, available 24 hours a day, and offers great variety. The use of personal computers for financial and career information will continue to rise. It is much more economical (money-wise and time-wise) for companies to post openings on the web than to mail out job announcements.

Examples of government websites on economic indicators include:

- **www.bls.gov** for consumer price index and inflation information.
- www.bog.frb.fed.us for information on the Federal Reserve.

Examples of business, organization, or media sites on financial planning include

- **www.personalwealth.com** for general goal setting information and investment planning.
- **www.thomsoninvest.net** for information on various topics, including the time value of money.
- **www.financenter.com/** from The FinanCenter, a source for questions and calculations regarding budgets, credit, savings, housing, automobiles, and investments.
- **www.moneyadvisor.com/calc** and **www.centura.com/formulas** for ways to compute the time value of money.
- **personalwealth.com** the Standard & Poors personal finance site with general information on financial planning.

Examples of websites offering career-planning benefits, and job information are:

- **www.ajb.dni.us** from the U.S. Department of Labor and state agencies.
- **www.fedworld.gov** (the FedWorld site) for information on federal job openings and agencies.
- **www.espan.com** for ESPAN's website on salary and relocation information as well as other job search services.

Examples of job boards of special interest to college graduates include the following:

- Alumni Network—**www.alumni-network.com**: Job openings for experienced and new graduates. Network with other students.
- CareerPark—**www.careerpark.com**: Focuses on accounting, investment, management, marketing, and sales jobs.
- JobDirect—**www.jobdirect.com**: Posts entry level positions. Candidates are alerted by email when a position opens.
- JobTrak—**www.jobtrak.com**: Comprehensive source with listings.
- Recruiter's Online Network **www.recruitersonline.com**. Recruiter listings in over 70 categories.
- The Job Resource—**www.thejobresource.com**. Founded by Stanford University students in 1996, for students. Companies are featured who are currently recruiting.
- Vault.com—**www.vaultreports.com** and **Topica.com**: Serve as ways to learn about what it is like to work for certain companies. For example, you could register for the Walt Disney World Gripes group through Topica to eavesdrop on what employees have to say.
- Monster Board—**monster.com**: A general job site popular with students. To show how big this site is, at last count there were 381,283 job postings. In its earlier days the number of job postings on the Monster Board jumped from about 16,000 to 50,000 in one year. It covers positions ranging from a taxidermy clerk to a chief operating officer.[9] Because this site has grown so huge, many recruiters and job seekers are migrating to niche sites such as **www.Dice.com**, **www.Techies.com**, or **Techjobbank.com** for jobs in technology, for example. For adventure seekers, **Coolworks.com** lists adventure jobs, such as park ranger. For high-paying jobs try **Sixfigurejobs.com**.

1. Why is personal financial planning considered a process?

   ■ Because it involves steps. The first step relates to goals, the second to plans, and the third to monitoring, evaluating, and revising. Plans need to be checked at least once a year.

2. Why is studying personal finance important?

   ■ It is important for a number of reasons including forces outside yourself—such as changes in the economy—and forces within yourself—such as the goals you have set and what you find satisfying.

CONSUMER ALERT

Be careful about posting your resumé on the Web because of privacy and security issues, especially regarding home phone numbers and addresses. Because of problems, some career coaches warn against posting electronic resumés. To avoid potential problems, discuss this issue with campus career services advisors and read about the security of the site before posting information. If you send a resumé directly to a company address in response to a job posting, you should be secure.

3. How has the U.S. economy been in the last few years?

   ■ The economy has been in an expansive period characterized by high growth and low unemployment.

4. What is the time value of money?

   ■ The value derived from the use of money over time to increase its total by investment and reinvestment. To give a practical example, if you know you will receive one dollar a year from now, what would you pay for it? You would not pay a dollar because you can take a dollar now, invest it, and whatever interest rate you receive would be the return on your dollar. If it is 4 percent, you would be willing to pay about 4 percent less ($.96) for the dollar because that way you are getting a 4 percent return on your investment. If you decided to invest your dollar today in an interest bearing account and received a 4 percent return, then you would have earned the same amount from that investment as from buying the dollar at a discount.

5. What is the most expensive benefit?

   ■ Health insurance.

6. How much is a good benefits package worth?

   ■ It can increase your compensation 30% or more.

SUMMARY

Parents can apply for their baby's Social Security number on the day of birth. So personal finance starts from day one.

Effective financial management requires planning.

The financial planning process involves three steps: setting financial goals; creating and activating action plans; and monitoring, evaluating, and revising plans.

Since personal finance takes place within the context of the general economy, it is important to understand the economic cycle, terms such as inflation and interest rates, and the Index of Economic Indicators. For example, a falling Index indicates a slowing down in the economy whereas a rising Index indicates prosperity.

Over the last twenty years the inflation rate has averaged three percent per year.

The key players in the economy are consumers, government, business, and media.

The economy has the following stages: expansion, recession, decline or depression, and recovery. In recent years the United States has been in an expansion stage.

The time value of money (the value derived from the use of money over time to increase its total by investment and reinvestment) can be estimated by using mathematical calculations. These formulas help individuals assess how to invest their money and whether to spend or invest now or to wait. Compound and simple interest can also be calculated.

Financial planning needs vary over the lifecycle.

Many factors influence financial situations, but none is more important than employment status.

Education increases potential earning power. A college graduate can expect to make twice as much as a high school graduate over the course of a lifetime.

Currently women earn 76 cents to every dollar men earn, but the gender gap is closing.

People between the ages of 45–54 have the highest median incomes.

Benefit packages can increase total compensation 30 percent or more. The most costly benefit is health insurance.

Most companies are either investigating or implementing interactive computer technology that will allow employees to access benefit information, and make changes.

The Internet is being used more and more for financial and job information and job searches.

REVIEW QUESTIONS

(See Appendix A for answers to questions 1–10)

Identify statements 1–5 as true or false.

1. Inflation and interest rates have little effect on personal financial planning.
2. Compound interest yields more return (money) than simple interest.
3. Level of education does not affect earning power.
4. The percentage of nonfamily households is decreasing.
5. People between the ages of 45–54 have the highest median incomes.

Pick the best answers to the following 5 questions.

6. Which of the following is not one of the three steps in the personal financial planning process?
   a. Setting financial goals
   b. Creating and activating plans
   c. Monitoring, evaluating, and revising plans
   d. Revenue sharing and reinvesting
7. __________ refer to what a person gives up in order to do something else.
   a. Inflation risks
   b. Opportunity costs
   c. Risk aversions
   d. The average propensities to consume
8. __________ refers to how readily an asset can be converted into cash.
   a. Wealth
   b. Credit
   c. Liquidity
   d. Churning
9. Which of the following is not one of the four key players in the U.S. economy?
   a. Brokers
   b. Consumers
   c. Government
   d. Business
10. The __________, the nation's central bank, has the responsibility of maintaining an adequate money supply.
    a. Federal Reserve System (nicknamed "The Fed")
    b. Securities and Exchange Commission
    c. Department of Treasury
    d. First National Bank

DISCUSSION QUESTIONS

1. What is the difference between simple interest and compound interest? Which one should consumers select, for example, when choosing a savings account? Explain your answer.
2. How can you use information about the state of the economy (inflation, interest rates, the Index of Economic Indicators, the Consumer Price Index) to influence the financial decisions that you make?
3. The United States economy has been in an expansion stage for many years, but how might people react if it suddenly took a downturn?
4. DeVane's uncle says he can't leave his job until he is fully vested. What does he mean by this?
5. How has this chapter made you rethink your financial situation and how you can improve it? What are your immediate financial goals?

DECISION-MAKING CASES

1. As a graduating senior, Jose has been offered a starting salary of $30,000. If Jose receives four percent raises each year, what will Jose's salary be in 5 years? Remember that each subsequent year will build on the past year. If the annual inflation rate is three percent, is he getting ahead?
2. Jennifer has inherited $5000 and wants it to earn money yet wants to keep it safe. If she invests in a treasury bond earning six percent interest a year, after two years with compound interest how much will she have in total?
3. A nutrition and fitness major, Simon, wants to open his own gym someday, but first needs to develop a personal financial plan. He has chosen the goal of owning a gym. What are the next two steps he should follow in the personal financial planning process? What specific decisions and actions (see Exhibit 1.1) would be involved in owning a gym?
4. Wiley has been told that a printed resumé mailed out to prospective employers may not be enough to compete in today's job market. What other types of resumés and what other types of methods are being used in job hunting? How do most people get their jobs?
5. Phil has been offered two different jobs, and he likes them equally well. The differences are in the salary and benefits packages and the perks. One offers a substantial cafeteria-style benefits plan with nearly everything, such as life insurance, extensive health and disability insurance, a dental plan, and perks like stock options, tuition reimbursement, and a company car, but the salary is $500 per year lower. The other offers minimal health insurance only. On the basis of the coverage in this chapter, what would you advise Phil to do?

GET ON LINE

Activities to try:

- Go to Monster Board (monster.com) or one of the other job-related sites targeted to students and locate your area of interest and see what jobs are posted. Summarize what you find.
- To check out government jobs go to www.hrsjobs.com for a database of federal jobs across the U.S. Provide a profile of desired job type, salary, and location, and this site will email applicable postings to you.
- More interested in corporate jobs? You can apply online directly to companies. To experience this, go to www.dell.com/dell/careers/index.asp and see what openings there are at Dell Computer, headquartered in Texas; or try Hewlett Packard at http://www.jobs.hp.com for jobs with their company around the world.
- For tips on preparing an electronic resumé go to www.excite.com/careers/careers_hub/top25 or www.eresumes.com. Report on how easy or difficult their instructions are to follow.
- To experience the financial planning process, go to FinancialEngines.com and enter your financial circumstances. The site tells you the odds that you will meet your financial goal and what specific steps you need to follow to make it happen.
- Visit www.money.com for general information and assistance on personal finance. Write a summary of what you find.
- For personal finance advice targeted to specific age groups, go to http://wsj.com, *The Wall Street Journal Interactive Edition's* Personal Finance Center. Enter in your age and see what advice comes up.
- Use a search engine such as Alta Vista (www.altavista.com) or Yahoo!(www.yahoo.com) and type in "personal finance"; see the variety of websites that come up. Find a site that interests you and explore its contents.

INFOTRAC

Each chapter in this text will provide an InfoTrac set of exercises to help you learn to research personal-finance related topics in the top journals, encyclopedias, and magazines available. To access InfoTrac, go to the address www.infotrac-college.com/wadsworth/access/htm. Use the passcode printed on your InfoTrac card to gain access to the online library.

Exercise 1: This chapter introduced the topic of personal finance. Using InfoTrac, type in he subject "personal finance." You will find reference book excerpts, over 500 periodical references (articles), subdivisions, and related topics. Click on view periodical references. Select one article and report on what it says. Does the article agree or disagree with the information given in the chapter?

Exercise 2: The chapter also covers career planning. Using InfoTrac go to the subject of career planning. You will find many articles on this topic, select one that interests you. What did you learn? Was the article helpful for someone beginning a professional career search?

chapter two

financial statements and budgets

Organizing Records
Making a Budget
Evaluating Net Worth

Did You Know?

- About 50% of Americans do not regularly balance their checkbook.
- There are over 500,000 financial planners in the United States.

Objectives

After reading Chapter 2 you should be able to do the following:

1. Understand the importance of keeping financial records.
2. Explain how to prepare an important numbers list and a budget.
3. Evaluate your net worth.
4. Evaluate your financial status by using three ratios.
5. Understand the process of evaluating and selecting financial advisors.
6. Explain how the Internet can be used to obtain financial information.

Chapter Overview

This chapter focuses on maintaining financial records, making budgets and net worth statements, and selecting financial advisors. It starts with record keeping because one of the most serious mistakes people make when managing their money is not keeping good records. Individuals who do not regularly balance their checkbooks (and about 50% of Americans do not) are probably not ready to move on to higher forms of finances such as investing. **Record keeping** is defined as the process of recording the sources and amounts of money earned, saved, spent, and invested over a specified time period. Budgets are based on financial records and show how your money flows over a period of time, such as a month or a year.

Record keeping
Process of recording sources and amounts of money, earned, saved, spent, and invested over a period of time.

Until you create a written record of where you have been, it is difficult to know where you are going. Keeping records of income and expenditures will provide the data needed to determine if progress is being made. In order to create written records, you need to find and organize financial information. By following the steps recommended in this chapter, you will lay a firm foundation on which to base future decisions.

The chapter contains case studies of two young single professionals and a married couple to show how budgets and net worth statements work. Successful financial planning requires the ability to identify strategies based on financial circumstances—how much money is coming in, how much money is going out, and what is left over. The case studies show how others manage their finances, and space is provided for you to write in your own monthly budget.

To aid in the organizational process, different types of financial planning software are described. Near the end of the chapter there is a discussion of how to evaluate financial advisors, followed by the Electronic Resources section with pertinent websites.

Objective 1:
Understand the importance of keeping financial records.

WHY KEEP GOOD FINANCIAL RECORDS?

As mentioned in the Chapter Overview, records give you a sense of where you have been and where you are going. They give you a basis on which to develop money management strategies.

Another reason financial record keeping is important is to help you plan for retirement. This was not an issue when few people lived past their productive work years, but now with improvements in water quality, nutrition, hygiene, and general health, life expectancy has risen. Along with this comes the necessity of making money stretch farther. Employer pension programs or retirement plans may be insufficient, or they may even be nonexistent in small businesses. If self-employed, you may be entirely on your own to build retirement funds in order to supplement your Social Security income. As increasingly many people want to retire early, this requires more planning to fill in the gap between, for example, age 55 and when pensions and Social Security begin. In order to calculate retirement income needs, many software programs and websites ask individuals to estimate their earnings over the next ten, twenty, or thirty years. Obviously this is not easy to do because it is difficult to estimate how much salaries will increase. The only way to approach estimates such as these is to base them on previous records such as average pay raises over the last five years.

The keeping of good records is useful in a crisis. For example, in one family with two elementary age children, the father suddenly died of a heart attack at age 48. Kay, his widow, found that within days of the funeral, bills had to be paid, insurance policies found, and so forth. Even though she worked full-time and was the major breadwinner, Kay had never handled the family finances. Her husband had left receipts and bills scattered on a desk at their house, nothing was labeled or organized, and creditors were asking for information that she had no idea how to find. Kay said it took nearly six months to straighten everything out, and it added stress to an already stressful situation.

Records Reflect Consumption

Financial records are also useful because they provide a picture of how money is spent. As noted in Chapter 1, a college student was spending over $1,000 a month on phone calls before her parents placed a $1,000 a month limit. Have you noticed the range of spending behaviors among students? Regarding consumption patterns, consider the differences between the following families:

> The Phelps family of four in Croton-on-Hudson, N.Y., has a cushy income of about $150,000 a year. For them, family togetherness means frequent boating outings on the Hudson River or foreign travel.
>
> The Kelly family [of five] of Springfield, Mo., manages on much less: $32,000 a year. The Kellys' version of family fun usually involves hunting, fishing, and playing softball.
>
> Two families, vastly different circumstances. Yet their purchasing priorities are strikingly similar and mirror those of the nation as a whole: They spend as much as they can on shared pastimes, hobbies and outings. And, they're cutting back on furniture, clothes and other nonessential household goods.[1]

Do your family's consumption patterns have anything in common with either of these families, or are they totally different? Have your personal consumption patterns changed over the last five years? Do you anticipate any changes in the next five years?

Organizing Records

So far the chapter has established that record keeping is important as a planning tool and as a revealer of consumption patterns. But what kind of records need to be kept, and for how long and where should they be kept? Exhibit 2.1 shows a suggested list of financial records and length of time each should be kept.

Besides the financial records shown in Exhibit 2.1, there are other kinds of documents that have financial or legal implications that you would also want to keep in a safe place. Their location should be accessible to you and also to your spouse, a parent, sibling, or an executor (also known as a personal representative, the person named in a will to carry out the directions and bequests in a will) in the event of an emergency or crisis. So the location of financial and legal records should be known to a select number of people that you trust. Exhibit 2.2 gives a list (also called an **inventory**) of typical items and a space to write in where they are located (file drawer at home, safe deposit box etc.). Kay, the widow described earlier, could have used such a list.

Inventory
Location list of financial and legal records.

If you are a young, single college student, it is likely that your parents have some of these records and as you start out on your own and establish a permanent residence, the records will transfer to you or you will begin your own set. In any event you should find out where your birth certificate is located and any other items in Exhibit 2.2 that pertain to you. Besides crisis events such as Kay experienced, there are other reasons why you may need these documents, such as having a copy of your birth certificate to get a passport or a copy of your undergraduate transcripts to apply for graduate school.

Safe-Deposit Boxes

For maximum security some records should be kept in a bank safe-deposit box. The smallest boxes (generally 2" by 5" by 18" long) cost about $15 to $25 a year. The largest boxes which are file cabinet size, cost about $100–$150 a year. Typically stocks and bonds certificates, property deeds, and marriage and birth certificates are kept in safe-deposit boxes. Other possible contents include:

* passports (unless used frequently, then these should be kept at home).
* citizenship or naturalization papers.
* divorce, child custody, and adoption agreements.
* military service papers.
* expensive or sentimental jewelry, coin collections.

EXHIBIT 2.1
Records to Keep

Documents, Records, Items	How Long You Should Keep Them Typically
Receipts and Warranties for major purchases	As long as you own them
Canceled checks	Three years or more
Bank Statements	Three years or more
Credit card receipts	Until you receive the statement showing the charge
Credit card statements	Three years or more
Monthly brokerage and mutual fund statements	At least until you receive the next statement, usually keep at least one year
Annual brokerage and mutual fund statements	Three years or more
Real estate documents (title, deed, mortgage contract, receipts for improvements or repairs	As long as you own the property and up to three years after the sale
Supporting documents for your income tax return (including income statements)	Three years after you filed the return
Income tax returns	Indefinitely*
Property tax receipts	At least three years
Wills	Permanently
Birth and Marriage Certificates	Permanently
Adoption papers	Permanently
Social Security data	Permanently
Stock certificates/bonds	As long as you own them

*An income tax preparer may be asked, for example, the date a pervious house was purchased. Income tax returns provide an accurate record of this sort of information.

- photographs or a videotape of possessions (usually of cars or home interiors), including written appraisals or sales receipts of valuable items.
- life-insurance policies.

Home Storage of Records

Documents used often are more likely to be kept at home. Tax records, checking account and credit card statements, homeowner's and auto insurance policies, pay stubs, debt statements (mortgage, car lease contracts, loans), and health care cost records should be safe but accessible.

For added security individuals may want home safes. Safes should be rated at least Class 350 for fire resistance by the Underwriters Laboratories, Inc. To protect computer disks, a Class 125 is recommended. Plain metal boxes are not recommended because the papers inside would burn from the heat of a fire.

Tax Considerations

For tax purposes other records to keep include income statements from employers and returned checks. All bank records for the current year—including statements, deposit slips, canceled checks, and evidence of securities transactions (stocks, bonds, and mutual funds bought and sold, dividends)—should also be kept. After completing taxes, last year's records should be labeled and stored, separate from the year in progress.

EXHIBIT 2.2
Inventory (List of Financial and Legal Documents)

Item/Document	Location
Auto/Other Insurance Policy(ies)	
Bank Statement(s)	
Birth Certificate(s)	
Brokerage Statement(s)	
Bond(s)	
Certificate(s) of Deposit	
Diplomas or transcripts	
Divorce Decree (s)	
Employee Benefit/Pension Records	
Health and Disability Insurance Policy(ies)	
Homeowner's Insurance Policy	
Life Insurance Policy(ies)	
Loan Document(s)	
Marriage Certificate(s)	
Mutual Fund Statement(s)	
Passport(s)	
Real Estate Deeds/Mortages	
Retirement Plan/IRA Records	
Social Security Card(s)	
Stock Certificate(s)	
Tax Returns – Past 3 Years	
Trust(s)	
Videotape or photographs of possessions	
Will(s)	
Other	

Since most audits occur between one to three years after a return is filed, most financial records should be kept for three years. The main exception to this rule is that the Internal Revenue Service (IRS) can go back six years if income is underreported by more than 25 percent. Thorough records will be useful in clearing up questionable items and arriving at corrections with a minimum of effort. The types of records that you should keep depend on the complexity of your finances and your return. For example, if you do not itemize but take the standard deduction, you will not need as many receipts and other proofs of expenditures. Details on taxes will be given in the next chapter.

Long-term Financial Records to Keep

Birth and marriage certificates, adoption papers, wills, and Social Security data should be kept permanently. Investment records should be kept as long as an investment is owned. Real estate records of the selling of property should be kept at least for three years after a sale due to a possible IRS audit.

Software, Technology, and Records

It was hoped that computers would simplify the number of financial and legal documents (paperwork) that we all have to juggle, but this has not been the case. Although computers have revolutionized the way finances are handled, we are still not at the point where a single number or code word will bring up all the records we need. Most of us are drowning in paperwork because computers have generated more paper, information, and record keeping options than ever before. Much of the paperwork produced circulates in the mail and arrives in our mailboxes in the form of bank statements and credit card bills.

One way to simplify financial record keeping is to use personal financial management software programs such as Quicken and MS Money. These programs keep track of expenses, investments, and bank and credit card balances; they prepare financial statements and budgets and print checks. Their most common use is to balance checkbooks. Each offers a wide variety of spin-off products, such as legal programs and retirement planning programs. For example, the Quicken Family Lawyer provides 74 legal documents, including a simple will and organ donation/refusal forms.

There are many competitors to the two programs mentioned. For example, WealthBuilder offers a sophisticated planning program with a portfolio manager to track investments. It allows a user to access online services to follow prices on securities and mutual funds and financial news updates.

Tax software programs are especially popular such as TurboTax, Kiplinger TaxCut, and Tax Saver. These products come in simple and deluxe versions. Many of the different types of software are designed to work with each other. For example, numbers on Quicken can be transferred automatically to TurboTax to make it easier to do taxes.

Since software prices vary considerably, it is worth investing your time to comparison shop. Software can be purchased in a number of ways, including by telephone, mail, fax, and the Internet; at retail stores; and through catalogs. An alternative to purchasing commercially available software is to design your own system of record keeping using a spreadsheet program to set up a budget and an expenditure program.

The main drawback to using software or setting up a system is that it takes time to learn how to use it. A rarer problem, but it happens, is that once the record keeping activity is established it can become too much fun. For example, a freelance writer got so involved in using Quicken that she was spending two hours every weekday morning tracking her expenses and investments. This cut into her income-producing writing time and whatever benefit she was deriving from tracking her money was being offset by spending too much time away from her writing.

Objective 2:
Explain how to prepare an important numbers list and a budget.

IMPORTANT NUMBERS LIST

Another useful tool in organizing financial records is an important numbers list such as the one shown in Exhibit 2.3. This type of list, gives company names, account numbers, phone numbers, and websites in case credit or debit cards are lost or stolen. If credit cards are lost or stolen it is especially important to have credit card numbers listed in one place that is easily accessible. For example, Courtney, a 20-year-old junior, reported in class that her mother called her at school and wanted to know why Courtney's credit card bill showed so many purchases at stores back home that she normally did not frequent. It turned out that someone had stolen Courtney's card when she was home for the holidays and was using it. Her numbers list came in handy when she called the credit card company to put a stop to the illegal use of her card.

Keep copies of numbers lists at home and give a copy to a spouse or parent, or let them know where the lists are kept. Another example of stolen credit cards occurred when Alyssa Lappert left her purse in a clothing store dressing room for a few minutes when she went out into the store to get other sizes. When she went to pay for her purchases, she realized she had been robbed. Alyssa called home. Her

EXHIBIT 2.3
Important Numbers List

Company Name	Account Number	Phone Number	Website
American Express	4955-123456-12345	800-528-4800	www.americanexpress.com
Discover	731 1234 1234	800-DISCOVER 800-347-2683	www.discovercard.com
First Union Visa	5663 1234 12341234	800-735-1012	www.firstunion.com/home-d.html
Bank of America MasterCard	5467 1234 12341234	800-732-9194	www.bankofamerica.com
Suntrust MasterCard	5468 1224 12341234	800-786-8787	www.suntrust.com
Citibank Visa	5240 1234 12341234	800-950-5118	www.citibank.com
Chase Manhattan Bank Classic MasterCard	5260 1234 12341234	800-633-0458	www.chase.com

Other Important Numbers

Company Name	Account Number	Phone Number	Website
Allstate Insurance	123456789	850-576-3986 (local claims office)	www.allstate.com
Blue Cross/Blue Shield	123456789	800-734-6656	www.bluecross.com
Suntrust ATM (do not include PIN)	5894 1234 1234 1234	800-734-6656	www.suntrust.com
Saturn VIN (Vehicle identification Number)	J4X23123456789	800-553-6000	www.saturn.com
Gateway CPU (hard drive)	Serial #0123456789 Customer #B1234567	800-GATEWAY 800-428-3929	www.gateway.com
Sony 25" TV	Serial #123456789	850-5262-8787 (local service center)	www.sony.com

Name	Function	Address/Phone Number
Jay Dowling	Insurance Agent	2000 Main St., 859-1725
Simon Crocker	Attorney	325 Anhinga Rd., 859-2222
Mary Sayer	Stockbroker	A.G. Edwards, 324 May St., 859-5666
Jane Boone	Accountant	Boone and Clarke, Inc. 888 S. Ride St., 859-7777
Dana Carson	Financial Planner	Moxley Financial, 24 7th St. 859-0361

Your List

Name	Function	Address/Phone Number

EXHIBIT 2.4
Sample List of Key Advisors

husband, using the numbers list called the credit card companies and told them about her loss. The cards were never found or returned, and the Lapperts replaced the cards immediately with ones with new numbers.

Another type of list is the one given in Exhibit 2.4. It shows a sample list of names and addresses of key advisors, such as stockbrokers, accountants, lawyers, and insurance agents.

Now that the importance of keeping records and numbers lists has been established, the next step is to use the information they contain to design money management strategies and assessment tools, such as monthly budgets and net worth statements.

Financial Statements

Financial statement
Assessment of current status of finances.

Budgets
Trace the flow of income and expenses over a period of time.

Balance sheets or **net worth statements**
List the value of assets and liabilities.

A **financial statement** is an assessment of the current status of an individual's, a family's, or a household's finances. The two main types of statements are **budgets** which trace the flow of income and expenses over a set period of time, and **balance sheets** (also called **net worth statements**) which list the value of assets and liabilities. A net worth statement reveals one's overall financial worth.

Michael's Monthly Budget		Your Monthly Budget
CASH INFLOWS		
Net Salary	$2,000.00	______
Interest	10.00	______
Total Cash Inflows	2,010.00	______
CASH OUTFLOWS		
Housing		
Rent	700.00	______
Utitities	100.00	______
Telephone	20.00	______
Cable TV	25.00	______
Auto Loan	300.00	______
Auto Insurance	75.00	______
Auto Maintenance/Gas	50.00	______
Clothing	175.00	______
Credit Card Payments	30.00	______
Entertainment	75.00	______
Food	200.00	______
Gifts/Contributions	20.00	______
Medical/Dental	20.00	______
Personal Care	30.00	______
Savings	25.00	______
Other Expenses	100.00	______
Total Cash Outflows	**1,945.00**	______
Monthly Summary		
Total Cash Inflows	2,010.00	______
Less: Total Cash Outflows	(1,945.00)	______
Surplus (Deficit)	$65.00	______

EXHIBIT 2.5
Michael's Monthly Budget

Budgets

The word "budget" strikes fear in the hearts of people because it sounds like scrimping, hardship, or sacrifice, but a budget simply shows an individual's or a family's projected income, expenditures, and savings over a period of time—usually a month or a year.

For budgets to work, individuals have to be committed to tracking where their money is going. Sometimes it takes a crisis like an over-the-top credit card bill, a bounced check, or a letter from an angry creditor to stimulate the need to rein in spending. Often people are afraid to take a hard look at where their money goes.

Exhibit 2.5 gives an example of Michael's monthly budget (cash inflow and outflow) and provides blanks for you to fill in with your own income and expenditures. Michael's budget is based on an annual salary income of $30,000, which—after taxes and Social Security are taken out—allows him to bring home (net pay) about $2,000 a month. He pays rent on his own apartment and saves $25 a month. His goals include paying off his car loan and saving more each month. Since in an average month he has $65 left over (see Surplus at the bottom of Exhibit 2.5), he could save more.

In reviewing your own expenses, do not be shocked if they are too high. Most people underestimate how much money they spend each month especially on miscellaneous items. If expenses far outweigh income, you have four choices:

1. Increase your income.
2. Cut back on expenses.
3. Sell something.
4. Try a combination of 1, 2, and 3.

For example, Jamie, a college student, chose option #1. She works in a restaurant during the school year; in the summer and over Christmas break she works extra nights and weekends at a department store to increase her income. Charlie, a 37-year-old accountant, chose option #3. He decided his new Lexus did not give him as much pleasure as he had hoped so he sold it to a friend and bought a Honda Accord instead, paid off his credit card bills, and pocketed what was left.

To work well, a budget should be flexible, bending with changing needs. It should not be a straitjacket on spending. Anything too rigid won't last.

A budget consists of three parts income (total income received), expenses (total expenditures made), and net income (net gain or net loss, figured by subtracting expenses from income).

Income can come from many sources. Here is a list of possible income sources:

* Wages, salaries, bonuses, commissions (*this category is the main source of income for most people*).
* Alimony and child support.
* Return on investments (interest and dividends from savings accounts, bonds, stocks, loans to others).
* Allowances or gifts (inheritances).
* Social Security benefits.
* Public assistance.
* Gains or losses from sale of assets.
* Other sources (rent income, royalties, loans, scholarships, tax refunds).

Fixed expenses
The same amount each time period.

Variable expenses
Expenditures over which individual has more control.

Expenditures can be broken down into two types: **fixed expenses** (usually the same amount each time period) and **variable expenses** (expenditures over which the

EXHIBIT 2.6
Fixed and Variable Expenses

Examples of typical fixed expenses:

- Auto (lease or installment payments)
- Cable television
- Contributions (giving a set, regular amount to a church or charity, such as United Way every week)
- Installment loans (appliances, furniture)
- Insurance payments
- Pension contributions (IRA, employer's plan)
- Rent or mortgage payments
- Savings and investments (regular fixed deposits)
- Taxes
- Child care (if same each month)

Examples of variable fixed expenses:

- Child care (after school, babysitting)
- Clothing, cosmetics, and accessories
- Contributions (irregular, different times to charities, schools, churches, organizations)
- Credit card payments
- Education (tuition, books, fees)
- Food and beverages (at home and out)
- Gifts
- Household furnishings and utilities
- Leisure and recreation
- Medical (dental, physicians, medicines/treatments)
- Miscellaneous (hobbies, newspapers, magazines, postage, etc.)
- Personal care (barbers, beauticians, dry cleaners)
- Savings and investments
- Tobacco, alcohol
- Transportation and gasoline
- Vacations

individual has considerable control). For examples of fixed and variable expenses, see Exhibit 2.6. Note that contributions and savings and investments can be fixed or variable. Child care can also be either or both. For example, there may be a set monthly charge of $500 for a 4-year-old's child care center, but the amount paid to babysitters when his parents go out on occasional weekend nights varies considerably.

Exhibit 2.7 shows how Dawn O'Brien, a college graduate with a double major in creative writing and psychology working as an assistant editor in a large city, kept a record of how she spent her money over the course of a year. She then took her typical yearly income (take-home net pay plus interest she earns on her savings account) and subtracted her expenses to figure what was left. The process can be diagrammed as shown in the box.

The General Format

Cash received (income) during a time period	Cash outflow −during the time period	Cash surplus = or deficit
Dawn's Case		
$32,601	−$31,362.00	= $1,239.00

Dawn's principle source of income is her job. Her monthly salary (or gross *income*) is $3,800, but she takes home only $2,700 a month (net *pay*) after deductions for Social Security and taxes are taken out. So Dawn has $2,700 to spend, save, or invest each month. Take-home pay is also called disposable income. She also receives about $200 a year in interest payments on her savings account, but she does not think of this as

EXHIBIT 2.7
Dawn's Annual Budget vs. Actual Spending Summary

	Monthly	Actual Spending												Annual Totals	
	Budget	Jan	Feb	Mar	Apr	May	Jun	Jul	Aug	Sep	Oct	Nov	Dec	Actual	Budgeted
Net Salary	2,700.00	2,700.00	2,700.00	2,700.00	2,700.00	2,700.00	2,700.00	2,700.00	2,700.00	2,700.00	2,700.00	2,700.00	2,700.00	2,700.00	2,700.00
Interest	17.00	15.00	15.00	15.00	16.00	16.00	16.00	17.00	17.00	18.00	18.00	19.00	19.00	201.00	204.00
Cash Inflows	2,717.00	2,715.00	2,715.00	2,715.00	2,716.00	2,716.00	2,716.00	2,717.00	2,717.00	2,718.00	2,718.00	2,719.00	2,719.00	32,601.00	32,604.00
Savings	550.00	600.00	550.00	550.00	550.00	550.00	550.00	550.00	550.00	600.00	550.00	550.00	550.00	6,700.00	6,600.00
Housing	200.00	200.00	200.00	200.00	200.00	200.00	200.00	200.00	200.00	200.00	200.00	200.00	200.00	2,400.00	2,400.00
Telephone	25.00	25.00	20.00	30.00	18.00	26.00	31.00	40.00	10.00	15.00	25.00	30.00	50.00	320.00	300.00
Food(away from home)	225.00	210.00	232.00	218.00	222.00	216.00	190.00	195.00	224.00	197.00	220.00	255.00	278.00	2,657.00	2,700.00
Clothing	200.00	98.00	187.00	120.00	156.00	185.00	177.00	215.00	222.00	187.00	295.00	256.00	451.00	2,549.00	2,400.00
Public Transportation	150.00	150.00	150.00	150.00	150.00	150.00	150.00	150.00	150.00	150.00	150.00	150.00	150.00	1,800.00	1,800.00
Auto Loan	300.00	300.00	300.00	300.00	300.00	300.00	300.00	300.00	300.00	300.00	300.00	300.00	300.00	3,600.00	3,600.00
Auto Maintenance/Gas	100.00	98.00	89.00	107.00	125.00	108.00	255.00	111.00	115.00	102.00	98.00	105.00	110.00	1,423.00	1,200.00
Student Loan	300.00	300.00	300.00	300.00	300.00	300.00	300.00	300.00	300.00	300.00	300.00	300.00	300.00	3,600.00	3,600.00
Auto Insurance	75.00		225.00			225.00			225.00			225.00		900.00	900.00
Entertainment	125.00	110.00	99.00	136.00	109.00	130.00	129.00	115.00	120.00	130.00	115.00	138.00	156.00	1,487.00	1,500.00
Gifts/Donations	100.00	95.00	50.00	70.00	75.00	83.00	155.00	93.00	80.00	50.00	75.00	125.00	400.00	1,351.00	1,200.00
Miscellaneous Expense	200.00	199.00	215.00	206.00	215.00	230.00	215.00	218.00	226.00	215.00	225.00	235.00	176.00	2,575.00	2,400.00
Cash Outflows	2,550.00	2,385.00	2,617.00	2,387.00	2,420.00	2,703.00	2,652.00	2,487.00	2,722.00	2,446.00	2,553.00	2,869.00	3,121.00	31,362.00	30,600.00
Cash Inflows	2,717.00	2,715.00	2,715.00	2,715.00	2,716.00	2,716.00	2,716.00	2,717.00	2,717.00	2,718.00	2,718.00	2,719.00	2,719.00	32,601.00	32,604.00
Less: Cash Outflows	(2,550.00)	(2,385.00)	(2,617.00)	(2,387.00)	(2,420.00)	(2,703.00)	(2,652.00)	(2,487.00)	(2,722.00)	(2,446.00)	(2,553.00)	(2,869.00)	(3,120.00)	(31,362.00)	(30,600.00)
Surplus (Deficit)	167.00	330.00	98.00	328.00	296.00	13.00	64.00	230.00	(5.00)	272.00	165.00	(150.00)	(402.00)	1,239.00	2,004.00

disposable income because she is leaving the money in to grow. When she has accumulated enough savings, she hopes to use other forms of investing that may pay higher interest rates.

Dawn is living at home with her parents for three years so she can pay off student loans and save for the downpayment and closing costs on a townhouse in the suburbs close to the city. This is why her savings rate is unusually high. She pays her parents $200 a month for living expenses that include her meals at home, her room and share of utilities, and laundry. This $200 is a fixed expense. Many of her other expenses vary, such as food away from home since she eats her lunch out every day and goes out after work and on weekends with friends. Dawn rides a commuter van (listed under public transportation in Exhibit 2.7) to her publishing job in the city, so that is a fixed expense of $150 a month, but she saves money on gasoline and a parking space during the week. At her parents' home she keeps a car, so she has gasoline and repair expenses for it.

Seeing where money goes for a whole year, as Dawn has done, is useful because some expenses fluctuate from month to month. Because of these fluctuations throughout the year, an annual statement rather than a monthly one (although a good place to start) gives the clearest financial picture.

The new townhouses in the area that she likes cost $130,000 and require a 10 percent downpayment of $13,000 and closing costs of around $6,000, so she needs to save $19,000 before she moves out of her parents' home into her own. She would rather rent an apartment with friends during these interim years, but she knows she can save a lot more by living with her family. Since the unit she likes has three bedrooms, during the first few years she figures she can rent out one of the bedrooms to a college student or a coworker and use the extra bedroom for a home office. The rent will help her meet the monthly mortgage payment. If she is promoted to an associate editor position with an increase in salary, she may be able to move in sooner than expected. Although Dawn feels the strain of trying to save so much each month, she likes the idea of building an investment through owning rather than renting.

Budget-Making Steps

Dawn went through several steps in the budget-making process. The steps in budget making consist of the following:

* *Setting goals.* Desired end states. Dawn wants a townhouse of her own and a promotion to associate editor.
* *Planning.* Deciding the actions and strategies that need to be taken. Deciding between two or more alternatives. A decision in Dawn's case was to save money by living at home with her parents.
* *Implementing.* Putting plans into action. Dawn saves part of her paycheck for the downpayment and tries to curb spending.
* *Evaluating.* Judging or examining the cost, value, or worth of the budget plan. Near the end of three years, Dawn will look back and evaluate her progress. Has she met her goals? If she has not already done so, Dawn would be well advised to develop a balance sheet (to be discussed next). As savings and investments build, it is useful to get a broader picture of one's total net worth.

Objective 3:
Evaluate your net worth.

THE PERSONAL BALANCE SHEET: NET WORTH

To make a net worth statement, subtract what is owed (debts) from what is owned to arrive at a figure. This is diagrammed as shown in the box.

Assets - Liabilities = **Net worth**

(What is owned) - (What is owed) = Net worth

Assets
What is owned.

Liabilities
What is owed.

Net worth
Assets minus liabilities.

Assets are *what you own*, such as a car fully paid for, and **liabilities** are *what you may owe* such as student loans.

Assets can be divided into three categories:

1. Monetary assets: liquid assets such as cash on hand, checking and savings accounts, money market accounts and certificates of deposit (CDs).

2. Tangible assets: items that can be sold for cash, such as cars, computers, houses, boats, and furniture. Usually these lessen in value as they age; exceptions include vintage cars and antiques. The tendency is to over value one's possessions. Ask, "Realistically, how much would I get today if I sold a certain item?"

3. Investment assets: items acquired that hopefully increase in value such as real estate, stocks, bonds, company pensions, and mutual funds.

Liabilities can be divided into two categories based on length of time:

1. Current liabilities are those that are due to be paid within a year. Examples of these are utility bills, taxes, insurance premiums, rent, medical/dental bills, short-term loans, and parking tickets. Some colleges and universities withhold diplomas until all debts (library fees, parking tickets) to the college or university are paid.

2. Long-term liabilities are those that go beyond a year, such as real estate mortgages (typically 15 or 30 years) and appliance, furniture, car, or education loans.

Although liabilities are usually thought of as negative (something to be avoided), when people get into a high tax bracket due to their wealth and investments, they may choose to reduce their tax burden by declaring liabilities (losses) on certain investments or business expenses.

The net worth statement of Brian and Kaytoria Wright—who have been married five years, do not have any children, and have a small house—is given in Exhibit 2.8.

Evaluating Your Net Worth

Using the formula, net worth = assets - liabilities, you can make a rough estimate of your net worth. Do not be discouraged by a low net worth figure or even a negative one if you are a fulltime student. This should change in a few years and this book is written to address the future as well as the present. Take comfort in the fact that government statistics reveal that the age group 55 to 64 has the highest net worth (see Exhibit 2.9).

For further interpretation net worth evaluations can be divided into four categories:

1. If your net worth is negative (meaning you owe more than you own), you should concentrate on getting out of debt. Pay off what you owe as soon as you can and get going on the plus side.

2. If your net worth is less than half of your annual income, keep spending in check and save and invest more. Work at increasing your net worth.

3. If your net worth falls between more than half of your annual income, but less than two or three years' annual income, then you are in fairly good shape especially if you are young.

4. If your net worth is over three years' worth of annual income, you are in very good shape for the future. Keep building.

EXHIBIT 2.8
Figuring the Wrights' Net Worth by Creating a Personal Balance Sheet

Brian and Kaytoria Wright
Personal Balance Sheet
December 31, 1999

Step 1
Prepare a list of all items of value (assets). Include bank account balances, real estate, investments, and the estimated current value of your possessions.

ASSETS		
Current Assets		
Checking Account	$540	
Savings/Money Market Accounts	2,360	
Cash Value of Life Insurance	980	
Total Current Assets		$3,880
Real Estate		
Current Market Value of Home		95,000
Investments		
Individual Retirement Accounts	6,430	
Mutual Funds	2,140	
Total Investments		8,570
Other Assets		
Market Value of Automobile	6,500	
Furniture and Fixtures	5,400	
Stereo and Video Equipment	2,100	
Computer	1,600	
Jewelry	1,950	
Total Other Assets		17,550
Total Assets		$125,000

Step 2
Prepare a list of all amounts owed to others (liabilities). Examples include credit card and charge account balances, auto loans, and amounts due on mortgages and other loans.

LIABILITIES		
Current Liabilities		
Medical/Dental Bills	$125	
Credit Cards	540	
Auto Loan	600	
Total Current Liabilities		$1,265
Long-Term Liabilities		
Home Mortgage	73,050	
Student Loan	5,000	
Total Long-Term Liabilities		78,050
Total Liabilities		79,315
Net Worth		45,685

Step 3
Subtract total liabilities from total assets to determine net worth.

If you are a college student, it is not unusual for you to fall into category one or two. If you are over thirty and working fulltime, then being in category one or two indicates the need to reassess your money management practices.

As you get older, financial planning becomes more crucial because more money is involved and you have fewer years to correct mistakes. Your net worth should steadily increase as long as your income outdistances your expenditures. For further analysis of your financial standing, use the personal financial planning software mentioned earlier or access financial planning advice on the Web (see the Electronic Resource section for suggestions) or consult with a financial planner or other finance professional.

EXHIBIT 2.9
Net Worth

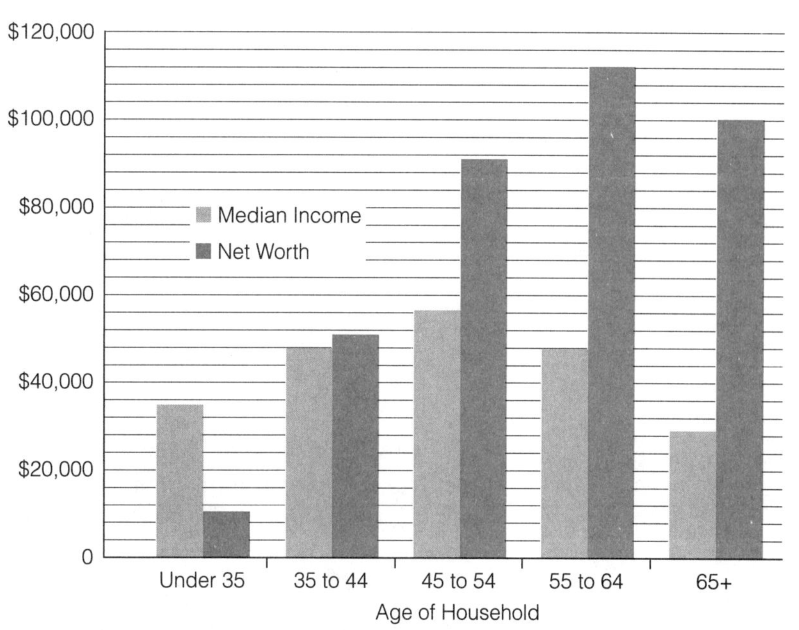

HOW SOLVENT ARE YOU?

Solvency
Ability to pay debts.

Insolvent
Person owes more than he or she owns.

The ability to pay debts is called **solvency**. If a person owes more than he or she owns then that individual is considered **insolvent**.

Three Financial Ratios

Objective 4:
Evaluate your financial status by using three ratios.

The three financial ratios (mathematical formulas) discussed in this section (solvency ratio, debt-to-total assets ratio, and debt-to-income ratio) are used to determine a person's financial status (beyond the net worth statement already presented) and to indicate the degree of solvency.

Solvency ratio
Extent to which an individual is exposed to insolvency.

Solvency Ratio The **solvency ratio** shows, in percentages, the extent to which an individual is exposed to insolvency. It indicates the potential to withstand financial problems by dividing total net worth by total assets. The calculations and an example, using the Wrights' net worth and asset figures from Exhibit 2.8, are as shown in the box.

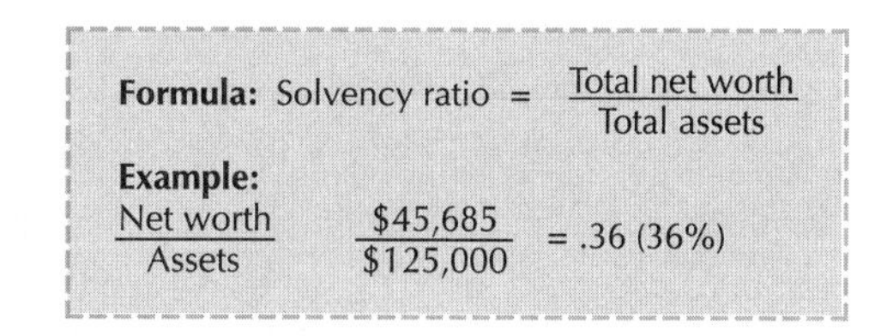
Formula: Solvency ratio = $\frac{\text{Total net worth}}{\text{Total assets}}$

Example:
$\frac{\text{Net worth}}{\text{Assets}}$ $\frac{\$45,685}{\$125,000}$ = .36 (36%)

Therefore the Wrights' solvency ratio is 36 percent. This means that Brian and Kaytoria could withstand around a 36 percent decline in the value of their assets before they would become insolvent.

Debt to total assets ratio
Liabilities divided by assets.

Debt-to-Total Assets Ratio Another way to determine solvency is to figure a **debt-to-total assets ratio**, defined as total liabilities divided by total assets.

For example, at the end of the year the Wrights (whose net worth was given in Exhibit 2.8) had total liabilities of $79,315 and total assets of $125,000. Their house is their greatest liability (they owe $73,050 on their mortgage), but it also accounts for some of their assets. Thus their debt-to-total assets ratio was .63 (or 63%) as shown in the box.

Formula: Debt-to-asset-ratio = $\frac{\text{Total debt}}{\text{Total assets}}$

Example:
$\frac{\text{Debts}}{\text{Assets}}$ $\frac{\$79,315}{\$125,000}$ = .63 (63%)

The 63 percent indicates that the Wrights are solvent. They own more than they owe. They can pay their debts because they have more assets than liabilities.

Debt to income ratio
Measures financial health by dividing loan and debt payments by income.

Debt-to-Income Ratio Another ratio that measures financial health is called the **debt-to-income ratio** or the **debt service ratio**. This ratio is calculated by dividing annual, or monthly, loan and debt payments by annual, or monthly, gross annual income.

As an example, let's figure the Wrights' debt-to-income ratio based on their gross annual income of $48,448 and annual loan payments of $13,500 ($10,800 for the mortgage loan and $2700 for other loans/payments). Their debt ratio amounts to 27 percent. The calculations are shown as follows:

Formula: Debt-to-income ratio = $\frac{\text{Annual debt repayments}}{\text{Gross income}}$

Example:
$\frac{\text{Debts}}{\text{Gross income}}$ $\frac{\$13,500}{\$48,448}$ = .27 (27%)

Brian and Kaytoria Wright did not know how to interpret 27%. Was it good or bad compared to how other people are doing? Since a *debt service ratio of 40 percent or less is viewed by many experts as a sign of sound financial health*, they are in good shape. Lenders are reluctant to extend credit/give loans to people whose ratio is over 40 percent. A more conservative figure is .36 or 36 percent, which some experts use as a financial yardstick because it allows for more flexibility. Even using the more conservative measure, Brian and Kaytoria are doing well.

The Wrights are solvent. They are on their way to improving their financial position. For now the Wrights feel confident about the way things are going, but in the long run they may seek outside advice. Who they could turn to for general financial advice is covered next and will be discussed further in upcoming chapters on banking, stocks, bonds, mutual funds, insurance, retirement, and estate planning.

Objective 5:
Understand the process of evaluating and selecting financial advisors.

SELECTING FINANCIAL ADVISORS

In choosing an advisor, what factors should the Wrights consider? They should find out about charges and credentials and gather recommendations from friends and relatives and trusted finance professionals.

Charges/Fees

Regarding charges and fees, they can be any of the following:

Fee-only. Advisors charge $50–$250 an hour for advice and to develop a plan. To maintain and update your plan, including buying and selling investments, a common arrangement is a yearly charge from 1 percent to 2 percent of your assets.

Commission-only. Advisors live solely on commissions, so if you buy a mutual fund or a life insurance policy through them they make a commission. For example, if you invested $5,000 into a mutual fund, the planner may take out 8.5% as a fee ($425), so that the amount actually put into the mutual fund would be $4,575. Because they receive greater commissions with certain types of funds or policies, they may not suggest the least costly alternative to the client. Some experts recommend avoiding advisors who earn commissions on the products they sell.[2] A counter point of view is that in the long run, if they want the client's return business and good word-of-mouth recommendations, they will consider the client's welfare above their own short-term gains.

Fee and commission-based Advisors charge an upfront fee for advice and a commission on any securities bought through them. *Most planners fall into this group* because it gives them the greatest flexibility.

Decisions about whom to choose should rest on the types of products and services desired and whether fees are justified. Good advice that saves money and makes a client more confident about the way money is invested and protected is worth a great deal. However, to get to this point there are more steps that need to be taken, especially the one to be discussed next, credential checking.

Credentials and Regulations

When selecting a financial advisor, another criterion is the advisor's credentials. *With over 500,000 people in the United States calling themselves financial planners*, there are outright fraudulent operators and many with limited skills.

Until 1997 financial advisors were regulated under federal law by the Securities and Exchange Commission (SEC). However, because there are so many advisors, studies have shown that the SEC was visiting some financial advisory firms only once in every 40 years.[3] Recently Congress changed the regulations so that state regulators became responsible for many of the SEC's former responsibilities. Under the National Securities Markets Improvement Act of 1996, the SEC regulates financial advisors with more than $25 million in assets under management. The regulation of the smaller firms falls to the states. The exception is that the SEC regulates investment advisors in Ohio and Wyoming because they have not adopted advisory laws as of the writing of this book.

Even in states with regulations, the credentials of financial advisors are still quite loose compared to credentialism in other fields. For example, "You have to show a higher level of competency to be a barber in some states than to be an investment advisor," said Denise Voight Crawford, securities commissioner of Texas and president of the North American Securities Administrators Association.[4]

So be forewarned. The term caveat emptor—"may the buyer beware"—is appropriate when selecting a financial advisor.

The knowledge gained from this book will help you separate the dependable from the unscrupulous. As alluded to earlier, a fundamental problem is that the advisor has an incentive to get you to change your investments (because this may be the main way they make their money), even if you do not really need to. A good advisor will explain why he or she is suggesting certain courses of action and will tell you upfront about the commissions he or she stands to earn on a mutual fund or an annuity.[5]

Professional designations and credentials for financial planners or advisors include all of the following (as of the writing of this book):

Certified financial planner (CFP). This is the best-known credential. A CFP has to pass rigorous examinations and be approved by the International Board of Standards and Practices for Certified Financial Planners (IBCFP).

Chartered financial consultant (ChFC). A ChFC has to pass tests through correspondence courses given by American College in Bryn Mawr, Pennsylvania.

Accredited financial counselor (AFC). An AFC has passed two tests and subscribes to the Association for Financial Counseling and Planning Education code of ethics.

Registered investment adviser (RIA). A RIA has registered with the Securities and Exchange Commission (call 800-732-0330 to check on this credential or to register complaints). The RIA designation means the adviser has paid a fee but has not necessarily passed any tests. The title is obtainable for $150, with no exam in most states.[6]

Mutual fund chartered counselor (MFCC). An MFCC passes a multi-part educational program and a final exam provided by the National Endowment for Financial Education and the Investment Company Institute.

Recommendations

According to a Roper Organization study, when people want financial advice, they most likely turn first to friends and relatives. After this they turn to finance professionals, such as bankers, lawyers, accountants, financial planners, real estate brokers, and stockbrokers. Some accountants and bankers have the CFP credential. Getting recommendations from several people whom you trust should narrow down your list of potential advisors.

Besides the criteria of charges and credentials already covered, you should find out what type of services and products he or she provides, what the track record is, how long the advisor has been in business, and the names of current customers with similar financial situations, so you can check their satisfaction levels.

As a final step, after visiting financial advisors, evaluate your comfort level. Do you feel comfortable discussing your finances with them? Do they understand your goals? How do you feel about the advice you are receiving?

Objective 6:
Explain how the Internet can be used to obtain financial information.

ELECTRONIC RESOURCES

Financial planning over the Web is a fast-paced, ever-changing option. It is not as specialized and personal as the advice you would get meeting one-on-one with a financial advisor, but it is free and easy to use with 24 hour access.

Examples of financial planning sites offering budgeting and record keeping advice are:

- **www.americanexpress.com/advisors**: American Express (see Exhibit 2.10)
- **www.aarp.org**: American Association of Retired Persons
- **www.fidelity.com**: Fidelity Insurance Co.
- **www.finance.com**: Citibank (See Exhibit 2.11)
- **www.jhancock.com**: John Hancock Mutual Life Insurance Co.
- **www.massmutual.com**: Mutual Insurance Co.
- **www.phoenixhomelife.com**: Phoenix Home Life Insurance Co.
- **www.prusec.com**: Prudential Insurance Co.
- **www.vanguard.com/**: Vanguard Investments/Mutual Fund Co.

Many of these offer computer generated financial plans based on the information you provide. They ask a few questions and then provide a profile. Five of the sites listed will be discussed to give you an idea of what financial websites typically offer.

One of the more comprehensive computer-generated financial plans is available at **www.jhancock.com**. This site starts with a detailed interview that requires the following information: tax returns, mutual fund, stock, and other investment documents, including IRAs; a list of potential colleges for your children; pension-plan information; 401(k) plan information; past credit card bills and bank statements; and insurance information. The areas covered include insurance and retirement planning, saving for college and general saving, plus psychological aspects of money such as risk tolerance. Although the site is operated by an insurance company, it does not sell specific investments or products. You can check a box asking for materials.

Two of the websites stressing retirement planning are offered by the American Association of Retired Persons (AARP) WebPlace at **www.aarp.org** and by the Vanguard website at **www.vanguard.com/**. The AARP site includes essays and articles about consumer rights for seniors and common legal questions.

General sites that are good for beginners are **www.finance.com** and American Express's "Create a Personal Financial Profile" (**www.americanexpress. com/advisors**). They work with information you supply to determine or evaluate your net worth, cash flow, net pay, retirement goals, and so on. Pages from these sites are given in Exhibits 2.10 and 2.11.

1. How can I start practicing good record keeping?

   ■ Balance your checkbook. Keep canceled checks, credit card statements, and receipts in one place, organized and labeled.

2. If I have records at school and some with my parents, how do I get organized?

   ■ Talk with your parents, find out what they have, then enter the information on an inventory list, such as the one in Exhibit 2.2.

3. How long should I keep financial records for income tax purposes?

   ■ Three years is sufficient for most records.

4. What records should I keep permanently?

   ■ Birth and marriage certificates, adoption papers, wills, and Social Security data. Depending on your circumstances, there may be others, such as military records, divorce papers, and bankruptcy filings.

5. How do I start a budget?

   ■ Figure out how much money you earn in a month and subtract your usual expenses to find out what you have left. To project into the future, estimate your starting salary per month, estimate rent and other large expenses, and design a budget around those estimates. See Exhibit 2.5 for an example of a monthly budget.

6. If I do not own anything but a used car (worth $4,500), clothes, and a computer (with no resale value), is it worth while to figure net worth?

   ■ Yes, because if you can understand net worth when your finances are simple, then you will be better prepared to figure it when your finances become more complicated. Any time you know your financial standing, you are less likely to

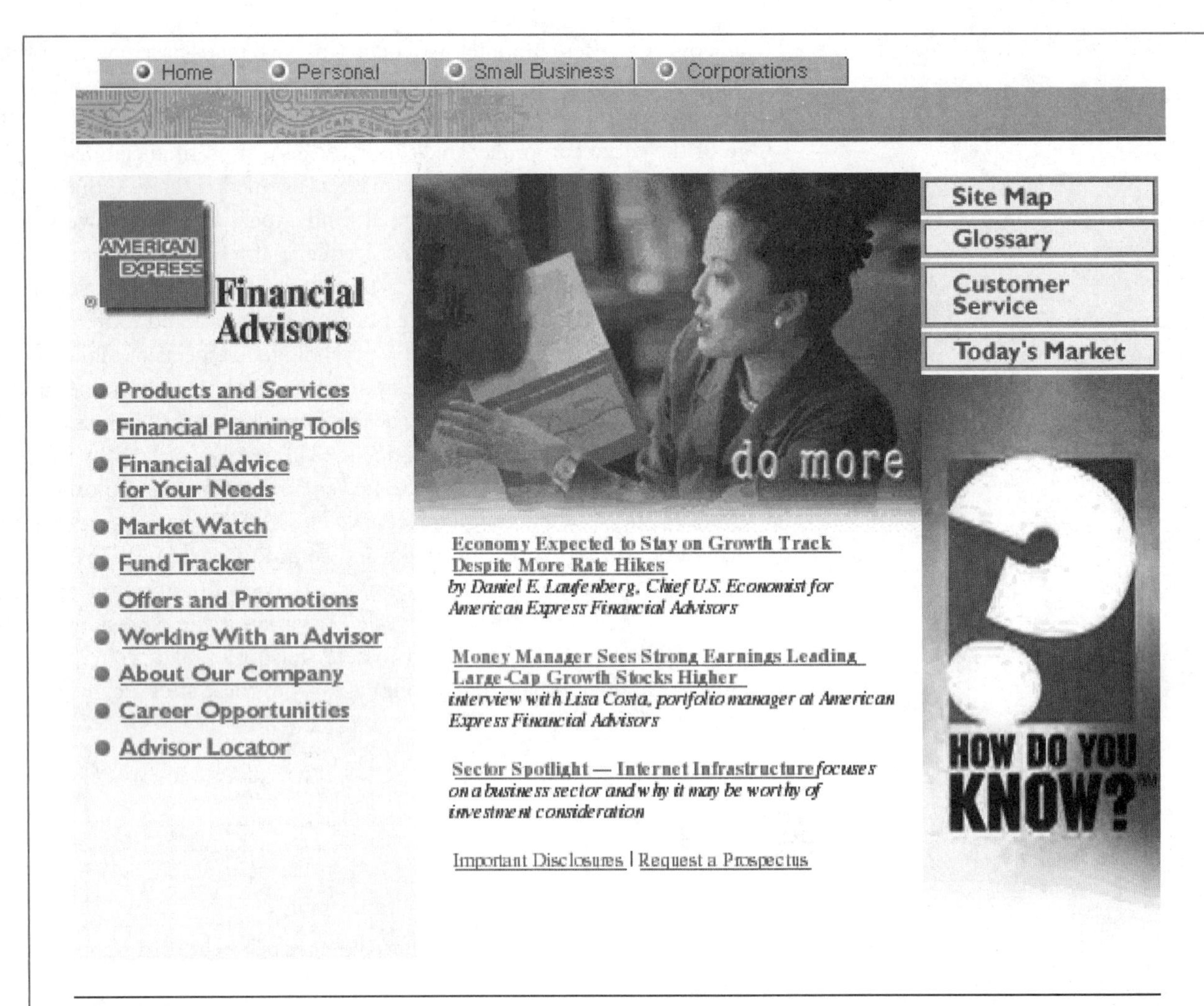

EXHIBIT 2.10 American Express Financial Advisors Web page.

drift into financial trouble. To figure net worth, subtract liabilities (what you owe) from assets (what you own). See the example in Exhibit 2.8.

7. What is the best known credential for a financial planner?

   ■ Certified Financial Planner (CFP).

8. How much will a financial planner charge?

   ■ This varies. An initial consultation may be free; after that fees go from $50 an hour and up. Most planners are both fee-and-commission-based.

EXHIBIT 2.11
Finance.com
Web page
Please visit the site for updated content.

SUMMARY

Assembling financial records and developing a system for their maintenance is an important first step in financial management.

The Internal Revenue Service (IRS) recommends that you keep most records for three years.

Making a monthly or a yearly budget and a net worth statement are means of determining money flow and current worth, and are useful for future planning purposes. Fixed expenses must be differentiated from variable expenses.

The net worth formula is Assets - Liabilities = Net worth.

Three ratios provide a yardstick of financial health and solvency. These are the solvency ratio, the debt-to-total assets ratio, and the debt-to-income ratio (also called the debt service ratio).

In choosing a financial advisor, keep in mind the term "caveat emptor," meaning "may the buyer beware." Factors to consider in selection are the advisor's services, products, fees, credentials, experience, and longevity, and the client's comfort level with the advisor and what he or she is proposing.

The best known credential for financial planners is the Certified Financial Planner (CFP).

The Securities and Exchange Commission (SEC) has a role in regulating some financial advisors; others are regulated by the states they operate in.

Financial planning software and Internet resources can help in the budgeting and record keeping processes. Users should be careful, however, about what they reveal over the Internet because there are very few privacy protections in place.

REVIEW QUESTIONS

(See Appendix A for answers to questions 1–10.)

Identify the following statements as true or false.

1. The Internal Revenue Service recommends that you keep records for ten years.
2. Budgets are the same as net worth statements.
3. Clothing is an example of a variable expense.
4. Solvency is your ability to pay off debts.
5. The debt service ratio is defined as the total value of all financial assets offset by net worth.

Pick the best answers to the following five questions.

6. Financial records are useful because
   a. they show where you have been.
   b. they give you a basis on which to develop money management strategies.
   c. they help you plan for retirement.
   d. all of the above are true.
7. People need to make financial plans now more than ever because
   a. life expectancy is declining.
   b. life expectancy is increasing.
   c. people are retiring later in life than ever before.
   d. Social Security takes care of only 90 percent of retirement expenses.
8. Technology has affected financial record keeping and information in the following way(s):
   a. Through financial planning software, such as Quicken and MS Money.
   b. Through financial planning websites.
   c. Through specialized software products, such as TurboTax and TaxCut.
   d. All of the above are true.
9. An inventory is a list showing
   a. where major personal and financial documents are kept.
   b. telephone numbers of financial planners.
   c. credit card numbers.
   d. addresses of executors.
10. A __________ lists the value of assets and liabilities.
    a. budget
    b. balance sheet—also called a net worth statement

c. commission
d. portfolio

DISCUSSION QUESTIONS AND PROBLEMS

1. How can some expenses be categorized as both fixed and variable? Give an example of an expense that can fit into both categories.
2. Why do people dread making budgets? Explain what a budget does or why it is useful.
3. How could a person have a negative net worth? If someone has a negative net worth, what can he or she do to turn it into a positive net worth?
4. What three ratios can be used to determine financial health and solvency? Which of these do you find the most useful and why?
5. It was suggested that the phrase "caveat emptor" is pertinent in the selection of a financial advisor. What does the phrase mean, and why is it relevant in the selection process?

DECISION-MAKING CASES

In this chapter two single working professionals—Dawn O'Brien and Michael—and a young married couple—Brian and Kaytoria Wright—were used as examples to show how to budget, figure net worth, and determine ratios of financial planning. Using the principles, formulas, and case studies, you should be able to do the following:

1. Brad has been turned down for a car loan. Even though he makes $26,000 a year, he owes $20,000 on his credit cards. What did the lender dislike when he figured Brad's debt service ratio?
2. Juanita is going through a divorce after 25 years of marriage. Her husband has always done the bill-paying and tax returns. What records does she need to locate to determine her financial standing? She would like to go to a financial advisor. What factors should she consider in her selection?
3. Becka's monthly expenses average $600 for rent and utilities, $220 for food, and about $180 for gifts, gasoline, clothes, entertainment, and miscellaneous. Her job covers her insurance, and her car is paid off. She brings home after taxes about $1550. Work out a monthly budget for her. As a goal, she'd like to build an emergency fund of 2–3 months of income. Right now she has no savings; how much can she realistically save each month?
4. Kimberlee has been rejected for a loan. The lender said that she owed too much money so she was a poor credit risk. Using the formula for the debt-to-assets-ratio, put in her current debt of $10,000 and her assets of $11,000 to see what the lender meant.
5. Dusty, age 28, works full-time for a computer networking company. He has paid off his student loan and is essentially debt free. Now, he would like to start building his money. Where could he turn for free financial planning advice? If he is willing to pay a fee, what are his options?

GET ON LINE

- Go to the American Express website (www.americanexpress.com/advisors), as shown in Exhibit 2.10, click on Financial Advice for Your Needs. Report on what you find. How easy was this site to use?
- Go to www.finance.com. Click on tools, select one, and report on what you find.

INFOTRAC

To access InfoTrac, follow the instructions at the end of Chapter 1.

Exercise 1: Type in the word "budget," select subject guide, and hit search. Then scroll down until you see Budget, Personal, click on See, Personal Budgets. This will bring up periodical references, subdivisions, and related subjects. Under subdivisions, click on management or planning. Select an article and report on what you find.

Exercise 2: Type in the words "net worth," select key words, and hit search. This will bring up articles on net worth. Select two and compare the advice they give on improving one's net worth.

chapter three

taxes

Minimizing Tax Liability

Did You Know?

* The average American pays between 20–40 percent of earnings in taxes.
* Over 40 million Americans hire tax preparers each year and the cost of assistance is rising.

Objectives

After reading Chapter 3 you should be able to do the following:

1. Describe the differences among progressive, regressive, and flat taxes.
2. Explain federal income tax fundamentals.
3. Understand tax audits, how often they occur, and what to do if audited.
4. Access tax preparation help, including online sources.
5. Discuss financial decisions and expenses that affect taxes.
6. Describe the benefits of tax-exempt and tax-deferred investments.

Chapter Overview

This chapter focuses on taxes. **Taxes** are the payments of money to local, state, or federal governments. Tax money is used to provide public goods and services, such as roads, schools, law enforcement, and public libraries. Failure to pay taxes can lead to various penalties including jail sentences. So taxes are not voluntary. As Benjamin Franklin said in 1789, in this world nothing is certain but death and taxes. The Internal Revenue Service (IRS) relies on citizens to report their income, calculate tax liability, and file tax returns on time.

Taxes
Required payments of money to governments.

Taxable income
Everything earned that can be taxed.

Besides taxes on income, taxes are placed on a variety of things, including property, investments, goods, and services. **Taxable income** is everything that you earn, including dividends from investments that can be taxed.

The U.S. Constitution gave Congress the authority "To Lay and Collect Taxes, Duties, Imposts and Excises" but did not specifically mention income tax. In 1913 the Sixteenth Amendment was passed; it stated, "The Congress shall have power to lay and collect taxes on income. . . ." The Underwood-Simmons Tariff Act (Section 2) provided for a one percent rate on taxable income. As the next section will show, taxes are now much higher than one percent.

This chapter covers various aspects of taxation with a particular emphasis on federal income tax and tax credits that affect students and their families, including child tax credits and education credits. As in all chapters, the chapter has sections of FAQs and Electronic Resources.

WHAT FEDERAL TAX PAYS FOR

The pie charts in Exhibit 3.1 show the relative sizes of the major categories of federal income and outlays. Note that the greatest amount of federal income (48.1%) comes from individual income taxes (what is paid in) and the greatest outlay finances Social Security, Medicare, and other retirement programs (37%) followed by national defense, veterans, and foreign affairs (18%).

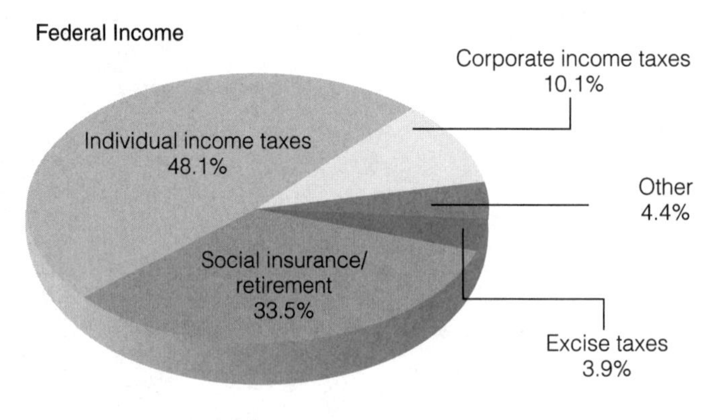

Source: Federal budget for fiscal 2001

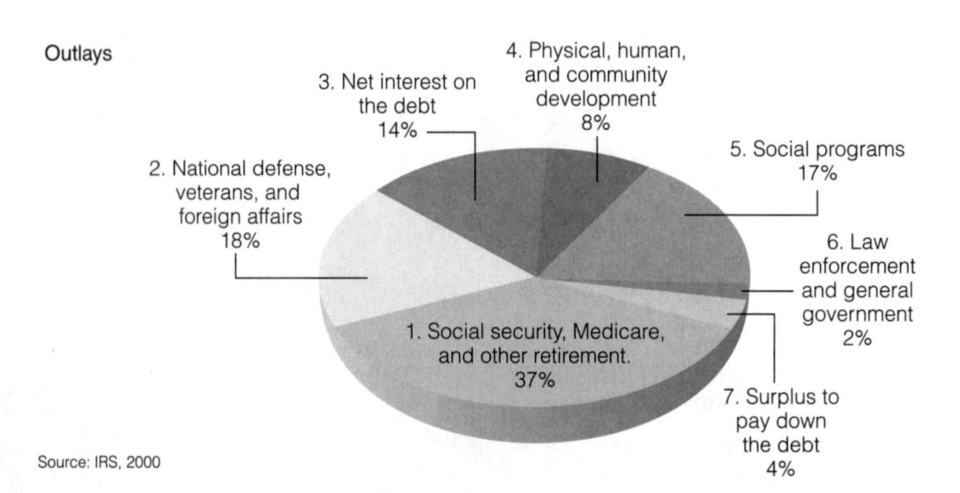

Source: IRS, 2000

EXHIBIT 3.1
Where the Money Goes
Major categories of federal income and federal outlays

TAXATION PHILOSOPHIES

Benefits-received philosophy
Those who receive benefits should pay taxes for them.

Ability-to-pay philosophy
Those with higher incomes pay proportionally more taxes than those with lower incomes.

Taxes are based on two different philosophies: benefits-received and ability to pay. The **benefits-received philosophy** means that those who receive the benefits of a particular public expenditure should pay for it. For example, a new bridge may require a dollar toll each time someone drives over it until the bridge is paid for. Similarly gasoline tax revenues can be used to pay for new highways. The **ability to pay philosophy** states that those with higher incomes should pay proportionally more taxes than those with lower incomes. The Internal Revenue Service (www.irs.gov) gives the following example regarding ability to pay:

> Jane makes a ton of money each year, and flies (on her private jet) to and from her houses in Miami and New York. Tough life. John earns a more modest salary and rents a small apartment. Jane and John do NOT pay the same amount in taxes. Their ability to pay differs vastly. John pays less, for his amount of income (wages, interests, profits) and assets (houses, cars, stocks, savings accounts) is less than Jane's.

Objective 1:
Describe the differences among progressive, regressive, and flat taxes.

Vertical equity
People in different income groups pay different amounts of taxes.

Progressive taxes
Those earning more, pay more.

State and federal income taxes are part of our vertical equity tax system, which is based on the ability-to-pay philosophy. **Vertical equity** states that people in different income groups should pay different amounts of taxes. State and federal income taxes are considered **progressive taxes,** meaning the more you earn, the more you pay. The progressive nature of the Federal Income Tax is shown in Exhibit 3.2. Note that in 1999 the highest tax rate was 39.6%.

The trend is toward the wealthiest Americans paying a considerably higher percentage of the nation's total federal income taxes. For example, the wealthiest in 1997—the one percent of Americans whose annual income was over $250,736—paid 33.2 percent of the nation's total federal individual income taxes up from 24.8 percent a decade earlier. This group earns 17.4 percent of total U.S. individual income.[1]

Regressive taxes
Approach that takes a smaller percentage of income as income rises.

In contrast, a **regressive tax** takes a smaller percentage from those with high income than from those with lower income. Regressive taxes are based on preestablished dollar amounts. For example, a sales tax of 7 percent remains the same, whether your income is $15,000 a year or over $500,000 a year. So sales taxes, which vary by state, are examples of regressive taxes.

In recent years to stimulate store sales and to support families with children, a number of cities and states around the country have offered no-sales-tax weeks (usually before the start of the school year) on clothing, backpacks, and footwear purchases.

Flat tax
Charges the same percentage of tax to everyone regardless of income.

Another new twist in taxation policy is called the flat tax. It was proposed by several candidates in the 1996 and 2000 presidential elections as a reform of the present federal income tax system. Some cities and states already have a version of it. A **flat tax** charges the same percentage of tax to everyone regardless of income. With a flat tax everyone would pay, for example, 25 percent of their income. There would be very few loopholes or exemptions. Paperwork would be simplified. Most people who are currently in a high tax bracket find the idea of a flat tax attractive, but it may put low income earners at a disadvantage.

How Much Is Average for Individuals and Families?

Taxes usually run between 20–40 percent of the average American's income. This range includes all taxes, not just the federal income tax rates shown in Exhibit 3.2.

Individual Return
If taxable income is:

Over:	But not Over:	Your tax is:	Of the amount over:
0	$ 25,750	15%	0
$ 25,750	62,450	$ 3,862.50 + 28%	$ 25,750
62,450	130,250	14,138.50 + 31%	62,450
130,250	283,150	35,156.50 + 36%	130,250
283,150		90,200.50 + **39.6%**	283,150

Head of Household Return
If taxable income is:

Over:	But not Over:	Your tax is:	Of the amount over:
0	$ 34,550	15%	0
$ 34,550	89,150	$ 5,182.50 + 28%	$ 34,550
89,150	144,400	20,470.50 + 31%	89,150
144,400	283,150	37,598.50 + 36%	144,400
283,150		87,548.50 + **39.6%**	283,150

Married Couples Filing Joint Return
If taxable income is:

Over:	But not Over:	Your tax is:	Of the amount over:
0	0	15%	0
$ 43,050	$ 43,050	$ 6,457.50 + 28%	$ 43,050
104,050	104,050	23,537.50 + 31%	104,050
158,550	158,550	40,432.50 + 36%	158,550
283,150	283,150	85,288.50 + **39.6%**	283,150

Married Couples Filing Separate Returns
If taxable income is:

Over:	But not Over:	Your tax is:	Of the amount over:
0	$ 21,525	15%	0
$ 21,525	52,025	$ 3,228.75 + 28%	$ 21,525
52,025	79,275	11,768.75 + 31%	52,025
79,275	141,575	20,216.25 + 36%	79,275
141,575		42,644.25 + **39.6%**	141,575

EXHIBIT 3.2
1999 Federal Income Tax Rates

Paying 40 percent in taxes may seem shockingly high, but it could be worse. In Norway, for example, the tax rate is about 60 percent. Another way to look at the tax burden of the average American is given in Exhibit 3.3. The "Tax Freedom Day" concept means that Americans work from January to May to support the government before they start making their own money. This exhibit also shows that the general trend is upwards, it is taking more days of work to earn enough to pay taxes.

In terms of families, experts estimate that the average American family pays about 32 percent of its income in taxes. That is more than the cost of food, shelter and clothing combined for the average family.

Tax Avoidance and Tax Evasion

Tax avoidance
Legitimate methods of reducing taxes.

Tax evasion
Illegal methods of reducing taxes.

Most people want to pay their fair share of taxes but not a cent more. This practice is called **tax avoidance** which refers to the use of *legitimate* methods to reduce one's taxes. Tax avoidance strategies may include buying tax-free municipal bonds, investing in tax-deferred IRA or Keogh accounts, shifting assets to children, establishing trusts, or making legitimate charitable contributions to generate tax deductions.

Tax evasion is the use of *illegal* methods to reduce taxes. Penalties can include the payment of back taxes with interest and civil and criminal penalties.

Marginal and Average Tax Rates

As already established, since the federal income tax is progressive, the rate at which a person is taxed rises with income. Connected to this are two important concepts in personal finance and taxes—the marginal tax rate and the average tax rate. The marginal rate is especially important because it tells how much extra earnings (from raises, bonuses, or investments) or a tax deduction (for example, from a charitable donation) are worth.

Marginal tax rate
Tax rate applied to the last dollar of earnings.

Average tax rate
Amount of taxes paid divided by income.

The **marginal tax rate** is the tax rate applied to the last dollar of earnings—or the highest rate at which a person pays. As shown in Exhibit 3.2, an individual taxpayer or married couple may be in the 15%, 28%, 31%, 36%, or 39.6% tax bracket and pay at one of those marginal tax rates. In contrast is the term **average tax rate**, or the effective tax rate, which is the amount of taxes paid, divided by the taxpayer's income.

To explain further, the following example is illustrated in Exhibit 3.4. Sierra is single and earned $35,000 in 1999. She took the standard deduction ($4,300) and claimed one personal exemption ($2,750), reducing her taxable income to $27,950. As

EXHIBIT 3.3
Working For Taxes
"Tax Freedom Day" is the last day average Americans have to work each year to cover their taxes. Here are past Tax Freedom Days and the number of calendar days needed to reach them.

Used by permission.

Year	Tax Freedom Day	as a Percent of income	No. of Days
1990	22–Apr	30.6%	112
1991	22–Apr	30.5%	112
1992	20–Apr	30.3%	111
1993	23–Apr	30.7%	113
1994	24–Apr	31.1%	114
1995	25–Apr	31.5%	115
1996	26–Apr	31.9%	117
1997	29–Apr	32.4%	119
1998	1–May	33.0%	121
1999	3–May	33.5%	123
2000	3–May	33.8%	124

Source: Tax Foundation, Washington, DC, 202-783-2760

shown in Exhibit 3.4, the first $25,750 was taxed at a 15 percent tax rate, and the remainder—$2,200—was taxed at the 28 percent tax rate. Thus Sierra paid $4,478.50. Her average tax rate—taxes paid, divided by taxable income—is 16 percent ($4,478.50 divided by $27,950), and Sierra's marginal rate is 28 percent (she paid 28% on her highest dollar).

TYPES OF TAXES

Taxes can be categorized into the four main types: taxes on earnings, taxes on purchases, taxes on property, and taxes on wealth.

Taxes on Earnings

Income taxes
Taxes on income, earned and unearned.

Income and Social Security taxes are the two main taxes on earnings. **Income taxes** are taxes on income, both earned (commissions, salaries, tips, and wages) and unearned (interest from savings accounts, stock dividends). Individuals and businesses are subject to income taxes. Employers withhold income tax payments and take Social Security taxes from paychecks. If you own your own business, you may be required to make estimated tax payments.

EXHIBIT 3.4
Sierra's 1999 Federal Income Tax Calculations

Taxable Income Range					
Mimimum	Maximum	Portion of Income Within Range		Tax Rate	Tax
0	25,750	$25,750	x	15%	$3,862.50
25,750	62,450	2,200	x	28%	616.00
62,450	130,250	0	x	31%	0
130,250	283,150	0	x	36%	0
283,150	———	0	x	39.6%	0
		Total $27,950			$4,478.50

End result: Sierra owes $4,478.50 in federal income tax.

Besides federal income tax, most states (the exceptions are New Hampshire, Tennessee, Texas, Florida and Nevada) have state income tax based on earnings. State income tax rates range from 1 percent to 10 percent. In South Carolina, for example, state income tax progresses on a graduated scale from 2 to 6 percent up to $10,000 and levels off at 7 percent over $10,000.

Taxes on Purchases

State and local taxes may be added to purchases in the form of sales taxes. The rates and exemptions (food and medicine) vary and not every state has a sales tax. Alaska, Delaware, Montana, New Hampshire, and Oregon do not have general sales taxes. As of the writing of this book, other states were considering removing sales taxes. **Excise taxes** are imposed on specific goods and services such as cigarettes, alcoholic beverages, tires, airplane tickets, telephone calls, and gasoline.

Excise taxes
Taxes on specific goods and services.

Taxes on Property

People pay taxes on property, including real estate, boats, cars, recreational vehicles, and business inventories. For example, an owner of a mountain cabin pays property taxes once a year. The amount of tax is based on the value of land and buildings and rates vary in each state and by community. To give you a ballpark figure, the property tax for a $150,000 home may run between $1,000–$5,000 a year, depending upon where you live.

Since the value of real estate holdings tends to increase, property taxes keep going up. Retirees worry about whether their fixed income will keep pace with ever-increasing property taxes. To help retirees and others, many states have **homestead exemptions**. For example, in South Carolina if you are 65 or older, totally and permanently disabled, or legally blind, you may apply for a $20,000 exemption of the fair market value of your property. A county tax assessor can give you the guidelines for homestead exemptions in your area. Exemptions do not happen automatically; the tax assessor has to be informed of age or disability changes that affect eligibility.

Homestead exemptions
Reductions in property taxes.

Taxes on Wealth

At the time of a person's death, an **estate tax** is imposed on the value of a person's net worth. This tax is based on the fair market value of the deceased person's property, investments, and bank accounts less allowable deductions and other taxes. Another related tax is **inheritance tax** which exists in about half of the states. Money and property that is passed down to beneficiaries is subject to taxation. These subjects will be covered more in-depth in the last chapter of the book.

Estate tax
Tax imposed on the value of a deceased person's net worth.

Inheritance tax
Taxes paid by beneficiaries of an estate.

In order to avoid or lessen estate and inheritance taxes, some people try to give their money away in the form of gifts to friends and relatives. State and federal governments realize that this is a possibility and, therefore, they have imposed **gift taxes.** Currently gift amounts greater than $10,000 in a single year are subject to federal tax. However, there are exceptions, such as amounts given for the payment of tuition or medical expenses. Since the laws are so intricate in these matters, an individual would be wise to seek advice from financial advisors before giving away large sums of money.

Gift tax
Federal tax a giver of a sizable gift may have to pay.

Because this book is written for students across the nation, the emphasis in the next section is on what is held in common—the federal income tax.

Objective 2:
Explain federal income tax fundamentals.

FEDERAL INCOME TAX

Who Pays?

If income reaches a certain amount, every citizen or resident of the United States and every U.S. citizen who is a resident of Puerto Rico is required to file a federal income

EXHIBIT 3.5
Filing Status, Age, and Gross Income

IF your filing is...	AND at the end of 1998 you were*...	THEN file a return if your gross income** was at least...
Single	under 65	$7,050
	65 or older	8,100
Married filing jointly***	under 65 (both spouses)	$12,700
	65 or older (one spouse)	13,550
	65 or older (both spouses)	14,400
Married filing separately	any age	$2,750
Head of household (see page 18)	under 65	$9,100
	65 or older	10,150
Qualifying widow(er) with dependent child	under 65	$9,950
	65 or older	10,800

* *If you turned 65 on January 1,2000, you are considered to be age 65 at the end of 1999.*
** *Gross income means all income you received in the form of money, goods, property, and services that is not exempt from tax including any income from sources outside the United States (even if you may exclude part or all of it).* ***Do not*** *include social security benefits unless you are married filing a separate return and you lived with your spouse at any time in 1999.*
*** *If you did not live with your spouse at the end of 1999 (or on the date your spouse died) and your gross income was at least $2,750, you must file a return regardless of your age.*
Source: IRS 1999 1040 Instructions booklet

tax return. About three-fourths of all Americans pay federal income tax. A single person or a couple (depending on age) does not have to file if their gross income was under a certain amount (see Exhibit 3.5)

Filing Status

Filing status
Marital or household status that affects tax rates.

One's **filing status** affects one's tax rate, as shown in Exhibit 3.2. There are four basic filing status categories.

Married Couples Filing Joint Return This category combines a husband's and a wife's incomes. For example, Mr. and Mrs. Brown have decided to file a joint return. Their taxable income is $25,330, and the amount shown where the income line and filing status column meet is $3,799 (see Exhibit 3.6). This is the amount they should enter on their Form 1040.

Married Couples Filing Separate Returns Each spouse is responsible for his or her own tax. Under certain circumstances a married couple may benefit from this filing status.

Individual Returns The filer has never married, is divorced, or is legally separated with no dependents.

Head of Household Return Filed by an unmarried individual or a surviving spouse who maintains a household (paying for more than half of the costs) for a child or a dependent relative.

EXHIBIT 3.6
Example of Mr. and Mrs. Brown Filing a Joint Return

1999 Tax Table

Use if your taxable income is less than $100,000. If $100,000 or more, use the Tax Rate Schedules.

Example. Mr. and Mrs. Brown are filing a joint return. Their taxable income on line 39 of Form 1040 is $25,330. First they find the $25,300–25,350 income line. Next, they find the column for married filing jointly and read down the column. The amount shown where the income line and filing status column meet is $3,799. This is the tax amount they should enter on line 40 of their Form 1040.

Sample Table

At least	But less than	Single	Married filing jointly	Married filing sepa-rately	Head of a house-hold
		Your tax is—			
25,200	25,250	3,784	3,784	4,265	3,784
25,250	25,300	3,791	3,791	4,279	3,791
25,300	25,350	3,799	**3,799**	4,293	3,799
25,350	25,400	3,806	3,806	4,307	3,806

Source: IRS 1999 1040 Instructions booklet

A much rarer filing status is Qualifying Widow(er) with dependent child.

Individuals who fall under several categories should determine which category is in their best interest. According to CCH, Inc., a publisher of tax and business-law information, 59 provisions of the tax code treat people differently depending at least in part on whether they are married or single. Although provisions may work against married couples, filing jointly is easier. According to the IRS, if you are married and file a separate return, you will usually pay more tax than if you file a joint return. For example, Exhibit 3.6 shows that Mr. and Mrs. Brown would have to enter a higher tax amount on their Form 1040 if they filed separately ($4,293) versus filing jointly ($3,799). An exception to this general rule is that filing separately may be wise if one spouse makes substantially less than the other and can claim itemized deductions that would otherwise be disallowed.

Another consideration is whether a married couple lives in a community property state (Arizona, California, Idaho, Louisiana, Nevada, New Mexico, Texas, Washington and Wisconsin). In a community property state, a couple can follow state law to determine what is community income and what is separate income. For details on community property states, see IRS Publication 555.

The basis for filing status changes comes from 1969 when Congress enacted a separate tax rate schedule to benefit single taxpayers. Congress figured a married couple could live more cheaply than two single people can. However, Congress made no special tax provisions for a man and woman living together who were not married. For example, in 1996 singles got a higher standard deduction, $4,000, on their federal income tax compared with $6,700 for married couples filing jointly. Thus an unmarried couple living together could claim a total deduction of $8,000 or $1,300 more than if they were married. Advocates of reform refer to this as a Marriage Penalty Tax.

Death of a Taxpayer

If a taxpayer dies before filing a return, the taxpayer's spouse or personal representative may have to file and sign a return for that taxpayer. The deceased taxpayer's estate or surviving spouse may owe money or receive a refund.

If Someone Cannot Pay

If a taxpayer can not afford to pay taxes the IRS is increasingly willing to make arrangements. The main options are paying in installments or agreeing to something more drastic, such as entering the "offer to compromise program." The IRS must be notified about the taxpayer's intentions and agree to a payment plan (by submission of Form 9465 with the tax return) or a compromise, which may involve giving the taxpayer time to sell a house or borrow the money.

Reaching the IRS

The federal income tax system has changed dramatically in recent years. Tax collection methods have been reorganized, and electronic filing—called e-filing—is encouraged. Exhibit 3.7 shows the main ways to access tax help and forms from the IRS.

To e-file, a taxpayer can find a list of participating software companies and instructions by going to www.irs.gov and clicking on "Electronic Services" or "On-Line Filing Companies" or "IRS e-file Partners. According to the IRS, the advantages of using e-file are

Quick and Easy Access to Tax Help and Forms

Note: If you live outside the United States, see **Pub. 54** to find out how to get help and forms.

PERSONAL COMPUTER

You can access the IRS's Internet Web Site 24 hours a day, 7 days a week at www. irs.gov to:

- Download Forms, Instructions, and Publications
- See Answers to Frequently Asked Tax Questions
- Search Publications On-Line by Topic or Keyword
- Figure Your Withholding Allowances Using our W-4 Calculator
- Send Us Comments or Request Help by E-Mail
- Sign up to Receive Local and National Tax News by E-Mail

You can also reach us using File Transfer Protocol at ftp.irs.gov

FAX

You can get over 100 of the most requested forms and instructions 24 hours a day, 7 days a week, by fax. Just call **703-368-9694** from the telephone connected to the fax machine. See pages 8 and 9 for a list ot the items available.

MAIL

You can order forms, instructions, and publications by completing the order blank on page 55. You should receive your order within 10 days after we receive your request.

PHONE

You can get forms, publications, and automated information 24 hours a day, 7 days a week, by phone.

Forms and Publications

Call **1-800-TAX-FORM** (**1-800-829-3676**) to order current and prior year forms, instructions, and publications. You should receive your order within 10 days.

TeleTax Topics

Call **1-800-829-4477** to listen to pre-recorded messages covering about 150 tax topics. See pages 10 and 11 for a list of the topics.

Refund Information

You can check the status of your 1999 refund using TeleTax's Refund Information service. See page 10.

WALK-IN

You can pick up some of the most requested forms, instructions, and publications at many IRS offices, post offices, and libraries. Some IRS offices and libraries have an extensive collection of products available to photocopy or print from a CD-ROM.

CD-ROM

Order **Pub. 1796**, Federal Tax Products on CD-ROM, and get:

- Current Year Forms, Instructions, and Publications
- Prior Year Forms, Instructions, and Publications
- Popular Tax Forms That May Be Filled in Electronically, Printed out for Submission, and Saved for Recordkeeping
- The Internal Revenue Bulletin

Source: IRS 1999 1040 Booklet

EXHIBIT 3.7
Quick and Easy Access to Tax Help and Forms

* the ability to file 24 hours a day, 7 days a week.
* the fast refunds—half the time, even faster with direct deposit.
* file now, pay later—with a credit card or direct debit you can wait to pay until April 17.
* less chance of receiving an error notice from the IRS because IRS e-file is more accurate than a paper return.
* filing federal and state tax returns together.
* an acknowledgment of IRS receipt within 48 hours.
* free and low-cost alternatives available.
* privacy and security are assured.
* chance of an audit is not greater than a paper return.

Forms, instructions, and publications can be picked up at post offices, libraries, and IRS offices. More taxpayers are filing returns by telephone and electronically each year. The IRS received about 33.6 million returns electronically in 2000, up from 29.3 million the previous year. Electronic filers who use a personal identification number assigned by the IRS do not need to mail in a paper signature form or send in W-2 statements.

If You Have Questions

Before contacting the IRS, have ready the following information:

1. The tax form, schedule, or notice to which your question relates.
2. The facts about your particular situation. The answer to the same question often varies from one taxpayer to another because of differences in their age, income, whether they can be claimed as a dependent, etc.
3. The name of any IRS publication or other source of information that you used to look for the answer.

Commonly Requested Forms

Which tax forms do most people use? Although there are about 700 forms, publications, and schedules, most people choose from three basic forms: Form 1040, Form 1040A, and Form 1040EZ. These three forms are given in Exhibits 3.8, 3.9, and 3.10.

Form 1040: *This is the most used form.* You are required to use Form 1040 (as shown in Exhibit 3.8) if your income is over $50,000 or if you can be claimed as a dependent on your parents' return and you had interest or dividends over a set limit. On the right hand side of Form 1040 notice the boxes to be checked regarding dependents. A **dependent** is an individual who relies on someone else for financial support. Most young adults (under age 21) are supported by their parents. If this is true in your case, your parents can claim you as a dependent. Various allowable expenses—such as medical costs, home mortgage interest, and real estate property taxes—can be used to reduce taxable income.

Form 1040A: *This form is simpler than Form 1040* it is designed for taxpayers with taxable income of less than $50,000 and no itemized deductions except for individual retirement account (IRA) deductions and certain tax credits, such as for child care and dependent care expenses. (see Exhibit 3.9).

Choosing between Form 1040 and Form 1040A depends on the type and amount of

EXHIBIT 3.8
The Most Used Tax Form

Form **1040** Department of the Treasury—Internal Revenue Service
U.S. Individual Income Tax Return **1999** (99) IRS Use Only—Do not write or staple in this space.

For the year Jan. 1–Dec. 31, 1999, or other tax year beginning , 1999, ending , 20 | OMB No. 1545-0074

Label (See instructions on page 18.) Use the IRS label. Otherwise, please print or type.
LABEL HERE
Your first name and initial | Last name | **Your social security number**
If a joint return, spouse's first name and initial | Last name | **Spouse's social security number**
Home address (number and street). If you have a P.O. box, see page 18. | Apt. no. | **▲ IMPORTANT! ▲** You must enter your SSN(s) above.
City, town or post office, state, and ZIP code. If you have a foreign address, see page 18.

Presidential Election Campaign (See page 18.)
Do you want $3 to go to this fund? | Yes | No | **Note.** Checking "Yes" will not change your tax or reduce your refund.
If a joint return, does your spouse want $3 to go to this fund?

Filing Status
1 ☐ Single
2 ☐ Married filing joint return (even if only one had income)
3 ☐ Married filing separate return. Enter spouse's social security no. above and full name here. ▶
4 ☐ Head of household (with qualifying person). (See page 18.) If the qualifying person is a child but not your dependent, enter this child's name here. ▶
5 ☐ Qualifying widow(er) with dependent child (year spouse died ▶ 19). (See page 18.)
Check only one box.

Exemptions
6a ☐ **Yourself.** If your parent (or someone else) can claim you as a dependent on his or her tax return, **do not** check box 6a.
b ☐ **Spouse**
c Dependents:

(1) First name Last name	(2) Dependent's social security number	(3) Dependent's relationship to you	(4) ✓ if qualifying child for child tax credit (see page 19)
			☐
			☐
			☐
			☐
			☐
			☐

If more than six dependents, see page 19.
d Total number of exemptions claimed
No. of boxes checked on 6a and 6b ____
No. of your children on 6c who:
• lived with you ____
• did not live with you due to divorce or separation (see page 19) ____
Dependents on 6c not entered above ____
Add numbers entered on lines above ▶ ☐

Income
Attach Copy B of your Forms W-2 and W-2G here. Also attach Form(s) 1099-R if tax was withheld.
If you did not get a W-2, see page 20.
Enclose, but do not staple, any payment. Also, please use Form 1040-V.

7 Wages, salaries, tips, etc. Attach Form(s) W-2 | 7
8a **Taxable** interest. Attach Schedule B if required | 8a
b **Tax-exempt** interest. DO NOT include on line 8a | 8b
9 Ordinary dividends. Attach Schedule B if required | 9
10 Taxable refunds, credits, or offsets of state and local income taxes (see page 21) | 10
11 Alimony received | 11
12 Business income or (loss). Attach Schedule C or C-EZ | 12
13 Capital gain or (loss). Attach Schedule D if required. If not required, check here ▶ ☐ | 13
14 Other gains or (losses). Attach Form 4797 | 14
15a Total IRA distributions | 15a | b Taxable amount (see page 22) | 15b
16a Total pensions and annuities | 16a | b Taxable amount (see page 22) | 16b
17 Rental real estate, royalties, partnerships, S corporations, trusts, etc. Attach Schedule E | 17
18 Farm income or (loss). Attach Schedule F | 18
19 Unemployment compensation | 19
20a Social security benefits | 20a | b Taxable amount (see page 24) | 20b
21 Other income. List type and amount (see page 24) | 21
22 Add the amounts in the far right column for lines 7 through 21. This is your **total income** ▶ | 22

Adjusted Gross Income
23 IRA deduction (see page 26) | 23
24 Student loan interest deduction (see page 26) | 24
25 Medical savings account deduction. Attach Form 8853 | 25
26 Moving expenses. Attach Form 3903 | 26
27 One-half of self-employment tax. Attach Schedule SE | 27
28 Self-employed health insurance deduction (see page 28) | 28
29 Keogh and self-employed SEP and SIMPLE plans | 29
30 Penalty on early withdrawal of savings | 30
31a Alimony paid b Recipient's SSN ▶ | 31a
32 Add lines 23 through 31a | 32
33 Subtract line 32 from line 22. This is your **adjusted gross income** ▶ | 33

For Disclosure, Privacy Act, and Paperwork Reduction Act Notice, see page 54. Cat. No. 11320B Form **1040** (1999)

Tax deduction
Items that reduce taxes.

Standard deduction
Deduction set by the government based on filing status.

income, the number of deductions, and the overall complexity of the tax situation. A **tax deduction** is the amount that a person or business can reduce from their taxable income. The more a person can deduct, the less he or she will have to pay. Most people, like Sierra earlier in the chapter, claim the **standard deduction** which is an alternative to itemizing deductions. The standard deduction is based on the main filer's status such as single taxpayer, head of household, married filing jointly, or married filing separately, as shown in Exhibit 3.2. The IRS reports that most individual income tax filers take the standard deduction.

EXHIBIT 3.8
continued

Form 1040 (1999) Page 2

Tax and Credits

34 Amount from line 33 (adjusted gross income) . . . 34

35a Check if: ☐ You were 65 or older, ☐ Blind; ☐ Spouse was 65 or older, ☐ Blind.
Add the number of boxes checked above and enter the total here . . . ▶ 35a

b If you are married filing separately and your spouse itemizes deductions or you were a dual-status alien, see page 30 and check here . . . ▶ 35b ☐

Standard Deduction for Most People
Single: $4,300
Head of household: $6,350
Married filing jointly or Qualifying widow(er): $7,200
Married filing separately: $3,600

36 Enter your **itemized deductions** from Schedule A, line 28, **OR standard deduction** shown on the left. **But** see page 30 to find your standard deduction if you checked any box on line 35a or 35b **or** if someone can claim you as a dependent . . . 36

37 Subtract line 36 from line 34 . . . 37

38 If line 34 is $94,975 or less, multiply $2,750 by the total number of exemptions claimed on line 6d. If line 34 is over $94,975, see the worksheet on page 31 for the amount to enter . 38

39 **Taxable income.** Subtract line 38 from line 37. If line 38 is more than line 37, enter -0- . 39

40 **Tax** (see page 31). Check if any tax is from a ☐ Form(s) 8814 b ☐ Form 4972 . . ▶ 40

41 Credit for child and dependent care expenses. Attach Form 2441 41

42 Credit for the elderly or the disabled. Attach Schedule R . . 42

43 Child tax credit (see page 33) . . . 43

44 Education credits. Attach Form 8863 . . . 44

45 Adoption credit. Attach Form 8839 . . . 45

46 Foreign tax credit. Attach Form 1116 if required . . . 46

47 Other. Check if from a ☐ Form 3800 b ☐ Form 8396
c ☐ Form 8801 d ☐ Form (specify) ______ 47

48 Add lines 41 through 47. These are your **total credits** . . . 48

49 Subtract line 48 from line 40. If line 48 is more than line 40, enter -0- . . . ▶ 49

Other Taxes

50 Self-employment tax. Attach Schedule SE . . . 50

51 Alternative minimum tax. Attach Form 6251 . . . 51

52 Social security and Medicare tax on tip income not reported to employer. Attach Form 4137 52

53 Tax on IRAs, other retirement plans, and MSAs. Attach Form 5329 if required . . . 53

54 Advance earned income credit payments from Form(s) W-2 . . . 54

55 Household employment taxes. Attach Schedule H . . . 55

56 Add lines 49 through 55. This is your **total tax** . . . ▶ 56

Payments

57 Federal income tax withheld from Forms W-2 and 1099 . . 57

58 1999 estimated tax payments and amount applied from 1998 return . 58

59a **Earned income credit.** Attach Sch. EIC if you have a qualifying child

b Nontaxable earned income: amount . . ▶ ______
and type ▶ 59a

60 Additional child tax credit. Attach Form 8812 . . . 60

61 Amount paid with request for extension to file (see page 48) 61

62 Excess social security and RRTA tax withheld (see page 48) 62

63 Other payments. Check if from a ☐ Form 2439 b ☐ Form 4136 63

64 Add lines 57, 58, 59a, and 60 through 63. These are your **total payments** . . . ▶ 64

Refund

65 If line 64 is more than line 56, subtract line 56 from line 64. This is the amount you **OVERPAID** 65

66a Amount of line 65 you want **REFUNDED TO YOU** . . . ▶ 66a

Have it directly deposited! See page 48 and fill in 66b, 66c, and 66d.

▶ b Routing number ▶ c Type: ☐ Checking ☐ Savings

▶ d Account number

67 Amount of line 65 you want **APPLIED TO YOUR 2000 ESTIMATED TAX** ▶ 67

Amount You Owe

68 If line 56 is more than line 64, subtract line 64 from line 56. This is the **AMOUNT YOU OWE**.
For details on how to pay, see page 49 . . . ▶ 68

69 Estimated tax penalty. Also include on line 68 . . . 69

Sign Here

Under penalties of perjury, I declare that I have examined this return and accompanying schedules and statements, and to the best of my knowledge and belief, they are true, correct, and complete. Declaration of preparer (other than taxpayer) is based on all information of which preparer has any knowledge.

Joint return? See page 18. Keep a copy for your records.

Your signature | Date | Your occupation | Daytime telephone number (optional) ( )

Spouse's signature. If a joint return, BOTH must sign. | Date | Spouse's occupation

Paid Preparer's Use Only

Preparer's signature | Date | Check if self-employed ☐ | Preparer's SSN or PTIN

Firm's name (or yours if self-employed) and address | EIN | ZIP code

Form **1040** (1999)

Form 1040EZ: As the name implies, this form is the easiest to fill out and the one most often used by college students. Form 1040EZ is appropriate if you are single or married and do not have any dependents or a lot of money. If your income is $50,000 or less and your interest income is $400 or less, then this is the form you will most likely want to use (see Exhibit 3.10).

For example, Jenna Stewart a college student, had a part-time job at a copy center. Since she was single, earned less than the amount needed to file, and had only $50 in interest income from her savings account, Jenna was able to use Form 1040EZ to obtain a refund of income tax withheld during the past year by her employer.

EXHIBIT 3.9
A Simpler Tax Return

Form **1040A** Department of the Treasury—Internal Revenue Service
U.S. Individual Income Tax Return (99) **1999** IRS Use Only—Do not write or staple in this space.
OMB No. 1545-0085

Label (See page 19.)
Use the IRS label. Otherwise, please print or type
L A B E L H E R E

Your first name and initial | Last name | Your social security number
If a joint return, spouse's first name and initial | Last name | Spouse's social security number
Home address (number and street). If you have a P.O. box, see page 20. | Apt. no.
City, town or post office, state, and ZIP code. If you have a foreign address, see page 20.

▲ IMPORTANT! ▲
You **must** enter your SSN(s) above.

Presidential Election Campaign Fund (See page 20.) Yes No
Do you want $3 to go to this fund? . . . .
If a joint return, does your spouse want $3 to go to this fund?
Note. Checking "Yes" will not change your tax or reduce your refund.

Filing status
1 ☐ Single
2 ☐ Married filing joint return (even if only one had income)
3 ☐ Married filing separate return. Enter spouse's social security number above and full name here. ▶
4 ☐ Head of household (with qualifying person). (See page 21.) If the qualifying person is a child but not your dependent, enter this child's name here. ▶
5 ☐ Qualifying widow(er) with dependent child (year spouse died ▶ 19). (See page 22.)
Check only one box.

Exemptions
6a ☐ **Yourself.** If your parent (or someone else) can claim you as a dependent on his or her tax return, **do not** check box 6a.
b ☐ **Spouse**
c **Dependents:**

(1) First name Last name	(2) Dependent's social security number	(3) Dependent's relationship to you	(4) ✓ if qualifying child for child tax credit (see page 23)
			☐
			☐
			☐
			☐
			☐
			☐

If more than seven dependents, see page 22.

No. of boxes checked on 6a and 6b
No. of your children on 6c who:
• lived with you
• did not live with you due to divorce or separation (see page 24)
Dependents on 6c not entered above
Add numbers entered on lines above

d Total number of exemptions claimed.

Income
Attach Copy B of your Form(s) W-2 here. Also attach Form(s) 1099-R if tax was withheld.
If you did not get a W-2, see page 25.
Enclose, but do not staple, any payment.

7 Wages, salaries, tips, etc. Attach Form(s) W-2. 7
8a **Taxable** interest. Attach Schedule 1 if required. 8a
b **Tax-exempt** interest. DO NOT include on line 8a. 8b
9 Ordinary dividends. Attach Schedule 1 if required. 9
10a Total IRA distributions. 10a 10b Taxable amount (see page 25). 10b
11a Total pensions and annuities. 11a 11b Taxable amount (see page 26). 11b
12 Unemployment compensation, qualified state tuition program earnings, and Alaska Permanent Fund dividends. 12
13a Social security benefits. 13a 13b Taxable amount (see page 28). 13b
14 Add lines 7 through 13b (far right column). This is your **total income.** ▶ 14

Adjusted gross income
15 IRA deduction (see page 30). 15
16 Student loan interest deduction (see page 30). 16
17 Add lines 15 and 16. These are your **total adjustments.** 17
18 Subtract line 17 from line 14. This is your **adjusted gross income.** ▶ 18

For Disclosure, Privacy Act, and Paperwork Reduction Act Notice, see page 53. Cat. No. 11327A Form **1040A** (1999)

What if Someone Does Not File in Time?

If someone cannot file a return by April 15, then he or she should file Form 4868, *Application for Automatic Extension of Time To File U.S. Individual Income Tax Return.* Form 4868 will give the taxpayer up to four more months to file Form 1040, 1040A, or 1040EZ. Interest on any unpaid tax is charged from the original due date of the return. Taxpayers do not have to file Form 4868 if they would prefer to use a credit card to get an extension of time. To pay by credit card, a person should call 1-888-272-9829 and follow the instructions.

EXHIBIT 3.9
continued

Form 1040A (1999) Page 2

Taxable income

19 Enter the amount from line 18. 19

20a Check if: ☐ **You** were 65 or older ☐ Blind ☐ **Spouse** was 65 or older ☐ Blind } **Enter number of boxes checked** ► 20a ☐

b If you are married filing separately and your spouse itemizes deductions, see page 32 and check here . . . . ► 20b ☐

21 Enter the **standard deduction** for your filing status. **But** see page 33 if you checked any box on line 20a or 20b **OR** if someone can claim you as a dependent.
- Single—$4,300 • Married filing jointly or Qualifying widow(er)—$7,200
- Head of household—$6,350 • Married filing separately—$3,600 21

22 Subtract line 21 from line 19. If line 21 is more than line 19, enter -0-. 22

23 Multiply $2,750 by the total number of exemptions claimed on line 6d. 23

24 Subtract line 23 from line 22. If line 23 is more than line 22, enter -0-. This is your **taxable income.** ► 24

Tax, credits, and payments

25 Find the tax on the amount on line 24 (see page 34). 25

26 Credit for child and dependent care expenses. Attach Schedule 2. 26

27 Credit for the elderly or the disabled. Attach Schedule 3. 27

28 Child tax credit (see page 35). 28

29 Education credits. Attach Form 8863. 29

30 Adoption credit. Attach Form 8839. 30

31 Add lines 26 through 30. These are your **total credits.** 31

32 Subtract line 31 from line 25. If line 31 is more than line 25, enter -0-. 32

33 Advance earned income credit payments from Form(s) W-2. 33

34 Add lines 32 and 33. This is your **total tax.** ► 34

35 Total Federal income tax withheld from Forms W-2 and 1099. 35

36 1999 estimated tax payments and amount applied from 1998 return. 36

37a **Earned income credit.** Attach Schedule EIC if you have a qualifying child. 37a

b Nontaxable earned income: amount ► and type ►

38 Additional child tax credit. Attach Form 8812. 38

39 Add lines 35, 36, 37a, and 38. These are your **total payments.** ► 39

Refund

Have it directly deposited! See page 47 and fill in 41b, 41c, and 41d.

40 If line 39 is more than line 34, subtract line 34 from line 39. This is the amount you **overpaid.** 40

41a Amount of line 40 you want **refunded to you.** 41a

► b Routing number ☐☐☐☐☐☐☐☐☐ ► c Type: ☐ Checking ☐ Savings

► d Account number ☐☐☐☐☐☐☐☐☐☐☐☐☐☐☐☐☐

42 Amount of line 40 you want **applied to your 2000 estimated tax.** 42

Amount you owe

43 If line 34 is more than line 39, subtract line 39 from line 34. This is the **amount you owe.** For details on how to pay, see page 48. 43

44 Estimated tax penalty (see page 48). 44

Sign here

Joint return? See page 20. Keep a copy for your records.

Under penalties of perjury, I declare that I have examined this return and accompanying schedules and statements, and to the best of my knowledge and belief, they are true, correct, and accurately list all amounts and sources of income I received during the tax year. Declaration of preparer (other than the taxpayer) is based on all information of which the preparer has any knowledge.

Your signature	Date	Your occupation	Daytime telephone number (optional) ( )
Spouse's signature. If joint return, BOTH must sign.	Date	Spouse's occupation	

Paid preparer's use only

Preparer's signature ►	Date	Check if self-employed ☐	Preparer's SSN or PTIN
Firm's name (or yours if self-employed) and address ►			EIN
			ZIP code

Form **1040A** (1999)

How to Avoid Common Mistakes

Mistakes may delay your refund or result in notices being sent to you. The most common mistakes are not doing the math properly and not signing the mailed-in forms. Not signing happens most often with married couples, when the preparer signs but forgets to ask his or her spouse to sign.

What about Dependents, Alimony, and Child Support?

The IRS checks for fraudulently claimed dependent exemptions (children that do not exist or elderly parents who have passed on). You can take an exemption for each of

EXHIBIT 3.10
The Simplest Tax Return Form

Department of the Treasury—Internal Revenue Service
Form **1040EZ** Income Tax Return for Single and Joint Filers With No Dependents (99) **1999** OMB No. 1545-0675

Use the IRS label here
Your first name and initial | Last name
If a joint return, spouse's first name and initial | Last name
Home address (number and street). If you have a P.O. box, see page 12. | Apt. no.
City, town or post office, state, and ZIP code. If you have a foreign address, see page 12.

Your social security number
Spouse's social security number
▲ IMPORTANT! ▲ You **must** enter your SSN(s) above.

Presidential Election Campaign (See page 12.)
Note. *Checking "Yes" will not change your tax or reduce your refund.*
Do you want $3 to go to this fund? ▶ Yes ☐ No ☐
If a joint return, does your spouse want $3 to go to this fund? ▶ Yes ☐ No ☐

Dollars | Cents

Income
Attach Copy B of Form(s) W-2 here. Enclose, but do not staple, any payment.

1 Total wages, salaries, and tips. This should be shown in box 1 of your W-2 form(s). Attach your W-2 form(s). 1
2 Taxable interest. If the total is over $400, you cannot use Form 1040EZ. 2
3 Unemployment compensation, qualified state tuition program earnings, and Alaska Permanent Fund dividends (see page 14). 3
4 Add lines 1, 2, and 3. This is your **adjusted gross income.** 4

Note. You **must** check Yes or No.
5 Can your parents (or someone else) claim you on their return?
Yes. Enter amount from worksheet on back. ☐ **No.** ☐ If **single,** enter 7,050.00. If **married,** enter 12,700.00. See back for explanation. 5
6 Subtract line 5 from line 4. If line 5 is larger than line 4, enter 0. This is your **taxable income.** ▶ 6

Payments and tax
7 Enter your Federal income tax withheld from box 2 of your W-2 form(s). 7
8a **Earned income credit** (see page 15).
b Nontaxable earned income: enter type and amount below.
Type | $ 8a
9 Add lines 7 and 8a. These are your **total payments.** 9
10 **Tax.** Use the amount on **line 6 above** to find your tax in the tax table on pages 24–28 of the booklet. Then, enter the tax from the table on this line. 10

Refund
Have it directly deposited! See page 20 and fill in 11b, 11c, and 11d.
11a If line 9 is larger than line 10, subtract line 10 from line 9. This is your **refund.** 11a
▶ b Routing number
▶ c Type: Checking ☐ Savings ☐
d Account number

Amount you owe
12 If line 10 is larger than line 9, subtract line 9 from line 10. This is the **amount you owe.** See page 21 for details on how to pay. 12

I have read this return. Under penalties of perjury, I declare that to the best of my knowledge and belief, the return is true, correct, and accurately lists all amounts and sources of income I received during the tax year.

Sign here ▶
Keep copy for your records.
Your signature | Spouse's signature if joint return. See page 11.
Date | Your occupation | Date | Spouse's occupation

For Official Use Only 1 2 3 4 5 6 7 8 9 10

For Disclosure, Privacy Act, and Paperwork Reduction Act Notice, see page 23. Cat. No. 11329W 1999 Form 1040EZ

your dependents who was alive during some part of the year for which you are filing a return. This includes a baby born during that year or a person who died in that year. The person must be your relative. The following people are considered your relatives:

* Your child, stepchild, adopted child; a child who lived in your home as a family member if placed with you by an authorized placement agency for legal adoption; or a foster child (any child who lived in your home as a family member for the whole year).
* Your grandchild, great-grandchild, etc.
* Your son-in-law, daughter-in-law.

EXHIBIT 3.10
continued

Form 1040EZ (1999) Page 2

Use this form if

- Your filing status is single or married filing jointly.
- You (and your spouse if married) were under 65 on January 1, 2000, and not blind at the end of 1999.
- You do not claim any dependents.
- Your taxable income (line 6) is less than $50,000.
- You do not claim a student loan interest deduction (see page 8) or an education credit.
- You had **only** wages, salaries, tips, taxable scholarship or fellowship grants, unemployment compensation, qualified state tuition program earnings, or Alaska Permanent Fund dividends, and your taxable interest was not over $400. **But** if you earned tips, including allocated tips, that are not included in box 5 and box 7 of your W-2, you may not be able to use Form 1040EZ. See page 13. If you are planning to use Form 1040EZ for a child who received Alaska Permanent Fund dividends, see page 14.
- You did not receive any advance earned income credit payments.

If you are not sure about your filing status, see page 11. If you have questions about dependents, use TeleTax topic 354 (see page 6). If you **cannot use this form,** use TeleTax topic 352 (see page 6).

Filling in your return

For tips on how to avoid common mistakes, see page 29.

Enter your (and your spouse's if married) social security number on the front. Because this form is read by a machine, please print your numbers inside the boxes like this:

9 8 7 6 5 4 3 2 1 0 Do not type your numbers. Do not use dollar signs.

If you received a scholarship or fellowship grant or tax-exempt interest income, such as on municipal bonds, see the booklet before filling in the form. Also, see the booklet if you received a Form 1099-INT showing Federal income tax withheld or if Federal income tax was withheld from your unemployment compensation or Alaska Permanent Fund dividends.

Remember, you must report all wages, salaries, and tips even if you do not get a W-2 form from your employer. You must also report all your taxable interest, including interest from banks, savings and loans, credit unions, etc., even if you do not get a Form 1099-INT.

Worksheet for dependents who checked "Yes" on line 5

(keep a copy for your records)

Use this worksheet to figure the amount to enter on line 5 if someone can claim you (or your spouse if married) as a dependent, even if that person chooses not to do so. To find out if someone can claim you as a dependent, use TeleTax topic 354 (see page 6).

A. Amount, if any, from line 1 on front ______ + 250.00 Enter total ▶ A. ______

B. Minimum standard deduction . . . . B. 700.00

C. Enter the LARGER of line A or line B here . . . . C. ______

D. Maximum standard deduction. If **single,** enter 4,300.00; if **married,** enter 7,200.00 . . . . D. ______

E. Enter the SMALLER of line C or line D here. This is your standard deduction . . . . E. ______

F. Exemption amount.
- If single, enter 0.
- If married and—
  - —both you and your spouse can be claimed as dependents, enter 0.
  - —only one of you can be claimed as a dependent, enter 2,750.00.

F. ______

G. Add lines E and F. Enter the total here and on line 5 on the front . . G. ______

If you checked "No" on line 5 because no one can claim you (or your spouse if married) as a dependent, enter on line 5 the amount shown below that applies to you.

- Single, enter 7,050.00. This is the total of your standard deduction (4,300.00) and your exemption (2,750.00).
- Married, enter 12,700.00. This is the total of your standard deduction (7,200.00), your exemption (2,750.00), and your spouse's exemption (2,750.00).

Mailing return

Mail your return by **April 17, 2000.** Use the envelope that came with your booklet. If you do not have that envelope, see page 32 for the address to use.

Paid preparer's use only

See page 21.

Under penalties of perjury, I declare that I have examined this return, and to the best of my knowledge and belief, it is true, correct, and accurately lists all amounts and sources of income received during the tax year. This declaration is based on all information of which I have any knowledge.

Preparer's signature ▶ Date Check if self-employed ☐ Preparer's SSN or PTIN

Firm's name (or yours if self-employed) and address ▶ EIN ZIP code

Form 1040EZ (1999)

- ✱ Your grandparent, great-grandparent, etc.
- ✱ Your brother, sister, half brother, half sister, stepbrother, stepsister, brother-in-law, sister-in-law.
- ✱ Your aunt, uncle, nephew, niece, if they are related by blood.

If there is a divorce, exemptions for children are usually claimed by the parent who had custody the longest during the tax year. The rules on taxation and children of divorce or separation are complicated, so any parent—custodial or non custodial—

should carefully check the IRS rules. Usually child support payments have no tax consequences, meaning that they are neither deductible by the payer nor included in the recipient's income. There are also child tax credits and tax credits for college expenses that will be addressed later in the chapter.

Alimony
Court ordered support to a former spouse.

Alimony, an allowance for support made under a court order and usually given by a man to his former wife after a divorce or legal separation, is deductible by the payer.

What About Other Tax Deductions?

Besides those categories already mentioned, like child support, the following *are not deductible* from federal income tax:

* Cosmetic surgery for improving appearance.
* Traffic violation fees.
* Attorney's fees for preparing a will.
* Life insurance premiums and death benefits.
* Workers' compensation benefits.

However, the following *are deductible:*

* The costs of home offices for many people who work at home.
* Income tax preparation fees.
* Mileage for driving to volunteer work.
* Up to $1,500 in student loan interest. The deduction is allowed for any loan in the first 60 months of its repayment schedule.

These categories are based on current tax codes and can change in any year. If, for example, you are considering plastic surgery, you should find out the latest tax codes as well as what your insurance covers. It may be important to document whether the surgery is strictly for appearance or if it is necessary for health reasons.

Withholding and Prompt Returns

Withholding
Amount employers deduct from employees pay and send to the federal government for income tax purposes.

Withholding, a national policy begun in 1943, requires an employer to deduct federal income tax from your pay and send it to the government. The withheld amount is based on the expected deductions and the number of exemptions that the employee has claimed on the W-4 form. After the year has ended, the employer sends W-2 forms to employees, reporting annual earnings and the amounts that have been deducted for federal income tax, Social Security, and state income tax, if applicable. A **refund** is due you when your employer deducts too much money from your paycheck and the government owes you the money back.

When can you expect a refund? If you file a complete and accurate tax return and are due a refund, your refund will be issued within 21 days (on average) if you filed electronically and within four weeks if filed by mail. It is not unusual for a refund to take six weeks.

Problems?

Problems not resolved through normal processes can be sent to the Problem Resolution Office. A caseworker will contact you within one week and will work with you to resolve the issue.

Objective 3:
Understand tax audits, how often they occur, and what to do if audited.

Tax audit
Detailed examination by the IRS of a tax return.

WHAT IF YOUR RETURN IS AUDITED?

Fewer than 1 percent of all individual income tax returns are audited each year. So the odds are very slim that you will be audited. A **tax audit** is a detailed examination of your tax return by the IRS. In most audits the IRS will request more information, such as receipts and canceled checks to support the entries on the tax form. As mentioned in the preceding chapter, well-organized records that go back three years should provide the answers the auditors need.

Who is most likely to be audited? The three categories of people most likely to be audited are people in cash businesses (occupations such as hairdressers, waiters, bartenders, and restaurant and gas station owners), certain professionals (doctors, lawyers, and accountants), and people taking unusually large deductions. People who have been audited in the past are likely to be audited again. And people living in certain locations—for example, in Las Vegas, Nevada—have a greater chance of being audited. The IRS may also audit if they receive a tip from individuals or from other government agencies that you are cheating on your tax returns. This accounts for 10 percent of referrals.

A red flag for auditors is whether or not your lifestyle matches your reported income. They may ask the following questions:

* How long have your owned your home? What did it cost?
* What are your mortgage payments?
* How often do you eat out? What's the average check?
* Where do you go on vacation? What do you spend?
* Which clubs do you belong to?
* Where do your kids go to school? What's the tuition?
* What year, make, and model are your cars? What are the payments?
* How much cash do you keep on hand?
* What's in your safe-deposit box?
* What do you spend on clothes? On hobbies?

For example, the IRS caught a Beverly Hills couple who claimed to live on $17,000 a year, yet deducted $28,000 in mortgage interest on their house. The audit turned up $250,000 in hidden income.[2]

There are three types of audits:

* *Correspondence audits*, for minor mistakes, are the simplest audits. A mailed inquiry is sent to you asking for clarification of a document or asking questions about your tax return. Usually when you send in the requested information, the matter is closed.
* *Office audits* require that you visit an IRS office on a specific date and time with the requested documentation to go over some aspect of your return.
* *Field audits* are more complex. An auditor visits you at home or at your business. This type of audit is usually done if the return is complicated and involves business operations.

What to Do if Audited

Do not panic. Your return may have been chosen at random and perhaps this is just a standard check. First, read the letter from the IRS carefully and see what it says you are to do. Maybe all you have to do is sign your return.

Audits make people nervous. Remember that IRS employees are only doing the job taxpayers pay them to do. If you had a certified public accountant (CPA) or attorney prepare your return, you should bring that person to the audit.

If you know one of your deductions is insupportable, come clean and pay the tax. If you dispute the results of the audit you can appeal and go through the courts. Some tax disputes have gone all the way to the Supreme Court.

Objective 4:
Access tax preparation help, including online sources.

SOFTWARE

Many different types of commercial software are available to help fill out tax returns. All of them have their pluses and minuses. The main thing is to find one that is easy for you to use yet covers every aspect of tax preparation that you need. Examples of the leading software offered by companies, all of which have websites, are TurboTax by Intuit, Kiplinger TaxCut by a unit of H & R Block, and TaxSaver by Microsoft.

Software prices vary considerably and sometimes there are rebates from the manufacturer, so comparison shop for the best price. Software is available for purchase online and at computer and office supply stores.

Tax Preparation Services

Is filling out a tax form too much work? Apparently a lot of people think so because *over 40 million U.S. taxpayers pay someone else to do their taxes.* Fees can range from around $40 to several thousand dollars. According to H & R Block, the average cost of federal, state, and local tax return preparation ran as follows in previous years:

1997	1998	1999
$71.69	$78.50	$84.57

Tax preparers' offices may range from a one-person shop to a national firm with thousands of offices, such as H & R Block.

If you hire a tax preparer you will still have to supply that person with information. Although most tax professionals are honest and competent, the IRS estimates that a few thousand out of the roughly 200,000 tax professionals in the country are incompetent, negligent, or fraudulent.[3] Even good tax preparers can make mistakes so hiring a preparer does not ensure that you will pay the correct amount. However, if you do not have the time, patience, or expertise, then perhaps a tax professional is what you need.

Volunteer Income Tax Assistance
Organization that helps individuals with filling out tax forms.

Free help may be available on campus or in the community through **Volunteer Income Tax Assistance (VITA)**. In this organization, people volunteer their time to help others fill out their tax forms and many offer direct e-filing. The service is free to those with limited or moderate income, the non-English speaking, the elderly, and the disabled. There is also AARP Tax Aide (www.aarp.org/taxaide) for older Americans with low or middle incomes.

If you work on your own, the IRS estimates that it takes two hours to complete Form 1040, 57 minutes for Form 1040A, and 22 minutes for Form 1040EZ. It is not unusual for someone to use a tax preparation service one year when their finances are complicated and to do it themselves the next. You have to weigh the costs and the benefits. With the growth in online services from the IRS and the computer literacy of the population, more people are filing their own taxes.

Objective 5:
Discuss financial decisions and expenses that affect taxes.

FINANCIAL DECISIONS AND TAXES

Your buying and selling decisions affect the amount of taxes you pay. As discussed earlier in the chapter, there are taxes on everyday items such as food and gasoline. This section focuses on expenses, college education decisions and child credits that may affect your taxes.

Owning a Home or Property

A home or property is an incredible tax-sheltered investment. While renting may be a good, temporary alternative until you are established, in the long run owning a home, given the tax benefits, makes more financial sense. Interest on mortgage payments and real estate property taxes are deductible and thus reduce your taxable income. Home mortgage interest ranks number one in dollar value in national returns.

If you own several homes, taxes become more complicated. Taxes are different on a first home that serves as a primary residence than on vacation homes. For example, Rose, a 77-year-old widow, had three homes. Two were in her native state of New Jersey (she lived in a big home on the edge of town and rented out a smaller one in the center of town), and the remaining residence was a vacation home in Florida. She was advised by her accountant not to sell them all at once due to heavy tax penalties. Her strategy was to gradually sell the big one and the Florida vacation house and within a few years to move to her smaller rental house in town.

Job-related Expenses, Including Moving and Home Offices

Certain work-related expenses (tuition, books, equipment, subscriptions, moving, computers, cars, telephones, and union or association membership dues) may be deductible. The rules on what is allowed change each year. The IRS is generally very tough on cars, computers, and telephones because these items fall in the gray area between business and personal use.

Usually job search expenses (i.e., résumés, telephone calls, and postage) are deductible if employment is sought in your current occupational category.

What about moving expenses? As long as your new job is at least fifty miles farther from your former residence than your former job, certain unreimbursed moving expenses are deductible. If you had no former workplace, your new workplace must be at least 50 miles from your old home.

Another category to be careful about is home offices. Generally you can deduct expenses related to a home office, if that office is used exclusively for business-related activities such as administrative or management activities. Many salespeople, consultants, teachers, and others can use the home office deduction.

> In a 1993 court case brought by the IRS, an anesthesiologist lost his bid for a home office deduction because he split his time between the office and three hospitals. The court said he had no principal place of business and the home office wouldn't qualify as one by default.[4]

There was an uproar over this decision that led to a new law, effective in 1999, that made more people eligible for home office deductions.

Travel and Entertainment Expenses

As more conventions and seminars take place at resorts, exotic locations (Hawaii, the Caribbean), and on cruise ships, travel and entertainment expenses have become more heavily scrutinized by auditors. The main guiding rule is whether or not the meeting was directly related to your business or profession. It is important to keep adequate records, receipts, and bills paid, including dates and times, who was entertained, and the purpose of the travel and entertainment.

> Unreimbursed business travel expenses for the self-employed are fully deductible, while such costs for employees are subject to the miscellaneous itemized deduc-

tion floor of 2 percent of adjusted gross income. However, the deduction for business meal and entertainment expenses is limited to 50 percent of the otherwise allowable amount...An annual deduction of up to $2,000 is allowed for attending a convention on a cruise ship, as long as all ports of call are in the United States or its possessions and the ship is registered in the United States.[5]

As with home office deductions, the best advice is to check with the IRS for the latest rulings at www.irs.gov.

USE OF CREDIT

The interest on home equity loans, also known as second mortgages, may be deductible in some states. Some people take out home equity loans on their primary or secondary homes to finance cars or to cover other debts.

Child Tax Credits

Tax credits
Items that reduce tax liability.

Tax credits are items such as child tax credits that reduce your tax liability. *For 1999 taxes a child tax credit of $500 was available for each qualifying child under the age of 17. This credit directly reduces the amount of the parents' tax liability.* If a family of four has a total tax liability equal to $4,000, the Child Tax Credit reduces this amount to $3,200 if both children are under age 17 through the 1998 tax year. If a family has more than two children under age 17, then they have to complete Form 8812.

Hope Credit and The Lifetime Learning Credit

You may be able to take Hope Credit for tuition and related expenses for yourself, your spouse, or dependents to enroll at or attend an eligible educational institution. The Hope Credit is a tax credit equal to 100 percent of the first $1,000 of tuition expenses for an eligible student's first two years of post secondary tuition and related expenses, and then 50 percent of the next $1,000 for students enrolled at least half-time (use Form 8863 to figure the credits).

The Lifetime Learning Credit is used for qualified tuition and fees paid after June 30, 1998, for courses to acquire or improve job skills, or for undergraduate and graduate-level courses at an eligible university or college. The maximum credit is 20 percent ($1,000 a year, up to $5,000) of qualified expenses. After 2002 the credit is slated to increase to $2,000 annually for a maximum of $10,000.

The Hope Credit and the Lifetime Learning Credit cannot be used in the same year for the same student. For 1998 taxes up to $1,000 of qualified student loan interest could be deducted. This amount is rising through the year 2001 and beyond. The IRS or tax advisors should be consulted for the latest rulings about these education-related credits.

INVESTMENTS AND TAXES

This section explores how investments impact taxes. Certain investments will reduce your taxes; others will increase it. Depending on your tax bracket, you may choose one form of investment over another because of its tax benefits. A financial advisor, such as a stockbroker, should be able to tell you in advance the tax advantages or disadvantages of certain investments so you can weigh the pros and the cons before investing in them.

Objective 6:
Describe the benefits of tax-exempt and tax-deferred investments.

Tax-exempt investments
Investments that are tax-free.

Tax-Exempt Investments

If you are in a high tax bracket (meaning you are expected to pay a lot in taxes due to high income), you may be interested in investing in **tax-exempt investments (or tax-free investments)**, such as municipal bonds which are issued by local and state governments and are not taxed by the federal government. They may pay lower interest rates than other types of bonds, but the tax savings may offset this.

Tax-Deferred Investments

Tax-deferred investments
Investments that are not taxed until later.

Tax-deferred investments put off the payment of taxes until a later date. Many people buy these with the idea of cashing them in during their retirement years when their income is reduced and, hence, they are in a lower tax bracket. Tax-deferred investments include all of the following:

* Retirement plans, such as IRAs, Keoghs and 401(k) plans (discussed in the section headed "Tax-Deferred Retirement Plans" and in Chapter 15).
* Tax-deferred annuities, usually offered by insurance companies (discussed further in the insurance chapters and in Chapter 15).
* Series EE U.S. Treasury Bonds issued by the federal government. The interest on these bonds is exempt from federal tax if it is used to pay tuition. (Chapter 13 on bonds will discuss these further.)

In summary, if you want to reduce your tax burden, then either tax-exempt or tax-deferred investments are worth exploring. As people age and their income grows, they become more interested in these types of investments as a way to avoid or defer taxes.

Capital Gains

Capital gains
Taxes on the profit from the sale of an asset.

Capital gains refers to profits from the sale of a capital asset, such as real estate, stocks, or bonds. Capital gains are tax-deferred, meaning you do not have to pay the tax on the gains until you sell the asset. Before selling a house or stocks and bonds, you should find out if there is a capital gains tax and whether it would be of maximum benefit to you to sell at a particular time.

Regarding houses, the latest ruling is that if you lived in the house you own at least two out of the last five years as your principal residence, the gain realized from its sale is tax-free up to $250,000 for singles and $500,000 for joint filers. If you sell your home before using or owning it for two years because of employment changes, health reasons, or other unforeseen changes, then the tax may be exempt if approved by the IRS.

Regarding stocks, the general rule is that you have to *hold a capital asset for at least one year* to get the 20 percent maximum rate for those in the 28 percent and higher tax brackets, or the 10 percent rate for individuals in the 15 percent tax bracket. Gains on stocks held for a year or less are taxed at your ordinary income rate, which could be as high as almost 40 percent. So a benefit of long-term investing is that the investor suffers a smaller tax bite.

Here is an example of how capital gains is calculated on a stock according to one financial advisor, The Motley Fool (www.fool.com/school/taxes). Suppose you bought 100 shares of XYZ stock at $25 each and paid a broker a $20 commission to make the purchase. Your total cost would be $2,520 (figured at $2500 plus $20 or $25.20 a share). If you sold 50 of the shares at $30 each a month later, the proceeds would be $1,500 less another $20 commission (the charge for selling) for a total of $1,480, or $29.60 a share. In this example your capital gain per share is $4.40 ($29.60 - $25.20). Because it was held less than a year, this gain will be taxed at the investor's usual tax rate.

Children's Investments and Taxes

Custodial account
Account that is created for a minor.

Parents can make investments on their children's behalf. They can list the children as owners through a **custodial account** wherein one of the parents serves as custodian for the child under the Uniform Gifts to Minors Act. This process is known as income shifting. It reduces the taxable income of the parents by shifting ownership of investments to children whose tax brackets are lower. However, once money is given to the child in this way, it is an irrevocable gift for the child's benefit or behalf. When the child turns 21 it is his or her money if it hasn't already been spent for college expenses or in some other fashion. Many parents set up custodial accounts to help save for their child's college education and at the same time reduce their taxable income.

There is a difference on taxation policies between children under 14 and those that are older. A child under 14 with investment income of more than $1,200 is taxed at the parents' top rate. For investment income under $1,200, the child receives a deduction of $600, and the next $600 is taxed at his or her own rate. Income shifting restrictions do not apply to children 14 and older.

If you are planning to file a return for your child who is under age 14, and certain other conditions apply, you may elect to report your child's income on your return. But you must use Form 8814, *Parent's Election to Report Child's Interest and Dividends*, in order to do so. If the parents make this selection, then the child does not have to file a return.

Tax-Deferred Retirement Plans

As mentioned previously, many working people use tax-deferred retirement plans—such as 401(k)s, and individual retirement accounts (IRAs)—as major tax strategies. The IRS allows people to deduct the contributions to these plans from current taxable income. This can have the benefit of lowering taxable profits by as much as $10,500 a year. The contributions and profits from the plans are not taxed until they are withdrawn, usually during retirement years when a person's income has dropped.

All working Americans may make an annual contribution of up to $2,000 per person or $4,000 per couple in an IRA. If you can afford it, you should try to put in the maximum amount. Like traditional and roll-over IRAs, Roth IRAs are limited to $2,000 per person or $4,000 per couple per year.

A self-employed person or a partnership can establish a Keogh or a SEP plan and contribute up to 20 percent of annual income, up to a maximum of $30,000. The 401(k) plan refers to a tax-deferred retirement plan sponsored by an employer. Keogh and SEP-IRA plans allow for a far higher amount than the $2,000 maximum limit set on other IRAs.

If you are interested in what your best retirement options are, you should consult with a financial professional and with the human resources department where you work because options vary so much by employer. Generally the higher your income, the higher the amount the employer sets aside for retirement. You can increase your financial security by setting aside money in IRAs, for example, that will supplement the retirement pension plans such as 401(k)s that your employer offers.

Estate and Gift Tax Transfers

If a person died in 2000, he or she could have left up to $675,000 to beneficiaries federal tax-free. This tax-free amount is on a graduated scale that increases to $1 million by the year 2006. There will be more on this subject in Chapter 15, but generally as any estate nears $500,000, estate and gift tax regulations should be investigated.

ELECTRONIC RESOURCES

As mentioned earlier in this chapter, the IRS website at **www.irs.gov**. is one of the best tax resources available. Access this site to:

* download forms, instructions, and publications.
* see answers to frequently asked tax questions.
* find news on the latest tax regulations.
* search publications online by topic or key word.
* figure your withholding allowances, using a W-4 calculator.
* send comments or request help via email.
* send in your return.

Privately maintained sites include the following:

1. **www.aicpa.org**, the home of the American Institute of CPAs. The site is geared to accountants, but anyone can use it to download forms and seek tax advice. The site also contains links to state revenue departments for questions about state taxes.

2. **www.taxresources.com** is maintained by accountant Frank McNeil in San Francisco. This is a thorough site with links to tax software sites, tax articles, and state tax sites. It is a starting point for people searching the Web for tax information.

3. **www.riatax.com** is the home of the Research Institute of America. The site features tax news and links to federal tax guides.

4. **www.1040.com** has information on regulations from the IRS.

5. **www.taxweb.com** features frequently asked questions, including how severance pay is taxed.

6. **www.quicken.com** contains calculators for difficult tax questions and tax chat rooms. It also offers tax preparation software. This site has a TurboTax Tax Estimator to find out if you are getting a refund next year from the IRS.

7. **turbotax.com** and **hrblock.com** (by Intuit and Block Financial) are websites that offer tax preparation advice.

1. What is a standard deduction?

   According to the IRS, the standard deduction is a dollar amount that reduces the amount of income on which you are taxed. The amount of the basic standard deduction depends upon your filing status (single, married, filed jointly, etc.). Each year the IRS publishes the standard deductions on their Form 1040 (see Exhibit 3.8) and in their instruction booklets.

2. To deduct charitable contributions, what form is needed?

   File Form 1040 and itemize deductions on Schedule A.

3. What are the criteria for being able to deduct moving expenses?

   The person must be an employee who moved at least 50 miles away from his or her former home. Also the person must have worked full-time for at least 39 weeks in the first 12 months after arriving.

4. How do education tax credits and refunds work?

■ Education tax credits are subtracted from individual tax returns, but they are non refundable. In other words if the credits are more than the tax, the excess is not refunded to the taxpayer.

5. Are tips taxable?

   ■ Yes, tips are subject to federal income tax. Tips include cash received, tips on charge cards, and any share of tips from tip-splitting arrangements.

6. What does a person do if he or she cannot file by April 15?

   ■ File an extension, Form 4868, or call 1-888-272-9829.

7. How does a day trader (someone who buys and sells stock very often) keep up with taxes?

   ■ Traders can physically record information, or they can use software programs designed to keep up with trades. More Web and software programs are being developed that will make managing trades and their tax consequences easier.

SUMMARY

Taxes are payments that governments require of individuals and organizations to support their activities. The largest percentage (48.1%) of federal income comes from individual income taxes, and the largest percentage (37%) of outlay goes to support Social Security, Medicare, and other retirement programs. Tax money also provides public goods and services, such as defense, roads, schools, law enforcement, and public libraries.

The average American pays 20–40 percent of income to taxes.

Progressive taxes are based on the idea that the more you earn, the more you pay. Examples are state and federal income taxes.

Regressive taxes take a smaller percentage of income as income rises. Sales taxes are regressive.

Flat taxes assess the same percentage from high and low income earners and allow fewer loopholes for the rich.

Tax avoidance is legal. Tax evasion is illegal.

Different types of taxes cover income, earnings, purchases, property, and wealth.

The most commonly used federal income tax form is 1040. Other often used forms are 1040A and 1040EZ.

For simpler returns and to speed up refunds, the trend is to file electronically. The IRS received about 33.6 million returns electronically in 2000, up from 29.3 million in 1999.

Most people claim the standard deduction rather than itemizing their deductions. The standard deduction varies by whether a person is a single taxpayer, head of household, married filing jointly, or married filing separately.

Tax records should be kept for at least three years.

Less than one percent of all individual income tax returns are audited each year. A tax audit is a detailed examination of your tax return by the IRS. There are three types of tax audits: correspondence, office, and field.

Over 40 million taxpayers a year hire someone to do their taxes.

The marginal tax rate is the amount of tax imposed on an additional dollar of income. It tells how much a tax deduction is worth and how much extra earnings will affect taxes.

The average tax rate (also called the effective tax rate) is the amount of taxes paid, divided by income.

Owning a home is one of the best tax shelters available.

Tax-exempt and tax-deferred investments help reduce current tax burden.

IRS rules change every year, including allowable amounts for deductions and credits. Federal income tax relief for parents and students paying college expenses is available from The Hope Credit and The Lifetime Learning Credit.

REVIEW QUESTIONS

(see Appendix A for answers to questions 1–10)

Identify the following statements as true or false.

1. Regressive taxes mean that as income rises, taxes rise. F
2. Tax avoidance is legal. T
3. Homestead exemptions are uniform across the country. F
4. Nearly 25 percent of the population gets audited by the IRS each year. F
5. An example of a tax-deferred investment would be an IRA. T

Pick the best answer to the following five questions.

6. _________ are the payments that a local, state, or federal government requires of individuals and organizations to support their activities.
   a. Taxes
   b. Tariffs
   c. Excises
   d. Appraisals
7. Who said, "In this world nothing is certain but death and taxes?"
   a. Theodore Roosevelt
   b. Calvin Coolidge
   c. Benjamin Franklin
   d. Donald Trump
8. In 1913 the _________ Amendment was passed; it stated, "The Congress shall have the power to lay and collect taxes on income."
   a. Tenth
   b. Thirteenth
   c. Sixteenth
   d. Twentieth
9. A child tax credit of $500 is available for each qualifying child under the age of _____.
   a. 14
   b. 17
   c. 20
   d. 21
10. Many people pay others to prepare their taxes. Free help for older Americans and those who qualify is available from AARP and from
    a. businesses such as H & R Block.
    b. welfare offices.
    c. Departments of Aging in each state.
    d. VITA (Volunteer Income Tax Assistance).

DISCUSSION QUESTIONS AND PROBLEMS

1. Tax assessments are based on either ability-to-pay or benefits-received philosophies. Give an example of how you experience both in daily life. Which philosophy do you think is fairer?
2. Sara and John Sullivan's baby was born on December 29, 1999. Could they claim her as a dependent on their 1999 tax return?
3. Why shouldn't you panic if you are audited by the IRS?

4. Why are certain professions, such as doctors and lawyers, audited more frequently than other professions?
5. What is capital gain and how does it affect the selling of stocks?

DECISION-MAKING CASES

1. Anthony graduated from college two years ago at age 23. He is single, has no children, and his current salary is $24,000 a year. He has a savings account and earned $100 in interest. He does not own a home or nor does he have an IRA. He does not plan to itemize deductions. The amount his employer withheld for federal income tax was $4,500.

   Questions:
   a. What tax form (1040, 1040A, or 1040EZ) should Anthony use?
   b. After selecting a form, can you determine Anthony's taxable income and total tax liability? For the answer use the information and exhibits given in the chapter or contact the IRS at www.irs.gov.
   c. Will Anthony probably get a refund?
2. Melissa is in the military and moves around a lot. She has been divorced for five years and her ex-husband has custody of their son, Cody. She has decided to take her income tax return to a tax preparer. What information will that person need?
3. Kevin, age 33, with one child and in a high tax bracket, wants to reduce his tax burden. What are five tax avoidance methods (investments or strategies) he can use?
4. Andrew and Kara are getting married in June and want to know if they should file their income tax return jointly or separately next year. Based on the chapter's coverage, what will the probable answer be, given that they both work and have similar incomes?
5. Robert began a new job at a company two hundred miles from the university from which he graduated. He wants to know what he can and cannot deduct from his income tax, including a home office, a computer at home, moving expenses, elective plastic surgery, travel and entertainment expenses, and charitable contributions. Which of these can he probably take off?

GET ON LINE

Activities to Try:

- For a "mentally taxing game," go to www.irs.gov and click on Braintaxer, select a player from the list given, then three questions will come up; answer them and hit submit. How did you score? Want to play again? Select another player.
- If you are planning to be a teacher or if you are interested in education, go to www.irs.gov and click on Taxpayer Help and Education, then TAXinteractive. There you will find a teacher tool kit and interactive games with tax information. Find the resource or game most interesting to you, explore it/play it, and report on what you find. In your opinion what age group is the material written for?
- Visit www.quicken.com/taxes and click on taxes. From there, click on the TurboTax Tax Estimator and find out if you are getting a refund next year from the IRS by entering your income amount. Or enter a hypothetical case, such as a software company CEO earning $500,000 a year. Or visit this site's chat room. What are taxpayers talking about?

INFOTRAC

To access InfoTrac, follow the instructions at the end of Chapter 1.

Exercise 1: Type in income tax and hit subject. This will bring up the subjects containing the words "income tax." This will include encyclopedia excerpts, reference book excerpts, periodical references, sub-divisions, and related subjects. Select one of the periodical references and report on what you find.

Exercise 2: Following the same steps as in Exercise 1, scroll down to Flat Income tax and click on See: Flat Tax to find periodical references on the issue. Choose two articles and compare what they say about the flat tax. Based on what you have read, what is your opinion of the flat tax?

chapter four
managing cash and savings

Building the Foundation of Wealth and Security

DID YOU KNOW?

* Most Americans save less than 5 percent of their incomes.
* One in four Americans has a savings bond.

OBJECTIVES

After reading Chapter 4, you should be able to do the following:

1. Explain the three steps in the cash management process.
2. Describe the new financial marketplace including online banking, ATM cards, smartcards, and debit cards.
3. Describe the various types of financial institutions and what they offer.
4. Understand how U.S. savings bonds can be used as part of a savings plan.

CHAPTER OVERVIEW

This chapter describes the different cash management and savings options available, including innovations in online banking, cards, accounts, and services offered by financial institutions. The chapter's purpose is to help you choose and use basic financial services effectively. It begins with a description of the cash management process, includes a section on Frequently Asked Questions (FAQs), and ends with Electronic Resources.

Objective 1:
Explain the three steps in the cash management process.

THE CASH MANAGEMENT PROCESS

Cash management
Routine, daily administration of cash and near-cash.

Cash management is the routine, daily administration of cash and near-cash (checking and savings accounts) to take care of an individual's or family's needs. The cash management process is made up of three steps, as shown in Exhibit 4.1:

1. *Awareness.* In this step individuals realize that money needs to be managed in a consistent way, that spending without a plan does not work.
2. *Analysis.* This is the stage in which individuals ask themselves, "What do I spend money on? Is it too much? Should I save more? What is really important to me?" An analysis is needed when people find they do not have enough money to meet financial goals, to handle anticipated life changes (moving, marriage, having children), or to deal with crisis events (layoffs, accidents), or if spending is out of control.
3. *Action.* This is the step in which individuals do something (open checking or savings accounts, talk with financial advisors, or change banks to get a better rate of return). For example, an action may be setting up a regular savings routine through automatic payroll deduction.

Part of the action step is making necessary adjustments. For example, to live within one's means, one should spend less than earned, save more, and invest what is saved. This sounds simple enough, but it is more easily said than done. As indications of the difficulties people have with spending less than they earn, consider that over 50 million Americans owe over $7,000 on their credit cards, and about a million people a year file for personal bankruptcy.

So the first guideline (spend less) is the most difficult. If someone brings home $2,000 a month, he or she can get caught in a cycle of spending it all each month and never getting ahead. This chapter focuses on how to get out of this rut by learning how to manage cash better and to consistently save.

Checking and savings accounts are familiar basic financial assets. They have the advantage of being *liquid.* As explained in Chapter 1, liquidity is the speed and ease with which an asset can be converted into cash, the most readily available form of financial resources. Because cash is so accessible with 24-hour ATMs (automated teller machines) and debit cards, it is becoming increasingly difficult to keep track of how much money one has. ATMS and debit cards are examples of ways to handle cash that will be described in this chapter.

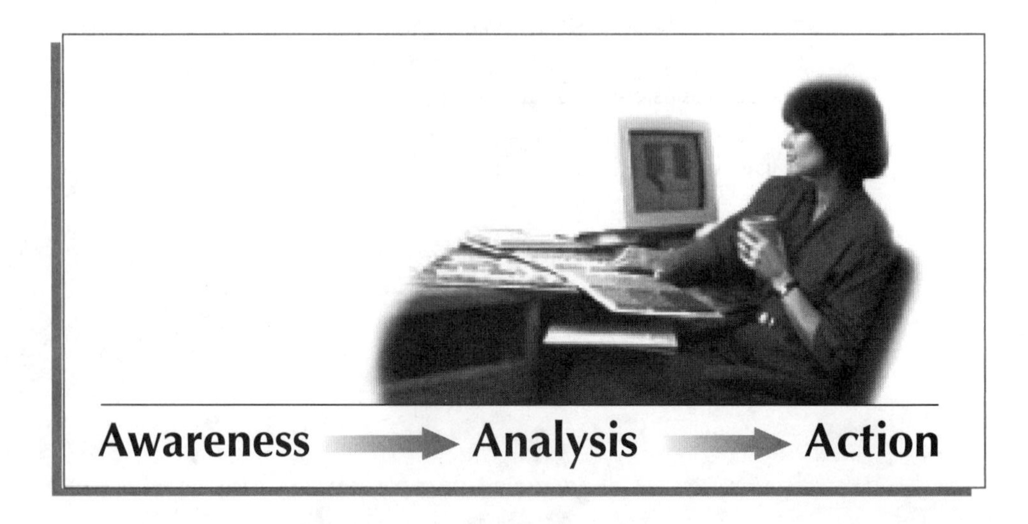

EXHIBIT 4.1
Cash Management Process

As noted in step three (action) of the cash management process, an individual may seek out financial advice and services. Exhibit 4.2 gives a list of today's financial services providers. Since there are so many providers offering the same or similar services it is up to the consumer to differentiate between them and pick the best choice. Speed, safety, ease of access, comfort level, cost, and interest are aspects to consider when choosing a provider and services.

Interest

Interest
The cost of using money.

Supply
Amount of money lenders are willing to lend.

Demand
Amount of money borrowers are willing to pay.

Certificate of deposit
Fixed-time deposit account.

Interest is the cost of using money. As defined in Chapter 1, intereste rate is the cost of borrowing money; it is what people pay in order to use other people's money. The rate of interest is determined by the supply and demand of credit. **Supply** refers to the amount of money lenders are willing to lend and **demand** is the amount of money borrowers are willing to pay.

The interest rate is expressed as a percentage and as an annual fee. For example, a one year **certificate of deposit (CD)**, a fixed time deposit, may offer a 6 percent interest rate. With a CD, money is lent to a bank for a predetermined number of months or years and the bank provides the lender a fixed rate of return (perhaps 6%). Because they are locked in to a set period of time, CDs are not as liquid as traditional savings or checking accounts. If lenders want their money back before the term is up, they will pay a penalty—usually a number of months' worth of interest.

You can be a receiver of interest (if you have a savings account or CD) or you can be a payer of interest if you have borrowed money for a purchase such as a house or furniture. Some interest rates are set for a period of time or they can fluctuate with inflation. Interest rates vary. While CDs may offer 6 percent interest, home mortgages may charge 8 percent interest and credit cards 12 to18 percent interest. Although these rates vary, the one thing they have in common is that most interest rates move together. A basic financial principle is that *when one part of the market adjusts interest rates, the other parts respond.*

Exhibit 4.3 shows the main cash management options available and the usual rates of return at the XYZ bank. Since options and rates change often, the latest update on options and interest rates should be obtained by calling or visiting financial

EXHIBIT 4.2
Providers of Financial Services

Providers*	Products and Services
Accountants	Preparation of income taxes, audits, financial planning, budgets, advice
Attorneys	Preparation of wills, estate plans, legal matters such as adoption, divorce, prenuptial agreements, bankruptcy
Banks, S&Ls, Credit Unions	Checking, share, money market, and/or savings accounts, traveler's checks, loans, mutual funds, safe-deposit boxes, ATMs, debit and credit cards, CDs, trusts
Brokerage Firms	In-person or online stocks, bonds, mutual funds, estate plans, financial planning, cash management accounts (CMAs), credit and or debit cards, 401k plans, Individual Retirement Accounts (IRAs), mortgages
Credit Counselors	Debt reduction, credit consolidation, financial planning, budgets, some offer 24-hour helplines
Financial Planners	Overall planning, money management, investment advice, life insurance, mutual funds
Insurance Companies	Policies for protection, annuities, retirement plans
Real Estate Companies	Buying and selling homes, investments, rentals, partnerships
Tax Preparation Services	Income tax returns, advice

*This exhibit serves as a general guide. In each category a provider may offer less or more than what is listed.

institutions or checking their home pages. Note in the exhibit that different levels of rates may be paid on accounts—the greater the amount deposited, the greater the interest rate.

Tiered rate
Interest-rate structure tied to a balance level.

An interest-rate structure by which the rate paid on an account is tied to a specified balance level is called a **tiered rate**. Also note in Exhibit 4.3 that Bank XYZ offers Prime 50 Savings for customers 50 and older. Banks like to have older customers because they leave large amounts of money in their savings accounts giving the banks a stable base of money to work with, but banks also like younger customers because they are more likely to build accounts and take out loans which banks make money on also. Because of this, banks and other financial institutions will offer a variety of accounts, services, and bonuses to attract different age groups.

How Much Should Be Kept in Checking and Savings Accounts?

Since most people's income goes directly into checking and savings accounts, they usually pay bills from these accounts. In a typical month people shift their money between their wallets and their checking and savings accounts sometimes setting aside money for other forms of saving. Take for example, the case of Peter and Cynthia Gray.

> Peter and Cynthia Gray deposit their combined monthly take-home income of $3,800 into their joint bank account. After paying their mortgage and other bills, their account usually carries a balance of $1200. Recently they started putting $200 a month into a mutual fund thus reducing their average checking account balance to $1000. Peter thinks this rate of savings is too low and they both know they have to be careful not to spend more than they have in their account.

The amount of money that should be kept in checking and savings accounts is a matter of preference. Some people get anxious if the balance drops too low, while others are comfortable living on the edge. If you are a person who gets nervous when

EXHIBIT 4.3
Savings Accounts from XYZ Bank

Account Name	Acct Type	Min. Balance To open	Balance Required to Avoid Charge	Monthly Service Charge	Interest Rate Tiering	% Rate	APY
SAVINGS ACCOUNTS							
Regular Savings	00/55	$100	$100	$5.00	Entire Balance / Compounded Daily	1.98%	2.00%
*Checkmate Savings	10	$100	$100	$5.00	$0 - 4,999.999	1.98%	2.00%
					$5,000 +	3.92%	4.00%
Variable Savings IRA	57	$25	N/A	N/A	Entire Balance / Compounded Daily	4.16%	4.25%
Roth Variable Savings IRA	12	$25	N/A	N/A	Entire Balance / Compounded Daily	4.16%	4.25%
Education Variable Savings IRA	14	$25	N/A	N/A	Entire Balance / Compounded Daily	4.16%	4.25%
**Prime 50 Savings	3	$100	$500	$5.00	$50-9,999.99	2.32%	2.35%
					$10,000–19,999.99	2.32%	2.35%
					$20,000–29,999.99	2.47%	2.50%
					$30,000–39,999.99	2.81%	2.85%
					$40,000–49,999.99	3.05%	3.10%
					$50,000–54,999.99	3.30%	3.35%
					$55,000 +	3.44%	3.50%

* Note: Tiered rates and APY. APY is the annual percentage yield which is the amount of interest earned on a yearly basis expressed as a percentage.

** Prime 50 Savings accounts are offered to customers age 50 and older at this particular bank.

the gas tank drops below 1/4 tank then you may be the kind of person who also does not like keeping a low bank balance. Your attitudes about maintaining a minimum balance have a lot to do with your risk tolerance personality as discussed in the first chapter.

If accounts drop too low, penalties are charged if a minimum balance is not maintained, or if a check bounces. Many financial institutions require the maintenance of a minimum balance to avoid checking fees. You can save more than $300 a year in fees by selecting a checking account with a minimum balance requirement that you can meet. For example, on one campus, students found they were paying $25 a month in fees at one bank and that by switching to another bank or a credit union they could reduce this to $5.00 or less per month. Regarding check bouncing, fees range from $15–$75 with a national average of $25 each time a check bounces. Banks may even charge $8–$25 if you deposit a bad check from someone else. Bank fees including ATM charges are on the rise so shop around.

To summarize, *cash management includes keeping abreast of your expenditures through maintaining a balanced checkbook, paying bills promptly, and establishing an ongoing savings program.* Being aware of cash management options and your cash management style are important first steps in the financial planning process. All of these will be discussed further in this chapter.

Why Maintain Cash Balances?

We all need cash for convenience. However, it is an expense to keep large sums in a wallet or in a non-interest bearing checking account because the money does not earn interest. Therefore cash management is part of the larger concept of opportunity cost (discussed in chapter 1) which is the value of what is given up when you decide on a certain course of action over another course of action. Since money deposited in a savings account or a CD earns interest, why would anyone have a noninterest-bearing account or carry a large amount of cash? The most obvious reason is that ready cash is needed to take care of small daily expenses. Checking accounts are more likely used for higher-priced items like rent, dentist, or utility bills. Some checking accounts pay interest others do not. To summarize, *cash is useful for small immediate needs and checks are useful for monthly bills or larger purchases.*

Why Are Savings Important?

Savings are an important resource in times of trouble and crisis. If you lose a job or if you are faced with unexpected expenses, savings can help tide you over. Thus savings serve as a cushion against financial emergencies. Savings can also serve as a way to accumulate funds to use to buy a car or to make a down payment on a house or to provide for education or retirement.

Savings Rates

Americans save on average less than 5 percent of their incomes each year. This is too low to save enough for retirement if the goal is to maintain current living standards during retirement. Many experts recommend a savings rate of 10 to 15 percent. The savings idea is that the more money put aside today, the more money there will be in the future.

As a nation we save far less than many other industrialized countries where savings average 12 to 18 percent of income. However, the figure of less than 5 percent may be misleading. Certain groups of Americans are saving far more than the national average. According to a recent study:

The 18-to-30 year olds who make up Generation X have something in common with their World War II-era grandparents: They save money. On average, X-ers save 16.2% of income, close to the 16.5% saved by 51-to-65 year olds, says the PaineWebber/Gallup Index of Investor Optimism.[1]

Another group saving more than the national average are the baby boomers (born between 1946 and 1964) because of their concerns about having enough money for retirement. Why Generation X-ers are such good savers is open to speculation.

Emergency Funds, Savings, and Basic Liquidity Ratio

Emergency fund
Fund kept in safe, readily available assets in the event of a financial crisis.

Basic liquidity ratio
Length of time an individual or a household can meet expenses based on liquid assets.

Because of potential crises, personal finance experts recommend setting up an **emergency fund** *of three to six months of after-tax income*. This amount is recommended because it should be enough money to take care of you and your family in case of job loss so that you can pay medical bills and take care of monthly needs. In other words an emergency fund should provide enough money to tide you over if you experience a financial emergency.

After a total loss of income the **basic liquidity ratio** reveals the number of months an individual or a household can meet expenses based on monetary (liquid) assets. In the example that follows, Stacee, a 25-year-old, has been laid off. She has $3,000 in monetary assets and her monthly expenses run $1200 for rent, utilities, gas, and groceries.

Formula: $$\text{Basic liquidity ratio} = \frac{\text{Monetary assets}}{\text{Monthly expenses}}$$

Example: $$= \frac{\$3{,}000}{\$1{,}200}$$

$$= 2.5$$

This ratio indicates that Stacee has sufficient money to support herself for two and a half months. Since a three to six months emergency fund is suggested, she would be well advised when she is employed again to build her fund up to that level.

Beyond the emergency fund someone starting out will want to build other forms of savings. It may take many years to build both an emergency fund and a savings fund. *The best way to reach your target is to take a regular amount out of each paycheck and put it into savings.* It is amazing how well this simple technique works. For example, if your money is earning 5 percent interest, compounded daily, saving $100 a month ($1200 a year) will produce a nest egg of $2523 ($123 interest plus $2400) in two years and $6,634 ($634 interest plus $6000) in five years.

Other ways to build an emergency fund include asking your bank to transfer a fixed amount each month from your checking account to your emergency fund, and depositing bonuses, tax refunds, or other unexpected income into your fund. For example, *the average tax refund is about $1,750.* A refund of this size could go toward paying off bills or into a number of different types of savings options, such as U.S. savings bonds, money market or savings accounts, or CDs. *The trend is towards depositing income-tax refunds directly into bank or savings accounts.* For example, the IRS reported that 15.4 million refunds were deposited directly as of May 2, 1997, up 53 percent from nearly 10.1 million a year earlier.[2]

COMPOUNDING: MAKING SAVINGS GROW

Savings grow based on the amount put in, the interest rate, the frequency that interest is compounded, and policies regarding deposits and withdrawals (how account balances are determined).

In chapter 1 the concept of the time value of money was discussed along with the related topic of compound interest.

Compound interest is the frequency that earned interest is added to the principal so that interest is earned on that amount as well as on the principal. The more frequently interest is compounded, the higher the future value. Savers should be interested not only in the **nominal rate of interest** (the stated annual rate of interest), but also in the **effective rate** (the actual rate of interest taking into account how often it is compounded). For example, if $100 is deposited in an account at 5 percent, the depositor will be credited with $105 at the end of the first year and $110.25 at the end of the second year. The extra 25 cents signifies the earned interest on the $5 in the first year. In this example interest is compounded annually, but more commonly interest is compounded continuously (called **continuous compounding**) throughout the day.

Nominal rate of interest
Stated annual rate of interest.

Effective rate
Actual rate of interest including compounding.

Continuous compounding
Compounding of interest continuously during the day.

Annual percentage yield
Amount of interest earned on a yearly basis expressed as a percentage.

Truth in savings act
Federal law requiring lenders to tell customers APY and other information.

Interest can also be compounded quarterly, half-yearly, or on some other time basis. *Since the more frequently interest is compounded, the more quickly money grows, continuous compounding is a benefit to the saver.* The **annual percentage yield (APY)** is the amount of interest earned on a yearly basis expressed as a percentage. The **Truth in Savings Act**, a federal law, requires that financial institutions tell customers the important terms of their accounts, including the APY, the interest rate, fees charged, and information about other features such as a required minimum balance.

Another factor affecting interest is how deposits and withdrawals are handled. Most financial institutions use the *day of deposit to day of withdrawal method* of computing interest. This method of computing interest is based on the exact number of days between when money is deposited and when it is withdrawn.

THE NEW FINANCIAL MARKETPLACE

Objective 2:
Describe the new financial marketplace including online banking, ATM cards, smartcards, and debit cards.

Today there are lots of cash management places and options that fall under the umbrella term of financial institutions. Before 1970 most people had a checking account from a bank and kept their savings in a savings and loan institution. Because of **deregulation** (fewer controls), in particular the **Depository Institutions Deregulation and Monetary Control Act of 1980**, the following happened:

Deregulation
Fewer controls, opening up of competition.

1. Many savings and loans closed.
2. There was more competition between remaining institutions.
3. Mergers occurred.
4. There was the development of nonbanks (new forms of banking including banks without main offices and branches called branchless banks).

For example, since 1980 credit unions have proliferated. Stockbrokers offer check-writing services, loans, and access to ATMs. Banks offer checking and savings accounts, CDs, traveler's checks, mutual funds, annuities, and insurance. Most recently some are placing stockbrokers or account executives in their banks. *The crossover among services offered is one of the most defining aspects of the new financial marketplace.*

Consumers benefit from the increased competition and from the multiservice, one-stop banking approach now available, but it can be confusing. It requires a more careful comparison of prices, charges, length of time money has to be invested for the full return (time deposits), interest rates, services, and convenience. **Time deposits** are accounts such as CDs with a maturity of at least seven days, from which you are not generally allowed to withdraw funds unless you pay a penalty. Cash management was a lot simpler when there was one bank in town and that was it.

Time deposits
Accounts with time limits, early withdrawals are penalized.

E-banking
Electronic access to cash and accounts.

Direct deposit
Paycheck directly deposited into accounts.

Electronic funds transfer
Making withdrawals, paying bills, depositing money or in some other way moving money electronically.

Electronic Banking/Electronic Funds Transfer

Another defining aspect of the new financial marketplace is **electronic banking** or **e-banking** which means 24-hour access through the Internet and to cash through an Automated Teller Machine (ATM) or **direct deposit** of paychecks or benefit checks into checking, share draft or saving accounts. Electronic banking is part of **Electronic Funds Transfer (EFT)** which means the use of computers and electronic signals to move money. Millions of American have discovered online banking, using their computers to monitor accounts, balance checkbooks, pay bills, apply for loans, and transfer funds between accounts. Banks may offer online banking service for free or charge $3–$10 a month for the right to pay bills electronically. One can also arrange to have a bank automatically pay recurring bills such as mortgage payments, auto leases or loans, or utility bills. This is especially useful for someone who travels a lot.

In most cases once an account is opened, you can initiate transactions through the bank's website. You enter a password and account number to log on. Onscreen options are given.

A type of e-banking is called supermarket banking which refers to the trend of one-stop shopping/banking. Consider the following excerpt from a newspaper article:

> Chris Blood, a computer programmer used to hate running errands during his lunch hour. First, he dashed to the bank. Then he headed to the grocery store to pick up a few items. By the time he finished waiting in lines, his lunch hour was over.
>
> For the past year, though, Mr. Blood has shopped at the Acme supermarket in Runnemede, N.J., near his job. At Acme, he buys bread and canned goods. He also visits the Mellon Bank on Acme's premises.
>
> "I come in every week to deposit checks," Mr. Blood said. "And I figure while I'm here, I might as well shop. It saves a lot of time."[3]

According to International Banking Technologies, an Atlanta bank consulting firm, there are more than 4,000 supermarket bank branches in the U.S.—up 27 percent from a year ago and up more than ten-fold since 1986.[4] Will conventional brick and mortar banks disappear? Probably not because there remains a place for them for more weighty transactions such as the purchase of CDs and the obtaining of loans. When people say "I am going to the bank" they usually mean they are going to a building housing a commercial bank.

EFT and Federal Payments

A provision of the Debt Collection Improvement Act of 1996 (DCIA) mandated that by 1999 all federal payments, with the exception of tax refunds, be made by EFT in lieu of paper checks. The reason for the legislation was to take advantage of the reduced cost, improved convenience, and security (Social Security checks won't be stolen from home mailboxes) of electronic technology.

Automated Teller Machines (ATMs)

Automated teller machines
Computerized, automated banking.

Automated Teller Machines (ATMs) are simple, fast, and 24-hour-accessible ways to access money and financial accounts. Because of their convenience, they are popular on campuses, in communities, and worldwide. If you have an ATM account in the U.S. you can get cash from machines in Ireland, for example, and nearly anywhere else in the world because most banks are members of ATM networks offering worldwide access.

Personalized identification number
Number assigned to customers for use at ATMs and for other forms of cash management.

ATM cards look like credit cards except that ATM cards have a special **personal identification number (PIN)**. The card and PIN allow the customer to make

CONSUMER ALERT

Usually banks or credit unions assign a PIN to you, but if you make your own choice don't pick a number that is easily associated with you—like your birth date, Social Security or telephone number, or home or office address. Also avoid repeating the same number (1111) or using consecutive numbers (1234). Never write the number directly on the ATM card or keep it in your wallet. When at an ATM machine, shield the screen from onlookers. If you become uncomfortable or suspicious during a transaction, cancel it, take your card, and leave. If your ATM or debit card is lost or stolen call your financial institution right away and follow up with a letter explaining what happened and when.

deposits and withdrawals, transfer funds between accounts, check their account balances, and make payments.

Banks vary in how much they charge customers to use cards. Generally there is no fee if you use your bank's own ATM machines, but transaction charges can range from $1.00–$5.00 if another bank's machine is used. Banks made over $2 billion from ATM surcharges in 1998 so naturally they are encouraging the use of ATMs. Legislative hearings have been held to discuss reducing surcharges.

Before signing up for a card, find out about the monthly or annual fees for the card, the "per use" fees when shopping or using the bank's or other banks' or credit unions' ATMs, and how fees can be avoided.

Always get receipts and record purchases in your checkbook to prevent overdrawing on your account. This will help you verify the accuracy of your bank statements as well as verify any unauthorized transactions.

Smartcards

Smartcards
Cards embedded with computer chips for a prepaid amount of money.

Smartcards are plastic cards embedded with a computer chip. A $20 card, for example, gives you $20 spending power. If a $1 soda is bought, the $1 will be deducted from the card making the card worth $19. If you forget the amount left on the card, you can check it with a portable card reader.

Smartcard promoters are relying on the idea that people do not like to carry coins or cash, they would rather carry a card. Smartcards come in three varieties:

1. A prepaid, disposable single purpose card such as telephone cards or library copy cards. When you have spent all the money encoded on the card, throw it out.

2. A prepaid, disposable bank card that you buy at a bank and can use at any retail outlet that has a terminal.

3. A reloadable card that you can load up again from a bank or ATM.

Why are the banks promoting smartcards? They produce income for the banks, who will charge a fee for this service (although initially the card may be free); the

CONSUMER ALERT

If you leave your card in the vending machine or library copy machine, you are out of luck if someone else picks it up and uses it. Remember to retrieve your card after each transaction.

merchant receiving your card may also be charged a fee. Smartcards are different from debit cards.

Debit Cards and Point-of-Sale Transfers

Debit cards
Cards that deduct purchase amounts from checking or savings accounts.

Debit cards or **cash cards** are plastic access cards that activate a system, such as an ATM machine or a point-of-sale transfer machine in retail establishments. The cards automatically deduct the purchase amount from checking or savings accounts. Thus *debit cards differ from smartcards with a prepaid amount and from credit cards for which you receive a monthly statement and have the option of paying all or part of the balance.* So what is the advantage of debit cards over credit cards? They are useful if you have a tendency to overspend on credit cards since debit cards restrict you to the amount in your account.

Debit cards can be used at retail establishments with point-of-sale (POS) transfers such as supermarkets, fast-food restaurants, gasoline stations, and convenience stores. They function in much the same way as ATMs. When you buy groceries, using your debit card, it is put through the POS machine at the checkout counter and the money is drawn from your account to pay for the groceries. Customers like using POS because they have to carry less cash, and retailers like it because they know that the customer has sufficient funds to cover the purchase. By federal law ATM and debit transactions must provide a paper record, and consumers must receive regular statements about EFT transactions.

Lost or Stolen Cards

As mentioned earlier, immediately reporting lost or stolen cards is important. Within two days of losing a card your liability for unauthorized use is $50. If you wait longer, your liability may go up to $500 within 60 days. Beyond that it can be limitless, you could lose all the money that was taken from your account. Consider what happened to Katherine Plowman:

> After her Master-Money card was stolen from her pocket in late December, it took Plowman a few days to realize the card was missing. She notified the bank, which immediately deactivated the card—but not before the thief had made five purchases worth $436. Counting the bank's overdraft charges and fees for two bounced checks, Plowman's account was debited an additional $260. The bank refunded the fees immediately, but didn't put the rest back into her account until it had completed an investigation. "We had to borrow money just to pay our rent," says Plowman.[5]

Future Trends

The trend is toward the development of one card to serve the function of a credit card, ATM card, debit card, and smartcard making it easier to keep track of money flow and to determine if there is an error or a loss. Dual-function ATM and debit cards are already common.

Eventually cards will be replaced by voice commands, codes, or scanning handprints, fingerprints, or eyes. For example, currently at the University of Georgia students with prepaid food plans do not use cards to gain access to the cafeterias. Instead they enter by placing one of their hands on a scanner. At other universities students gain access to their dormitories not by using keys or cards but by using hand scanners.

Bill presentment
Bills sent and paid for electronically.

Another trend is **bill presentment**. In this system, bills are delivered to customers electonically and paid for online without any paper exchanged. The main use is for recurring bills such as auto loans, credit card and telephone bills.

Another trend is the growth of connecting directly to a bank's website as distinct from using proprietary software. As more households go online and as more banks develop their sites this will continue to rise in use.

Objective 3:
Describe the various types of financial institutions and what they offer.

FINANCIAL INSTITUTIONS

Financial institutions
Multi-purpose institutions offering banking and other financial services.

Financial institutions are businesses with services, such as checking and savings accounts, car loans, home mortgages, credit cards, investment guidance, and retirement counceling. They offer the following benefits.

- *Safety*: You can be protected from theft, loss, and fire.
- *Convenience*: Cash can be accessed quickly.
- *Cost Savings*: Financial institutions offer cheaper ways to cash checks and buy money orders and other services than some other businesses.
- *Security*: In a federally insured institution, money is protected by the federal government *up to $100,000 for each depositor.*

Federal Deposit Insurance Corporation
Government insurance of bank and S & L accounts.

Most banks and savings and loan companies are insured by the ***Federal Deposit Insurance Corporation*** *(FDIC) and most credit unions by the National Credit Union Administration (NCUA). In the same bank a family could have $400,000 insured by the FDIC if two parents and their two children had separate accounts.* Since there are some financial institutions that are not federally insured, it is important to find out if they are insured by the FDIC or the NCUA. If they are, they post signs with their logos or you can contact the FDIC through the Internet to check if a bank is insured through them (see the Electronic Resources section).

Despite the advent of the electronic transfer of funds, over 60 billion paper checks are written annually according to the Consumer Bankers Association.[6] And in spite of deregulation which allowed the proliferation of services and institutions and electronic access, *most people still keep their money in traditional banks, savings banks, savings and loan associations, and credit unions because they are familiar and convenient.* These financial institutions tend to offer similar products so they will be summarized briefly before being discussed more in-depth in the section that follows.

CONSUMER ALERT

Although they may be purchased at banks, products such as mutual funds, annuities, and trusts are probably not protected by the FDIC, unlike checking and savings accounts. This has confused many investors. A survey conducted by Prophet Market Research and Consulting revealed that nearly a third of bank brokers failed to inform customers that mutual funds lack FDIC insurance and 15 percent never mentioned the possibility that customers could lose money on these investments. Bank brokers are often given incentives such as free trips or prizes for signing customers up for mutual funds, so they may not be as forthcoming as they should be about the risks involved.

Never invest in something you do understand and never assume that because a product or service comes from a bank or other financial institution, it is secure. Ask for proof or evidence.

1. Commercial Banks are financial institutions that operate under federal and state laws and offer a full range of products and services from checking and savings accounts to trust and loan assistance. They are the only institutions that offer non interest-paying checking accounts.

2. Savings and Loan Associations (S&Ls) are financial institutions, operating under federal and state laws that use depositors' money to finance consumer loans primarily mortgage loans. They generally pay a slightly higher interest rate on savings accounts than do commercial banks. Since deregulation, S&Ls offer other financial products and services as well. Similar to S&Ls are mutual savings banks (MSBs), which are savings institutions that are legally permitted in 17 states primarily in the Eastern U.S. In MSBs savings depositors own the institution and share in the earnings.

3. Credit Unions are non profit, member-owned cooperatives that operate under federal or state laws. Most offer loans, credit cards, and checking (share drafts) and savings (share) accounts that are insured by the federal government. Their members usually have something in common such as the same employer, church, or union.

4. Brokerage Firms offer banking services in the form of central asset accounts, which are coordinated money management plans that combine checking accounts, debit or credit cards, and money market funds with a traditional brokerage account. An example is Merrill Lynch's CMA account which requires a minimum investment of $20,000 and costs $100 a year. Other brokerage firms may require higher minimum balances, but charge no fees. One of the advantages of using a brokerage firm is that money can be shifted easily between accounts.

Commercial Banks

Of these four main types of financial institutions, traditionally *commercial banks have been the largest*. They are often called full-service banks because they offer safe-deposit boxes, traveler's checks, check-cashing privileges, savings accounts—including club accounts—credit cards, consumer loans, and trust services. Club accounts have special purposes or time limits such as vacation or Christmas clubs. Weekly or bi-monthly deposits are designated toward a specific savings goals such as $1,000 for Christmas shopping. The problem with club accounts is that they often pay less interest (in the past many paid no interest) than regular savings accounts.

As noted previously, only commercial banks can offer noninterest-paying checking accounts. With a checking account you use checks to withdraw money that you have deposited. Regular checking accounts do not pay interest, negotiable order of withdrawal (NOW) accounts (to be discussed later in the chapter) do pay interest. Also as mentioned before, another thing to be careful about is minimum balances that need to be maintained in order to avoid loss of interest. **Overdrafts** are checks written against insufficient funds. A bank may close an account or refuse to open an account for someone who has a history of overdrafts.

Overdrafts
Checks written against insufficient funds.

Banks also review every morning large transactions that occurred on the previous day so that, for example, if a check was written for $5,000 or over and it did not go into a certificate of deposit (CD) or anything else that could be expected, the bank will call the customer to make sure he or she wrote the check. This is to protect customers from any unauthorized use of their checks.

The average family spends about $200 a year on bank fees so it is a sizeable enough amount to warrant shopping around. Checking accounts may cost $5 to $20 a month, plus a $1 charge every time you write a check or use an ATM. Some banks charge $5 a month if savings accounts are inactive or drop below a certain amount. Certifying a check may cost $15 to $20, and there may be a fee for traveler's checks.

A supply of checks may cost $10 or more—the fancier the check designs the higher the rate. To reduce this, look for banks that do not charge for checks or checking accounts (some even pay interest on checking accounts), or print your own checks by using Quicken or a similar software program or through Internet websites.

Saving and Loan Associations (S&Ls)

Since deregulation it is getting increasingly difficult to differentiate savings and loan associations (S&Ls) from commercial banks. They both offer many of the same products except that S&Ls cannot offer noninterest-paying checking accounts. Another difference is that S&Ls usually pay slightly higher interest rates on savings accounts than commercial banks.

Their main function traditionally was as the name implies providing savings accounts and loans. If you are considering setting up a savings account or will need a home improvement or car loan in the future you may check out S&Ls to see what their rates are like compared to banks and credit unions.

Savings and loans may charge upwards of 10 percent to lend money, but pay considerably less for the use of your money (low interest rates, 2.2 to 2.7 percent, on savings accounts, close to what savings are losing to inflation). What can you do? If you have a considerable amount, one option would be to switch savings to money market mutual funds that charge low or no sales fees and pay a higher interest rate. However, these funds are not insured by the government and they require a minimum investment.

If you keep a savings account, shop around for the highest interest rates as well as for a convenient location or Internet access. If you can not easily access your money, a slightly higher interest rate is not much of a bargain.

Credit Unions

Credit Unions are nonprofit cooperatives owned by their members. Over 70 million people belong to credit unions in the United States. At a credit union savings accounts are called share accounts. Credit union members have share draft accounts rather than checking accounts and share certificate accounts rather than certificate of deposit accounts.

Essentially credit unions serve many of the same functions as banks because laws and economic forces have allowed banks and credit unions to enter one another's traditional areas of business while continuing to offer their specialized services to the public. Over the years there have been many turf fights and it remains to be seen how all financial institutions will fare in the future. Currently credit unions are limited in that they are able to expand their loan portfolios and deposit categories only in the consumer marketplace. Commercial banks have the ability to make large leaps in real estate lending, corporate financing, and financial service transactions. Credit unions are one twelfth the size of banks meaning that banks still have more than 12 times the amount of assets that credit unions do.

The strengths of credit unions are their low share account charges and loan rates and that they usually pay higher interest rates than commercial banks or S&Ls. So credit unions are attractive to most consumers as a choice in the financial services marketplace because of their competitive rates. They are also secure if they are federally insured.

Brokerage Firms

To the general public, brokerage firms are relative newcomers to banking and related services. One of their advantages is, as mentioned earlier, that a customer can easily shift money from one account to another within the firm's offerings. For example, money in a central asset account could be shifted with a telephone call to an IRA, stock, or mutual fund. At the end of the month, the customer receives a statement summarizing all transactions. This simplifies record keeping. Exhibit 4.2 shows the wide range of services brokerage firms provide.

Financial Institution Summary

To conclude this section on credit unions, commercial banks, S&Ls, and brokerage firms, the main advice is to shop around for the best rates, products, convenient locations, and services you can find. With deregulation and the growth of electronic options, the need to comparison shop is more important than ever before with the goal of keeping fee charges as low as possible.

An advantage to settling on one financial institution is simplicity. Another advantage is that establishing a relationship (having an account for several years) will help when applying for a loan. Loyal customers are usually offered lower interest rates than are offered to individuals the financial institution does not know.

WHAT TYPES OF DEPOSIT ACCOUNTS ARE AVAILABLE?

A great variety of accounts are available and many have been discussed earlier (share accounts offered by credit unions, time deposits such as CDs), but it is useful to have a separate section on accounts distinct from the general discussion of financial institutions since so many accounts are not limited to one type of financial institution and many are available electronically. *Making Sense of Savings* a publication of the Federal Reserve System, divides accounts into four types to be discussed next.[7]

Checking Accounts

A personal check directs the institution to subtract money from your checking account. Many institutions let you withdraw or deposit funds at an automated teller machine (ATM) or pay for purchases at stores with your ATM card.

The person who writes the check is called the **drawer or payer**. The **drawee** is the financial institution at which the account is held, and the **payee** is the person or company for whom the check is written. Exhibit 4.4 shows a sample check.

A regular checking account (frequently called a demand deposit account) does not pay interest; a **negotiable order of withdrawal (NOW) account** does pay interest or dividends.

Depending on the type of account, the balance in the account, and the institution, checks may be free or there may be a charge. Some institutions charge a fee for every transaction, such as for each check written or for each withdrawal from an ATM.

Although a checking account such as a NOW account that pays interest, is attractive, it is important to look at fees for both interest-bearing and no-interest accounts. Consider how many checks a month you normally write and how much money you normally keep in an account before deciding which is the best choice.

Banks compete by offering packages of products and services. Exhibit 4.5 shows how extensive a checking account can get. Note in the advertisement that free

Drawer or Payer
Person who writes a check.

Drawee
Financial institution with the account.

Payee
Person or company for whom a check is written.

NOW account
Type of checking account that pays interest or dividends.

EXHIBIT 4.4
Sample Check

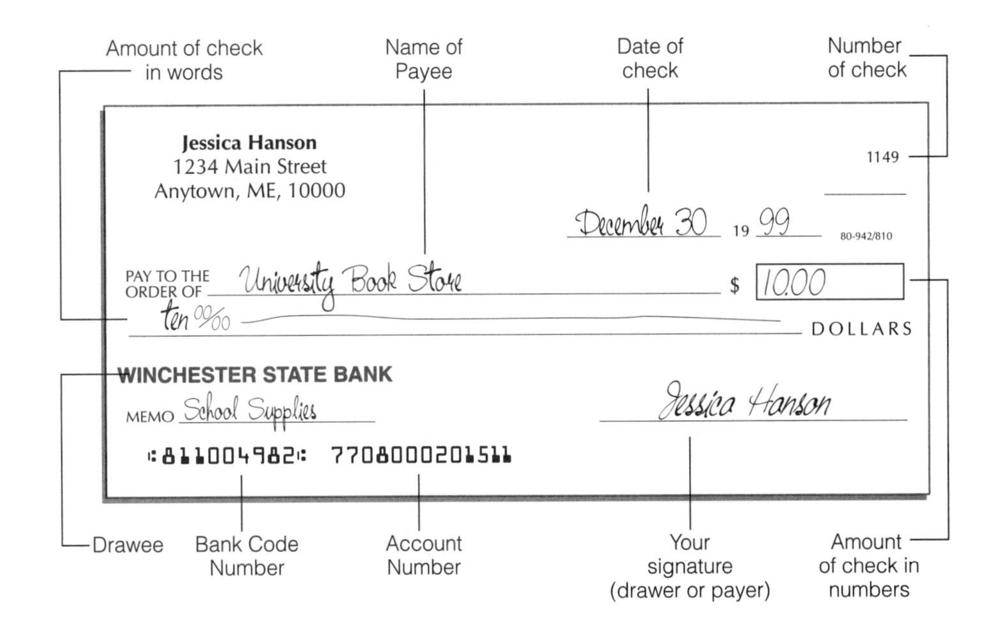

checking is available if a minimum of $100 is maintained. Also notice all the other features including a loan service, insurance, a MasterCard, a VISA card, and sports bag.

Money Market Deposit Accounts

Money market deposit account
Type of account that offers a higher rate of interest than most checking, savings, or NOW accounts.

A specialized interest-bearing account that allows you to write checks is called a **money market deposit account (MMDA)**. This type of account often pays a higher rate of interest than regular checking, savings, and most NOW accounts. Institutions can afford to do this because MMDAs usually require a higher minimum balance, such as $2500.

Exhibit 4.6 shows a bank advertisement offering money market accounts and CDs. Note that with MMDAs the APY goes up as the minimum balance increases. This is also true for CDs but in the case of CDs they also go up the longer they are invested.

One of the drawbacks of MMDAs besides the high minimum balance is that many are not as convenient to use as a checking account. For example, you may be limited in the number of transfers, usually three to six, you can make per month. As with checking accounts, most institutions impose fees and penalties if the balance drops below a minimum amount on the MMDAs.

Savings Accounts

With savings accounts you can make withdrawals, but you do not have checks. The number of withdrawals or transfers you can make on the account each month is limited as is the case of MMDAs.

Passbook savings account
A savings book record of deposits and withdrawals in an account.

Statement savings account
A statement of withdrawals and deposits in an account.

Many institutions offer passbook savings and statement savings accounts. The most familiar is a **passbook savings account** in which you use a record book called a passbook to record your deposits and withdrawals. The passbook should be brought to the institution each time you make deposits and withdrawals. As the name implies, a **statement savings account** means that the institution mails you a statement that shows your withdrawals and deposits for the account. *Passbook savings accounts often pay a slightly higher interest rate than statement savings accounts* because in passbook savings accounts people tend to leave in larger amounts of money because they cannot

EXHIBIT 4.5
Advertisement for a Checking Account. Used by permission.

easily get to the money. A typical statement savings account may have $100 to $1,000 in it, but a passbook savings account typically has more. So if someone is saving $100 a month for two years to finance a major vacation he or she may be wise to put the money into a passbook savings account to gain the greatest amount of interest possible on an insured account. As Exhibit 4.3 shows, institutions may require a minimum balance to open and maintain an account.

Checking and statement savings accounts can be linked offering overdraft protection. If the checking account runs out of money, then the bank can go into the savings account if the customer signed paperwork allowing them to do so.

Time Deposits (Certificates of Deposit)

As mentioned earlier, time deposits are often called certificates of deposit or CDs. For a specified term, from several days to several years, they offer a guaranteed rate of interest, which is usually higher than interest rates paid on checking and savings accounts. Another good feature of bank CDs is that they are federally insured up to $100,000. Also the more money you put in (for example, $50,000 versus $1,000), the higher the annual percentage yield. There may be prizes or bonuses such as toasters

EXHIBIT 4.6
Advertisement for Special Accounts

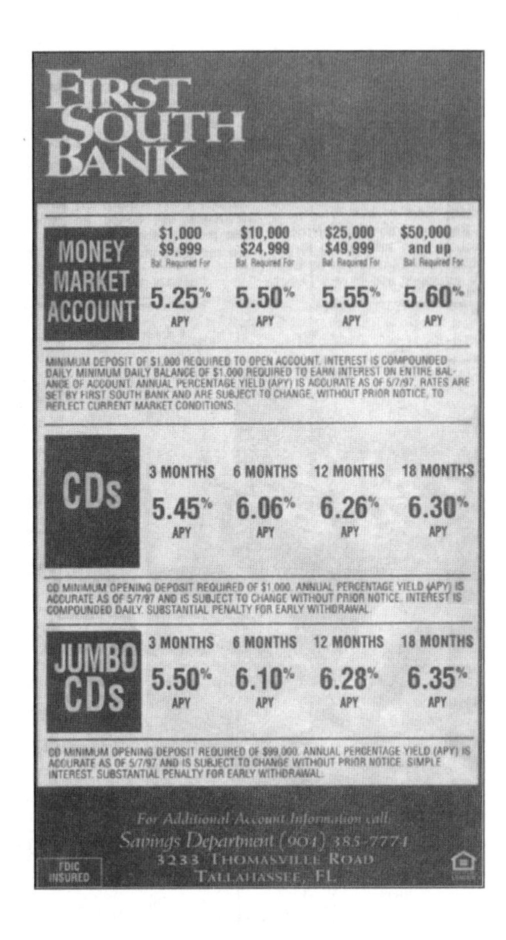

or turkeys, attached to opening a new CD. To determine if these are worthwhile, estimate the approximate price of the bonus item and decide if it adds substantially to the offered interest rate.

Maturity
A designated future date.

Principal
Initial deposit or investment.

Roll over
When a CD or other form of investment moves from one investment to another.

If you keep money in the CD the agreed upon amount of time, it is called waiting until **maturity**. In some cases, you can withdraw the interest you earn even though you may not be permitted to take out any of your initial deposit (the **principal**).

If a CD is withdrawn before it matures, a penalty is frequently charged. When a CD is getting close to maturity, the institution will notify you and encourage you to renew it. If you do not notify your institution about what you want to do, usually the CD will **roll over** (meaning continue) for another term.

What Type of Account Is Right for You?

Selecting the right account depends on your needs. Perhaps you prefer a basic, no frills account that gives you a limited number of services for a low, set price, or you may want something with more services. In any case find out the fees, services, and penalties before you sign up and consider the risk involved in tying your money up at a set rate. A general strategy to use is that if interest rates are falling you should consider long-term instruments (CDs) that "lock in" earnings at current high rates. Besides the accounts already discussed, two other options are: Money Market Mutual Funds (MMMFs) and Asset Management Accounts (AMAs). Exhibit 4.7 summarizes the main points about the different types of accounts.

Money market mutual fund
A money market account that is in a mutual fund.

A **money market mutual fund (MMMF)**is a money market account that is in a mutual fund investment company rather than a commercial bank. The company pools together the cash of thousands of investors and passes the benefits, relatively safe and high returns, to the investors. The drawbacks are that check writing is discouraged

EXHIBIT 4.7
Comparing Different Kinds of Accounts

Type of Account	Is there a minimum investment?	Will I earn interest?	May I write checks?	Are there withdrawal limitations?	Are fees likely?
Regular Checking	Yes	No	Yes	No	Yes
Interest Checking (NOW)	Yes	Yes	Yes	No	Yes
Money Market Deposit Account (MMDA)	Yes	Yes, usually higher than NOW or savings	Yes	Yes	Yes
Savings	Yes	Yes	No	Yes	Yes
Certificate of Deposit (CD)	Yes, some as low as $100, but typically $1,000 and up	Yes, usually higher than MMDA	No	Yes, usually no withdrawals of principal until the date of maturity	Yes, if you withdraw principal funds before the date of maturity
Money Market Mutual Funds (MMMF)	Yes, $1,000-$2,500	Yes	Yes, most firms	Yes, $200 or more is customary	Yes, typically one-time annual fee
Asset Management Account (CMA)	Yes	Yes	Yes	No	Yes, typically one-time annual fee

since checks must be for large amounts ($200 or more) and the funds are not insured by the FDIC. A definite plus is the tax benefits most MMMFs offer.

To open an account, an investor would contact a mutual fund investment company. Examples of such companies are Dreyfus, Fidelity, Kemper, and T. Rowe Price. They advertise in newspapers, magazines, and on television or can be contacted over the Internet by using their name.com.

Asset management account
Multi-purpose account for handling cash, checking, credit, and investments.

An **asset management account** (also known as a central asset account or **cash management account, CMA**) is a multipurpose package of a number of transactions/services all in one account. These can include checking, credit cards, debit cards, ATM access, loans, stock brokerage accounts, and money market mutual funds. There is usually a required minimum amount to open the account and fees (opening and monthly maintenance charges) vary.

Where do you get one? They are offered through companies, retailers, financial institutions, mutual funds, and brokerage firms. Brokerage firms like to offer AMAs or CMAs because money can be shifted easily from the accounts to stock and bond purchases or vice versa and dividends can be put into the cash management part of the account. This saves the customer and the stockbroker time and allows for greater flexibility.

Checks for Specialized Needs

Before leaving the subject of institutions and accounts, there are four kinds of checks with specialized functions to be discussed.

Cashier's checks
A guaranteed check purchased by a customer drawn on a bank's general funds.

Cashier's checks are made out to a specific payee drawn on a bank's general funds. To obtain this type of guaranteed check, you pay a financial institution the amount of the check and have a bank officer prepare and sign it. A fee for this service is charged.

Certified checks
Personal checks on which the bank has guaranteed payment.

Certified checks are personal checks drawn on your account on which the institution has imprinted the word "certified," signifying that your account can cover it. A fee for this service is charged.

Money orders
Types of checks purchased for use instead of personal checks.

Money orders are offered by the U.S. Postal Service, financial institutions, and grocery stores. Money orders are written for small amounts and constitute a type of checking instrument. Because of fees charged for this service, if you write more than three checks a month you would probably be better off opening a checking account than using money orders.

Traveler's checks
Types of checks pre-purchased for set amounts for travel.

Traveler's checks are issued by large financial institutions such as American Express or VISA, and sold through financial institutions and groups such as the American Automobile Association (AAA). A fee is usually charged. If you are a member of the group or have a certain type of account the fee may be waived.

Joint versus Separate Accounts

When opening an account, ordering traveler's checks, or renting a safe deposit box, you have to decide if you are signing by yourself or if you are signing with another person, signifying shared ownership. A third option that some couples select is to open a joint account for bill paying and to maintain separate accounts for each person for gifts and other purchases.

Objective 4:
Understand how U.S. savings bonds can be used as part of a savings plan.

SAVINGS BONDS: A SAFE WAY TO SAVE

Besides savings accounts and certificates of deposit, another savings option is U. S. savings bonds. The U.S. Secretary of the Treasury says that *about one in four Americans owns a savings bond.*[8] Why are they so popular? A lot of people especially *conservative savers who are low-risk takers like government-backed instruments.*

> More than 55 million Americans collectively own $186 billion in savings bonds, commonly available through banks and payroll savings plans. With the smallest denomination of a Series EE bond selling for just $25, they've been a popular savings vehicle for children's college education or as holiday and birthday gifts for young relatives.[9]

How do you purchase U.S. savings bonds? Besides banks and payroll savings plans, U.S. savings bonds can be purchased online at www.savingsbond.gov. The Treasury Department has sold them for more than 50 years. U.S. Savings Bonds are appealing because of their safety, patriotism (investing in America), and affordability.

Identification numbers (bond serial numbers, issue dates) should be kept in a safe place separate from the bonds. What if bonds are lost or stolen? The bond owner can apply free of charge for replacement to The Bureau of the Public Debt, Parkersburg, WV, 26106-1328.

CONSUMER ALERT

The critical words to watch for in the wording of the title of joint accounts and safe deposit boxes are "or" and "and." With the word "or" either spouse can empty the joint checking or savings account or close out the safe deposit box without the other person's knowledge. With the word "and" both signatures are needed to close an account or a safe deposit box. In divorce or separation situations, it is not unusual for one person to empty out boxes or accounts. With the "and" word either person can take an account down to a $1.00 or empty the box, but they cannot close the account or end the safe-deposit box arrangement at the bank without the other person's knowledge. Therefore the word "and" does not provide total protection, but it at least alerts the other person that the spouse is making significant changes.

The three main types are Series EE, Series HH, and Series I bonds. Interest earned on these bonds is exempt from state and local income taxes. The bonds are sold through financial institutions qualified as savings bonds agents or through employer-sponsored payroll savings plans. The registration on a savings bond (owner's name) is conclusive of ownership. Bonds may be registered in the names of individuals in one of three ways: single ownership, coownership, or beneficiary form. When bonds are purchased as gifts, the purchaser should be sure to have the correct Social Security number and the correct spelling of the recipient's name.

Series EE

Series EE bonds are Treasury securities that earn interest through periodic increases in value for up to 30 years. The purchase price of a bond is 50 percent of its face amount; for example, a $50 bond costs $25. Bonds are available in $50, $75, $100, $200, $500, $1,000, $5,000, and $10,000 denominations. There are two maturity dates for Series EE bonds. Original maturity is 17 years after the issue date and final maturity is 30 years after the issue date. Bonds stop earning interest at final maturity.

Series HH

Series HH bonds come in denominations of $500, $1,000, $5,000 and $10,000. So the lowest amount that can be invested in Series HH bonds is $500. They pay interest semiannually at a fixed rate and because of this the interest generally is subject to federal income tax. The rate is set for the first 10 years that the bond is held and can change when the bond enters an extended maturity period of an additional 10 years. Series HH bonds are issued and redeemed by Federal Reserve Banks and can be processed by qualified financial institutions.

Series I

Series I bonds (the "I" stands for inflation) were introduced in September 1998 and come in denominations from $50 to $10,000. These bonds guarantee a real rate of return no matter what happens to the inflation rate. After purchase they must be held for at least six months. To get the full value they should be held for five years. They are tax-deferred meaning that federal income tax on earnings is not paid until the bonds are redeemed. As of the writing of this book they were paying an attractive 7.49 percent.

ELECTRONIC RESOURCES

Government websites relevant to this chapter include

* the Federal Deposit Insurance Corporation at **www.fdic.gov**, where you can find out whether or not a bank is federally insured.
* the Treasury Department's Bureau of Public Debt website at **www.savingsbond.gov**. Besides selling savings bonds, this site answers frequently asked questions about bonds, provides forms, and has a kid's page.

Business and organization websites are extensive. General information on banking can be obtained from **www.aba.com** (American Bankers Association) and for savings from **www.asec.com** (American Savings Education Council). Samples of banking software can be seen at **www.microsoft.com/catalog/** (Microsoft Money) and **www.intuit.com** (Quicken Financial Network).

Most major banks have websites and some are branchless banks that operate only on the Internet (Security First Network Bank, Net.B@nk, CompuBank, and First Internet Bank of Indiana). The OnLine Banking Report at **www.onlinebankingre-**

port.com contains links to more than 240 U.S. banks with online services and has an online magazine. Bank Online at **bankonline.com** lists links to banks in specific regions of the country and has an interactive map. De Armas Internet Banking at **ddsi.com/banking** provides links to national and international banks.

The Gomez Advisors (The Internet Banker Scorecard) at **www.gomez.com** provides a consumer-oriented list that ranks bank websites according to their ease of use, customer confidence, costs, on-site resources, and relationship services. In 1999 some of their highly ranked banks included Wells Fargo, Bank of America, Bank One, and Citibank. For their latest top ten list go to their website.

Besides the banks already mentioned, two others who post CD and money market yields are

- Atlanta Internet Bank at **www.atlantabank.com**.
- 1st Source Bank at **www.pawws.com/1stSrc**.

For example, Sara James found that her local bank was offering 5 percent CD rates whereas two of the listed national providers offered over 6.5 percent. Carmen King has $10,000 of her savings in one of these sources in a money market account that yields 5.06 percent annually compared to the national average of 2.64%. Nationwide institutions offer better rates because

> "We don't have the overhead of a bank with branches," explains Mitchell Caplan, TeleBank's president. "We can pass through our savings in the form of higher rates."[10]

These providers offer CDs and money market funds that are federally insured (although this should be confirmed before an investment is made). They may set minimum entry levels such as $1,000 which may initially have to begin by mail and later move onto the Internet. If you send a check and while the check is in transit the rates change, usually the institution will honor the rate you were initially quoted (assuming your check arrives within seven days of the quoted rate).

One of the drawbacks of going online is the loss of personalized service, but unfortunately today even some "full-service" banks are charging customers if they want to talk to a teller. Another drawback is security. To learn more about security on the Internet contact Rutgers University Network Services at **www-ns.rutgers/edu/www-security/index.html**.

1. Should I buy government bonds or put the money into the stock market?

   Putting risk and security issues aside, money that you expect to keep invested five or more years would have a better chance of growing in the stock market. As long-term investments, stocks outperform bonds. This advice is based on historical analysis that shows that from 1802 to 1997 (195 years) the stock market offered an average nominal annual return of 8.4 percent per year, compared to 4.8 percent for long-term government bonds and from 1966 to 1997 (a thirty year period) stocks rose 11.5 percent vs. 7.9 percent for long-term government bonds.[11] However, savings bonds make sense for those seeking a guaranteed return on their money. The investment chapters will cover more strategies and ways of comparing different types of investments.

2. I'm graduating and moving across country. How can I use the Internet to find banks in that area and compare rates?

■ The Electronic Resources section will give you a number of leads. Most sites offer the ability to click on the state you are interested in. Another option is to go online and contact the Chamber of Commerce in your new location — they can give you bank information or they will contact banks and the banks will send you information. A third option is to use a general search engine such as www.yahoo.com or try www.banx.com.

3. Why are there fewer banks than there used to be?

■ There are a lot of reasons including mergers, ATMS, the globalization of the economy, and the growth of online services.

4. Why do Americans save so little?

■ They would rather spend their money partly as a cultural tradition. We live in a consumption society with many more choices than is common in the rest of the world and we have a higher standard of living than most other countries.

SUMMARY

Cash management is a three-step process including awareness, analysis, and action.

Americans save on average less than 5 percent of income per year. This is a lower savings rates than in many other industrialized nations.

An emergency fund equal to three to six months of take-home pay should be set up.

With deregulation, financial institutions have become more competitive and non-banks have developed. More services and products are available electronically and at a greater variety of sites and outlets.

Interest rates may be tiered meaning that the more money put into the account, the higher the interest rate. In the case of CDs, the higher the amount invested and the longer it is invested, the higher the interest rate.

Debit cards, credit cards, and smartcards serve different functions, but each is a substitute for cash.

Future trends include the development of one all purpose card, bill presentment (receiving and paying bills online), and directly accessing bank websites as opposed to using proprietary software.

The four main kinds of financial institutions are commercial banks, savings and loan associations, credit unions, and brokerage firms. An advantage to establishing a relationship with a financial institution may be lower interest rates when applying for loans.

Different kinds of accounts are checking, NOW, money market deposit, savings, club, time deposits (CDs), money market mutual funds, and asset management accounts. Checking and savings accounts can be linked to provide overdraft protection.

Cashier's checks, certified checks, money orders, and traveler's checks are types of checks for specialized needs.

The three types of U.S. savings bonds are Series EE, Series HH, and Series I.

REVIEW QUESTIONS

(See Appendix A for answers to questions 1–10)

Identify the following statements as true or false.

1. Cash and checking and savings accounts are the most liquid, most easily accessible forms of financial resources.
2. Americans rank high as savers compared to other industrialized nations.

3. Continuous compounding is a benefit to the account holder.
4. Consumers have been hurt by financial institution deregulation.
5. For the safety of a PIN, avoid repeating the same number (2222) or using consecutive numbers (5678).

Pick the best answer to the following five questions.

6. Which of the following is not a step in the cash management process?
   a. Awareness
   b. Analysis
   c. Appraisal
   d. Action
7. _______ refers to the amount of money lenders are willing to lend.
   a. Supply
   b. Demand
   c. Yield
   d. Premium
8. Using the Basic Liquidity Ratio formula (monetary assets divided by monthly expenses), how many months does Bryan have enough money for given that he has lost his job, has $2,000 in his checking account, and his monthly expenses run $1,250?
   a. 1 month
   b. 1.6 months
   c. 2.3 months
   d. 2.9 months
9. ______ or cash cards are plastic access cards that activate a system such as an ATM machine or a point-of-sale transfer machine in a retail establishment. These cards automatically deduct the purchase amount from checking or savings accounts.
   a. Debit cards
   b. Smartcards
   c. Credit cards
   d. Money market cards
10. Most accounts in banks and savings and loans are insured by the Federal Deposit Insurance Corporation (FDIC) up to
    a. $10,000.
    b. $25,000.
    c. $100,000.
    d. $1 million.

DISCUSSION QUESTIONS AND PROBLEMS

1. How has electronic banking revolutionized the way people access money? Give examples. In what ways do you use e-banking?
2. What do you think about banks charging a customer for talking with a teller? Why are banks charging for what used to be a free service?
3. Generation X-ers appear to be better than average savers. Why would 18-to-30 year-olds be saving more than some other age groups in the population?
4. What are the advantages and disadvantages of banks, S&Ls, and credit unions? Which one(s) do you use and why?
5. U.S. savings bonds are owned by one out of four Americans. Why do so many people have them? Where can you get savings bonds?

DECISION-MAKING CASES

1. Josh is a freshman enrolled at a university. He is getting a lot of offers in the mail for checking accounts, debit cards, and smartcards. What advice would you give him? How should he compare offers?

2. Brett has received a $2000 tax refund. His expenses are taken care of for the next three months until he graduates. Should he put the tax refund into a savings account or a 3-month CD or savings bonds? He wants the choice that will give him the most interest and allow him to get the money out in three months.
3. Nichole and Jeremy are getting married. They have lived and worked separately for years and put money into separate accounts. Should they open a joint account and pool all their money? What factors should they consider? If they go with a joint account, how should the title be worded to provide protection to both of them?
4. Karen is setting up a savings account at a Savings and Loan Association.
   a. According to which federal law should she be told about interest rates, fees, and the annual percentage yield? What else can she expect on a regular basis because of this law?
   b. One of the reasons Karen is setting up the account is to build an emergency fund since she works in an industry known for laying off workers. So far she has $1,000 saved up, but her monthly expenses run $1,800. How much more does she need to save to have at least a three-month emergency fund? If she saves $400 each month, how many months will it take her to build up her emergency fund to this level? Assume no interest is earned on the account in calculating your answers.
5. Mark and Rona Lee are having their first baby. As a present, Mark's aunt wants to give the baby a $50 Series EE Savings Bond. How much will it cost her? How long will the child have to keep the bond before it can be cashed in for its full value?

GET ON LINE

Try these activities:

- Go to www.americanexpress.com/banking, a site run by American Express that offers interest rates on deposits. Compare its rates to the national average on the opening page and compare other services. Report on what you find. Other choices here are www.citibank.com, the site of Citibank, and www.wingspan.com, the site of Wingspan, part of First USA Bank, a division of Bank One in Chicago. All offer online banking.
- For more general sites that note which banks and other financial institutions are offering the highest rates on savings accounts and certificates of deposit (CDs), go to www.bankrate.com or www.banx.com or www.bankrate.com/brm/default.asp and find the following: mortgage rates, rates on one-year and five-year CDs, and MMDA accounts.
- Use a general search engine such as www.yahoo.com or www.excite.com and find two banks that offer online banking, compare rates and services.
- Go to www.pathfinder.com/money and find the calculators. Pick a savings goal (for example, $3,750) and use the calculators to find out how much you need to save each month to meet this goal.

INFOTRAC

To access InfoTrac, follow the instructions at the end of chapter 1.

Exercise 1: Type in the word "banking," select the subject guide and hit search. Then limit your search to a subtopic like ATMs or e-banking, select an article and report on what you find.

Exercise 2: Type in the word "savings", select subject guide and hit search. This will bring up articles on savings. Select two and compare the advice they give on how to save more or where to put savings.

chapter five
managing credit and loans

Avoiding Debt
Getting Ahead

Did You Know?

* Over a trillion dollars a year are spent by using credit cards.
* The typical cardholder carries 8 to 10 credit cards with an average monthly balance of $3,900.

Objectives

After reading Chapter 5, you should be able to do the following:

1. Discuss the pros and cons of credit.
2. Describe how to acquire and manage credit.
3. Identify the types, sources, and costs of credit.
4. Identify the types, sources, and costs of loans.
5. Distinguish between the two main types of bankruptcy.

Chapter Overview

This chapter focuses on credit, loans, and bankruptcy. It starts with a definition of credit and moves on to the pros and cons of credit, including online credit. Sources of credit and loans are described. There is a special section on college students and credit.

When people go deeply into debt one alternative is credit and debt counseling, another is filing for bankruptcy. Types of bankruptcy are described near the end of the chapter. As in all chapters, there are sections on Frequently Asked Questions (FAQs) and Electronic Resources.

CREDIT AND CREDIT CARDS

Credit
Receiving goods, money, and services based on an agreement between the lender and the borrower.

Credit cards
Cards used to purchase something or to get cash now with the promise of future payment.

Credit refers to a situation in which something is gained today (goods, money, services) based on an agreement between lender and the borrower. It is a broad term that encompasses loans, credit cards and other types of credit.

Credit cards are conventionally thought of as plastic cards that are used to purchase something or to get cash now with the promise of future payment. The first credit cards were offered in 1973. Exhibit 5.1 shows how credit cards fit in the evolution of money. The newest development is online credit. The big names in the credit card industry—such as American Express, VISA, and Mastercard—are being challenged in terms of services and products by new players—such as NextCard and First USA—for positions in online credit. Forrester Research projects online credit to be a $167 billion business by 2003.[1]

Computers made the first plastic cards possible because of their ability to transmit information instantly electronically. Now, people can go online use to compare credit card fees and services. In the future the online credit market will not be limited to plastic and its varied forms. A number of companies are developing ways to market micropayments—payments so small that it would be inefficient to process credit cards. So, the trend is toward online credit and paperless/cardless exchanges. More on this subject will appear throughout the chapter and especially in the Electronic Resources section.

The number of credit card holders, the number of credit cards held, and the amount of credit card spending have vastly increased over the last few decades. According to the *Statistical Abstract of the United States*, since the 1970s credit card spending has grown from about $100 billion dollars annually to over $1 trillion annually. Issuers of cards extend credit because they have confidence in the holder's ability to pay.

WHAT ARE THE PROS AND CONS OF CREDIT?

Objective 1:
Discuss the Pros and Cons of Credit

There is no question that credit cards are popular. The typical cardholder carries 8 to 10 credit cards with an average monthly balance of $3,900.[2] But what are the good and bad reasons for using credit?

The Pros

Convenience. Probably the main benefit of credit is convenience. With credit cards you do not have to carry large sums of cash or wait to buy things.

Taking advantage of sales or interest-free promotions. If you do not have the cash or the total amount of money readily available, you can still take advantage of sales or promotions.

Accumulating rebates, frequent flyer points, gifts, and other benefits. For signing up for a credit card, you may receive a sports bag or a T-shirt. When you use a card

CONSUMER ALERT

"Buy Now: Pay Over the Next Three Months: No Interest Charged" is a good deal if you pay the bill within the time allowed. If not, some stores will charge you interest from the first day of purchase.

EXHIBIT 5.1
Evolution of Money in Its Various Forms of Exchange

Adapted from and used with permission from MasterCard.

Prehistory • *Agricultural products*
Money was what you grew or hunted. Everything from grains to game was traded, but is was hard to make change.

700 B.C • *Gold coins*
Coins represented a great leap forward in portability and convenience.

1661 • *Modern paper money*
During a coin shortage, Sweden became the first European country to issue printed paper money.

1700 • *Checks and bank drafts*
Checks let people move money readily, access it without carrying it, and make exact payments. Currency was no longer a prerequiste for payment.

1973 • *Credit cards*
Credit cards made purchasing even more convenient, as banks began communicating transaction information through computers.

1988 • *ATMs*
ATMs proliferated worldwide, providing consumers with convenient, secure access to money and account information.

1991 • *Debit cards*
Debit cards extended card functionality to the deposit account, making it much more convenient for consumers to shop and pay bills.

1996 • *Chip cards*
Chip cards, sometimes called "smart cards," offer the latest technology to support a variety of new uses beyond traditional payment methods.

frequently enough, many companies offer airline miles, vacations, televisions, and other benefits. For example, a movie video and music store chain may offer cards that can accumulate points for free CDs or tapes. The rebate feature is growing in popularity.

Identification. Credit cards can be used for identification for check cashing, for renting an apartment, and making reservations.

For emergencies. Automobile repairs or medical expenses may crop up unexpectedly.

For big ticket items. Many people cannot afford to pay cash for a new car or a computer. Credit allows them to have the use of the product now and gradually pay for it.

Simplifying record keeping. The detailed monthly statements sent by credit companies provide a useful record of spending.

For insurance benefits. Major credit cards may offer some of the following benefits:

* Best value guarantee. This feature means that if a card is used to buy an item—for example a television—and the same item goes on sale within 60 days, you may be entitled to a refund on the difference.
* Buyer's assurance plan. The original manufacturer's warranty may be extended on some items for an additional period of time.
* Car rental loss and damage insurance. Your card may offer additional personal car insurance policies that cover loss and damage to rental cars.

CONSUMER ALERT

Consider the value of the rebate in terms of the card's total cost. How much do you have to spend in order to qualify for meaningful prizes? Is the card still a good deal? Does it have a low interest rate? Another cautionary note is that in some cases awards can be changed or rescinded without notice.

* Purchase protection plan. Some items that you buy by using a card are protected against accidental damage and theft for a period of time after purchase.
* $100,000 travel accident insurance. This covers death and dismemberment for you and family members while you are traveling on a common carrier (airline, bus, ship, train) if you used your credit card to purchase the ticket.

The Cons

Overextending. People overuse credit and go deeply into debt. *If over twenty percent of monthly take-home pay goes to making debt repayments, the consumer is overextended.* If twenty percent is too much, what is an appropriate level? Ten to 15 percent is recommended as a reasonable amount. An explanation of this is given in a later section titled "How Much Credit Is Affordable?"

Interest and finance charges are high. The higher the interest rate and finance charges and the longer the loan is held, the higher the end cost. There are also **teaser rates** on credit cards which means the company signs the customer up at, for example, 5.9% interest one year and then hikes the rate much higher after the year is up.

Teaser rates
Lower rates on signing that jump to a higher rates later.

Add-on fees. If a borrower misses a payment or falls behind there may be a penalty fee for late payment. Or if borrowers go over their limit on a credit card there may be a fee. Other types of fees will be covered later in this chapter.

Lost cards. Under the Truth in Lending Act, cardholders are liable for lost or stolen credit cards up to $50 per card. Report missing cards immediately.

Loss of privacy. Fraud and ripoffs can easily occur. Guard personal and financial information carefully. Examine your credit statement and report any suspicious charges. Do not give your credit card number out over the telephone or Internet to *anyone* whom you did not contact first. Save all purchase and ATM receipts to compare to statements.

Loss of freedom. If you get too weighted down with credit payments (still paying for December spending in April), then you do not have the money to spend on new purchases or to take care of emergencies. Debt is a burden.

Establishing Credit

In order for creditors to lend you money you need to be perceived as a good credit risk. A number of factors will enter into their decision, such as present earnings, earning potential, past track record on paying bills, and net worth. Factors such as these can be categorized under the five Cs of credit worthiness—capacity, capital, character, collateral, and conditions—as shown in Exhibit 5.2. It is not unusual for someone to score high in one of the Cs and low in another. Banks and other lending agents consider the whole person when weighing these factors.

To establish a strong credit rating you should

* open a checking and savings account.

EXHIBIT 5.2
Credit worthiness

Lenders determine borrower's credit worthiness by checking the 5 C's of credit.

Capacity refers to the current and potential income available to make repayment.
Capital is a measure of financial net worth (assets – liabilities).
Character is an assessment based on past credit history and reliability.
Collateral is property used to secure repayment of a loan.
Conditions refer to the general state of the economy at the time of the credit application.

* Use credit by opening one or two charge accounts. For example, get a retail card or a VISA card and make a few credit purchases. Make your payments when due.
* pay off existing loans such as student loans.
* get a steady job.
* buy or rent and stay for a year or more, thereby establishing residency. Stability is an asset when applying for credit.

Objective 2:
Describe how to acquire and manage credit.

Opening an Account

Credit accounts are opened by completing an application. Sometimes applications are sent in the mail or obtained directly from a store, bank, or company or over the Internet (see the Electronic Resources section). The usual questions asked center around your address, where you go to school or work, sources of income, and the name of your bank and your account number.

Once the form is filled out, it is subject to a credit investigation. The investigation involves contacting credit references. This can include a call to banks or a more thorough investigation through a credit bureau. The purpose of the credit investigation is to verify information on the credit application and to determine credit worthiness.

Credit Bureaus

Credit bureau
Reporting agency that collects, stores and sells credit information to potential lenders.

A **credit bureau** is a type of reporting agency that collects, stores, and sells information about individual borrowers and, for a fee, releases the information to financial institutions that request it. A bank, for example, may pay a small fee to obtain a short, one month summary report on a person's finances when a person applies for a $1,000 loan. The bank may pay $50 or more for a full-blown, total financial report if the person is buying a house, for example. The higher fee is usually passed on to the client when they close on the house. The reports come over the computer to the banker in a matter of seconds. A sample credit profile from Equifax is given in Exhibit 5.3.

Equifax is one of the three largest national bureaus. The other two are Trans Union Corporation and Experian. Each may be contacted by mail, telephone, or via the Internet.

Equifax
P. O. Box 740241
Atlanta, GA 30374
(800) 685-1111
www.equifax.com

Trans Union Corporation
Consumer Relations Center
P. O. Box 390
Springfield, PA
19064-0390
(800) 916-8800
www.transunion.com

Experian
P.O. Box 2104
Allen, TX 75013
1-888-397-3742
www.experian.com

EXHIBIT 5.3
Sample Credit Profile – JavaScript version. Used with permission.

SAMPLE CREDIT PROFILE

This is a sample profile. Click the links to see more information about each section.

Personal Identification Information — Social Security Number 123-45-6789
Your Name — Date of Birth Day Month Year
Your Current Address
City, State, Zip
Previous Address(es)
Your Previous Address
Last Reported Employment Your Position and Employer

Public Record Information

Bankruptcy filed on 02/97 in City of Atlanta with case or other ID number 123456789012341 With Liabilities of $5,000, Assets of $60,234, Exempt Amount of $60,000 Type of Personal, Filed Individual, and status Voluntary Ch-7

Lien filed on 03/9:Fulton CTY:Case or Other ID Number-32114:Amount-$26667:Class-State-Released 07/93:Verified 07/93

Satisfied Judgment filed on 07/94:Fulton CTY:Case or Other ID Number-898872:Defendant-Consumer:Amount-$8984-Plaintiff-ABC Real Estate:Satisfied 03/95:Verified 05/95

Collection Agency Account Information

Pro Coll (800) XXX-XXXX

Collection Reported 05/96:Assigned 09/93 to Pro Coll (800) XXX-XXXX Client ABC Hospital:Amount-$978:Unpaid:Balance $978:Date of Last Activity 09/93:Individual Account:Account Number 787652JC

Credit Account Information

Company Name	Account Number	Whose Account	Date Opened	Last Activity	Type of Account and Status	High Credit	Terms	Balance	Past Due	Date Reported
Macy's	12341234	Joint	02/94	01/97	Revolving -90 Days Past Due	$4235	50	$243	$0	02/97
Citibank Visa	12341234	Joint	02/97	05/97	Revolving - Pays as Agreed	$10000	50	$243	$0	06/97
Rich's - FACS	12341234		02/94		Lost or Stolen Card					02/97

Additional Information

Foreclosure reported 02/97 by Firm Name verified on 02/94

Checking Account reported 02/97 opened on 04/83 closed for reason: non-sufficient funds with amount $500

Companies that Requested your Credit File

02/15/97	AR Sears	02/13/97	ACIS 712000003
02/13/97	Richs/Facs	02/11/97	Equifax – Update
01/16/97	Macy's	01/13/97	Equifax – Disclosure
01/18/96	AM Macy's	10/16/96	PRM – Citibank Visa

THE FOLLOWING INQUIRIES ARE NOT REPORTED TO BUSINESSES:

PRM - This is a promotional inquiry in which only your name and address were given to a credit grantor so you could be solicited with an offer such as a credit card. (PRM inquiries remain on file for 12 months.)
AM or **AR** - These inquiries indicate a periodic review of your credit history by one of your creditors (AM and AR inquiries remain on file of 12 months.)
EQUIFAX, ACIS or **UPDATE** - These inquiries indicate Equifax's activity in response to your contact with us for either a copy of your credit file or a request for research.
PRM, AM, AR, INQ. EQUIFAX, ACTS and **UPDATE** inquiries do not show on credit files that businesses receive, only on copies provided to you.

For More Information

Each of the credit-reporting agencies listed allows consumers to obtain a copy of their report by mail with a check or money order. Equifax and Experian permit phone orders, using a major credit card. On their websites they have order forms and information. If you have been denied credit, employment, insurance, or rental housing in the past 60 days based on information in credit files, then you are legally entitled to a free copy of your report; otherwise in most states there is a charge of $8. Since these fees and rules change, before sending any money visit the credit agency's website or call them.

It is possible that there are errors on a report. A common error is mistaken identity. This occurs if you have a common name such as John Smith. Other errors include the nonrecording of late payments or ex-spouse's data. It is also possible that if one credit-reporting bureau has a mistake, the others will also. It would be wise to contact all three bureaus to clean up a report since lenders may request a report from any one of the three.

If you are denied credit or find an error, contact the credit bureau immediately. Request a copy of the corrected file. Many of the rules governing credit reporting procedures were formalized by a 1996 amendment of the Fair Credit Reporting Act enacted by the Federal Trade Commission (see the Electronic Resources section for their website).

How Do You Manage Credit?

To summarize, the best ways to manage credit are to

1. use it only when necessary.
2. know what the agreement says.
3. make payments on time.
4. *pay off credit card charges in full each month. Although only 36% of cardholders do so*, it is a goal to shoot for so you can avoid interest charges and fees.
5. inform creditors if you cannot make all or part of your payment. When you call them first, creditors are more understanding.

College Students and Credit

Managing credit is a part of life for most college students since nearly all have cards. Consider the following:

* an estimated 80 percent of colleges and universities permit some form of on-campus credit card solicitation.
* 77 percent of full-time undergraduates have credit cards.
* Only 44 percent of college students clearly understand the term budget.
* Only 34 percent of college students clearly understand the concept of buying on credit.
* The average credit card limit for a student is $6,122.
* The average outstanding balance for a student is $2,226.
* 54 percent of college students expect their parents to help support them after graduation.[3]

Affinity cards
Credit cards offered by institutions and organizations that lend their names for a percentage return.

In addition some schools earn sizable fees (typically 0.5 percent to 1 percent of the amount charged) by lending their names to **affinity cards** that are marketed to students and alumni. Affinity cards are credit cards offered by institutions and organizations such as professional sports teams as well as universities.

College students have been targeted by credit card companies because students spend a lot of money and companies want to establish brand loyalty before graduation. Although many students are responsible, it is not unusual for students to overextend because of rising tuition, room, board, and fees, spending on comfort and luxury, image/appearance consciousness, ignorance, and naivete.

Regarding this last point, a survey by the Consumer Federation of America revealed that 78 percent of college juniors and seniors did not know that the best way to figure out the cost of a loan was to look at the interest rate.[4] Lenders take advantage of this ignorance by burying the interest rate in small print, but writing ***"CONGRATULATIONS, YOU HAVE BEEN PRE-APPROVED FOR $10,000 OF CREDIT"*** in big, bold print. Be wary of anything that sounds too good to be true.

To illustrate how easy it is for a college student to get in debt, read the following story that appeared in the Free Press of the University of Southern Maine:

> When University of Southern Maine sophomore Carrie Heselton received her first credit card, she felt like she won the lottery.
>
> "It just came in the mail one day, and I was like 'Yeeee-hahhhhh!' she says. Carrie, then 19, bolted 10 steps to the nearest pay phone, activated the card and rushed to L.L. Bean.
>
> With no money and no job, Carrie had filled out the credit card application in the campus center two months prior because her car tires were so old she needed the free tire gauge AT&T Universal offered.
>
> She never thought she'd actually get the card. "It felt like I found $800 in the gutter to do with as I pleased," she says.
>
> "I went nuts. It was like, 'Oh, I really love that dress, I have to have it. Oh hey! I have a credit card!' Or, 'does anybody want to go to lunch? You don't have any money? Oh hey! I have a credit card!..."
>
> She figured she'd pay the bill with the money she made waitressing over the summer. But she says she wound up needing it for school because she didn't get the financial aid package she expected.
>
> AT&T eventually found Carrie's parents and began calling them. When her mom and dad found out Carrie was issued a credit card they were infuriated with AT&T—not Carrie—for issuing her the card in the first place.
>
> When creditors finally tracked Carrie down, they began hounding her over the phone. She's been threatened with bad credit, lawsuits and even jail time...[5]

In the aftermath of the case Carrie was sued for $1600—twice what she spent. An attorney at the university's Student Legal Services worked with Carrie to clear up her credit problems. He said he saw credit card problems all the time.

Unfortunately debt doesn't stop at graduation. It follows graduates for years directly through repayment and indirectly through the shaping of career choices. Consider the following:

> Desiree Saylor, a senior at the University of Texas, says she had hoped to work in academia as a physical anthropologist. "If I didn't have this debt, I'd go for it," she says. But she has $16,000 in debt and has decided on the higher-paying career of occupational therapy. Even the children of the wealthy find career options affected by their debt. The two sons of Joseph Biden, who makes six figures as a U.S. senator, have amassed $196,000 in student debt between them. The younger worked at a homeless shelter between college and graduate school, Sen. Biden says, but 'with his debt, there's no possibility he could take a job there.' With a third of all medical school graduates owing more than $75,000, few can work in low-paying health clinics.[6]

To summarize this section, three trends are students are being pursued by credit card companies, debt can build-up in college, and debt is affecting students' career choices. The worst case scenario is for a student to incur debt while in college, quit, and never get a degree.

Women and Credit

Although legislation such as the Equal Credit Opportunity Act has attempted to resolve inequities, women who have not established a credit history in their own name may find themselves having difficulty obtaining credit and loans.

Women who wish to establish credit should:

* use their own legal name when filling out a credit application.
* open a checking or savings account in their own name.
* notify creditors if there is a marriage or divorce or name change.
* take out a small loan and pay it off quickly.
* check to see that the information reported to the credit bureau is in the wife's name as well as the husband's.

How Much Credit Is Affordable?

Debt safety ratio Proportion of monthly consumer credit obligations to monthly take-home pay.

Since as stated earlier more than 20 percent of monthly take-home pay allocated to credit is too much, how much credit can individuals comfortably handle? The answer to this is similar to how much risk a person can handle, it depends on the individual. Some people don't like to owe anything and others are comfortable owing a lot. It also depends on future earning capacity. But in general, experts suggest that *most people can handle 10 to 15 percent,* which is expressed as a **debt safety ratio**, the proportion of total monthly consumer credit obligations to monthly take-home pay.

To illustrate how the debt safety ratio works, if you bring home $2,000 a month then by using the 10 to 15 percent estimate, you can comfortably owe $200 to $300 a month. The calculations are shown as

$$\$2{,}000 \text{ X } .10 = \$200 \text{ or } \$2{,}000 \text{ X } .15 = \$300$$

The lower the percentage the easier it is to manage so that 10 percent can be thought of as low, 15 percent as medium, and 20 percent as high. Exhibit 5.4 provides a chart to help you figure monthly consumer credit payments based on monthly take-home pay for each of the categories—low, medium, and high.

EXHIBIT 5.4
Consumer Credit Guidelines Based on Ability to Repay

	Monthly Consumer Credit Payments		
Monthly Take-Home Pay	Low Debt Safety Ratio (10%)	Medium Debt Safety Ratio (15%)	High Debt Safety Ratio (20%)
$ 750	$ 75	$ 112	$ 150
1,000	100	150	200
1,250	125	188	250
1,500	150	225	300
2,000	200	300	400
2,500	250	375	500
3,000	300	450	600
3,500	350	525	700
4,000	400	600	800
4,500	450	675	900

Objective 3:
Identify the types, sources, and costs of credit.

TYPES AND SOURCES OF CREDIT

There are three types of consumer credit: open-ended, installment, and noninstallment credit.

Open Ended Credit

Open-ended credit
Extended in advance of transactions, borrowers are pre-approved up to a limit.

The most commonly used type of consumer credit is called **open-ended credit**. This is a form of credit extended in advance of any transaction such as in credit cards or charge accounts. In open-ended credit, borrowers do not reapply for each transaction, they are preapproved for a credit limit up to a maximum amount such as $5,000. The main sources of open-ended credit are banks or other financial institutions—such as credit unions, S&Ls, and brokerage firms whose cards are referred to as bank credit cards—and retail establishments.

Bank credit cards
Versatile cards that can be used almost anywhere for cash or purchases.

Bank credit cards (VISA, Discover, MasterCard) are versatile because they can be used in a variety of places whereas retail cards are usually linked to a particular store or a network of stores. Bank credit cards can be used for a **cash advance** which is a loan that can be obtained by the cardholder at any participating financial institution. These loans begin to accrue interest immediately and require no formal application.

Card issuers may also offer many other benefits, such as insurance travel protection, emergency services, discounts on telephone calls, rebates, and gifts (as mentioned earlier). Shop around as well as consider your consumption patterns in order to make the best choice.

Also, as mentioned previously, a **credit limit** is set which is the maximum amount that can be charged to a single card or account. If card holders go a few dollars over their limit, they probably won't hear anything, but if they go 20% or more over the limit they will probably hear from the lender.

Credit statement
Monthly summary of credit transactions.

Each month a person with open account credit receives a **credit statement** that summarizes credit transactions. If the amount is paid in full then interest and fees are avoided. This is a smart use of credit because you get the benefits without incurring costs.

Installment versus Noninstallment Credit

Installment credit
An agreement that requires a fixed number of regular payments of principal and interest.

Noninstallment credit
Rare type of agreement in which a single payment is due at the end of the loan period.

Other types of credit include installment and noninstallment credit. **Installment credit** on an installment loan requires the repayment of amounts owed in a specific number of equal payments such as in a car loan when $500 per month is owed for 48 months. Payments are regular and fixed. Besides cars, furniture, appliances, audio equipment, computers, and jewelry are often bought on installment. **Noninstallment credit** refers to a single-payment system such as owing $2,000 at the end of a year. In consumer credit, noninstallment is rare compared to installment or open ended credit.

CONSUMER ALERT

Department stores and clothing shops notoriously have high interest rates on their credit cards. Read the fine print, they may be charging 21 percent or 22 percent interest versus a bank credit card that may only charge 5.9 percent or 6 percent interest.

CONSUMER ALERT

Cash advances may be a very expensive way to get cash. Most cards charge a higher interest rate on cash advances than on purchases. Read the fine print on the agreement.

COSTS OF CREDIT

A number of factors contribute to the total cost of credit. This section covers how credit cost expenses are figured including interest rates and fees.

What Is a Good Interest Rate?

A good interest rate is the lowest one you can find that also offers the services and benefits that you want. The annual rate of interest charged on credit cards ranges from about 4 to 21 percent. Bank credit cards tend to have lower interest rates than retail cards. Some states have **usury laws** that limit interest rates to a certain maximum.

Usury laws
State laws that limit interest rate.

Prime rate
Rate that banks use to determine loans.

Most bank cards have variable interest rates tied to the **prime (or base) rate**. This is the rate that a bank uses to determine loans to individuals and small or medium-sized businesses. To keep up with changes in the prime rate, card issuers adjust the interest rate quarterly and have minimum and maximum rates. So a card may be based on a prime rate of +10 percent, with a minimum of 13 percent, and a maximum of 20 percent. These percentages have to be disclosed to the person who is signing up for a card. For an existing card, the issuer by law has to disclose rate changes. If terms or fees are increased, you may want to switch to a lower cost card.

As mentioned earlier, you can avoid interest charges by paying charge accounts off in full each month. There is usually a 25-to-30 day **grace period** in which you can do this. Remember that interest rates are applied to any unpaid balance from previous months as well as current purchases, but interest on cash advances begins immediately (there is no grace period).

Grace period
Amount of time to pay for purchases without incurring finance charges.

Additional Fees

Lenders like to add on fees because this is how they can make money beyond interest charges. Besides a fee that may be charged if you go significantly over your limit or are late with your payments, there may be other fees. For example, some card issuers require an annual fee of around $15–$100 a year for the privilege, prestige, or benefits associated with using their card.

Another type of fee may be charged for cash advances or small loans. A transaction fee of $2 per cash advance or 2 percent of the amount obtained in the transaction, whichever is more, may be charged.

A final type is an overrun fee imposed if you exceed your cash limit. Consider these fees when choosing cards.

Annual Percentage Rate (APR)

The actual or true rate of interest paid over the life of credit (or a loan) is called the **annual percentage rate (APR)**. According to the Truth in Lending Act, lenders must disclose the rate of interest that they charge and their method of computing finance charges. It is your right to be told the dollar amount of charges and the APR on any credit or loan for which you are applying.

Annual percentage rate
Interest rate paid per dollar per year for credit.

CONSUMER ALERT

Regarding the grace period, although 25 to 30 days is typical, some cards limit the grace period to 15 days. This is a very short turnaround time.

One student, Bridget, was so frustrated by the complexity of her retail credit card statement and keeping up with payments on her account between semesters that she went to the credit department of the store and asked the clerk what she needed to do to pay the card off completely. The clerk pointed to the appropriate amount on the statement, Bridget paid off the card with a check and went straight home and cut up her charge card. Now she uses only one bank card and receives only one statement per month.

The APR is one of the most important concepts covered in this chapter because (except for the possible add-on fees) the lower the APR the lower the cost of credit. The simplest way to explain how APR works is by example. If a $1,000 loan has an APR of 12 percent that means that the finance charge will be $120; the total cost to the borrower will be $1,120 if paid at one time in a lump sum at the end of a year. If the APR was 15 percent on the same loan, the finance charge would be $150, and the total cost to the borrower would be $1,150 if paid in one lump sum at the end of the year. So an APR of 12 percent is better than an APR of 15 percent.

In the case of credit charges that are paid in monthly installments, the actual interest cost or finance charge would be more complicated than this example because monthly purchases and balances would most likely vary. But the basic principle to keep in mind is that the lower the APR the lower the cost of credit.

Average Daily Balance (ADB) Method

Lenders have to tell consumers not only the APR but also the method they use to compute the finance charges. Most banks and retailers charge one of four variations of the **average daily balance (ADB) method**. In this method finance charges are computed by applying interest charges to the average daily balance of the account over the billing period, including or excluding purchases or returns made during that period. The calculations of the four variations of the average daily balance method are given in Exhibit 5.5.

Average daily balance method
A way finance charges are determined, there are four variations.

1. Average daily balance excluding new purchases
   In this method, the cardholder pays interest only on any balance left over from the last month.
2. Average daily balance including new purchases with a grace period
   In this method, the balance calculation includes the balance from the previous month and any charges made during the billing cycle. A grace period allows for the exclusion of charges made during the cycle if the balance from the previous cycle was zero.
3. Average daily balance including new purchases with no grace period
   In this method, the balance from the previous month and any charges made during the billing cycle are included in the balance calculation even if the previous month's balance was fully paid off.
4. Two-cycle average daily balance including new purchase
   This method eliminates the grace period on new purchases made during the current billing cycle and retroactively eliminates the grace period for the previous month each time the account carries a balance. This is the worst method for consumers because a credit card holder has to pay off his or her balance for at least two months to avoid finance charges.

EXHIBIT 5.5
Four methods used to calculate average daily balance on a credit card billing statement

CONSUMER ALERT

While some card issuers expect a minimum payment of five percent of your monthly bill, some go as low as two percent. Are they being nice? No, this is profitable for them because they are stretching out your repayment and getting more money. If you have a number of credit card statements each month, a good strategy is to pay off your highest-interest cards first and to try to pay more than the minimum balance.

Cardholders paying off their accounts each month avoid these charges. If a cardholder cannot pay the entire balance, then he or she can pay an amount equal to or greater than the minimum monthly payment specified on the statement.

How to Find the Best Deal

To summarize the points covered thus far, look for the following credit card features:

* Rate of interest (APR), the most important feature.
* Rebates and gifts.
* Fees.
* Length of grace period.
* Method of calculating balances.

If you pay off your account balance each month in full, then rate of interest is irrelevant, but you will want a long grace period and no annual fees. If, however, you are like most people and carry a balance then get a card with a low rate of interest on unpaid balances.

CONSUMER CREDIT LEGISLATION

As mentioned in the beginning of this chapter, over $1 trillion is charged on credit cards. Since so much money is tied up in credit, the potential for fraud is enormous. Over the years a number of consumer credit protection laws have been passed to reduce fraud and stealing from consumers and merchants. Three Acts already mentioned in the chapter are the Truth in Lending Act, the Fair Credit Reporting Act, and the Equal Credit Opportunity Act. Of particular note is the Truth in Lending Act's **right of rescission** which provides the right to cancel a contract or an agreement that has been signed. Section 125 of the Act provides the "cooling off period" which gives consumers three business days to rescind a credit transaction.

Right of rescission
The right to cancel a contract or agreement after it is signed.

MANAGING LOANS

As mentioned in the introduction, the subject of credit includes loans, but loans are unique enough in terminology, types, sources, and costs to deserve their own section. **Loans** are sums of money lent at interest for big-ticket items—such as sports utility vehicles, boats, and college educations—or for personal loans. In **personal loans** (also known as **cash loans**), a person borrows cash then uses it to make purchases or investments or pay off debts. A typical use of a personal loan is to get money to pay for a vacation.

Loans
Sums of money lent at interest.

Personal Loans
The borrowing of cash that is repaid.

CONSUMER ALERT

Never cosign a loan without careful thought because if the borrower defaults, you will have to repay the loan. Not only could this be very costly, but it could adversely affect your credit standing.

Collateral
Property acceptible to secure a loan.

Cosigner
Person who agrees to repay a loan if the borrower defaults.

Defaults
Unpaid loans.

Liens
Legal rights to take and hold property or sell it for repayment of a loan.

To help ensure repayment of a loan, lenders often want secured loans which require collateral or a cosigner. **Collateral** is property (an RV, a car) acceptable to secure a loan. A **cosigner** is another person who agrees to repay the loan if the borrower does not do so. Parents often cosign loans for their children. A person who does not repay a loan **defaults**.

Liens are legal rights to take and hold property (i.e., homes, farms, cars) or sell the property for repayment of a loan. When a loan contract is paid up, liens are removed.

Objective 4:
Identify the types, sources, and costs of loans

TYPES OF LOANS

Loans can be divided into two main types based on what individuals take out loans for, and what kind of repayment schedule is used. In the first category, individuals take out a variety of loans such as

* auto loans.
* durable goods other than autos (computers, appliances, furniture, mobile homes, airplanes).
* personal (or cash) loans.
* consolidation loans which put several loans into one.
* home equity loans.
* home equity line-of-credit loans.
* home improvement loans.
* education loans.

On the list the most diverse category is the "durable goods other than autos," not only in terms of the breadth of loans taken out but also in the cost and length of repayment. For example, a mobile home repayment schedule may be seven to ten years, whereas a television or appliance may have a 6–12 month payback schedule.

Home equity loans
Borrowing against the equity in homes.

Home equity loans are second mortgages on homes that require a fixed amount of money to be paid back in a limited time frame such as five to ten years. Home equity line of credit loans operate as open ended credit wherein borrowers can write checks up to a credit limit. Often the interest rate is fairly low for both types of loans and another advantage is that the interest paid is usually deductible from income tax. The main negative is that the borrower could lose the home through failure to repay the loan. This rarely happens, however, because in most cases the lender would rather have the money paid back than take the house (the exception would be a house that has greatly escalated in value).

Generally people take out home equity and home equity line-of-credit loans for major purchases or home improvements. On a home equity loan, points (each point is one percent of a loan, so for a $100,000 loan one point would be $1,000) can be paid up front. By paying points, a person can buy-down (reduce) the cost of a mobile home, a house, or a lot (raw land).

In the second category, loans are distinguished by repayment schedules: installment, noninstallment, and open ended credit. Installment loans involve monthly payments. Noninstallment loans usually involve one lump sum due at the end of the loan period. Open ended credit can be paid partially or in full each month. Rates may also be fixed or variable.

Sources of Loans, Including Ease of Access

The main sources of loans are

1. Banks and S&Ls: all types of loans to people deemed to be average and better credit risks; lower rates to their own customers.
2. Brokerage firms. If you have an account, you can borrow money from the firm related to the value of the securities held in your account.
3. Consumer Finance Companies which specialize in small loans (Household Finance Corporation (HFC) and Beneficial Finance Corporation (BFC)), take on more risk therefore rates are high.
4. Credit Unions: lend to members only, often have the lowest rates.
5. Sales Finance Companies or Captive Finance Companies that are business-related lenders such as Ford Motor Credit, General Motors Acceptance Corporation (GMAC), and General Electric Credit Corporation (GECC).
6. Life Insurance Companies: lend on the cash value of the insurance policy, can be used for different kinds of loans including home equity loans.

Applying for a loan has never been easier. You do not even have to go in person any more since applying can be done over the telephone or the Internet.

Other sources of loans include the federal government, pawnshops, check cashers, consumer finance companies, and relatives and friends. Terms of loans should be clearly defined and agreed upon by the borrower and the lender. Loans can come from multiple sources. For example, education loans are available from colleges, the federal government, and banks and other financial institutions.

Education Loans

As college costs rise, more students are taking out loans and many have a difficult time paying them back. According to congressional testimony by David Longanecker of the Office of Postsecondary Education, as of November 1997, 35.4 percent of the federal borrowed loan money has been paid off. Further, he testified that since 1959 the federal government had made $285 billion in higher education loans. The yearly default rate dropped during the 1990s mostly because the Department of Education began removing schools from loan eligibility if too many students defaulted.

CONSUMER ALERT

When a consumer finances a car at a dealership, typically the car salesperson and the financial salesperson make a commission. This commission cost is passed on to the consumer. Because of the commissions and the money to be made from loans, the car dealership will push the consumer to finance the car through them. In many cases they do not offer the best interest rates so the car buyer would be wise to shop around for better rates such as at a credit union or bank before signing up for a car loan.

Currently a one-year federal student loan is capped at $2,625 to $5,500 for undergraduates depending on which year of school the student is in.[7] Given the high cost of a college education, this amount of money is often not enough to fill the gap between the money the student has and the cost of the school attended. Parents and students can borrow the rest of the money through the federal PLUS program (Parent Loans for Undergraduate Students) or private loans. Private loans can come from banks or other lenders or through college financial-aid offices.

In keeping with the main subject of this chapter, the focus of the following discussion will be on educational loans and strategies for paying them back. For a discussion of how to save for a college education see Chapter 7, and for information on Education IRAs (individual retirement accounts) see Chapter 15.

The main application form is the Free Application for Federal Student Aid (FAFSA), which can be filed during a student's senior year in high school. The web address for FAFSA, as well as other financial-aid resources, are given in the Electronic Resources section near the end of the chapter. The FAFSA calculation determines the parental contribution which is the expected amount parents and the student are capable of paying each year. This calculation is based on family income that subtracts taxes and on an "income protection allowance" that exempts a portion of income based on the family's size and number of children in college. Next, the calculation considers parents' and students' assets (savings and investments). The FAFSA formula excludes equity in the family home (which is the paid off part of the house), but many colleges add it back in when figuring nonfederal aid. The following are examples of how FAFSA works:[8]

- If a single parent has a $25,000 income and $2,500 in assets (excluding his or her home) and the child has $255 in income and $175 in assets, it is assumed the parent could contribute $1,285 and the student $430.
- If parents with $50,000 income and $27,000 in assets (excluding their home) have two children with one entering college and if the child entering college has $4,000 in income and $2,500 in assets, then the parents' expected contribution is $5,380 and the student's is $1,800.

Naturally as income and assets rise the calculations become more complicated. The bottom line is, understandably, the less money you and your parents have, the more aid you are likely to get.

One of the most usual types of federal loans is the Stafford loan. The standard application form requests only the most basic information about the student borrower. No information is required about a student's credit history or income. The Stafford loan carries a low interest rate—between 8.25 percent and 8.72 percent as of the writing of this book.

The loan process begins by filling out applications and contacting college financial aid offices. These offices offer information about their own and state and federal loan programs that require repayment and scholarship and grant programs that do not require repayment. Other college costs and financial-aid information are available from magazines, financial institutions, high school guidance offices, books, and over the Internet. Loan repayment begins after graduation.

Most of the students reading this chapter already have their loans and are beginning to wonder how they are going to repay them. Here are some strategies:

1. Have an exit counseling session with the financial aid office (or bank) before you leave school to help you understand your rights and responsibilities. If you have a federal Stafford loan, you can register online for exit counseling.

2. Make repayment part of your monthly budget. The most common repayment method is a 10-year level repayment plan in which you pay equal monthly installments until the loan is paid off. If you owe less than $10,000 this is probably the plan that you want. Other plans go as long as 30 years or are based on a graduated payment scale (as your income goes up so do your payments).

3. Review all loan statements, have a clear picture of what you owe and how the repayment process works.

4. If you are not receiving statements or are having trouble making payments, ask your lender for help or for a deferment.

5. It is rare, but some employers will pick up student loans as a benefit (perk) of employment with them.

Costs of Loans

This section goes back to the general topic of loans and the costs thereof. When you are shopping for loans, lenders will explain the repayment schedule options and the costs of loans so that you do not have to do calculations yourself. However, to show you how one of the most basic methods works the APR calculations for a single-payment loan, using the simple interest method, will be covered here. In the example the APR is 12 percent; it is shown in bold print throughout so you can follow it through the calculations.

APR Calculations for Single-Payment Loans Using the Simple Interest Method

Banks, credit unions, and S&Ls often use this simple interest method. The formula is as follows:

$$I = PRT$$

I = Interest or finance charge
P = Principal amount borrowed
R = Rate of interest
T = Time

Here is an example of using the formula. Ben took out a $1,000 loan for one year at a simple interest rate of **12 percent** (also written as **.12**). His interest charges would be

$$\$120 = \$1{,}000 \times \mathbf{.12} \times 1 \text{ year}$$

At the end of a year Ben would owe $1,120.

To calculate the APR for a single payment loan, using the same figures already given:

$$\text{APR} = \frac{\text{Average annual finance charge}}{\text{Average outstanding loan balance}} = \frac{\$120}{\$1{,}000} = \mathbf{12\%}$$

The APR is 12 percent. In the simple interest method of loan calculation the simple rate of interest and the APR are always the same.

Debt Warning Signs and Debt Collectors

Credit cards and loans are useful tools, but overuse can lead to problems. The warning signs of overextension are

* having no idea where your money is going.
* paying off only the minimum each month on credit cards.
* postponing paying bills.
* living paycheck to paycheck.
* having no savings.
* experiencing discomfort about financial trouble .
* being turned down by stores when you apply for credit.
* receiving letters, faxes, or phone calls from debt collectors.

How far can debt collectors go? They can use aggressive tactics to get people to pay up, but they cannot threaten violence; use obscene language; threaten arrest or property seizure without going through proper procedures; harass with repeated, anonymous, or collect calls; call before 8 A.M. or after 9 P.M.; misrepresent themselves; or contact debtors by postcard (because this is a violation of privacy). They can charge fees (usually a percentage of what is owed), contact you in person and within the time guidelines given, and ask other people where you work or live.

If a person receives a letter from a collection agency, he or she must acknowledge receipt by returning a letter within 30 days. The agency will respond by mail a second time before taking further action.

CREDIT PROBLEMS AND SOLUTIONS

When finances get out of control, people may turn to credit counseling to resolve their credit problems or they may file for bankruptcy. Bankruptcy is an extreme end result.

Credit and Debt Counseling

Someone having financial troubles may turn to professional help. People have financial problems that run the gamut from someone too busy to remember to pay the utility bills (so the electricity is turned off) to someone slowly sliding into debt. Credit and debt counselors have seen it all.

National Foundation for Consumer Credit
Non-profit organization that provides credit and debt counseling.

One of the best known nonprofit organizations that offers credit and debt counseling is the **The National Foundation for Consumer Credit (NFCC)** with 1,300 offices in the United States, Puerto Rico and Canada. The NFCC has a toll free number at 800-388-2227 (English-speaking) and another at 800-682-9832 (Spanish-speaking). Consumers can locate the closest NFCC member office or visit their web page at www.nfcc.org or check the telephone book under credit counseling.

NFCC provides free, or low cost help in setting up budgets, dealing with creditors, creating debt-management plans, and consolidating debts. You may wonder how NFCC can offer these services for free or how a person in debt can pay even a small fee. The answer is that the local credit counseling offices are supported by contributions from banks, credit unions, merchants, creditors, consumer finance companies, and community minded organizations such as United Way.

Bankruptcy is discouraged by credit counselors. They would suggest ways that an individual could consolidate debts and stop the reckless use of credit cards. The way debt consolidation works is that a counselor could take an individual's revolving credit accounts that charge 18 percent and consolidate them into one 12 percent credit union loan. In this way the individual can save money plus have the convenience of paying off one monthly bill instead of many.

Sometimes debt consolidation has a reverse psychological effect because the debtor sees that this is less and in only one place and goes out and gets more credit cards. Talking regularly with a credit counselor helps keep the person on track. Many offices now offer 24-hour telephone counseling services as a convenience for people who cannot come in for appointments. The average NFCC client profile reveals that the client is most likely to be a woman, age 35.4 years, married or never married with children, with lower middle income, and in debt $2,300 with 11 creditors.

Alternatives to NFCC include family and friends, other credit and debt counseling services such as Consolidated Credit Counseling Services, and the different financial advisors listed in Chapter 2 including bankers and managers of other financial institutions, financial planners, attorneys, accountants, and Employee Assistance Programs (EAPs) at workplaces.

Bankruptcy

Personal bankruptcy
Legal recourse open to insolvent debtors.

Personal bankruptcy is a form of legal recourse open to insolvent debtors in which they petition a federal court for protection from creditors and arrange for the liquidation of their assets. The typical defaulter or bankrupt has 20 or more cards, an average of $40,000 in card debt, and an annual income averaging about $35,000.[9]

The rate of personal bankruptcies has tripled since 1981. Each year over 1 million Americans file for personal bankruptcy and one in 70 families declares bankruptcy.[10] Some are credit abusers, but many are average working people who intend to pay their bills but do not. Crises can play a part, such as unexpected medical treatments, divorces, accidents, loss of employment, and natural disasters. Under stress many people rush into bankruptcy without looking into alternatives.

Filing for bankruptcy is not to be taken lightly because it *remains on credit reports for seven to ten years*, affecting an individual's ability to buy a car, get a mortgage, or rent an apartment. Future employment may also be affected if a person files for bankruptcy. About 25 percent of U.S. employers look into credit backgrounds before hiring; many of these employers are retailers.[11] Employers cannot look into your credit past without telling you first. The Federal Fair Credit Reporting Act requires that employers obtain a form from prospective employees authorizing them to look into credit pasts.

One usually tries to avoid filing for bankruptcy because of what it does to one's credit record, possessions, expenditure patterns, psyche, and future employment opportunities. However, declaring bankruptcy is not a permanent handicap. There are laws that prohibit discriminating against a person who files for bankruptcy. For example, an individual cannot be denied a driver's license because of filing for bankruptcy.

Objective 5:
Distinguish between the two main types of bankruptcy

The two main types of bankruptcy open to individuals are stipulated in Chapter 7 and 13 of the federal bankruptcy code. Many assets are turned over to an appointed trustee, who will sell these assets to pay debts, then make partial payments to creditors. Under the federal bankruptcy laws certain property is exempt. Once the trustee has sold everything allowed and paid off as much of the debt as possible, then the remainder of what is owed is discharged (erased) and an individual can start over again. In 1996, 70 percent of bankruptcy filers sought maximum relief under Chapter 7 of the federal bankruptcy code, asking the court to erase all of their dischargeable debt.

Chapter 13 This can be used if you have a regular income. It provides a method for repaying debt over a period of time according to a court-approved plan. Essentially it sets up a debt-consolidation plan with legal safeguards that allows you to stretch out bill payments. To be eligible for this choice, an individual has to have unsecured debts of less than $250,000 and secured debts of less than $750,000. Any individual who is eligible can file for Chapter 13. This includes sole proprietors of small businesses and those living on welfare, fixed pensions, investment income, and Social Security.

Most experts say that Chapter 13 bankruptcy is a better choice than Chapter 7 because under Chapter 13 you can keep some of your assets and have more control over how the debt is repaid since you develop a plan that is approved by the bankruptcy judge. *Creditors and future lenders tend to look more favorably on a Chapter 13* because the debtor is responsible to pay back what is owed within a maximum repayment period of five years.

To better understand how bankruptcy affects families, read the following story about the Williams family:

> It's not easy to be restrained, especially when children are involved. Two of Shana and Edward Williams's three daughters ages four, seven and 10, have asked for bicycles this year. But the 27-year-old San Antonio legal assistant and her husband opted for a Chapter 13 bankruptcy petition, which poses even more restrictions on their spending than Chapter 7. Under the plan, the Williamses can regain a clean credit record in seven years, but must pay back some of their old debt, at a reduced rate, over an extended period of time. They have little money left over after living expenses.[12]

The article went on to explain that part of the reason for the bankruptcy is that Mr. Williams temporarily lost his job.

Another type of bankruptcy is Chapter 11 which is generally used to reorganize businesses. An individual may use this option if he or she is involved in a business. The benefit of this choice is that the business can continue to operate while creditors are repaid through a court-approved plan.

Hiring an Attorney

Most people hire an attorney to handle their bankruptcy. Forms are available in bookstores or from a court clerk (to find the local bankruptcy court look in the telephone book under federal government listings), but generally it is advisable to hire an attorney in bankruptcy cases. An attorney can help an individual decide which type of bankruptcy is best, suggest other options, and negotiate with creditors.

Costs include the following:

1. *Attorney fees* may range from $400 to over $1,000 depending on the complexity of the case. This will be the biggest cost.

2. A *filing fee* of about $175 is due when a bankruptcy file is petitioned. Under certain circumstances this can be paid in installments.

3. *Trustee fees* are established by the bankruptcy judge or by a U.S. trustee in certain districts.

In some cases out-of-court agreements can be reached thus avoiding having bankruptcy on your record.

Steps

If an individual decides to declare bankruptcy, the first step is to file a petition and schedules at the clerk's office of the federal bankruptcy court. The petition includes a list of creditors, income sources, property, and living expenses. The following documents are needed:

- Deeds, mortgages.
- Any papers related to past bankruptcies (a person can file for Chapter 7 again after six years have passed).
- Copies of tax returns for the past two years.
- All legal papers, summonses.
- Credit card bills, medical and dental bills, and all other outstanding debts.
- Savings records.
- Loan papers.

One of the benefits of filing for bankruptcy is that it will stop the calls from bill collectors and halt creditor lawsuits, at least temporarily. So, if the bank is threatening to foreclose on a home or repossess a car, a person is given a breathing space once a petition has been filed. The process start to finish takes about four to six months.

ELECTRONIC RESOURCES

There is a tremendous amount of information on the Internet regarding the topics covered in this chapter. The following websites are some of the options available.

For more information about student loans, contact any of the following:

- The Federal Student Aid Information Center at **www.ed.gov/prog-info/SFA/StudentGuide/** or try **www.fafsa.ed.gov**.
- For help if you have defaulted on a student loan go to **www.ed.gov/offices/OPE/DCS**.
- To calculate repayment estimates, figure out interest expenses and pull together a monthly budget or to find answers to FAQs go to **www.salliemae.com** or **www.usagroup.com/students/faqs.htm**.
- For Stafford loan information go to **www.nelliemae.com** or **www.ed.gov**, the website of the U.S. Dept. of Education that has details on Stafford loans for college students.
- **www.ed.gov/proginfo/SFA**. This is a financial aid primer from the U.S. Department of Education that includes links to FAFSA.

Business, Organization, and Media sources regarding student loans and financial aid include **www.finaid.org**, the Financial Aid Information Page sponsored by the

CONSUMER ALERT

Unfortunately there are a lot of scholarship and financial aid search scams. Information is available for free online and from colleges, the federal government, libraries, and guidance counselor offices.

National Association of Financial Aid Administrators. It contains scholarship and grant information as well as warnings about fraudulent search firms.

- **www.collegeboard.org**. The College Board website contains a database of scholarship sources called Expand Scholarship Search. A similar database through a software program called Fund Finder is also available from the College Board, libraries, and guidance counselors' offices. In addition most colleges have their own Web pages with information.
- **www.fsatweb.com**. This is a free, searchable scholarship database.

To learn more about credit rights, you can email **consumerline@ftc.gov** for more information on the Fair Credit Reporting Act enacted by the Federal Trade Commission that protects your credit rights or call the FTC at 202-326-2222.

On the subject of bankruptcy the American Bankruptcy Institute has a home page at **www.abiworld.org**. Another source for the latest on filing for bankruptcy and bankruptcy issues or to join a discussion group is **www.bankrupt.com**.

The American Bar Association (ABA) has a home page at **www.abanet.org**. Lists of attorneys who specialize in bankruptcy can be found on the ABA home page or from West's Legal Dictionary at **www.wld.com**. The ABA home page also suggests ways to get a lawyer even if you cannot afford one. Attorneys can file a bankruptcy petition over the Internet.

Banks and other financial institutions and credit card companies have their own Web pages listing options and interest rates. *Money* magazine and *Kiplinger's Personal Finance Magazine* and their websites offer lists of the issuers of the lowest credit card and loan rates in the country on a regular basis.

FAQs

1. When I graduate, I will owe $8,000 in student loans. What is the best way to pay them off?

   ■ The best way is in a lump sum if you have it, but most people find that consistently making monthly payments—probably through a 10 year level repayment plan—is the way to go. The sooner you pay the loan off the sooner you can get that money working for you in savings and investing.

2. How much is too much credit?

   ■ If you are paying over 20 percent of take home pay a month to credit, that is too much.

3. I owe $5,000 on my credit cards and my utilities were turned off because I forgot to pay my last three bills. I do not want to tell my parents. I want to handle it myself. I have a job but I can't seem to keep up with money. Where can I go for help?

   ■ Call the National Foundation for Consumer Credit at 800 388 2227 or contact Consolidated Credit Counseling Services at 800-728-3632 or **www.debtfree.org** for a credit counseling office near you or check the telephone book under Credit Debt Counseling Services. NFCC or CCCS offer free or low-cost counseling and will help you get out of debt and get the utilities turned back on.

SUMMARY

Credit card spending is rising. The numbers of credit card holders and credit cards are also rising, as is the use of online credit.

The main types of credit are installment (same, regular payments), noninstallment (single payment), and open account (credit up to a specified maximum amount). When people talk about credit, usually they are referring to open account credit cards such as a VISA or MasterCard.

The main benefit of credit is its convenience. The main problem is overextension. Debt is a burden.

Before signing for a credit card, potential borrowers should check the interest rate expressed as the annual percentage rate (APR), finance charges, and fees. Of these, the most important way to figure out the cost of credit or a loan is to look at the interest rate. In a survey of college juniors and seniors 78 percent did not know this.

Creditors lend money to those whom they think can repay it.

The five C's of credit are capacity, capital, character, collateral, and conditions.

To establish credit, several steps are recommended, including opening a checking and savings account and paying bills on time.

Credit bureaus gather and supply information to creditors. The three largest national credit bureaus are Equifax, Experian, and Trans Union Corporation.

Women and college students may have special credit problems.

For students three trends are notable: the rise in student loan volume, the proliferation of credit cards, and the effect on career choices due to debts incurred while pursuing an education.

Bank credit cards are versatile, usually have a lower APR rate than retail cards, and can be used for cash advances. The majority of bank cards have variable interest rates tied to the prime (or base) rate.

There is usually a 25-30-day grace period in which to pay bills although some credit cards allow only 15 days.

Lending agencies may use any of four variations of the average daily balance (ADB). The ADB affects how much a person pays.

Loans can be divided into two main types, depending on what the loans are for and the repayment schedule. Loans may be for a variety of purposes including educational loans. Major loan sources include banks and S&Ls, credit unions, consumer finance companies, sales finance companies, and life insurance companies.

Over 1 million Americans file for bankruptcy, and one in 70 families declares bankruptcy each year.

The two main types of personal bankruptcy are Chapter 7 (the type most chosen) and Chapter 13. Most people filing for bankruptcy hire an attorney.

About 25 percent of U.S. employers check into job applicants' credit histories. Under the federal Fair Credit Reporting Act, they need permission from applicants to do this.

The National Foundation for Consumer Credit (NFCC) and Consolidated Credit Counseling Services (CCCS) offer free or low-cost advice on budgeting and debt-management and debt-consolidation plans.

REVIEW QUESTIONS

(See Appendix A for answers to questions 1–10)

Identify the following statements as true or false.

1. The use of credit is decreasing.
2. The typical credit cardholder has 1 or 2 credit cards.

3. The higher the annual percentage rate (APR), the lower the cost of credit.

4. Most people can handle 10 to 15 percent of annual income allocated to credit debt.

5. In the U.S. over one million bankruptcies are filed each year.

Pick the best answers to the following five questions.

6. Probably the main benefit of credit is
   a. getting rebates and prizes.
   b. paying less for credit than if cash were used.
   c. increased privacy over the use of cash.
   d. convenience.

7. Which of the following is not one of the five C's of credit worthiness?
   a. Capacity
   b. Capital
   c. Class
   d. Character

8. Many of the rules governing credit reporting procedures were formalized by a 1996 amendment of the ______________________.
   a. Fair Credit Reporting Act.
   b. Equal Credit Opportunity Act.
   c. Federal Trade Reporting Act.
   d. Interstate Commerce Act.

9. Only ___ of cardholders pay off credit card charges in full each month.
   a. 10 percent
   b. 20 percent
   c. 30 percent
   d. 36 percent

10. __________ are credit cards offered by institutions and organizations, such as professional sports teams and universities.
    a. Debit cards
    b. Affinity cards
    c. ATM cards
    d. Debt-Safety cards

DISCUSSION QUESTIONS

1. How do people fall into the credit trap? List five ways they can manage their credit better.

2. Credit is not always negative. What are some of the good reasons for using credit?

3. Describe some of the consumer credit legislation. For example, what does the Truth in Lending Law cover?

4. Why would someone go to see a consumer credit counselor? What can the counselor do?

5. If someone is going to file for bankruptcy what should they consider? Who should be consulted? If the filing is initiated, how long will the procedure usually take?

DECISION-MAKING CASES

1. Jeremy, a college junior, received a credit card application in the mail. He does not have a job and owes over $8,000 in student loans, yet the credit card issuer is offering him $2,000 in preapproved credit with a teaser rate of 6 percent for one year that will climb to 18 percent interest. He could use the money. Should he get the card? If so, how will he pay his credit card bills? If he needs a card, how could he find a better rate?

2. Michael and Sabrina are college friends. Sabrina, a full-time student, wants Michael, a part-time student with a full-time job, to cosign for a personal loan for a skiing trip in Colorado that she wants to go on. He hesitates, why?

3. Using the debt safety ratio, figure out how much consumer credit debt the Philbins can comfortably handle. They bring home $4,000 a month and are debt-free. Specifically they are wondering if they can afford $378-a-month payments for a new car.
4. Ben and Jessie bring in $22,500 a year, but have run up credit charges and other bills amounting to over $40,000. They have a ten year old car and rent an apartment. They are thinking of declaring bankruptcy and starting over. Should they do this? If so, whom should they consult?
5. Janese spends about 22 percent of her take-home pay on credit card bills. Is this too much? What percentage is recommended as a reasonable amount of credit for most individuals to handle?

GET ON LINE

Here are activities to try:

- Go to the American Express site at www.americanexpress.com and see what types of credit that company offers. For example, American Express recently debuted Blue, a card containing a smart chip designed to provide customers additional security in their Internet-based transactions when used in combination with a smartcard reader. Report on what you find out about Blue.
- Forrester Research was mentioned in the chapter. This is a firm that tracks consumer trends including online credit. Their reports can be found at www.forrester.com. Summarize what you find at their site keeping in mind that the site is designed for business and media sources.
- Compare the interest rates and services of different credit cards by visiting their websites such as the American Express one above. The website www.quicken.com/banking_and_borrowing/credit_cards/ offers advice on shopping for the right card. Compare this information with the information from the companies offering cards.
- Go to the Federal Trade Commission website www.ftc.gov and summarize what they have to say about credit.
- Select one of the student loan and financial-aid websites listed below and report on what you find:
  - www.ed.gov/prog-info/SFA/Student Guide/
  - www.ed.gov/offices/OPE/DCS
  - www.salliemae.com
  - www.nelliemae.com
  - www.usagroup.com/students/faqs.htm
  - www.usagroup.com/loancons/lcmain.htm

To access InfoTrac, follow the instructions in Chapter 1.

Exercise 1: Under the topic of credit cards you will find articles on affinity cards, discounters, and online credit. Of particular note are the many research articles in the *Journal of Money, Credit, and Banking*. General audience magazine articles are available from *Kiplinger's Personal Finance Magazine*, *Time*, and *U.S. News & World Report*. Select an article from the Journal and an article from a general article magazine, compare how they are written and compare what they say.

Exercise 2: Under bankruptcy there are over 100 articles and 24 subdivisions such as bankruptcy law, cases, Chapter 11 and Chapter 13. Select one article on bankruptcy and report on what you find.

part two

purchasing decisions and insurance

chapter six
housing and transportation

Purchasing Shelter and Vehicles

DID YOU KNOW?

* 67 percent of American households own their own homes, and home equity accounts for 23.5 percent of household net worth.
* About 30 percent of new cars are leased.

OBJECTIVES

After reading chapter 6, you should be able to do the following:

1. Analyze the costs and benefits of renting or owning a home.
2. Discuss the steps and finances involved in buying a home.
3. Discuss the finances involved in owning and selling homes.
4. Compare the costs of buying versus leasing vehicles.
5. Determine the cost of operating vehicles.

CHAPTER OVERVIEW

This chapter provides practical information about renting, buying, owning, and selling housing and about leasing, buying, and operating vehicles. Exhibit 6.1 shows why these are such important areas of expenditure to cover. *Housing represents 32 percent, transportation 18 percent, and utilities 5 percent of a typical family's budget. A home is the single, most expensive purchase* made by most people.

A record number of American households own their own homes. According to RFA Dismal Science, a consulting firm in West Chester, Pennsylvania, 67 percent of American households own their homes, and home equity accounts for 23.5 percent of household net worth. Many heads of households depend on appreciating home prices to help finance children's college educations or their own retirements.

The chapter begins with a discussion of the pros and cons of renting versus buying. Conventional wisdom held that buying housing was always better than renting, but buying is not always the best bet, at least in a strictly financial sense, experts say.[1] Two significant factors affecting potential financial gains are location and how long you plan to live in the house. These and other factors affecting housing decisions will be explored in this chapter. Next the specific financing options and the steps involved in buying a home will be covered, followed by the costs involved in maintaining and selling homes. Then the chapter switches to a discussion of the costs involved in buying, leasing, and operating vehicles. As in all chapters, FAQs and Electronic Resources are given near the end.

RENTING VERSUS BUYING A HOME

Everyone needs a place to live so one of the most basic life decisions to make is whether to rent or buy housing. Since most people rent before buying, renting will be discussed first.

Objective 1:
Analyze the costs and benefits of renting or owning a home.

Rent
Payment for use of property.

Security deposit
Payment required in advance to cover wear on unit.

Rent is the cost or payment for using property. Besides a first month's rent, the main cost prior to moving in is a security deposit. A **security deposit** is required by the landlord in advance to cover wear and tear of the unit beyond what is normally expected. Typically this is the same amount as or a slightly lesser amount than a month's rent. So if rent is $1,000 a month then the renter pays $1,000 for the first month's rent and a security deposit of $1,000, for a total cost of $2,000 before moving in. There may also be a nonrefundable pet deposit of about $100–$300.

In competitive markets property managers may offer lower security deposits in order to attract renters. Since tenant turnover is costly (loss of rent from vacant units, painting, recarpeting), property managers may encourage renters to stay by returning part of their security deposit if they sign on for another year, or by reducing the rent slightly.

Lease
Legal document between renter and landlord.

Sublease
Property can be leased to another.

A **lease** is a legal document between the renter and the landlord, defining the rights and obligations of both. An important part of the lease is whether or not the renter can **sublease** which means the property can be leased by the original tenant to another. Disagreements about leases and the return of security deposits are common especially in college towns.

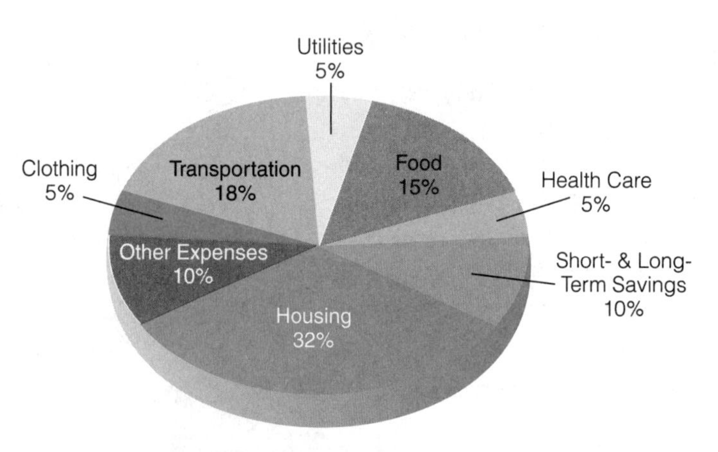

Spending allocations
Although every family's budget is different, this chart shows the average percentage in main categories of expenditures.

EXHIBIT 6.1
A Typical Family Budget

Tenant Rights

Tenant rights vary by state and community. Most of them center on the idea of fair treatment and the landlord's responsibility to provide a habitable property.

If a rental unit is left clean and undamaged, then tenants have the right to a prompt return of their security deposits. To ensure that this happens, a renter should give proper notice of intent to move out (usually at least one month's notice), make lists of all damages and defects before moving into the unit and before moving out and have the landlord sign them, and try to resolve any differences with the landlord first. If this does not happen, consider going to the university student services attorney or to small claims court to obtain a refund if there is a dispute.

Owning versus Renting

After spending several years renting, an individual may start to think about owning rather than renting. In the short run renting is more economical, but in the long run owning usually is more economical because of tax deductions on mortgage interest, tax-free capital gains, and **appreciation** (increase in the home's value). These however, may be offset by maintenance and repairs (expenses to be discussed later in the chapter).

Appreciation
Increase in value.

According to Jane Bryant Quinn, financial columnist, "A house may not be your best investment in the decade ahead. If the eighties and early nineties taught us anything, it's that real estate doesn't always go up."[2] If owning a home is not such a wise investment, why do most people buy homes? According to the National Association of Realtors, the main reason first-time homebuyers give is that they are "tired of paying rent." Second on the list is the tax advantages, and third is the desire for a larger home.[3] These reasons and other pros and cons of owning and renting are given in Exhibit 6.2. Exhibit 6.3 gives an example of the financial aspects of owning versus renting during the first year of occupancy.

EXHIBIT 6.2
The Pros and Cons of Renting versus Buying

	Pros	Cons
Renting	Apartment extras (pool, fitness centers)	No tax deductions
	Few responsibilities	Transient neighbors
	No maintenance or repairs	No gain from rising value of property
	No yard work	Little storage
	Low moving-in costs	Most alterations cannot be made
	Mobility, less commitment	Usually no garage
	Close neighbors give sense of security	Usually no laundry hook-ups in unit
	Manager to turn to for repairs	Rents go up with little notice
Buying	Build equity	Tied down
	Pride of ownership	Down payment and move-in costs
	Monthly payments remain stable	Costs of repairs and maintenance
	Can borrow against equity	Limited liquidity
	More room, more storage, yard, garage	Commitment on time, energy, money
	Income tax deduction for mortgage interest and property taxes	Yard work
		Homeowner Association fees
	Stronger credit rating	Property taxes
	Laying down roots in a community	

EXHIBIT 6.3
First-Year Cost Comparisons

		Buying				Renting
Purchase price		$100,500		Monthly rent		$700
Down payment (10%)		$10,000		Security deposit		$350
Closing costs (2.5%)		$2,500				
Mortgage rate		7.5%				
Loan term		30 yrs.				
Factor from Table A		6.9921				
Total loan amount (number of thousands)	X	$90,000				
Purchase price – downpayment						
Monthly payment		$629				

First Year Analysis

Up-Front Costs		Buying				Renting
						$350
						.0500
Interest lost	=		$750		=	$17.50
Taxes saved	–		$250		–	0
Total Up-Front			$500			$17.50
Annual Costs						
Number of months	X	12		12	X	
Monthly payment		$629		$67.50		
Total payment			$7,548.00		=	$8,100

		Buying				Renting
Insurance			$500			$250
Utilities			$2550			$2200
Real Estate Taxes			$1500			0
Maintenance			$1000			0
Total payment	X	$7,548				
Factor from Table D		89.01				
	+	$6,718				
Property taxes		$1,000				
	X	$7,718				
Marginal tax rate		28%				
Income Taxes Saved			$2,168			
Initial loan amount		$90,000				
Factor from Table C		.0092				
Principal Repaid			$828			
Purchase price		$155,000				
Factor from Table B		.0200				
Appreciation (2%)			$2,000			
Total Annual Cost*		$7,559				$10,850.00
Total Up-Front		$500				$17.50
Total Housing Cost*		$8,059				$10,867.50

*The Total Annual Cost is figured to:

$7,548	$12,548	$10,387	$ 9,559
+ 500	– 2.161	– 828	– 2,000
2500	$10,387	$9,559	$7,559
1000			
1000			
$12,548			

EXHIBIT 6.3
continued

Table A: Monthly Payment per $1,000 Borrowed

Interest Rate (%)	Loan Term (Years) 15	30
5.0	$7.9079	$5.3682
5.5	8.1708	5.6779
6.0	8.4386	5.9955
6.5	8.7111	6.3207
6.75	8.8491	6.4860
7.0	8.9883	6.6530
7.5	9.2701	**6.9921**
8.0	9.5565	7.3376
8.5	9.8474	7.6891
9.0	10.1427	8.0462
9.5	10.4422	8.4085
10.0	10.7461	8.7757

Table B: Interest Earned/Appreciation

Number of Years	2%	3%	4%	5%	6%	7%	8%
1	0.0200	0.0300	0.0400	0.0500	**0.0600**	0.0700	0.0800
2	0.0404	0.0609	0.0816	0.1025	0.1236	0.1449	0.1664
3	0.0612	0.0927	0.1249	0.1576	0.1910	0.2250	0.2597
4	0.0824	0.1255	0.1699	0.2155	0.2625	0.3108	0.3605
5	0.1041	0.1593	0.2167	0.2763	0.3382	0.4026	0.4693
6	0.1262	0.1941	0.2653	0.3401	0.4185	0.5007	0.5869
7	0.1487	0.2299	0.3159	0.4071	0.5036	0.6058	0.7138
8	0.1717	0.2668	0.3686	0.4775	0.5938	0.7182	0.8509
9	0.1951	0.3048	0.4233	0.5513	0.6895	0.8385	0.9990
10	0.2190	0.3439	0.4802	0.6289	0.7908	0.9672	1.1589

Table C: Percent of Loan Repaid after Given Number of Years (30-Year Loans)

Interest Rate (%)	Number of Years 1	2	3	4	5	10	20
5.0	1.48%	3.03%	4.66%	6.37%	8.17%	18.66%	49.39%
5.5	1.35	2.77	4.27	5.86	7.54	17.46	47.68
6.0	1.23	2.53	3.92	5.39	6.95	16.31	46.00
6.5	1.12	2.31	3.58	4.94	6.39	15.22	44.33
6.75	1.07	2.21	3.43	4.73	6.12	14.70	43.51
7.0	10.2	2.11	3.27	4.53	5.87	14.19	42.70
7.5	**0.92**	1.92	2.99	4.14	5.38	13.21	41.09
8.0	0.84	1.74	2.72	3.78	4.93	12.28	39.52
8.5	0.76	1.58	2.47	3.45	4.51	11.40	37.98
9.0	0.68	1.43	2.25	3.14	4.12	10.57	36.48

Table D: Percent of Payment Going to Interest after a Given Number of Years (30 Year Loans)

Interest Rate (%)	Number of Years 1	2	3	4	5	10	20
5.0	77.10%	76.51%	75.91%	75.28%	74.63%	71.04%	61.67%
5.5	8.203	79.67	79.09	78.49	77.87	74.38	65.01
6.0	82.93	82.41	81.86	81.29	80.69	77.32	68.03
6.5	85.26	84.77	84.25	83.72	83.15	79.93	70.77
6.75	86.31	85.83	85.33	84.81	84.26	81.11	72.05
7.0	87.28	86.82	86.33	85.83	85.30	82.23	73.26
7.5	**89.01**	88.59	88.14	87.67	87.17	84.26	75.51
8.0	90.51	90.12	89.70	89.26	88.80	86.06	77.56
8.5	91.81	91.44	91.06	90.66	90.22	87.65	79.42
9.0	92.92	92.59	92.24	91.86	91.47	89.05	81.11

Objective 2:
Discuss the steps and finances involved in buying a home.

STEPS AND FINANCES INVOLVED IN BUYING A HOME

Step #1: Determining Financial Readiness

The first step in home ownership involves building savings for a down payment and figuring out how much house a person can afford. There are a number of ways to do this.

1. To get a general idea of the maximum amount of house that can be afforded multiply annual household income by two and one-half times. For example, if a dual career household brings in $50,000 a year they can afford to spend about $125,000 for a house.

Mortgages
Loans to purchase real estate.

PITI
Principal, interest, taxes, and insurance.

2. A more accurate method, used by lenders in estimating **mortgages** (loans to purchase real estate in which the real estate itself serves as collateral), is based on two conditions. The buyer must be able to handle the monthly payments of principal, interest, taxes and insurance (**PITI**), and have an adequate down payment.

So a buyer will need money up front in the form of a down payment as well as the proven ability to meet monthly payments. A typical down payment is 10–20 percent of the purchase price; if a house costs $100,000, the down payment would be $10,000–$20,000 and the mortgage would be on the remaining amount. The range of down payment options will be discussed in Step #3.

3. Another method is the *28/36 qualifying ratio*. In this ratio monthly gross income is multiplied by .28 (28%) or .36 (36%) to show the range that lenders would consider as appropriate in order to determine monthly mortgage payments.

	28 Qualifying Ratio	36 Qualifying Ratio
Monthly Income	$3,500	$3,500
	x .28	x .36
	$980	$1,260

Given this example, a potential buyer with a monthly income of $3,500 could afford monthly payments (including insurance and property taxes) of between $980 and $1,260. Using the 28/36 qualifying ratio helps potential buyers see how much house they can afford, but lenders will not lend 36 percent if the buyer has large debts (auto or educational loans or credit card debts).

Homebuyers may not want to spend the maximum amount that they can afford based on these formulas for housing because they would rather have more money for

other expenses. Kim McGrigg of Consumer Credit Counseling Service Southwest in Phoenix, Arizona cautions, "You may qualify for a more expensive home than you can reasonably afford. By doing a budget analysis, you will better know how much you can comfortably pay on a monthly basis."[4] She recommends that you should figure out how much house you can afford before beginning to shop.

How Much Do Homes Cost?

Besides how much house a person can afford, another concern is how much houses cost in the area where a person lives. In 1999 the median price for a U.S. home was $132,000, but prices varied considerably depending on area of the country. The median is the midpoint—half are sold for more and half for less. According to the National Association of Realtors and RFA Dismal Science, the fastest-rising home prices are in the West (Seattle-Bellevue-Everett in Washington, San Francisco and other areas in California) and in New York City and Middlesex-Somerset-Hunterdon, New Jersey.

Step #2 Determining Preferences

There are other things to consider besides financial readiness and location when home shopping such as

* Type of housing desired. The main types of housing are:

1. Single-family dwellings: houses detached from other units.
2. Multi-units: duplexes and townhouses.
3. **Condominiums**: individually owned units in a building with several other units with monthly fees for operating the building and common areas, such as swimming pools and lobbies. The monthly fees are called **homeowner's fees**. The owner of a condominium has title to the unit.
4. **Cooperative housing**(co-ops): units, usually apartments in cities, are owned by the co-op not directly by the buyer. The buyer owns shares in the building as a whole with the right to lease a certain unit. Monthly operating fees are charged, similar to those for condominiums.
5. **Manufactured housing**: units fully or partially assembled in a factory and moved to the living site. This category includes mobile homes.

* Commuting distance.
* Service availability (water, sewer, roads, cable tv) .
* Taxes (as shown in Exhibit 6.3).
* Size and Quality.
* Zoning laws and covenants including what can and cannot be done to and at the home (i.e., whether or not business can be conducted, what improvements such as fences and additions are allowed), and what schools are zoned for that home.

Condominiums
The homeowner has title to his or her unit and jointly owns common areas.

Homeowner's fees
Monthly fees paid by home owners.

Cooperative housing
Units owned by the co-op, not the homeowner who owns shares in the building.

Manufactured housing
Units assembled at a factory and moved to a building site.

CONSUMER ALERT

Do not assume that a nearby school is zoned for the home you are considering; check with the school board to learn which schools are zoned for your area. Even if you do not have children or plan to have them, school zones are important because they affect property values.

Step #3 Finding a Home, Making an Offer

Real estate agent or Broker
Offers services to buyers and sellers of property.

Real Estate Agents, Realtors, and Buyer's Agents Locating a home can be done on one's own or through the services of a **real estate agent or broker**. Usually the seller pays the real estate broker's commission (around 6 percent of the selling price of the house) so the buyer can use the services of the agent without any direct cost involved. Real estate agents who are members of the National Association of Realtors can use the registered trademark of Realtor.

A real estate agent or realtor can provide the following services to someone in the market for a home: show homes, present an offer to the seller, negotiate a final price, assist in the obtaining of financing, and represent the buyer at closing

Buyer's agent
Represents the buyer.

Another type of professional that may be consulted is a **buyer's agent**. This person represents the buyer not the seller. He or she may want an upfront retainer fee for services to be rendered which include showing a large selection of homes and representing the buyer in negotiations and at the closing.

Home warranty
Provides additional protection to the buyer.

A potential buyer should inquire about a **home warranty** that provides additional protection for the buyer. Home warranties can be purchased from insurance companies or arranged by a realtor or buyer's agent. Sometimes sellers buy a home warranty and offer it to potential buyers as an incentive. Home builders may offer a one-year home warranty on a new home.

Another professional that homebuyers may consider using is a certified home inspector. A home inspector can assess the condition of the home by checking mechanical systems, interior spaces, appliances, and exterior conditions. The inspector will take only a few hours to do the assessment and will charge a few hundred dollars. The time and expense are worth it, especially in assessing an older home. If problems are found, such as cracks in the foundation or a leaky roof, the seller should offer to pay for repairs or reduce the price.

Making an Offer

Once the purchaser has decided on a home, he or she makes an offer directly or through the real estate agent, realtor, or buyer's agent. A little haggling over price and counteroffers may be necessary.

Binder or earnest money
Security deposit by the buyer as evidence of a serious offer.

Escrow
Earnest money held by the mortgage lender until closing.

If buyers are sure they want a particular home, the usual practice is to put down a **binder** (also known as **earnest money**), a portion of the purchase price deposited as evidence that the purchase offer is serious. It can range from a few hundred dollars to several thousand dollars.

Earnest money is held in **escrow** (held by the mortgage lender) until closing day. If the offer for the home is accepted, then the earnest money is figured into the closing costs (see Exhibit 6.4). If the offer is not accepted, the binder is returned. In rare cases—usually due to the buyer's failure to complete the purchase—it is kept by the seller (see the Consumer Alert box).

As soon as the binder is accepted, the buying process speeds up and goes to contract. Going to contract means that the buyer and the seller outline the terms of the sale, including the down payment. You may want a real estate attorney to look over

CONSUMER ALERT

Do not put down a binder unless you are sure this is the home that you want. In some cases the binder may be kept by the sellers if they were inconvenienced by keeping their house off the market by a less-than-"earnest" offer.

the contract. This is the first legal document that the buyer and the seller sign. The conditions of the sale that it outlines include a free and clear title (explained in the next section), mortgage contingencies (a mortgage has to be approved by a lender), and termite inspections and other required inspections.

A closing date will be written on the contract. Sellers like to move quickly so they can get their money, but the buyer may need three to five weeks to get the finances in order.

Step #4 Getting Financing

Once the home is selected and the price agreed upon, then buyers will need a mortgage loan unless they pay the full price in cash or make a trade. Less than 10 percent of home buyers pay cash for a home. The type and the amount of the mortgage will depend on several factors including the down payment which may be 5 percent, 10 percent, or 20 percent of the purchase price. The main exception to such a down payment would be a qualified veteran of the armed forces who may obtain loans with a zero percent down payment.

The higher the down payment the lower the amount of the loan and subsequently the lower the monthly payment, as indicated in Exhibit 6.5. See Exhibit 6.6 for general estimates of how much cash will be needed given varying amounts of purchase price, 5 percent down payments, and closing costs.

Typical closing cost estimates are given in Exhibit 6.6. Note on Exhibit 6.4 that one of the closing costs is for a title search and insurance. The **title** is the legal right of ownership of property. In property transactions the title is transferred from the seller to the buyer. The buyer—usually with the help of an attorney—determines if there is a clear title to the property (no claims, no other owners). If there is any question, a title search and the purchase of insurance can protect the new owner. In title searches attorneys examine the **abstract** (a detailed written history of the property's ownership) and may issue a certificate of title.

Title
Legal right of property ownership.

Abstract
Written history of property's ownership.

Deed
Document that transfers property ownership.

Lenders often require that the new owner buy title insurance in case the title is later found to be faulty. Also due at closing will be a fee for recording the **deed**, the document that transfers ownership of property from the previous owner to the new owner.

Sources of Loans

In the past, financing a home was simple. You walked down the street to the bank, where the banker greeted you by your first name and if you qualified, you received the loan. Today you can go to the following sources for mortgages: banks, brokerage firms (if you have an account at a certain level), credit unions, mortgage finance companies, insurance companies, private investors, pension plans, savings and loan associations.

Access can be in person or through toll-free telephone calls or the Internet. Many financial institutions offer mortgage loan prequalification checks over the telephone or the Internet.

Amortization
Process of paying off a loan through a series of payments.

The term "mortgage" comes from the concept of **amortization** which is the process of gradually paying off a loan through a series of payments to the lender. For example, a monthly payment of $800.00 can be broken down into the portion for principal repayment and a portion for interest payments. Over time the outstanding loan balance reduces until the home is paid off in 15 or 30 years.

Types of Loans

A home buyer can save thousands of dollars by taking the time to explore mortgage options. Different types of mortgages are described next.

EXHIBIT 6.4
Closing Costs Checklist

Every mortgage lender charges borrowers differently, and usual closing costs between buyers and sellers vary greatly by location. The following is a checklist of typical charges.

Common Charge	Cost	Who Pays?	Comments
Earnest money (example)	$1,000	buyer	returned on purchase of house
Points charged by lender	______	______	______
Loan assumption fee	______	______	______
Loan origination fee	______	______	______
Prepaid mortgage insurance	______	______	______
Appraisal	______	______	______
Credit report	______	______	______
Property survey	______	______	______
Initial mortgage property	______	______	______
Property inspections	______	______	______
Recording fees	______	______	______
Prepaid homeowner's insurance for 1st year	______	______	______
Prorated property taxes	______	______	______
Attorney's fees	______	______	______
Title search and insurance	______	______	______
Closing taxes	______	______	______
Down payment (this may be less earnest money paid)	______	______	______
If condo or co-op, prorated maintenance for remainder of month	______	______	______
Other	______	______	______
Total amount needed at closing by buyer	$ ______		

Fixed rate, conventional mortgages. These are fixed-rate, fixed-term, fixed-payment loans, usually carried for 15 or 30 years. The virtue of this option is its predictability (8 percent for 30 years). Fifteen-year mortgages have a 15-year mortgage payback, which saves homeowners thousands of dollars in interest payments. The downside is that monthly payments will be much higher than for thirty year mortgages. Tax benefits versus saved interest should be weighed when selecting a 15-year mortgage over a 30-year mortgage. Since fixed, conventional mortgages are a popular choice, the pros and cons should be weighed carefully:

Pros	**Cons**
Housing costs are known.	If interest rates drop, borrower cannot benefit.
Budgeting is easier.	Monthly payments may be higher than for ARMs.
Mortgage rate is stable, does not go up if interest rates go up.	Closing costs may be higher.

Adjustable rate mortgages (ARMs) These have interest rates that fluctuate with the movement in the economy. Less is paid initially, often as much as two or three percent less (called a teaser rate) than the going (fixed) conventional mortgage rate; but if interest rates increase, so does the payment. Most have a payment cap so the interest cannot increase more than 5 percent above the original interest rate. ARMs can be adjusted yearly or after three years or five years. In other words ARMs can change at preset intervals, depending on whether interest rates have increased or decreased. The appeal to the consumer is the low initial teaser rate, but the consumer

EXHIBIT 6.5
The Down Payment Effect

Effect of Down Payment Size on Monthly Payment for a $100,000 Home (Figured on 7 Percent Mortgage Loan for 20 Years)

Down Payment	Amount of Loan	Monthly Payment
$5,000 (5%)	$95,000	$736.53
10,000 (10%)	90,000	697.77
15,000 (15%)	85,000	659.00
20,000 (20%)	80,000	620.24
25,000 (25%)	75,000	581.74

should keep in mind how high the final interest rate may climb. It is important to shop around for ARMs and understand all the fine print. There is more risk and variability in ARMs than in fixed-rate, conventional mortgages.

Convertible mortgages These are adjustable loans that can be converted to a fixed rate at certain times during the course of the loan.

Two-step loans These are adjustable loans with a built-in rate increase after five or seven years. Initially these are tempting, but they could be difficult for the last 25 years of the loan.

FHA and VA loans These are offered if the lender determines that a person qualifies for Federal Housing Administration (FHA) or Veterans Administration (VA) government-backed loans. Both types allow for very low down payment, or in the case of VA loans, no down payment at all.

Owner financing These loans are offered by the property owner, who will take all or part of the mortgage at a rate higher than average and for two or three years. The balance of the note, known as the balloon, will be due in full at the end of the loan. Then the buyer would pay off the owner and switch over to a bank or mortgage lender. If this option is selected, be sure that this switch is possible, and try to get the longest possible repayment period from the owner.

Assumable mortgages. These allow buyers to assume the original mortgage from the seller of the home. For example, if the original owner is paying only a 7 percent mortgage rate on a two-year-old home, the buyer may want to assume that rate (and the amount on the mortgage that was paid up to the point of sale) rather than finance the home at the current rate of 9 percent. In this case the down payment will be larger because the home has increased in value and the principal of the loan will have been reduced somewhat over the two years of ownership.

No income check, no asset check loans These types of loans are rare and may be a choice selected by people with uncertain or poor credit ratings. Lenders will provide

EXHIBIT 6.6
Examples of Cash Needed for Home Purchase

This chart provides an estimate of how much cash is needed to cover the down payment (figured at 5%) and typical closing costs.

Purchase price	Down Payment 5%	Estimated Closing Costs	Approximate Total Cash Needed
$ 75,000	$3,150	$1,500	$4,750
$100,000	$5,000	$2,000	$7,000
$125,000	$6,250	$2,500	$8,750
$150,000	$7,500	$3,000	$10,500

a loan based on tax returns and other proofs of ability to pay. Since the lender is taking a chance, interest rates are high.

As a final note, there are other more obscure types of loans besides the ones described in this section. The majority of home buyers get the first type described: fixed-rate, conventional mortgages. After a loan source and type are selected, a loan application is filled out. Sometimes there is an application fee that includes an appraisal fee and a credit report fee.

Step #5: The Closing

Closing
Meeting in which real estate is transferred from seller to buyer.

When the loan is approved, the buyer moves toward **closing**, a meeting in which occurs the transference of real estate from a seller to a buyer, also known as the settlement. Fees are associated with the closing, documents are signed, and the title is passed. Throughout the loan and closing process, realtors, mortgage professionals, and loan officers assist buyers and sellers with the options and the paperwork. To get ready for closing, the buyer needs to get mortgage insurance; set aside money for closing costs as shown in Exhibit 6.4 (lenders can provide an exact total); set a date for the final walk-through to see that the home is in the agreed upon condition; and consider having a home inspection by a professional and getting a home warranty.

Mortgage insurance will be required by the lender if the buyer makes a down payment of less than 20 percent of the home's purchase price. This insurance will insure the first 20 percent of a loan in case the buyer defaults. Mortgage insurance is available, if you qualify, from a private mortgage insurance (PMI) company (the largest is Mortgage Guaranty Insurance Corporation), the Federal Housing Administration (FHA), and the federal Department of Veterans Affairs (VA).

Closing costs (fees) are negotiable. How closing costs are divided between seller and buyer varies from location to location and by individual preference.

Points
Fees paid to the lender at the closing.

One type of closing costs, called points, is shown in Exhibit 6.4. **Points** are fees charged by the lender, separate from interest but designed to give the lender more money. A point is one percent of the total loan. So two points on a $100,000 loan would be $2,000 due at the closing. Because this is an additional cost, usually the more points the borrower assumes, the lower the interest rate. An astute buyer can save money on closing costs. According to Martin Atkinson of Fleet Mortgage, "Ninety percent of customers shop interest rates, but they fail to shop closing costs and points and other things that are negotiable."[5] Occasionally all closing costs are dropped in order to promote sales. Closings usually take place in the office of the mortgage lender or an attorney.

Objective 3:
Discuss the finances involved in owning and selling homes.

OWNING A HOME

Once a home is purchased, the expenses and mortgage options do not end. Depending on the type of home and location, there may be homeowner association fees or condominium or co-op fees.

Refinancing may be desirable if interest rates drop considerably from when the home was bought. Other expenses include property taxes, utilities, and renovations.

Maintenance and repair costs are estimated to run about two percent of the cost of the home per year, but this varies greatly, depending on owner preferences, age and condition of the home, and the owner's financial resources. These ongoing expenses may seem to discourage home ownership, but it should be remembered that the value of a home changes continually, and so does a person's equity with a home as long as the home has a mortgage.

Another estimate of expenses comes from a study commissioned by *The Wall Street Journal* that revealed that almost every house, no matter how recently or expertly built, will require substantial sums to maintain. The study found that the cost of keeping a typical home up to current standards for 30 years is almost four times the purchase price. In fact it was suggested that it may actually be cheaper to buy a new or fully remodeled home every ten years rather than dealing with repair problems.[6]

Property Taxes

Fair market value
What a willing buyer would pay a willing seller for property.

Assessed value
A percentage of fair market value.

Communities depend on the revenue gained from real estate property taxes to provide for numerous community services, including schools. Local government officials determine a **fair market value** for an owner's home and land. The fair market value is what a willing buyer would pay a willing seller. Then an **assessed value** is determined, which is a percentage of fair market value. The percentage varies by community. For example, a $200,000 home at 60 percent would yield an assessed value of $120,000. Property with an assessable value of $120,000 may be taxed at a rate of 30 mills (each mill represents one-thousandth of a dollar). This would result in a tax liability of $3600 ($120,000 x 0.030) due yearly. Multimillion dollar houses in prime areas may be assessed over $100,000 a year in property taxes.

Before buying land or a home, contact the tax assessor to find out the typical yearly tax rate. Previous owners can also show records of past taxes. If the assessed value of the property is too high, the owner can appeal to the tax assessor for a reduction. Once the home is owned, property taxes may increase periodically. For example, on undeveloped land if the city puts in a sewer the property taxes will probably go up.

Utilities: Reducing Energy Costs

The chart in Exhibit 6.1 shows that utilities amount to 5 percent of the typical family's budget. Within the utility category, the largest expense is usually home-heating costs. Average annual home-heating costs by area of the country are given. As may be expected the Northeast and Alaska lead the nation in home heating costs.

A homebuyer can ask to see the previous owner's utility bills to get an idea of energy costs for that home. In addition most utility companies will come to the home and give a free home energy audit. This audit may reveal the need to replace appliances, furnaces, or water heaters with more energy-efficient models.

In warm climates yearly air-conditioning costs can run as high as heating costs. Ceiling fans, awnings, and shade trees can cut air- conditioning costs by as much as 50 percent.[7]

Refinancing Mortgages

Refinance
Revising a payment schedule to reduce monthly payments and to modify interest charges.

If interest rates drop while someone owns a home, the owner may want to **refinance** (revise the payment schedule) the loan. Refinancing works best when there are *at least two percentage points difference* between the old and new loan to offset costs charged by lenders in refinancing a loan. Banks, mortgage companies, and other lenders will show a homeowner the cost savings and charges involved in refinancing a loan. Whether refinancing is a good option or not depends on how much difference there is in interest rates, how long the owner plans to stay in the home, and what refinancing costs there are. Besides refinancing, there are other financial options, including paying off the mortgage early, taking on a second mortgage, and signing up for a reverse mortgage.

Paying Off Mortgages Early

To speed up the process of owning a home free and clear, a homeowner may want to pay off the mortgage early. Some mortgages have a prepayment penalty, so a homeowner should check to see if that is the case. Otherwise paying off early may be an option to consider for peace of mind. This decision should be weighed carefully in terms of taxes and a determination whether the money spent on paying off the loan would be better spent in another way.

Second Mortgages and Home Equity Loans

Second mortgages and home equity loans were covered in the previous chapter. To recap, homeowners take out second mortgages to get money in a lump sum for home improvements or other expenses, such as children's college education or to buy a new car. It is, just as the name implies, another mortgage on property; it has a secondary claim to the primary mortgage. Second mortgages are considered more risky than primary mortgages because in the case of default, the first mortgage has priority of claim over the second. Because of this, secondary mortgages tend to have higher interest rates and shorter payback times. Second mortgages are repaid in monthly installments.

A variation on the second mortgage is the home equity loan. Home equity loans are secured by independent appraisal of the property value. The loans may be used as a line of credit, which may be drawn down as a check or a credit card. Since the home serves as collateral in a second mortgage or a home equity loan, if the owner defaults on the loan, he or she could lose the house. A new twist in the home equity loan market is called **125 loans**. These loans allow the borrower (the home owner) up to 125 percent of the value of the home. So the attraction is the ready cash or line of credit. However, 125 loans should be avoided because they charge much higher interest rates than other types of home equity loans, not all the interest is tax deductible, the house cannot be sold until the 125 loan is repaid, and the house can be lost if payments are not kept up.

125 loans
Home equity loans that allow the homeowner up to 125 percent of the home's value.

Reverse Mortgages

Homeowners age 62 and older can apply for **reverse mortgages**, also called **home equity conversion mortgages**, which provide tax-free income in the form of a loan that is paid back (with interest) when the home is sold or the homeowner dies. Getting money this way is often a preferred choice if the elderly person is house rich and cash poor. Reverse mortgages are available from government agencies and private lending institutions. More details on these will be given later in the book.

Reverse mortgages
Arrangement wherein owner borrows against home equity.

SELLING A HOME

Mortgages are held on average only five to seven years because whenever a house is sold the mortgage is paid off and the finance cycle begins again. When homeowners decide to sell, the first thing they usually do is set a price. Next they get the house into marketable condition and decide whether to sell it themselves or use a real estate agent. Since a person's home is such a valuable asset, a lot of care should be taken in putting the house up for sale.

Setting the Price

Sellers should find out the sale prices of homes that were recently sold in their neighborhood. They should compare house size, condition, appeal and other features that affect desirability. A more comprehensive assessment can be gained from realtors or a certified appraiser. An estimate of the current value of a home is called an **appraisal**.

Appraisal
Estimate of home's current value.

Marketing a Home

Most people want a house in move-in condition with curb appeal. This means that the front of the house should be as attractive and as well kept as possible. An open-house event can stir interest in a home. Advertisements can be placed in newspapers, brochures, through the Internet, and on television. Computer listings are becoming more and more popular because pictures and descriptions can be displayed 24 hours a day. The Electronic Resources section at the end of the chapter lists websites that offer these.

Seller or Real Estate Professional?

About 10 percent of home sales are made by owners each year.[8] It is not unusual for homeowners to try selling the house by themselves for a few months and then turn it over to a realtor.

As mentioned earlier real estate agents charge around 6 percent of the house selling price which is paid by the seller. They provide many services, including suggesting a selling price, advertising, contacting potential buyers and other real estate professionals, conducting open houses and house showings, screening potential buyers for financial ability, and helping set up financing. There are discount brokers that charge less than 6 percent but offer fewer services. Other companies, such as Help-U-Sell Real Estate, charge a flat fee or 1 to 2 percent of the price.

Many of the other steps involved in selling homes have already been covered in the buying homes section (financing, closing costs). Usually the seller will be expected to pay the real estate agent, the transfer of taxes, the balance of the mortgage plus fees, and whatever costs are specified in the contract. Regardless of whether or not the seller makes money on the sale, he or she will have to report the sale on the federal income tax return for the year in which the sale took place.

Objective 4:
Compare the costs of buying versus leasing vehicles.

TRANSPORTATION AND VEHICLES

The chapter changes gears now and describes leasing, buying, operating, and selling vehicles. An interesting trend is that Americans are keeping their cars longer than ever, and that means they are spending more on maintenance and repairs. The median age of the U.S. vehicle fleet is almost eight years, the oldest it has been in more than 40 years.[9] People are keeping vehicles longer due to the rising cost of new cars.[10]

As the chart in Exhibit 6.1 showed, transportation takes up about 18 percent of the typical family's budget. For many people their car comes second only to their house in terms of amount of purchase price. A basic decision is whether to lease or buy. In 1985 only 5 percent of all cars were leased. In 1996 this number soared to 30 percent.[11] About three million leased cars were returned to dealers in 1997, swelling the used car market.[12]

Leasing

Leasing
Contractual agreement outlining monthly payments, security deposit, and terms of lease.

Leasing is a contractual agreement outlining monthly payments, a security deposit, and terms and conditions of the lease. The concept of leasing versus buying vehicles is analogous to renting versus buying housing. The *advantages* of leasing are several:

* Up front, leasing may require less money than buying because a security deposit may amount to only two monthly payments, whereas buying a car may require a large down payment.
* Monthly payments may be less for leasing than for owning.

* Leasing may be paid for by a business or may be favored for business purposes.
* No trade-in or resale worries follow leasing.

The *disadvantages* are fewer, but not trivial::

* The vehicle is not owned.
* There may be extra costs for mileage, repairs, turning the vehicle in early or moving to another state. *The average car is driven over 15,100 miles per year, but many lease contracts allow only 10,000–12,000 miles per year.*[13]

There are three types of leases:

1. The most common type of leasing is closed-end or "walk-away" wherein the leaser returns the vehicle at the end of the lease period and pays any additional fees for damage or mileage.
2. Open-end or finance leases may require the leaser to pay the difference between the expected value of the leased automobile and the amount for which the leasing company sells it. In this type there may be an end-of-lease payment.
3. Single-payment lease allows a customer to obtain a discount on the motor vehicle rental agreement.

Buying a New Vehicle

Buying a new vehicle presents lots of options. The first step in a car search is getting information. The April issue of *Consumer Reports* lists test results on new cars. *Car and Driver* and *Road and Track* also review cars.

List prices and auto guides are available over the Internet, telephone, in books, in magazines such as the December issue of *Kiplinger's Personal Finance* or *Edmund's New Car Prices* or through auto pricing services available through Consumer Reports Auto Price Services and Nationwide Auto Brokers.

Consumer Reports Auto Price Services
P.O. Box 570
Lathrup Village, MI 48076
(800) 933-5555

Nationwide Auto Brokers, Inc.
17517 West Ten Mile Rd.
Southfield, MI 48075
(800) 521-7257 (toll-free)
(313) 559-6661 (Michigan only)

CONSUMER ALERT

Because leasing is a fairly new option, there have been misunderstandings and unscrupulous dealings. For example, some elderly people who leased a car thought they had bought it. Be sure you know what you are paying for and what you are getting. Present laws do not fully protect leasers.[14] Partial protection is offered under the Consumer Leasing Act, which is administered by the Federal Trade Commission (FTC). The Federal Reserve Board has approved a simple, one-page form explaining leasing (monthly payments, charges, etc.) to replace the dense contracts that often mislead car-lease customers. Consumers can obtain a copy from the Fed's public affairs office (202) 452-3204.

Price Bargaining, Buying Services, Set Price Dealers

Once potential buyers have decided how much they can spend, and have selected a model and options, then they should compare prices at several dealers or consider buying through a car-buying service offered through credit unions, motor clubs (such as the American Automobile Association), auto brokers (businesses offering cars just over dealer's costs), churches, warehouse clubs (such as Sam's), and community organizations.

Invoice price
Car cost to dealer.

Sticker price
Price listed on the car.

Before negotiating, consumers should find out the car's **invoice price**, which is what the car cost the dealer. The **sticker price or suggested retail price** is the price of the car, including options, "market adjustment," and transportation charges. The astute buyer negotiates from the bottom up, in other words from the invoice up, and does not pay for unwanted extras or extended service warranties. Information about invoice and sticker prices can be obtained from a variety of sources, including financial magazines and their websites, *Consumer Reports*, and IntelliChoice (www.intellichoice.com).

The value of car-buying services and auto brokers is that they can search nationally for the car desired and compare prices through computer searches, saving the buyer time, energy, and usually money. The negative is that the buyer does not get to see the car before purchase unless he or she can find the same make and model on a local dealer's lot. In the late 1990s an estimated 20 percent of all vehicles were bought through a buying service or a broker. Buyers who are flexible (willing to buy a tan car instead of a white car) and comparison shop can save money. Using the information gathered about invoice price from the sources mentioned earlier, the buyer has an estimate of how low a dealer can go.

Set price dealers, such as Saturn use a no-haggling car-selling method. Prices are presented to customers to accept or reject. One of the reasons set-price dealers have grown in popularity is that a lot of people are uncomfortable with bargaining and are aware of the potential for ripoffs (see Consumer Alert box).

For both set and bargained prices another consideration is trade-in value of the previously owned vehicle. It may reduce the cost enough to make the new price attractive, or it may be better to sell the old vehicle oneself.

Sales Agreement

Before the sale is completed, a buyer has to sign a sales agreement, which is a legal document outlining the details of the purchase. As in any legal document, customers should be careful what they sign and make sure everything expected is in the sales agreement.

Financing

Paying cash is one option, but most people buy vehicles on credit because of the high cost. Financing sources include dealer financing, banks, credit unions, consumer

CONSUMER ALERT

To avoid buying a vehicle at an unfair price, watch out for the following practices: lowballing (a very low price is quoted but there are add-on costs at the end) and highballing (a high amount is offered for a trade-in but the extra amount is made up in an increased new car price).

finance companies, and other financial institutions locally and accessed by toll-free numbers and the Internet.

As discussed in Chapter 5, the best indicator of the true cost of credit is the annual percentage rate (APR). So a buyer should shop around for the lowest APR. Traditionally, credit unions have offered especially low APRs on car loans. However, rebates with car purchases may alter the total cost of the loan. All charges, rebates, and the APR have to be considered in order to figure out what is the best deal.

Dealer financing is designed to make signing up for credit easier since the financing is set up right at the place the car is bought. A variety of options are available from Chrysler Financial Corporation, Ford Motor Credit Company, and General Motors Acceptance Corporation, including credit plans with balloon-payment loans, variable-rate loans, and special loans for recent college graduates. Balloon-payment loans are illegal in some states. The way they work is that smaller payments are made during the first 48 months, after that the buyer has the option of (1) continuing to make payments, (2) returning the car to the dealer for a fee, and (3) selling the car, paying off the loan, and keeping the profit, if any.

BUYING A USED CAR

High-priced new cars are out of the price range of many people who turn to the used car market. Sources of used vehicles include all of the following:

* New car dealers selling trade-ins and previously leased cars.
* Used car dealers specializing in used car sales. These can range from a dealer with a few cars to huge national corporations such as CarMax, with tens of thousands of cars. The large companies offer a service in which the buyer goes in and tells the salesperson how much money he or she has to spend. This information is fed into the computer and a list comes out stating which cars are for sale at that price and where they are located on the lot.
* Private individuals selling their own cars.
* Others including auctions, dealers that sell cars used by businesses, auto rental companies, and government agencies.

Fair prices can be determined by checking *Edmund's Used Car Prices* and the *Kelley Blue Book*, which is the common name for the National Automobile Dealers Association Official Used Car Guide. The blue book, updated monthly, is available at libraries, dealerships, and financial institutions, such as banks and credit unions. Edmund's and the blue book are also available on the web (see Electronic Resources section near the end of the chapter). Another source is the *black book*, a weekly publication of car auction sales published by National Auto Research of Gainesville, Georgia.

Objective 5:
Determine the Cost of Operating Vehicles.

Cost of Operating Vehicles

The costs involved in operating a vehicle can be divided into fixed and variable expenses.

CONSUMER ALERT

The Federal Trade Commission (FTC) requires businesses that sell used cars to place buyer's guide stickers on car windows. This disclosure states whether the car is under warranty or being sold "as is." If bought from private individuals there are few guarantees. To be safe, buy from a reliable source and have the car's condition inspected by a trained mechanic that you choose, not the seller.

Fixed	**Variable**
Depreciation	Gasoline, oil
Insurance	Maintenance, repairs
Interest on car loan	License, fees, registration
Parking, tolls	
Taxes	

Each year the American Automobile Association (AAA) conducts a Driving Cost Survey. The most recent report revealed that a motorist driving 15,000 miles per year paid $6,908 in auto-related expenses. Each year the costs rise, primarily due to higher ownership costs such as insurance, depreciation, taxes, license, and registration. The factors that affect auto related expenses include the following:

* Driving costs involving the size of vehicle and number of miles driven.
* **Depreciation**, which is the loss in the vehicle's value due to time and use. *This is usually the largest fixed expense.* Depending on the type of car and age and driving record of the driver, insurance can be higher than depreciation. Federal tax guidelines assume a vehicle depreciates 52 percent during the first two years of an expected lifespan of five years. For a $21,000 car , this means it will depreciate $10,920 in the first two years.
* Insurance—personal injury and property damage.
* Variable operating costs, including gasoline, oil, tires, maintenance, and repairs. The life of a car can be extended through proper maintenance.
* Depending on tax laws, some vehicle costs may be deductible if the vehicle is used for business. For the latest rules and regulations contact the Internal Revenue Service.

Depreciation
Loss of the vehicle's value due to time and use.

ELECTRONIC RESOURCES

For government housing and transportation information, visit **www.fedstats.gov.**, a general site from the U.S. government for a variety of agencies, including those dealing with housing and transportation.

For vehicle information there are a number of websites. The trend is toward using the Internet to research prices or to look at new car models. "Analysts estimate about 15% of new-car buyers do their research on-line and as many as 2% make their purchases from dealers they find on the Net."[15] websites include:

* CarPoint: a site produced by Microsoft Corporation at **carpoint.msn.com**. Specific model cars' performance data, a calculator to figure out how much car you can afford, and a tour of cars using "surround video" are a few of the options this site offers.
* Edmund's Automobile Buyer's Guides: packed with information on new and used car prices and the steps in purchasing at **www.edmunds.com**.
* Kelley Blue Book: gives trade-ins and used car pricing information at **www.kbb.com**.

For home information there are also many websites, and the trend in real estate is to offer home previews on the Web. This is especially helpful if you are being transferred and buying or renting from out of town. The Internet is giving borrowers access to a raft of information—from calculators to figure mortgage payments to home pages listing rates, and including applications from various lenders.

To comparison shop mortgage rates, you would choose one of the sites listed next and enter the type of loan you want, how much you want to borrow, and a few other facts. The sites list the best loans offered by the lenders it tracks. Then you can handle the loan yourself by applying online. At the most competitive sites the fee should not exceed 0.5 percent which compares favorably to a bank or a mortgage broker who may charge 1 to 1.5 percent. Examples of mortgage rate websites are

* **www.bankrate.com** for Bank Rate Monitor.
* **www.eloan.com**.
* **homeadvisor.msn.com**.
* **www.homepath.com** for Fannie Mae.
* **www.hsh.com** for HSH Associates.
* **www.iown.com**.
* **msn.com** for the Microsoft Network.
* **www.mbaa.org** for the Mortgage Bankers Association.
* **priceline.com** for Priceline (also has travel and other prices).
* **www.quickenmortgage.com** for Quicken Mortgage.

The website for the National Association of Realtors, which offers comprehensive national as well as regional home pricing and trend information, is **www.realtor.com**. Using this website, you can electronically search for a home as you can with other sites, including **www.prudential.com**, described in the Get On Line section. Another home search website is **www.homes.com** of Homes and Land Magazine with sale and rental listings in 14,000 communities.

1. When is it time to buy a house? I have been renting for six years and have a full-time job and about $6,000 in savings.

   ■ It is time to buy a house when you think you are ready emotionally and financially. Most first-time homebuyers have grown tired of renting. In the long run buying is a good option. Renting makes more sense if you do not plan to live in the house for more than three years. Use the formulas given in the chapter to see how much house you can afford, and visit houses in neighborhoods that you like, to see what is available.

2. Does it make sense to prepay a mortgage since the interest paid is tax-deductible?

   ■ A mortgage represents a tax deduction, but tax deductions do not put money into your pocket, they reduce expenses. So prepaying a mortgage shortens the term of the loan, thereby saving the owner of the house thousands of dollars of interest. This amount of savings would have to be compared to savings in tax deductions to see if prepayment is the right step. Another consideration is what the homeowners would do with the money (for example, $50,000) if they did not pay off the mortgage. If they invested the $50,000 wisely, the money would grow and might be a better investment than prepaying the mortgage.

3. Do home values beat inflation?

   ■ In fast-growing markets prices often outpace inflation, but increasingly most homes in most markets stay even with inflation. This answer may change if the economy changes radically.

4. Do I need a lot of money to buy a house?

   ■ The majority of Americans can afford a house. Lower-income homebuyers can start out with modestly priced homes, apply for a state low-income program for first-time buyers, or rent to own through the Federal Home Mortgage Corporation.

5. Is a fixed, conventional 30-year mortgage always best?

   ■ The mortgage choice depends on personal financial circumstances and interest rates. Most people pick a fixed, conventional mortgage, but an adjustable-rate mortgage (ARM) may be a better choice if interest rates are high or you need the lower rate initially to afford payments. At this writing about 30 percent of borrowers were choosing ARMs because the least expensive ARMs were at 6.3 percent, compared to fixed rates at 8 percent.

6. Do you always need a 20 percent down payment?

   ■ No, the usual amounts are 10 percent sometimes 20 percent and 5 percent is not uncommon, especially for inexpensive, starter homes. Military veterans can get a Veterans Administration loan for zero down. The Federal Housing Administration may require as little as 3 percent down.

7. Are home equity and second mortgages the same thing?

   ■ This is a trick question because they have a lot in common and the terms are often used interchangeably. Home equity loans are a variation of a second mortgage. In both, your home is at stake and you receive a loan by using your home as collateral. Traditional home equity loans were based on a formula that allows up to 80 percent of the value of your home minus the amount still owed on the mortgage. Today there is a new variation called 125 loans that can go as high as 125 percent of the value of the home.

8. Should I lease or buy a car?

   ■ That depends on many factors; use the guidelines given in the chapter to make a decision. Generally leasing has the advantage of little or no down payment, and typically you get a new car every three years. Whether or not leasing is right for you depends partly on what kind of car you want and how many miles you drive each year.

9. How do I find out if a car dealer is offering me a good price?

   ■ Go to the Internet to compare prices or go to a bank or credit union and look up suggested prices for new and used cars in the Kelley Blue Book or other books that they have.

SUMMARY

Sixty-seven percent of American households own their homes, and home equity accounts for 23.5 percent of household net worth.

In the short run renting is more economical, but usually in the long run owning a home is more economical because of appreciation and tax advantages.

Housing takes 32 percent, transportation 18 percent, and utilities 5 percent of the average family's budget. Housing is the biggest expense for most people.

Different types of housing include condominiums, in which a unit is owned but the owner has access to common areas with neighbors and cooperatives, in which an owner has shares in a building as a whole with the right to lease a certain unit.

A particularly popular way to figure how much home a person can afford is the 28/36 qualifying ratio. This means that lenders prefer buyers to spend no more than 28 percent to 36 percent of their monthly gross income on housing.

The steps in buying a home include determining financial readiness and housing preferences, finding a home, making an offer, and closing. Closing costs, including points, are negotiable. A point is one percent of the total loan. So two points on a $100,000 loan would be $2,000.

Home financing sources and mortgage loan types should be considered when buying and also when already owning a home. Refinancing a loan is suggested when there is at least a two-percentage point difference between the original loan and newly available mortgage rates.

Most people have fixed, conventional mortgages, but at this writing about 30 percent had adjustable-rate mortgages (ARMs). Fixed rates were about 8 percent, compared to the least expensive ARMs at 6.3 percent.

Leasing vehicles is growing in popularity. About 30 percent of new cars are leased.

New and used vehicle sources of information and purchase and loans abound. Buyers should be wary of two pricing practices: lowballing and highballing.

Both fixed and variable costs are involved in vehicle ownership, including the rapid depreciation of vehicles.

REVIEW QUESTIONS

(See Appendix A for answers to questions 1–10.)

Identify the following questions as true or false

1. Condominiums are units, usually apartments in cities, that are owned by a corporation, not directly by the buyer.
2. The title is the legal right of ownership of property.
3. Most homeowners have conventional mortgages.
4. The largest fixed expense in owning a vehicle is usually depreciation.
5. More people than ever before are leasing vehicles.

Pick the best answers to the following questions.

6. What percent of a typical family's budget goes for housing?
   a. 15
   b. 32
   c. 40
   d. 50
7. What percent of a typical family's budget goes for transportation?
   a. 5
   b. 10
   c. 18
   d. 27
8. Given the 2½ times rule, if your income is $28,000 a year, how much home can you afford?
   a. $56,000
   b. $60,000
   c. $62,800
   d. $70,000
9. About ____ percent of home sales are made by owners each year.
   a. 10
   b. 25
   c. 40
   d. 50
10. In the 1990s an estimated ____ percent of all vehicles were bought through a buying service or broker.
   a. 5
   b. 10
   c. 20
   d. 30

DISCUSSION QUESTIONS AND PROBLEMS

1. Describe the pros and cons of renting versus owning a home.
   Why do most people buy their first home?
2. What is the importance of the APR (annual percentage rate) in buying a vehicle?
3. Which costs more for the average family: housing or transportation? What are the typical percentages for each category?
4. Why does Kim McGrigg of CCCS recommend that people think twice before they spend the maximum amount that they can on housing?
5. Why would someone want to lease a car instead of buying it?

DECISION-MAKING CASES

1. Al Miller bought a small lakefront cabin in a rural area twenty years ago. Now his cabin is surrounded by million dollar mansions. His property taxes have climbed to over $20,000 a year and he is retired living on a fixed income. Can Al do anything to reduce his property taxes or does he have to sell?
2. Pier Bateson and his landlord are having a dispute over Pier's getting his security deposit returned. What tenant rights does Pier have?
3. Skye and her fiancé are house hunting. They would like to start out in a condominium or co-op. What should they consider in making this decision?
4. Marc wants a sports utility vehicle and has a model picked out. How can he search for the best price?
5. Jenn thinks she will lease a car instead of buying one because she does not want to deal with reselling or trading in a car. Because she is a sales representative, she typically drives 25,000 miles a year. How may this affect her leasing versus buying decision?

GET ON LINE

Activities to Try:

- Go to **www.e-loan.com**. Since interest rates change all the time, go to the website and report on the current mortgage rates that you find, home equity rates, and auto rates. For auto rates compare the APRs for 5-year loans and 3-year loans.
- Find **pathfinder.com/money/features/auto** for an interactive website comparing buying versus leasing. Try completing the worksheet and report on how well it works.
- Choose a car you would like to own (Honda Civic, BMW Z3) and visit the following two websites to compare invoice prices and retail prices: **carpoint.msn.com** and **www.autobytel.com**.
- Go to **www.prudential.com** and click on buying a home and selling a home. Report on the information that you find. Also click on "Search for a home" to find a database of homes for sale.
- Visit Fannie Mae's website—**www.homepath.com**—and click on "Home Starter Path." Follow the guidelines and report on what you find. For example, would a first-time home buyer find this site useful?

To access InfoTrac follow the instructions at the end of chapter 1.

This chapter discussed buying homes (versus renting) and vehicles (versus leasing).

Exercise 1: To learn more about renting, go to Rents (Property) and you should find about 25 periodical references, 12 subdivisions, and 9 related subjects, including tenant rights. Under the periodical references search for an article that compares buying versus renting a house. What are the pros and cons of renting versus buying? Summarize the key points in the article. Another option is to run a key word search, for example on "real estate" or "mortgages." Report on the trends in housing and mortgage rates. Are interest rates going up or down?

Exercise 2: To find out more about leasing versus buying cars, try "leasing cars" or "buying cars." Or run a key word search on "leasing," "cars," or "automobiles." What are the current issues regarding leasing versus buying vehicles? Are there any new government rulings (for example, by the Federal Trade Commission) on leasing? What are the trends in car financing?

chapter seven
costs of providing care

PUTTING PEOPLE FIRST

DID YOU KNOW?

* Middle-income parents spend nearly $300,000 to raise a child from birth to age 18.
* Disney's "The Ultimate Fairy Tale Wedding" costs $100,000.

OBJECTIVES

After reading chapter 7, you should be able to do the following:

1. Discuss the costs of childbirth, adoption, and raising children, including child care.
2. Describe the different types of money gifts and investments for children.
3. Discuss ways to save for children's college education's.
4. Discuss prenuptial agreements and the costs involved in weddings, divorces, child support, and widow(er)hood.
5. Discuss Social Security and Medicare.
6. Discuss elder care and living expenses.
7. Distinguish between legitimate and less-than-legitimate charities.

CHAPTER OVERVIEW

This chapter is a departure from the previous one on houses and transportation. Rather than focusing on big-ticket, tangible items, this chapter emphasizes the intangibles involved in caring for others. Many of the costs to be discussed involve services rather than products. Some of the expenses are crisis-oriented; others can be planned for. Regardless of their origin, many life expenses take a sizable portion of individual and family budgets and deserve a place in a book about personal finance. For example, people save for a wedding just as they save for a house or any other large expense. The purpose of this chapter is to put the "person" back into *personal* finance.

According to author Suzie Orman, the people you may want to provide care for include those who helped you enter the world, those to whom you gave life and those who have guided your passage through life.[1]

This chapter adds to this list others that you may want to help through other forms of giving such as charity. The chapter also provides an explanation of government programs that provide support, such as Social Security and Medicare. Other forms of providing care and protection will be discussed in future chapters on insurance and estate planning.

The chapter begins with the costs of childbirth, adoption, and raising children, including child care. This is followed by different forms of money gifts or investing for children and how to save for a college education. Since Chapter 5 covered college costs and loans, the emphasis in this chapter is on how to save for a child's college education rather than dealing with the expenses incurred while in college.

The financial side of weddings and marriage, including prenuptial agreements, will be covered along with divorces, child support, and widow(er)hood. Then government programs are covered. The chapter concludes with elder care and living options and a discussion of charitable giving. As in all chapters, there are FAQs and Electronic Resources.

Objective 1:
Discuss the costs of childbirth, adoption, and raising children, including child care.

COSTS OF CHILDBIRTH

Newborns come into this world with a considerable price tag. The routine prenatal care, delivery, and one postpartum visit by the physician, plus hospital costs, range from $6,500 to nearly $10,000 for a vaginal delivery—and $5,000 to $6,000 on top of that is charged for a Cesarean section. Costs vary considerably by area of the country and by services preferred by the mother such as a private room versus a shared room. Typically the mother and baby are released from the hospital 24 hours after a vaginal delivery birth.

Because of the high cost of childbirth, the necessity for health insurance is evident. Some employers have been cutting back on their coverage, so it is wise to learn the following ahead of time:

* What is covered (will it cover a birthing center as well as a conventional hospital? a midwife? prenatal care and tests?).
* The copayments or deductibles.
* The number of days in the hospital.
* The rules about whether or not the mother must take childbearing education classes.
* The pediatric care for the newborn in the hospital.

Costs of Adopting

Adoption offers so many options, costs, and sources that it is a difficult subject to discuss. This section gives general information, but it is such a changing area that readers should see this as only a starting point.

A directory of agencies that places healthy American-born children and *The Adoption Fact Book* can be obtained from the National Council for Adoption, 1930 17th Street NW, Washington, DC 20009-6207. There is a charge for the book, which is a 300-page comprehensive guide covering state regulations, lists of resources, and a discussion of issues.

Children from foreign countries or United States-born children with special needs are increasingly of interest to prospective adoptive parents. *Most parents adopt children through agencies*, some adopt through lawyers and doctors. The Internet offers information on agencies. For example, by typing in "adoption.org" on a general search engine thousands of websites come up. You can narrow the search by selecting a subcategory, such as Eastern European adoptions, which produces over 50 agency names. Public adoption agencies charge no fees or small fees, whereas private adoption agencies sponsored by religious or charitable groups charge about $9,000, including attorney's fees.[2]

Nearly all states have fee waivers or subsidy programs for parents who choose special needs children to encourage the adoption of these children. Adopting children from foreign countries involves the expense of travel and fees that can range from several thousand to $20,000.[3]

All of the figures in this section on adoption are estimates. It is not unusual for an adoption to run over $30,000, especially if travel, medical care of the child, or the expenses of the birth mother are involved.

Costs of Raising Children

According to the U.S. Department of Agriculture, the first year of a child's life will cost middle-income parents more than $9,000 and upper-income parents as much as $14,000. The same source estimates that from birth to age 18, *middle-income parents will spend nearly $300,000 and upper-income parents nearly $430,000.* These figures do not include college.

When people think about having children they tend to think about health care and college costs, but there are many other expenses that add up. For example, it is estimated that a baby will use 1,750 diapers per year at a cost of about $400 per year. Infant formula can cost upwards of $1100 for a six-month supply. For older children summer camp and afterschool care—along with music lessons, computers, and sporting equipment—can run into the thousands and tens of thousands of dollars. On average, *housing is the biggest part of the total cost* of raising a child. In the preschool years the next-highest expense categories are child care and education. For young teens the highest expense after housing is food.

Costs of Child Care

Child care costs vary by area of the country and the type of child care chosen. Choices include in-home care by a relative (usually a parent or grandparent), family child care homes, child care centers, or in-home care by a paid, nonfamily member.

Costs of care can vary by the age of the child and the number of children in the same setting. The ratio of children to staff and the annual child-care-teacher turnover rate should be considered. As in everything else you get what you pay for, so high quality care is going to cost more.

Dependent-care assistance plan
Employer sponsored plan to reimburse parents for child care costs.

Employers may offer on-site child care or child care subsidies to reduce the costs to parents. Many employers offer a **dependent-care assistance plan**, which allows workers to put away as much as $5,000 a year in pretax dollars. The $5,000 can be used to reimburse a parent for child care costs. Usually, however, if the money is not used in a year, it will be forfeited. If an employer does not offer a dependent-care account or if child care expenses exceed $5,000, then parents may qualify for dependent care tax credit on their income tax. A tax advisor or the Internal Revenue Service (IRS) should be consulted on this matter.

Child care costs can run as much per year as college tuition. Nationwide, families pay an average of about 7.5 percent of annual pretax income on child care. *The average American family spends about $4,600 each year for the care of one child.*[4] Many pay much more than this, especially those living in cities. For example, in New York a child care center's fee is much higher than the national average, and a full-time, in-home nanny can cost over $400 a week, adding up to more than $20,000 a year.[5]

Child care costs weigh heavily especially on the working poor, who pay the highest percentage of their income—some as high as 25 percent— for child care. Their ability to stay off welfare often hinges on finding affordable child care. Some states offer subsidized child care for low-income families that reduces costs to the families that qualify.

If parents select in-home care, they are required by law to pay Social Security and Medicare taxes if their child care giver receives more than $1,000 a year. Parents must provide W-2 forms for the child care worker and the Internal Revenue Service. There may be state taxes as well. Many parents prefer the ease of paying a child care center because W-2 forms and taxes are taken care of by the center.

Au pairs
Young Europeans who live-in and provide child care.

In many families child care is a mix of child care centers, baby-sitters, and au pairs or nannies. **Au pairs** are usually young Europeans that live in the home and provide as much as 45 hours of child care a week as part of a year-long cultural exchange. Because they are non immigrants, parents do not pay Social Security. In 1999 the average cost to parents was approximately $210 to $250 a week plus room and board. There are eight authorized au pair agencies that arrange placement and charge fees of around $4,000 for health insurance, travel, training, and local services. So there are start-up costs and weekly costs. Parents have to weigh the advantages and disadvantages of all the child care options. Some make more financial sense than others especially based on the number and ages of the children—plus there are tax and quality of care considerations.

CHILDREN, YOUNG ADULTS, AND TAXES

Tax relief takes the sting out of some of the childcare costs. Parents need to look into the federal child care tax credit and employer-sponsored programs. Usually reimbursable expenses include payments for nursery schools, child care centers, day camps, and babysitters. Certain things do not count, such as overnight camp and transportation to and from the child care center. In addition to federal tax breaks, roughly half of the states offer tax breaks for child care.

> Typically, though, higher-income employees will find flexible-spending accounts more advantageous, while lower-paid employees will get greater benefit from the tax credit.
>
> Here's why. The credit, which reduces the amount owed to the IRS dollar-for-dollar at tax time, equals as much as 30% of the first $2,400 in child-care expenses if you have one child, or the first $4,800 if you have two or more children.
>
> The actual benefit, however, depends on how much you earn. Low-income families get the full 30% credit—a maximum of $1,440 for two or more children. Families with an adjusted gross income of $28,000 or more, however, get only 20% of their expenses up to the limit. That's a maximum credit of $480 for one child and $960 for two or more children.
>
> By comparison, the savings on a flexible-spending account depend on your tax rate. Someone in the 15% bracket would save a maximum of only $750 by using a child-care account, while someone in the 39.6% bracket could save as much as $1,980.[6]

Kiddie-tax
Tax filed by parents for the investment income of children under age 14.

Since figures and amounts change as guidelines change, parents and tax advisors need to keep up-to-date. What may be allowable one year may not be the next.

Another aspect of children and taxes is the "**kiddie tax**." At age 14 a child is no longer taxed at the parents' rate, but at the child's own rate. Up to age 14 (at this writing) the first $700 of unearned income (interest and dividends) is tax-free to children. The next $700 is taxed at the child's rate, which would usually be 15 percent, the lowest tax bracket. So when the child becomes 14, the parents should talk with their tax accountant or financial planner about readjusting their investment strategy for saving for the child's college education. The kiddie tax rule was enacted to discourage parents from shifting assets to young children in order to reduce their own tax burden. In most states at age 14 a child can get working papers and a job. Also at age 14 the child can set up an individual retirement account, which can be used as a way to save for college.

Other important tax birthdays are ages 19 or 24 because at age 19 the federal government says you are no longer a dependent on your parents' tax returns unless you are still in school. If you are still in school, then you can be counted as a dependent until the age of 24.

Allowances

As part of the socialization process, children learn how to handle money. Giving an allowance may be part of this process. Neale Godfrey, president of Children's Financial Network, says children like an allowance system because it gives them independence. They learn what money can and cannot do. She says, "it's about empowering kids."[7]

Opinion is divided over whether allowances should be linked to chores or grades or whether it should be a set amount each week, regardless of work or academic performance. Each family has to come up with an allowance system they are comfortable with.

One formula is to give the child a dollar for each year of his or her age, so that a 10-year-old would get $10 a week. It could be all discretionary (the child spends what he or she wants), or it could be divided into savings and spending.

Another question is *when* to start an allowance. This decision should be based on the child's interest and maturity. Some experts suggest that children as young as three can handle an allowance; others suggest waiting until ages seven or eight or even later; and others say the best age is when they start school, around ages five or six. Some children are not ready until 11 or 12.

What Do Children Spend Money On?

According to a Texas A&M study by James McNeal, children under 12 spend about 33 percent of their money for food and drink, 25 percent for play and toys, 15 percent is saved, and a significant portion goes to apparel.[8] As children age more goes to apparel. Business is interested in what children spend because it is estimated that the under-12 set has $20.2 billion to spend per year. Teenagers earn over $100 billion a year and spend it mainly on clothing, food, and entertainment.

Parents and Schools

What can parents do to help children learn how to manage money? Besides giving them an allowance, they can show children by example, such as taking them along when shopping or going to the bank. Later (usually around the ages of 11 to 13), parents can talk about or show how they invest in stocks and mutual funds. Younger children are most interested in stocks they can relate to such as McDonald's, Disney, Wrigley, Apple, or Coca-Cola.

Schools are partners in the financial learning process. In preschool and kindergarten, children learn the difference between pennies, nickels, dimes, and quarters. For example, young children are taught that dimes (even though they are smaller) are worth more than nickels. By age 7 and 8 children can read price tags and make sure they get the correct change.

Parents can expect that their children will make purchasing mistakes. For example, new toys will break or disappoint in some other way. This is part of the consumer socialization experience. Allowances or money gifts for birthdays or other occasions should include some money that the child can spend any way that he or she wants.

Objective 2:
Describe the different types of money gifts and investments for children.

MONEY GIFTS OR INVESTMENTS FOR CHILDREN

Parents, aunts and uncles, and grandparents often want to give money gifts to children. For a newborn the gift may be a bank with his or her first dollar in it. For young children the gift may be a piggy bank or savings account with a starting amount. For more long-term giving there are many options:

* Series EE savings bonds can be purchased for a minimum of $25 from a bank or online from the federal government.
* With a savings account a child can take the money out at any age. For this reason for larger amounts many parents and relatives prefer other methods, such as trusts or investments covered under the Uniform Transfer to Minors Act (read on for descriptions of trusts and this Act).
* A stockbroker can buy five or ten shares of a company that a child is familiar with, then the giver can show the child how to follow the price of the stock in the newspaper or a parent can do a direct investment over the telephone or the Internet. A teenager may like an index fund or a mutual fund. Most mutual funds charge at least $1,000 to open an account so this may be too steep, but there are less expensive no-load funds (mutual funds with no sales charge) that may be appropriate. Examples of funds requiring less than a $1,000 investment are the Strong Asset Allocation Fund and the Strong Total Return Fund, the David L. Babson Growth Fund and Babson Bond Trust, the Sentry Fund, and Century Shares Trust.
* If someone wants to give a lot of money to a child, then that person should see a lawyer about setting up a trust. In a trust an age can be designated for when a child gets the money and it can be stated as a sliding scale (so much at age 21, more at age 25, and the rest at age 30).
* A tax law provision allows people to give away as much as $10,000 a year, tax-free, to a child or anyone else. Many elderly people—especially grandparents or great-grandparents—take advantage of this provision in order to reduce estate taxes.
* Another option for substantial gifts ($50–$100 a month for ten years to build a college fund) is to invest in **custodial accounts** through the **Uniform Transfer to Minors Act (UTMA)**, or in some states the **Uniform Gifts to Minors Act (UGMA)**. In custodial accounts money is not put directly in the child's name; it is held in custody by an adult— usually a parent, grandparent, or family friend, or sometimes a bank. The custodian is responsible for investments and record keeping for the account. The money can be used to pay for summer camp, a new bike or car, or college, or held until later for another expense. Usually it cannot be spent for routine upkeep, such as food, clothing, or housing. If someone is interested in this, he or she could start a mutual fund, for example, by asking for the appropriate papers, filling out the child's and custodian's name and mailing them back. Brokerage firms, banks, and mutual funds can help with custodial accounts.

Custodial accounts
Accounts created for minors usually at brokerage firms, banks, or mutual funds.

Uniform Transfer to Minors Act or Uniform Gifts to Minors Act
Provide simple ways to transfer wealth to minors without a formal trust.

If the custodian dies or becomes unable to manage the account or chooses to no longer serve in this capacity, a successor custodian may assume his or her responsibilities. Most states allow a child over age 14 to designate a successor. For children under

14 the child's guardian usually designates a custodian.

At age 18, or in some states age 21, the child takes over the money. So although the parents may be hoping that this money will go to pay for college, once the child hits the **age of majority**—18 or 21, depending on state laws—the child can spend the money on anything he or she wants. Because of this, some parents prefer not to put too much money into custodial accounts. Also, as mentioned under the kiddie tax discussion, tax considerations affect how much should be put into these accounts.

Age of majority
Age when young adults can access custodial accounts in their name.

Objective 3:
Discuss ways to save for children's college educations.

Saving for College

When adults are asked what they are saving up for, the usual answers are a house, their children's college educations, and their retirement. Even the most financially secure parents worry about how to fund their children's educations. The first step is to estimate how much college will cost in the years that the child will attend. A general estimate and a college tuition calculator are given in Exhibit 7.1 (other calculators are available on the Web).

If it seems impossible to think about saving huge sums for each child, parents should not panic because—given inflation—their incomes will rise from what they are today; investments should grow; and schools, financial institutions, and the govern-

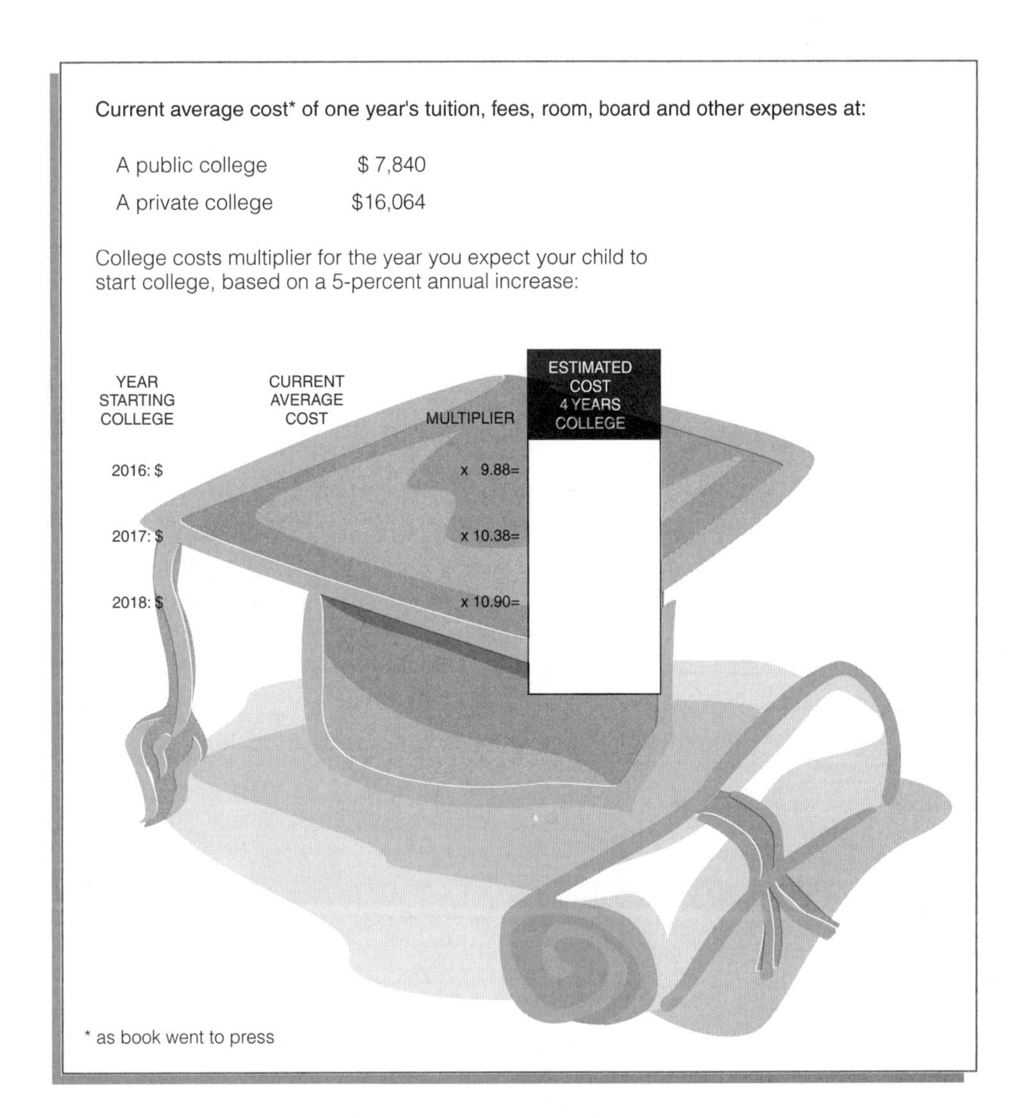

Current average cost* of one year's tuition, fees, room, board and other expenses at:
A public college $ 7,840
A private college $16,064
College costs multiplier for the year you expect your child to start college, based on a 5-percent annual increase:
YEAR STARTING COLLEGE
CURRENT AVERAGE COST
MULTIPLIER
ESTIMATED COST 4 YEARS COLLEGE
2016: $
x 9.88=
2017: $
x 10.38=
2018: $
x 10.90=
* as book went to press

EXHIBIT 7.1
Estimated Future Cost of College

ment offer financial aid in the form of grants and loans. Most schools assume a student will also contribute to his or her own educational expenses through summer jobs and working part-time during the school year.

Given that college funds will be needed, what is the best way to save? The previous section showed some ways money can be set aside for children's futures. The next sections will show more ways. The best general advice for parents is that they should *start as early as possible putting together a college savings plan.*

Investing for College

Most experts agree that some of the college fund should be invested in stocks or mutual funds for the greatest growth potential. Bonds are another choice because the timing of maturity dates of bonds can coincide with dates of the child's enrollment in college. As mentioned earlier, because of tax considerations any savings or investing plans begun when children are young should be reassessed when they turn 14.

Time horizon factor
Number of years before money is needed.

The **time horizon factor** (number of years until the money is needed) is part of the tuition calculator estimates. The farther away from actual need, the more risk (more stocks and stock-based funds) can be taken. Nearer to college the proceeds from the stocks or funds could be put into a money market fund, for example. If the stock market is rising, some experts would suggest that you leave the rest of the investment fund in stocks and draw out slowly during each of the four years of college. At any stage of the college savings program, it is wise to *diversify the investment portfolio because of the ups and downs in the economy* (do not put everything into a single bond or stock or a savings account).

According to Kenneth Kelgon, a financial planner in Lansing, Michigan, the most common mistake he sees is that college money is invested too conservatively. He says, "Parents think this is the college money, so they really should be conservative and put the bulk of their money in bonds and money-market funds. But if you've got enough time, that is the wrong thing to do."[9]

> Historically, stocks have done better in the long run than bonds, so they should be your best bet for making college money grow. Suppose, for instance, you sock away $200 a month. If you plunk the money in bonds that earn 7% annually, you will have $81,598 after 18 years. But if you will stash the money in stocks that gain 10% a year, you will end the period with $109,438, some 34% more.[10]

Another mistake parents make is to pull money out of retirement accounts to pay for college. This is a mistake because if a parent is in the 28 percent bracket, he or she could lose close to 70 cents out of every dollar to lost aid, taxes and penalties. For example, if a parent is under age 59½, he or she will have to pay a 10 percent tax penalty for withdrawing money early. Experts would suggest instead that the money be borrowed from the government or a financial institution at a low annual percentage rate (APR) rather than withdrawing from retirement funds.[11]

Tax Laws/Legislation

Given the evolving nature of tax laws and financial-aid policies, many of the amounts and guidelines given in this section regarding saving for college may change considerably. In the 1990s putting money aside became simpler with the advent of the Education IRAs and Roth IRAs. The Education IRA allows parents to put away $500 for tuition per year per child, and the Roth IRA up to $2,000 per year. When a child goes to school, the money is withdrawn, and there are no taxes due on any interest that the money has earned. There will be more details on IRAs in the retirement chapter.

Financial Aid: How to Access?

Chapter 5 covered many of the student loans available. The main aid form is the Free Application for Federal Student Aid (FAFSA), which can be submitted as early as January 1. The form is available from high school guidance counselors or college financial-aid offices, by calling (800) 801-0576, or by downloading it from www.ed.gov/offices/OPE/express.html.

Over $31 billion is available in aid, and 60 percent of American students receive some financial aid in the form of grants, work-study, or loans or some combination thereof.[12] Individual states offer programs—such as Georgia's Hope Scholarships and Florida's Bright Scholars—that provide tuition waivers, tuition reductions, and other forms of financial aid to qualified students attending state-funded institutions. High-school guidance counselors and college financial-aid offices can provide information about these programs.

Prepaid-Tuition Plans and College Savings Plans

Prepaid tuition plans or college savings plans State-offered tuition savings plans.

Most states offer special tuition savings plans to encourage parents to save for their children's education. There are two varieties: prepaid-tuition plans and college savings plans. Of the two the most typical is the **prepaid-tuition plan**, in which parents make regular payments (or pay a lump sum) into a fund that invests the money and agrees to help protect them against rising college costs. Prepaid plans pay a rate of return equal to the inflation rate at public universities in your state.

College savings plans are state-sponsored accounts in which money is invested by the state, and participants share in the earnings of the account. In college savings plans the rate of return depends on how good a job the state's money managers do with the funds they invest. Both plans offer tax advantages, and in most states you have a choice of only one. Usually the plans are applicable to state-funded universities and colleges. If the child goes out of state or to a private college, the money can be removed from the fund (sometimes with a penalty), or it can be transferred to a sister or a brother.

Prepaid-tuition plans and college savings plans are an option for people who have difficulty disciplining themselves to save in other ways. As in all savings decisions, investors should examine the prospectus and understand the plan, including the tax advantages and growth potential.

Objective 4: Discuss prenuptial agreements and the costs involved in weddings, divorces, child support, and widow(er)hood.

BEFORE MARRIAGE, INCLUDING PRENUPTIAL AGREEMENTS

Before marriage couples should discuss money. An inventory should be taken of assets and liabilities. A decision should be made about credit cards (separate or jointly held) and about whether to have a joint banking account or separate ones. Other decisions involve in whose name major assets—such as cars, houses, savings accounts and investments—will be held. The couple should familiarize themselves with state laws (available by contacting the local Bar Association or Legal Aid Society) since each state has its own laws governing marital and separate property and what happens when the two are mingled. Community property states (Arizona, California, Idaho, Louisiana, Nevada, New Mexico, Texas, Washington, and Wisconsin) have their own rules regarding ownership arrangements and estates.

Benefits and insurance policies (health, life, disability, auto, property) should be examined. If both the husband and wife have group plans then money may be saved by dropping one plan and adopting the other. Often, one plan offers better coverage at a lower cost or the premium for you as a couple is lower than your combined premiums as singles. If you have two cars, you will probably save money if you use the same insurer.

Much of the advice given earlier about establishing goals and emergency funds, organizing records, budgeting, and taxes should be reexamined at marriage. Taxes usually go up with marriage: called the **marriage penalty**. In most cases a married couple filing jointly will owe more tax than when they filed as singles. The government is working on this inequity that started out as being marriage-friendly, based on the idea that one of the spouses earned most of the money. If only one spouse is working or if one spouse earns considerably more than the other, newlyweds may experience a tax cut. But for most couples, with both working, a marriage penalty is more common.

Marriage penalty
Effect of a tax code that makes married people pay more than two single people.

The marriage penalty may not always exist. In recent presidential campaigns candidates have suggested doing away with the marriage penalty and inheritance taxes. So both of these tax situations may change.

If there are children, their futures should be considered. Wills, especially details about guardianship, should be examined. Inheritances and taxes will be affected by marriage.

In recent years there has been quite a bit of publicity about **prenuptial agreements (prenups)**, which are legal contracts. Each partner has his or her own lawyer negotiate and review the contract that spells out in advance of the marriage who owns what as well as how things will be divided in the event of a divorce. Although some are contested, most courts uphold prenuptial agreements. In the event of a marital break-up, a prenuptial agreement can save thousands of dollars in fact-finding and legal fees. A doctor or business owner may, for example, stipulate that in the event of a divorce, his or her practice or business or future earnings will not be affected, instead a lump sum will be offered to the spouse.

Prenuptial agreement
Legal contract between a future husband and wife about the splitting of assets in the event of a divorce.

The most likely people to have prenups are those with high salaries, significant assets, and children. For example, a prenup may be necessary if you have children in order to protect their inheritances. If the marriage fails, you will probably not want your assets to be used for child support payments to children from the spouse's previous marriage(s). Couples can work out their own prenuptial agreements, but as previously mentioned each should have his or her own lawyer. Each prenup should be evaluated on its own merits. A person may hesitate before signing for any of several reasons: it lacks romance, or it offers a poor settlement or does not offer some income if the marriage ends, or it does not have a time limit (many end at five years into the marriage), or the individual feels pressured to sign.

Marriage is not only for the young or the middle-aged. When seniors remarry there are special considerations regarding pensions, Social Security benefits, health care, estate planning, and taxes. Pension benefits may be forfeited on remarriage.

At any age if spouses have vastly unequal incomes or assets and different money philosophies, deep resentments can occur that do not bode well for the marriage. Marriage is a legal, financial, and emotional arrangement.

Weddings

Currently the average wedding costs about $18,000. In the U.S. about 2.4 million weddings are held each year costing a total of over $32 billion.[13] The wedding reception costs the most, followed by the engagement ring. A wedding cake and punch in the church hall or the parents' home can run less than $200. Compare this to elaborate catered dinners that can run over $200 per guest in Los Angeles, Boston, or

New York. January is the least popular wedding month, June the most popular for weddings.

Besides the wedding there are the related costs of gifts and honeymoons. Although newlyweds make up only 2.6 percent of all U.S. households, they account for 75 percent of all fine china, 29 percent of tableware, and 21 percent of jewelry and watches sold every year.[14] Wedding gift registries (in stores and online)—originally for china and crystal—have expanded into all kinds of goods from outdoor camping equipment to a federal registry, through the Department of Housing and Urban Development, that offers an interest-bearing savings account toward the purchase of a new home.

According to a survey in *Brides* magazine, the average American couple spends $3,200 on a nine-day domestic honeymoon and $3,800 on a foreign trip.[15] The most popular destination is Hawaii, followed by Florida. A new twist is a destination wedding where the entire wedding party meets at an exotic location to attend the wedding. An example of this is "The Ultimate Fairy Tale Wedding" at Disney World that costs about $100,000. The wedding party and guests get the Magic Kingdom to themselves, the bride-to-be rides in a carriage to meet her Prince Charming on a white stallion, and white chocolate cream cake is served in glass slippers while Tinkerbell flies overhead, scattering pixie dust.[16]

MARRIAGE

Dating offers couples the opportunity to find out if they have compatible saving and spending habits. Marriage will unite a couple's finances. First-time brides are on the average 25 years old. First-time grooms are on the average 27 years old. Early decisions should be made about who will pay bills and when each should be consulted about major purchases. Besides financial matters already discussed, if a husband and wife do not have life insurance, they may want to get it in order to protect each spouse. Extra homeowner's insurance may be needed on wedding and engagement rings.

If the bride takes the husband's name, or if they choose to hyphenate their last names the couple may need to change bank accounts, credit cards, driver's licenses, income tax forms, insurance policies, passports, social security cards, and voter registration cards.

DIVORCE

Divorces are emotionally wrenching experiences and it is difficult for many people to settle down enough to gather financial information. Besides assets of the marriage, future pension payments, stock options, profit-sharing plans and other deferred benefits count as divisible assets.

First marriages last on the average 11 years, whereas remarriages that end in divorce last about seven years. Divorce numbers are highest for men aged 30–34 and for women aged 25–29. States vary in their divorce laws (most notably community property versus noncommunity property states). Usually any property accumulated during a marriage will be divided. This will involve documenting what was brought into the marriage, by whom, and what was obtained after the marriage. A thorough accounting of all assets and liabilities will have to be made. The three basic questions to be settled are these:

1. How much is there to be divided?
2. What portion is liquid and income-producing?
3. How should it be split?

The purpose of the divorce settlement is to emerge from a divorce with one's quality of life and lifestyle intact. One's standard of living has to be considered as well as that of dependents and future costs such as college educations. Debt is divided as well as assets. Child support, alimony, custody of the children, and visitation are also part of divorce.

About 15–16 percent of divorced wives get alimony. Of these women 25 percent never see any of it, according to The National Center for Women and Retirement Research (NCWRR). Short-term alimony (two years of support until she can complete her college degree) is more common than long-term alimony.[17]

Rather than alimony, the trend is towards lump-sum, one-shot settlements so that each partner can move on unencumbered. In some cases the house they lived in will be sold and the proceeds split rather than allocating the house to one person or the other.

Attorneys, Mediators, Certified Divorce Planners

Divorce attorneys can charge $10,000 and more for a divorce between a middle-class couple with children, a house, and other assets.[18] If there are a lot of disagreements and the case goes to trial, the fee escalates. An alternative is divorce mediation which runs anywhere from a few hundred dollars for drawing up a separation agreement to $5,000 for a final divorce.[19] Generally divorce mediation is more agreement-oriented, less traumatic, faster, and less costly than litigation. There are also Certified Divorce Planners that help with financial settlements.

If there is considerable wealth and if children are involved, it is wise not to agree on a settlement without consulting financial and legal experts. Some mediators and Certified Divorce Planners have private practices. Some attorneys are also mediators or can recommend one as well as Certified Divorce Planners. A settlement is forever. The effects may be felt for a lifetime and if there are children and grandchildren the effects can extend into generations.

Financial Effects of Divorce

Both men and women will feel the financial effects of divorce, but, on average, women's income drops 27 percent after a divorce whereas men's income increases 10 percent.[20] Note that this is an average; a man with six young children, who divorces his unemployed wife, may end up paying high child support payments and alimony and, hence, have a vastly decreased income, compared to the national average.

In divorce proceedings an ex-husband's future earnings and pension fund are usually considered, especially if the ex-wife has no pension of her own. Another problem is who should get the house or whether the house should be sold and the profits divided. Sometimes divorcees end up with a house they cannot maintain and less cash than they need. Consider the following:

> Sara Fera, 54, a real estate sales rep in Marietta, Ga., now regrets that she traded away 25% of her ex-husband's profit-sharing plan in return for his share of their house when they divorced 11 years ago. She estimates that a booming stock market has ratcheted his plan balance to nearly $1 million. Meanwhile, the house, which Fera sold in 1987, netted her only $25,000 after the real estate commission and taxes. "It seemed fair to me at the time," laments Fera. "Now it just looks dumb."[21]

Child Support

There are three main types of child support:

1. Temporary: When parents first separate or file for divorce, the judge may order one parent to pay temporary child support.

2. Permanent: When the divorce is finalized, the court issues permanent child support (although this is a misnomer since circumstances may change the amount and payment schedule).

3. Family support: Some states allow child support and alimony to be combined into one payment called **family support**. There are tax considerations involved with this type. Family support is taxable income to the receiving parent and a deduction for the paying parent, whereas child support is neither a taxable income nor a deduction.

Family support
Child support and alimony combined in one payment.

Child support is designed to meet the minimum financial needs of the child such as food, clothing, education, and shelter. It is not set up to meet all the costs of raising a child. A **child support gap** exists, which is the difference between what the children cost to raise and what the parent receives. Although children deserve support, less than half of all women awarded child support get the full amount.[22] Each parent is assumed to provide support based on ability to pay which is based on gross income, adjusted gross, or net income. Generally speaking, gross income includes everything whereas adjusted gross and net income allow certain expense deductions, such as taxes and mandatory retirement contributions. Because child support often ends at age 18, it is a good idea to have an agreement stating who will pay for the child's college education. It is also wise to have the paying spouse purchase a life insurance policy covering the term of the child support payments or the college expenses.

Child support gap
Difference between what the parent receives and how much it costs to raise a child.

Usually child support payments are handled by circuit courts. The courts are dedicated to ensuring that children receive the financial support accorded them. To accomplish this, there are active enforcement programs to help a custodial parent collect overdue child support payments. A parent may qualify for these programs if child support payments are overdue or if he or she is not currently under contract for collection services with the appropriate child support department in his or her state. States rely on a standardized formula to determine a minimum level of child support payments, but the courts may choose to award more.

If payments are not received in a timely manner (monthly, semi-monthly, bi-weekly), the delinquent parent can be taken to court if payments are consistently late or stop. There is usually no charge to the custodial parent; however, a fee may be assessed to the delinquent parent to cover administrative costs. One of the results may be that a new payment plan will be set up. Another result may be to set up a direct deposit plan for child support and/or alimony payments. This service eliminates the hassle of lost or stolen checks, the inconvenience of going to a bank, and the worry of making regular payments while away on vacation or business.

Estate Planning in Second Marriages

Estate planning in a second marriage can be spelled out in a prenuptial agreement, especially in terms of how the marriage will affect the inheritance of children from the first marriage. In particular a later-in-life marriage should have wills or living trusts clearly stating what children and the spouse should receive.

Two possibilities requiring attorneys that may be considered in a second marriage are a Q-tip trust which passes income to the surviving spouse for life and preserves the principal for the children upon the spouse's death, and a "sprinkling" trust which grants a trustee discretion over the distribution of income from a trust.

Widow(er)hood

Becoming a widow (female) or widower (male) affects a number of financial decisions. Soon after the funeral or memorial service, the surviving spouse will be asked to settle the financial affairs of the deceased. Depending on the size of the estate, this can take a short amount of time and little expense or years in court and a lot of expense. The process may involve probate which will include a determination of the authenticity of the will. If there is no will, the estate will be distributed to heirs according to state law. If there are no children, the money usually goes to the surviving spouse, but in some states the deceased's parents get a share.

The surviving spouse should consult an attorney to help with the procedures and paperwork involved in probating the estate. The exception to this is if the estate was only a few thousand dollars, the surviving spouse could handle everything on his or her own under each state's Small Estate Statutes. Under these a court clerk helps with the paperwork. For more information about the Small Estate Statutes and probate procedures, the local Bar Association or Legal Aid Society can be contacted.

In addition the surviving spouse should contact the following:

* The Social Security Administration (800-772-1213) regarding benefits and the Department of Veterans Affairs (if the deceased was in the military).
* Insurance companies (see the next section about life insurance).
* Employers regarding benefits.
* Banks or brokerages where there were accounts.
* Creditors.
* The Internal Revenue Service regarding taxes.

In some cases children may be eligible for benefits through Social Security or through the Department of Veterans Affairs.

The widow or widower should be prepared at each contact or each meeting with professionals with a set of questions that relate not only to settling the deceased's estate, but also to gain advice about his or her own financial future. More on estate planning will be covered in the final chapter.

Collecting Life Insurance

The widow(er) should contact the agent who sold the policy, and the agent will send back a claim form. Then the widow(er) fills out the claim form and sends it and a copy of the death certificate back to the agent. Usually the money will be sent within a month.

Guidelines for Surviving Divorce and Widow(er)hood

Here are a few guidelines regarding how to survive divorce and widow(er)hood:

* Know what is rightfully yours.

CONSUMER ALERT

Divorcees and widows can fall prey to hot tips and unscrupulous financial salespeople. Crooks read the obituaries and divorce notices to locate victims. If the pitch sounds too good to be true (offering a sure return of 30% a year), it probably is a rip-off. Divorcees and widows should find their own financial advisors based on good recommendations, stability in the community, and reputation.

* Take precautions before the crisis strikes.
* Take it slowly, get good advice from trusted professionals.
* Adjust your lifestyle.
* Protect current and future assets.
* Update benefits, wills, and insurance.
* Create a new life: time and positive action will help.

Divorce and widow(er)hood are emotionally as well as financially stressful. According to Alexandra Armstrong, a financial planner in Washington, DC, "I've had dozens of professional women arrive in my office after a husband has died, or while they're in the throes of an ugly divorce, and they have no idea what insurance policies they have or what brokerage accounts they hold."[23] Such ignorance can lead to a poor divorce settlement or a loss of money from unclaimed benefits or from lost documents or bank accounts.

Objective 5:
Discuss Social Security and Medicare.

SOCIAL SECURITY

When you work, you pay taxes into Social Security (a federal government program); and when you retire or become disabled, you, your spouse, and your dependent children receive monthly benefits that are based on your earnings. Your survivors collect benefits when you die. So although people generally associate Social Security with the elderly, depending on your circumstances, you could collect from Social Security at any age. About 44 million Americans, one out of every six, collect some kind of Social Security benefit.

Out of every dollar you pay in Social Security and Medicare taxes

* 69 cents goes to a trust fund that pays monthly benefits to retirees and their families and to 8 million widows, widowers, and children of workers who have died.
* 19 cents goes to a trust fund that pays for some of the cost of hospital and related care of all Medicare beneficiaries.
* 12 cents goes to a trust fund that pays benefits to people with disabilities and their families.
* about 1 cent goes to pay for administering Social Security. (This adds up to 101 cents due to rounding off numbers.)

Money not used to pay benefits and administrative expenses is invested in U.S. government bonds, a very safe investment.

Your first link to the Social Security system is your number. The number is needed in order to get a job and pay taxes. Newborns are assigned numbers at birth.

Social Security's Internet address is www.ssa.gov. For specific information contact one of the 1300 Social Security offices around the country, or for a free estimate of your earnings and benefits write to the Social Security Administration, ATTN: Reports Clearance Officer, 1-A-21 Operations Bldg., Baltimore, MD 21235-0001.

Retirement Benefits

Although as mentioned, anyone—given certain circumstances—can collect Social Security, the majority of recipients are retired. The full Social Security benefit has traditionally begun at age 65, but beginning in the year 2003 the age at which full benefits are payable will increase in graduated steps from 65 to 67 over the course of many years. For example, a person born in 1950 will be eligible to receive full retirement benefits at age 66. Anyone born in 1960 or later will be eligible at age 67.

What about early retirement? Reduced benefits are available beginning at age 62. Starting at that age one would get 80 percent of the following benefits; at age 64, 93 1/3 percent of the full benefit. Beginning in the year 2003, the reduction will be greater as the full retirement age increases.

What about late retirement? There are two-fold gains from delaying retirement. First, extra income will usually increase "average" earnings, and the higher the average earnings, the higher the benefits. Second, a special credit is given to people who delay retirement. This credit is a percentage added to the Social Security benefit. The credit varies, depending on date of birth.

Benefits for Divorced Men and Women

Divorced persons can be eligible for benefits based on their ex-spouse's records. According to the rules, the ex-spouse must:

- have been married for at least ten years to the person claiming the benefit
- be at least 62 years old
- be unmarried
- not be eligible for an equal or higher benefit on his or her own Social Security record, or on someone else's Social Security record.

If an ex-spouse collects Social Security, does it affect the amount of any benefits payable to other family members or to that person directly? The answer to this is no. What about multiple marriages? Could someone collect on two or three spouses if she had been married to each for ten years or more? The answer to this is no. One benefit amount will be offered based on one's own earnings or one of the marriages—whatever is the higher amount.

There are also benefits to divorced widows and widowers. In order to qualify, two of the main rules are that the ex-spouse must be at least 60 years old (or 50 if disabled) and have been married for at least 10 years.

Some people who receive Social Security benefits have to pay taxes on their benefits; usually this is the case if their income is substantial. To find out more about this and other rules, contact the Social Security Administration.

MEDICARE

Medicare Federal health insurance program for elderly and many people with disabilities.

Medicaid Federal health care program for low income people.

Medicare *is the U.S.'s basic health insurance program* for people 65 or older and for many people with disabilities. Someone who gets Social Security at age 65 will be automatically enrolled in Medicare.

Medicare should not be confused with **Medicaid**, which is a health care program for people with low income and limited assets. Some elderly, poor people qualify for both Medicaid and Medicare. A free booklet on Medicare is available at Social Security offices.

CONSUMER ALERT

Note the importance of being married ten years in order to claim the benefit. Given this, a person would be unwise to divorce at nine years if this benefit will be needed (a person has not put in enough Social Security payments from his or her own working years).

Medicare has two parts:

1. Hospital insurance, including in-patient hospital care, skilled nursing facility care, home health care, and hospice care.

2. Medical insurance that helps pay for doctors' services, outpatient care, tests, home health visits, and other services.

For more details on Medicare, call their hotline at 800-638-6833. Many nursing homes have both Medicare and non-Medicare patients, which leads into the next discussion on the costs of elder care and living arrangements.

The average life expectancy today is 72 for a man and 79 for a woman. With parents living longer, the role of children has changed. Even if parents are self-sufficient today, in a few years they may need care. Eventually children have the responsibility of arranging for their care and financial affairs. When to step in and what to do are some of the most difficult decisions families make. Often aging parents are the first to notice that they need some help with their finances (confusion over bills, insurance payments). The first step is to determine needs and goals and then to find information. For example, a daughter or son (or other designated person) should find out about sources of income, assets, debts, insurance (particularly medical insurance), advisors, and where documents are kept.

Objective 6: **Discuss elder care and living expenses.**

LIVING ARRANGEMENTS AND CARE OPTIONS

The following comments will be explained within the context of children taking care of parents because this is typically what happens, but elders may seek living and care options for themselves or with the help of people other than children. Most elders do not want to move, and fortunately the growth of the home health care industry has made this option more viable than in the past.

Continuous daily care of parents include the following options:

Independent Living. Parents stay in their own home and take care of their own needs or move to a retirement community where healthy retirees live independently. Children may drop in once or twice a week to mow the lawn or do housework or to drive parents to the grocery store or to appointments.

Home care. Parents live in their own home or with friends or children, but require care. Home health care professionals may be hired or the children may be the sole caregivers.

Assisted-Living Community (also called Continuing Care Retirement Community). Parents live independently in a facility that provides some additional support such as meals or housekeeping. For example, each person may have a separate, private apartment and join others for lunch and movies and exercise classes. Nursing care or a nursing home facility may or may not be available. Usually parents start out in an apartment and then if their health declines they move to another section of the retirement facility that offers more care.

Adult Day Care Services. Parents go to a community-based program for part of each day or from 8 to 5, usually when their adult children are working. There are nearly 3,000 adult day care centers in the U.S.[24] They provide as a minimum recreation or activities, interaction, and a hot lunch.

Intermediate or Skilled Care Facilities. These facilities provide round-the-clock professional care. This category includes nursing homes. State departments of health or aging can provide lists of facilities, as can local doctors, hospitals, friends, religious personnel, and social workers.

In most cases the family serves as a major support system when it comes to providing care and assessing the options. The parents' physical and mental health and preferences should be considered along with the family's time, energy, and financial resources.

Finding Help

Community resources, such as area chapters of the Alzheimer's Association, and nationally based resources, such as the American Association of Retired Persons (AARP) are available to assist with elder care. The AARP offers publications on retirement planning and making health care decisions. They can be reached at 404-888-0077.

The U.S. Administration on Aging provides a toll-free ElderCare Locator Information Line at 800-677-1116. This is an especially useful resource when a parent is no longer fully independent but does not need a nursing home. The ElderCare Locator will refer the caller to a community agency that tells how to find local services such as adult day care, senior-citizen lunch groups, home-delivered meals, home health care, chore services, financial aid legal services, hospice programs, and transportation. This service can also recommend a care manager who will visit the parent, assess needs, and suggest services. Care managers can also be found through referrals from doctors and hospital social service offices or by calling the National Association of Professional Geriatric Care Managers at 520-881-8008. More resources can be found in the Electronic Resources section.

Costs, Including for Nursing Homes

Costs can be kept down by using community services, such as Meals on Wheels and by providing in-home care. Moving in with children is another option. According to a survey by the AARP, the number of parents living with children over age 55 doubled, to 4 percent, between 1992 and 1996.[25] If elders stay in their own home or if they move in with children, sometimes remodeling is necessary. Modifying is less expensive than adding on.

> Access is the biggest problem. Doorways may not be big enough for walkers or wheelchairs, the bath is not convenient, and there may be stairs (Lawrence Murr quoted by Sandra Dallas). [26]

Contrary to popular belief, *most elders do not live in nursing homes.* Even in nursing homes the trend is toward providing short-term care and release while focusing on sicker patients who need post hospital care. Most people can get by if in-home care includes home nursing and housekeeping services.

Regarding costs, *nursing home care can cost as much as $60,000 a year,* whereas home care can go as high as $35,000. The national average for nursing homes was $31,000 a year in the 1990s, or $86 a day.[27] Medicare picks up only a small fraction of these costs, and there are many stipulations about accessing these funds for nursing home care. About 25 percent of the Medicare population needs supervision and assistance.

An elder person's resources or insurance may cover all or some of the cost of care, or it may totally deplete their resources in a short time. In anticipation of long-term care costs, many people in their 50s and 60s are purchasing long-term care insurance (to be covered in Chapter 9). Another option is for the elder to get monetary support from adult children.

> Most people want to take care of their parents, and they do an admirable job at it. In approximately 80% of cases, any help the elderly require comes from fami-

ly members. And the National Council on the Aging reports that up to 11.8% of the U.S. workforce is involved in providing care for elders—a figure expected to jump dramatically as the number of aged people surges in the years ahead.[28]

Objective 7:
Distinguish between legitimate and less-than-legitimate charities.

CHARITIES

Caring for others goes beyond caring for family members. Americans donate nearly $6 billion a year to the 400 largest charities in the U.S. In recent years the Salvation Army has received the largest donations.

According to *Money* magazine, there are 626,226 charities vying for contributions and an additional 30,000 join their ranks each year.[29] So the question is, "How do you decide which ones deserve your dollars?" Your interests provide one guide because charitable giving is driven by the desire to help others in a way you feel is appropriate. Another approach, suggested by *Money* magazine, ranks charities on the basis of how efficiently dollars are spent. For example, in 1996 they rated the American Red Cross number one because an average of 92 percent of its income went to programs. Other magazines, organizations such as the Better Business Bureau (BBB), government consumer-protection agencies, and the state attorney general's office provide lists of charities and warnings about ones that are spending too much on fundraising and administration and too little for the people they are supposed to be serving. Any unethical or illegal tactics should be reported to the police, the BBB, or the state attorney general's office. Also charities can be checked by using the Web (see the Electronic Resources section for websites).

In order to distinguish between charities you want to give to and those you do not, ask yourself three questions:

* Is this the group I want to give to? Do I know their exact name and address and what they do?
* Do they use their money wisely? To find out, request a copy of the charity's annual report, budget, audited financial statements, and list of board members, or check them out through the BBB or other organizations, lists, or agencies.
* Do I endorse the charity's goals and methods?

Tax Benefits

Donating to charity can minimize taxes. To qualify for a deduction, charitable contributions must be made to U.S., state, or local governments exclusively for public purposes or to certain qualifying nonprofit charitable organizations. Not all non profit organizations qualify; the IRS determines whether donations to a particular organization are deductible.

Check with the IRS (www.irs.ustreas.gov/prod/bus_info/ed/)or with the organization in question, such as Goodwill, about whether donations to that particular organization are tax-deductible. Usually donations to political causes are not deductible.

CONSUMER ALERT

Many fraudulent charities use names that are very similar to legitimate charities. Be wary of any unsolicited phone calls asking for contributions.

For tax purposes you will need evidence showing your donation. *Never give cash, use a credit card or write a check and make the check out to the charity, not to an individual.* For donations under $250, a canceled check is all the record you will need. For donations over $250 you must obtain a contemporaneous written acknowledgment of the donation from the charity. Donations made through trusts, estates, or the giving of property are more complicated and should not be made without advice from a tax advisor or attorney.

Giving and Receiving Benefits

Even with all these cautionary notes, few things are more satisfying than giving to people and causes you believe in. Take the time to plan giving so that your dollars do the most good.

ELECTRONIC RESOURCES

This chapter covered many topics with resources on the Web. You can gain more information about college costs, admissions procedures, and financial aid from the College Assist website at **www.edworks.com/**. Also colleges and universities maitain their own websites. For a general search try Yahoo's Universities list at **www.yahoo.com/Education/Universities.**

For legal information regarding adoptions, marriages, and divorces check the American Bar Association home page at **www.abanet.org**.

The last two topics of elder care and giving to charities are loaded with websites. Six charity watchdog sites are:

* **www.give.org** (the National Charities Information Bureau). This organization monitors charities that receive $500,000 and more a year.
* **www.bbb.org** (Council of Better Business Bureaus).
* **money.mag.com** (*Money* magazine's data on how efficiently the nation's 100 largest charities use money).
* **www.nonprofits.org** (Internet Nonprofit Center) has a IRS database of all 1.2 million nonprofits; use it to find out if a group is registered as a charity.
* **www.contact.org** (A Nonprofit website Directory offers individual charity Web pages).
* **www.interaction.org** (the Interaction site offers information when a crisis breaks about relief charities).

FAQs

1. What do weddings cost and what is the highest cost in a wedding?

   ■ On average, weddings are running about $18,000, and the highest cost is the reception.

2. How can you tell a legitimate charity from a fake or quasi legal one?

   ■ Investigate before you give. Find out what the charity does with its money. Hesitate before giving to an organization you have never heard of before. Check them out on the Web (see the Electronic Resources section). Beware of phone solicitations.

3. How can I give a money gift to my nephews and nieces?

   ■ There are lots of ways to give money to children. The least expensive choices are cash or savings bonds in their names. More expensive choices include stocks and mutual funds, trusts, and custodial accounts.

4. Are state-sponsored tuition savings plans a good idea?

   ■ Yes, they are for people who need a forced savings plan in order to save for their children's education and for parents who are reasonably sure that their children will go to state-supported colleges and universities. These plans offer tax advantages. Before deciding whether to commit to a college plan, parents should examine the prospectus to read about the growth potential of the plan and tax advantages.

5. How does someone determine that aging parents need help with their finances?

   ■ Aging parents often are the first to notice they need help with paying bills or locating documents. There may be a crisis event, such as an automobile accident, that brings up insurance issues or a mundane occurrence, such as an unpaid bill, or confusion over a financial matter. In other situations the children may notice first that they have to take steps so that their parents are protected. Awareness and communication are key.

SUMMARY

The high cost of childbirth necessitates the need for health insurance.

Most adoptive parents adopt children through agencies.

Middle-income parents spend nearly $300,000 and upper-income parents nearly $430,000 to raise a child from birth to age 18, not counting college expenses.

Child care in the early years can cost as much as college tuition. Tax relief takes the sting out of child care costs.

There are a variety of ways to give money gifts to children (savings bonds, savings and custodial accounts). Tax aspects should be considered. Up to $10,000 can be given per child tax-free a year. In custodial accounts money is not directly put in the child's name; rather it is held in custody by an adult, bank, or trust company.

Two kinds of state-sponsored tuition savings plans are prepaid-tuition plans and college savings plans.

Prenuptial agreements are legal contracts specifying in advance of the marriage who owns what and how things will be divided in the event of a divorce.

About 15–16 percent of divorced wives get alimony.

On average, women's income drops 27 percent after a divorce, whereas men's income increases 10 percent.

Child support may be temporary, permanent, or coupled with alimony and called family support.

Reduced Social Security benefits are available starting at age 62. Full benefits start at age 65. In the year 2003 the minimum age for full benefits will begin moving slowly upward until it reaches age 67.

Medicare is the U.S.'s basic health insurance program for people 65 or older and for many people with disabilities. It should not be confused with Medicaid, which is a health care program for people with low income and limited assets.

Living and care options for elders include independent living, home care, assisted-living communities, adult day care services, and intermediate or skilled care facilities, including nursing homes.

Americans donate billions of dollars a year to charities. For tax purposes evidence of contributions should be kept.

REVIEW QUESTIONS

(See Appendix A for answers to questions 1–10)

Identify the following questions as true or false

1. Food is the biggest expense in raising children.
2. Child care costs can run as much per year as college tuition.
3. Au pairs are the least expensive child care option.
4. Most women get alimony in a divorce.
5. Experts agree that children's allowances should be based on their performance of household chores.

Pick the best answers to the following five questions.

6. The age of majority varies by state, but it usually is
   a. age 16 or 18.
   b. age 18 or 21.
   c. age 19 or 23.
   d. age 25 or 30.
7. At age _____ a child is no longer taxed at the parents' rate.
   a. 10
   b. 12
   c. 14
   d. 18
8. While a student is still in school, a parent can claim the child as a dependent on tax returns up to the age of
   a. 18.
   b. 21.
   c. 24.
   d. 30.
9. An option for giving substantial money gifts is to invest in __________________ through the Uniform Gifts to Minors Act.
   a. a custodial account
   b. a Q-tip account
   c. an AAFCS life account
   d. life insurance
10. Most experts agree that some of a college fund should be invested in ____________ for the greatest growth potential.
    a. corporate bonds
    b. government bonds
    c. real estate
    d. stocks and mutual funds

DISCUSSION QUESTIONS

1. When considering having a baby, what health insurance questions should be asked? In other words what should be covered?
2. If people are going to adopt, what avenues are open to them? Through what type of organization are most babies adopted?
3. What are some of the ways parents can set aside money for their children's college education?
4. If someone is getting divorced for the first time after two years of marriage and no children, how will the assets from the marriage probably be divided up?
5. In order to settle things after the death of a spouse, what should a widow(er) do?

DECISION-MAKING CASES

1. Lauren, a single parent who works full-time, has a three-year-old named Jasmine. Lauren is moving to a new city to take a promotion and has to make child care arrangements. What are her choices, and what should she find out from her employer?
2. Christian, a carpenter, is considering a state-sponsored prepaid tuition plan for his two-year-old son, Brian. What are the pros and cons of prepaid tuition plans? Using the calculator (multiplier) in Exhibit 7.1, imagine that Brian will be starting at a public college in 2016; what is the estimated cost for four years of college?
3. Brooke is planning her wedding and wants to estimate how much it and the honeymoon will cost. She does not want anything extravagant, but she wants to be realistic about what things will run. How much is average?
4. Jesse, owner of three copy stores, is marrying Angel. This is Jesse's second marriage and Angel's first. He has a daughter, Brianna, from his first marriage and would like to protect her and his businesses so Jesse wants a prenuptial agreement. Angel thinks it is unromantic. What should they do?
5. Aaron's mother is in declining health at age 80 and can no longer live on her own. He works full-time and lives alone. What are the options for her care?

GET ON LINE

Regarding charities, visit **www.give.org**, click on the National Charities Information Bureau (NCIB), and go to their reports on charities. Does the American Heart Association meet all NCIB standards? Select another charity; does it meet NCIB standards? Go to the website's Frequently Asked Questions section and scroll down until you find a question you are interested in. What is the question and what is the answer?

Web sites on elder care and housing include the following:

- **www.caregiver911.com**, which provides links to caregiver resources and to sites with information on chronic illnesses that affect the elderly.
- **www.n4a.org** for a nationwide database carrying lists of local providers of care and services for the elderly or www.nfacares.org for the National Family Caregivers Association.
- **www.aahsa.org** offers information on choosing a nursing home, continuing care retirement community, or assisted-living facility.
- at Extendedcare.com you can fill out a questionnaire about an elderly person's condition and get a list of appropriate nearby resources. **Careguide.com** also provides resource lists.

Instructions: Select two of these websites and compare what they offer. Suppose that you have to take care of an aging parent or grandparent with health or daily living problems; which of these websites provides the most valuable information?

To access InfoTrac follow the instructions at the end of Chapter 1.

Exercise 1: This chapter addresses a number of subjects related to the personal side of personal finance. After accessing InfoTrac, run a search on marriage or divorce. For example, if you select "marriage" for the subject search, many subcategories will come up such as Marriage Law or Prenuptial Agreements. Choose one of these subcategories and choose one of the articles related to marital finances and report on what you find.

Exercise 2: InfoTrac has many articles on college education. Choose a subject within college education such as college education costs. This will elicit over 150 articles. Scroll down the articles and find an article of interest to you and write a summary report. Be sure to document the article (author, title, source, date, page numbers) or print it out and attach it to your summary report. To give personal context to this subject, how do you feel about the cost of a college education? Do you think students should pay for college themselves, have help from their families, or a combination? If you have children or plan to have children, how will you save for their college educations?

Exercise 3: Under the subject "elder care" go to the economic aspects. You should find over 20 articles. Scroll down and select an article of interest to you, document it (see instructions in Exercise #2) or print a copy of the article, and summarize what the article says. Note that there are journal articles, general audience articles, and government publications on this topic.

chapter eight

property, liability, and automobile insurance

Providing Security and Protection

DID YOU KNOW?

* Almost one in twelve dollars in the U.S. economy is spent on insurance.
* There are 35 million auto accidents a year in the U.S.

OBJECTIVES

After reading Chapter 8, you should be able to do the following:

1. Understand the fundamentals of insurance and risk management.
2. Describe the four steps in the insurance management process.
3. Identify the different types of property and liability insurance.
4. Identify the different types of automobile insurance.

CHAPTER OVERVIEW

This chapter introduces the subject of insurance and risk management. It is the first chapter in a three-part series on insurance. The next chapter covers health, disability, and long-term care insurance. The third chapter covers life insurance.

In this chapter insurance and personal risk management are explained; the four steps in the insurance management process are described; and specifics about property, liability, and automobile insurance are given. Near the end of the chapter is an Electronic Resources section and a Summary.

Objective 1:
Understand the fundamentals of insurance and risk management.

INTRODUCTION TO INSURANCE AND PERSONAL RISK MANAGEMENT

Sound financial plans include protection from major risks that can threaten financial security. Failure to plan for risks may keep individuals from reaching their goals. Insurance is the main way people protect themselves and their belongings.

Insurance is a financial arrangement in which individuals concerned about potential hazards pay **premiums** (payments) to an insurance company that reimburses them in the event of loss or injury. *Insurance transfers the costs of risk from the individual to a larger group*, the insurer or the insurance company, which is better able to pay for losses. Insurance companies can afford to do this because they profit by investing the premiums they receive. **Actuarial risk** is the risk an insurance underwriter (an insurance company) takes (automobile accident, premature death) in exchange for premiums.

Insurance
Financial arrangement between individuals and insurers to provide protection from loss or injury.

Premiums
Payments to an insurer.

Actuarial risk
Risk an insurance underwriter takes in exchange for premiums.

Personal risk management is the identification and evaluation of risks faced by individuals and families and the selection of the most appropriate techniques or mechanisms for treating such risks. Because it is multifaceted, it can be thought of as a process.

Personal risk management
Process of identifying and evaluating risk faced by individuals and families.

Risk was defined in Chapter 1 as the possibility of experiencing harm, suffering, danger, or loss. In insurance **pure risk**, the threat of loss existing without the possibility of gain, is the primary concern. For example, if a car is totaled in an accident, you may have insurance to replace the car. Pure risk is distinguished from another kind of risk, **speculative risk** which involves the possibility of gain as well as loss such as investing in stocks and mutual funds.

Pure risk
Risk that has a threat of loss without possibility of gain.

Speculative risk
Risk that has the possibility of gain and loss.

Regarding risk, a person could try to avoid it (avoiding high-risk occupations, not smoking), reduce it (burglar alarms, smoke detectors), assume it (emergency funds, savings), or transfer the risk to others through insurance. Before issuing a policy, an insurance company must determine if the requirements are met for an insurable risk.

Sources of risks are called **exposures**. Driving an automobile exposes you to more risk than if you stayed home. **Perils** are events that cause financial loss, such as car accidents. An example of exposures and perils can be diagrammed as

Exposures
Sources of risks.

Perils
Events that cause financial loss.

item	*exposure*	*peril*
car	driving	accident

Individuals, before taking action, consider the potential of exposures or perils. For example, someone may refuse to fly in small airplanes, fearing that they are more likely to crash than large jets. This person is judging the likelihood or severity of loss. Rather than avoiding risk, another way an individual may try to handle risk is through insurance. Risks commonly associated with insurance include

Property Risks: Individuals that own, rent, or use property are exposed to the risk that the property may be damaged, destroyed, or stolen.

Liability Risks: Individuals can be held legally liable for a number of reasons, ranging from personal acts that cause bodily harm or property damage to others to liability from the negligent use of boats or cars. U.S. society has become increasingly litigious, which means that individuals are often held financially liable for damages resulting from a vast array of circumstances/events. **Liability** is a claim on the assets of a business, organization, or individual. Since this is a personal finance book, the emphasis will be on what liability claims may mean for individuals and families. Liability insurance is usually associated with automobile or homeowner's insurance policies. The way it works is that it compensates others for losses you caused (for example, if you are at fault in an auto accident).

Liability
Claim on the assets of a business, organization, or individual.

Personal Risks: These include potential losses involved with the health and well-being of individuals from such things as unemployment, injuries on the job and medical problems, or the premature death of the family breadwinner (the primary revenue provider). For example, the family breadwinner who purchases a $300,000 life insurance policy does so to protect his or her family against economic loss due to death. This risk has to do with (1) the possibility of loss and (2) uncertainty about when the loss will occur.

To summarize, *the connection between risk management and insurance is that most people rely heavily on insurance as the main way to handle risk.* Through buying insurance they transfer their personal risk to the insurer.

The Insurance Industry

Over one million people are employed by 35,000 insurance companies. Some of the well-known insurers are State Farm, Allstate, Aetna, Prudential, and Citigroup. Almost one in every twelve dollars spent in the U.S. economy goes to pay for insurance.[1]

Agent
Represents the insurer and acts on the insurer's behalf.

Nearly all insurance is sold through insurance agents who work on commission, although this is changing with the advent of buying insurance online. An **agent** represents the insurer and has the legal authority to act on the insurer's behalf. Independent agents represent several insurers. Exclusive agents represent only one insurer or group of insurers. When you pay premiums, part of it goes to the agent in the form of a sales commission.

Pooling
Spreading of loss of a few over an entire group.

A basic concept in the insurance industry is **pooling**, the spreading of losses incurred by a few over an entire group. Average loss is substituted for the actual loss. For example, if in a community one $200,000 house per year burns down and 1000 people pay homeowner's insurance (including the owner whose house burned down), their pooling of money will take care of the one house that burns down. Each person may pay $200 a year for the insurance but this is better than being at risk for having to pay for a whole new house if your house is the one that burns down.

Another basic concept in the insurance industry is that of risk transfer. This is a process in which risk is transferred to the insurer, who has a greater ability to pay than the individual. For example, in a car accident the liability lawsuit will usually be transferred to the insurer (a large company with immense assets) instead of the car owner (an individual with limited assets).

Typical Coverage

People buy insurance to protect themselves and those they care about and their assets. Insurance is available for nearly everything, including race horses' legs and pianists' fingers, but the more typical types of insurance are for property (including homeowners and renters), liability, automobiles, health, disability, long-term care, and life insurance.

Other Insurance Fundamentals

Policy
Contract between individual or group and insurer.

An individual joins a risk-sharing group (an insurance company) by signing a contract called a **policy**. The insurance company that holds the contract agrees to assume a certain amount of risk for a fee called the *premium* paid for by the insured or the policyholder. Each policy can be broken down into parts. Exhibit 8.1 shows the five basic parts of insurance contracts that are described next.

Declarations
Description of what is insured.

1. **Declarations** describe what is being insured, the name of the insured, the premium to be paid, and policy and time limits.

EXHIBIT 8.1
Building Blocks of Insurance Policies

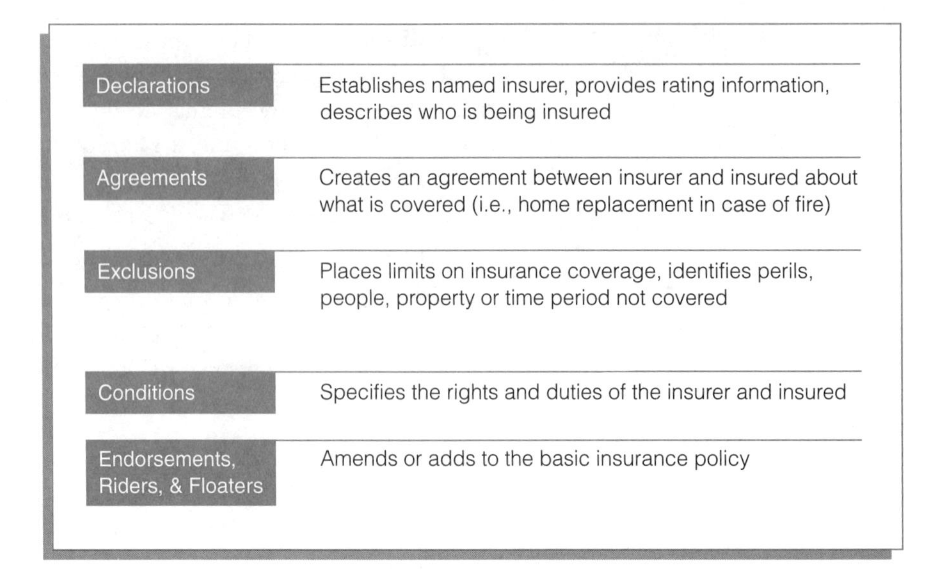

Declarations	Establishes named insurer, provides rating information, describes who is being insured
Agreements	Creates an agreement between insurer and insured about what is covered (i.e., home replacement in case of fire)
Exclusions	Places limits on insurance coverage, identifies perils, people, property or time period not covered
Conditions	Specifies the rights and duties of the insurer and insured
Endorsements, Riders, & Floaters	Amends or adds to the basic insurance policy

Agreements
Description of extent of coverage.

Exclusions
Items insurer will not pay for.

Conditions
Obligations imposed on policyholder and insurer.

Endorsements
Amend or add to a policy.

Deductible
Amount policy holder pays toward a loss before insurance coverage begins.

Coinsurance
Sharing of an insurance risk between two or more insurers.

2. **Agreements** describe the coverage provided; for example, in a homeowner's policy the insurance company may agree to replace a house if destroyed by fire.
3. **Exclusions** list items for which the company will not pay, such as flood damage.
4. **Conditions** delineate a set of obligations imposed on both the insurance company and the insured in the policy. For example, the company will not replace a house or business destroyed by fire if the homeowner or business owner purposely set the building on fire.
5. **Endorsements** (riders and floaters) amend or add to a basic insurance policy (expensive jewelry, collections).

Depending on the type of policy there may be a **deductible**, the amount you pay toward a loss before the insurance coverage begins. For example, a deductible of $100 on a health or dental plan means that the person being treated will pay for the first $100 of the bill, and the insurance company will pay for amounts above that in a given time frame. In general higher deductibles mean lower premiums.

Another part of an insurance contract may be a statement about coinsurance, the sharing of an insurance risk. **Coinsurance** is common when claims could be of such size that it would not be in one company's best interest to underwrite the whole risk. Usually the underwriter is liable up to a stated amount, and the consumer's liability is for amounts above that limit. What this means for consumers is that they should read policy clauses about the amount of coverage (full or partial) and whether coinsurance is involved or needed particularly in regard to such hazards as fire or water damage.

PLANNING A PERSONAL INSURANCE PROGRAM

Individuals confronted with thousands of companies and policies eventually have to make decisions about what insurance to purchase. These decisions depend on how much protection is needed, how much can be afforded, and how much is covered by employers.

Insurance coverage should be based on individual's and family's needs and goals. According to the nonprofit National Insurance Company Organization (NICO), more than nine in ten Americans purchase and carry the wrong types and amounts of insurance coverage. Usually this is because an individual's insurance program has not kept pace with needs, goals, resources, and lifestages (an elderly woman still paying for maternity coverage). The following sections provide general guidelines based on lifestage.

Early Stages

A single or married worker starting out needs to investigate employer benefit packages to see what insurance is offered and fill in the gaps with outside coverage for assets, such as homes, cars, and personal property and for increased personal coverage (if necessary) in life and health insurance. People who are self-employed, between jobs, not working or not covered by a family or school policy should investigate group coverage offered through social, religious, or professional associations or businesses such as the Automobile Association of America (AAA). Group policies usually offer lower rates than individual policies.

At age 30 drivers are viewed as adults by the auto insurance industry. As long as an individual's driving record is good, car insurance rates should substantially decrease. Insurers do not always keep up with birthdays, it is up to the insured to find out if he or she qualifies for a reduced rate at age 30.

Families with dependent children especially need sound insurance coverage. As a minimum, families should have homeowner's or renter's insurance, life and disability insurance for the main wage earner(s), and adequate health care (including maternity benefits if children are expected).

Later Stages

When an individual's or a family's income and possessions increase and children grow older, insurance needs change. Teenagers mean more drivers, more cars, and higher insurance rates. More life insurance is needed, especially if there are dependent children, and provisions should be made for the care of aging parents. Health insurance needs to be revisited with long-term care and retirement benefits taken into account.

As one approaches midlife, a person's entire financial situation should be evaluated, including insurance coverage. For example, there may be less need for life insurance. Another way to look at midlife to later life is that a successful investment program, in a sense, helps the family become self-insured (protected).

Objective 2:
Describe the four steps in the insurance management process.

THE INSURANCE MANAGEMENT PROCESS

Although the description of the "earlier" and "later" stages is useful as a broad overview of insurance over the lifecycle, ultimately insurance has to be tailored to the individual or family. One way to approach this is to go through the four steps in the insurance management process: awareness, analysis, action, and evaluation (see Exhibit 8.2).

Awareness

First, individuals have to become aware that insurance is needed before they can move on to the next step. Awareness may occur through an advertisement; an insurance

EXHIBIT 8.2
The Insurance Management Process

AWARENESS → ANALYSIS → ACTION → EVALUATION

salesperson's call; an announcement of an open-enrollment period at work (offering new or extended policies); a disaster that befalls a friend, relative, or community; or different events, such as moving, the birth or adoption of a child, a job change, marriage, or divorce. For example, in coastal communities people may not pay much attention to their homeowner's policy until a hurricane is approaching; then suddenly they want to know if they are covered and for how much.

Analysis

Second, individuals should ask themselves "Is there a peril? What do I need to protect? How much protection do I need?" Usually insurance goals are linked to life changes, such as launching a career, having children, or making a major purchase. For example, a person may want to purchase additional insurance to protect a new car or a diamond engagement ring.

Analysis revolves around five fundamental questions:

1. What should be insured (perils, property, location)?
2. Who should be insured (people covered)?
3. What kind of insurance is needed?
4. From whom?
5. For how much and how long (amount, cost, and time)?

Regarding the fourth question, one way to sort through the choices is to ask friends, relatives, colleagues, financial advisors, and employers for their recommendations. It is also important to then examine the ratings of insurance companies, such as the ratings given by A. M. Best in Exhibit 8.3. The preferred ratings are A++, A+, and A. Do not buy insurance from a company with an A. M. Best rating of less than A-. In addition such magazines as *Consumer Reports*, *Money*, and *Kiplinger's Personal Finance* regularly rate customer satisfaction with various insurance companies.

Question number five regarding how much insurance one needs, has two inherent pitfalls: underinsurance and overinsurance. A person is underinsured if he or she seriously underestimates the cost of something, such as insuring a $200,000 house for only $80,000 or having a life insurance policy on the major breadwinner for only $50,000 when he or she earns over $100,000 a year. Overinsured means when someone has too much insurance on something like an out-of-date computer or old used car that current income could cover replacing.

As part of a final analysis, one should not overlook the fine print in insurance contracts. Know what you are buying. Read the policy and ask questions about what you do not understand. Besides the agent ask your human resource advisor at work, an attorney, a friend, or relative to explain the policy to you. What you want to do is avoid the following:

> Just ask Pamela Rogers-Amy and Sara Harrington. Both had big losses from fires. Both were sure they would be fully covered by their insurance. Both now say they will never shop for homeowner's insurance in the same way again. "People don't know what they've bought," says Ms. Harrington, whose Lexington, Mass., house was struck by lightning in 1995 and reduced to a shell by the resulting fire. "It is a cliché to say you have to read the fine print, but you do."[2]

Action

Third, in this step action is taken by seeking more information and selecting insurers and policies. Since there are so many insurance companies and each of them offers a

EXHIBIT 8.3
Rating Systems of Insurance Rating Agencies

	A. M. Best	Standard & Poor's Duff & Phelps	Moody's	Weiss Research
Superior	A++ A+	AAA	Aaa	A–
Excellent*	A A –	AA+ AA AA –	Aa1 Aa2 Aa3	A A – B –
Good	B++ B+	A+ A A –	A1 A2 A3	B B – C –
Adequate	B B –	BBB+ BBB BBB –	Baa1 Baa2 Baa3	C C – D –
Below Average	C++ C+	BB+ BB BB –	Ba1 Ba2 Ba3	D D – E –
Weak	C C – D	B+ B B –	B1 B2 B3	E E –
Nonviable	E F	CCC CC C, D	Caa Ca C	F

* Do not buy insurance from a company listed below this level.

variety of policies, this can be an overwhelming task. To simplify things once you have narrowed down the choices, some experts suggest that you have only one insurance agent and buy broad coverage.[3 & 4] Know the name of your agent and the insurance company that insures you. They may be different from the name of the insurance agency or Third Party Administrator. Broad coverage means that you would buy all-inclusive policies such as life insurance and major medical coverage. Everyone needs some form of health insurance, regardless of the state of the person's health.

Examples of types of insurance that you probably do not need include:

- life insurance on children (unless the child provides significant income to the family, such as a child actor) or life insurance for a single adult with no dependents.
- specific disease insurance, such as cancer insurance (unless the disease runs in the family and there is a stronger-than-average possibility).
- hospital indemnity insurance that covers extended hospital stays (most people are in and out in one to three days).
- flight insurance.
- rental car insurance (your regular car insurance should cover this; an exception may be renting a car when abroad).
- vacation policies (you may need cancellation insurance if it is likely you may cancel on a cruise, for example, but otherwise your regular insurance should cover loss or medical expenses).
- accidental death and dismemberment that covers reimbursement for medical costs resulting from accidents (you would be better off purchasing a comprehensive health policy and life insurance).

To narrow down the list of companies and broad policies needed, cost and coverage comparisons between companies should be made. This is an important step because one company could charge triple what another one does for the same coverage. Online websites have made it far easier to make cost comparisons. The Electronic Resources section gives addresses of websites that will do the comparisons for you on automobile insurance, life insurance, etc. Although price is important, coverage is too. The online sources will provide information about coverage as well. Keep in mind that the difference in cost between "bare bones coverage and a policy that will cover everything is not usually large," says Judith W. Lau, a Wilmington, Del., financial advisor.[5]

As mentioned earlier, group plans—because of the increased purchasing power of a group—usually offer less expensive coverage than an individual can obtain alone. Group plans, in particular for health, disability, and life insurance, are available from employers and also from professional or service organizations.

Once minimum coverage goals are met, then additional coverage should be sought, especially if income has increased and there are dependent children. Long-range goals may include the establishment of a consistent insurance purchase routine and the development of a trusted relationship with an insurance agent.

Another part of this step is keeping copies of insurance records. Store extra copies in a safe-deposit box or with a friend or attorney. Keep all copies of insurance policies (past and present) on financed items until the loan is satisfied and you have received your title or deed. It can be very hard to go back and find these records.

If you decide to change coverage or companies, make sure the new coverage is in effect before canceling the old policy. Also be aware that if you do not maintain coverage on a financed car or mortgaged home, the financial institution will purchase insurance to protect its own interest. You will be charged for this very expensive coverage, which may not adequately cover your needs.

When purchasing insurance, get a detailed receipt and never pay cash if it can be avoided. Try to pay insurance premiums by check or money order made out to the company or premium finance company.

Evaluation

In this last step an individual looks back on the decisions and actions made earlier. If there was a car accident or some other form of trouble, was the insurer prompt at filing claims? Did the insurer live up to expectations? Is the cost of the policy still appropriate?

> Each insurer offers its own mix of coverage and pricing, depending on the market it is targeting, and each has its own personality when it comes to claims. "Some companies seem to try to help you. Other companies play hardball," says Robert Hunter, director of insurance for the Consumer Federation of America in Washington.[6]

In a final evaluation or review a person can feel that insurance was a waste of money. If nothing happened and insurance was paid into but not accessed in any way, could the money have been better spent? Undoubtedly yes, but when the insurance was purchased, the purchaser had no way of knowing nothing was going to happen. A positive way to look at insurance is that having it relieves the anxiety or fear of financial disaster that could result from—or even lead to—an accident or another peril. How do you put a price on peace of mind?

Denied Coverage?

A person who applies for an insurance policy may be turned down, especially for health, life, or disability insurance if there is an existing medical problem, or for homeowner's insurance if he or she lives in a high-risk area. If a person is denied coverage, he or she can

1. Ask the insurer why he or she was turned down.
2. Request a copy of one's medical information file. You can request a copy of your medical information file by writing to the Medical Information Bureau at P.O. Box 105, Essex Station, Boston, MA or by calling them at 617-426-3660.
3. Try other companies.
4. Check for availability and costs before buying a new house. For example, flood insurance may not be available if the house is in a high-risk area, and this is important because over 90 percent of all disasters in the U.S. are flood-related. An alternative is to pay premiums for Federal Flood Insurance to protect your home and belongings.
5. Check with the state department of insurance to find out what the problem might be.

Objective 3:
Identify the different types of property and liability insurance.

Property insurance
Pays for losses to homes and personal property due to a number of causes.

Liability insurance
Pays for losses from negligence resulting in bodily injury or property damage to others for which the policyholder is responsible.

PROPERTY AND LIABILITY INSURANCE

Property insurance pays for losses to homes, autos, boats, recreational vehicles, and other personal property due to theft, fire, vandalism, natural disasters, or other causes. **Liability insurance** pays for losses from negligence of individuals that results in bodily injury or property damage to others. Liability insurance usually pays the damages awarded by the court and attorney fees and other defense costs.

Since there is considerable overlap between property and liability insurance and because they are usually sold together, they will be discussed in the next section. The discussion begins with a major part of property and liability insurance: homeowner's insurance.

Homeowner's Insurance

A homeowner's policy protects you from the cost of damage to your home from fire, falling trees, lightning strikes, and other perils. *The most important feature of homeowner's insurance is the replacement-cost coverage.* This is the coverage that would pay to rebuild your house if it were completely destroyed. You should insure your home for at least 80 percent of its replacement cost. Purchase enough insurance to protect your home's structure and your personal possessions, not the land your house is on. An online source or an insurance agent can calculate replacement-cost coverage. Arriving at an accurate figure is crucial because you do not want to be under- or overinsured.

> "A lot of people get confused between replacement value and market value, which tends to fluctuate more," says Bryan Freeman, assistant vice president of fire and casualty underwriting at State Farm Insurance Co. in Bloomington, Ill. "If your home is in a high-priced subdivision of town, the land value may far exceed the replacement cost."[7]

As another example of the importance of location, on the East Coast replacement value can be as much as 30 percent higher than market value.[8] So if you live in an area where property and home values are rising rapidly, this should be considered when assessing replacement value.

Another consideration is if there have been major renovations or improvements to the home or the property. If this is the case, a homeowner policy should be updated to reflect the higher replacement value. For example, if coverage was for $120,000, then a $20,000 family room was added, the replacement cost should be increased to $140,000.

Also the homeowner should be aware of any local ordinance or law affecting the home. Check with the agent because the insurance company may not be responsible for paying the cost of upgrading your home to meet ordinance or building code changes. The agent must offer coverage to bring your home up to standard, and you must sign a rejection form if you do not want it.

When buying a house that requires financing (as most do, versus paying cash outright), the buyers are required by the lender to purchase insurance to protect the lender's financial interest in the property. If homeowners do not have any insurance and their $100,000 house burns down, they are out $100,000. If they had homeowner's insurance for the correct replacement value, the company would give them the money to rebuild.

Homeowner Policy Forms The Insurance Services Office (ISO) is an industry association that provides standard policy forms. The ISO Homeowners Insurance Program uses the following forms (summarized in Exhibit 8.4):

1. HO-1: Basic.
2. HO-2: Broad.
3. HO-3: Special.
4. HO-4: Contents Broad (covers renters' contents and liability).
5. HO-6: Unit Owners (used for condominiums or cooperatives).
6. HO-8: Modified (used for older, urban homes).

The numbering of this list seems unusual, but it is accurate. HO-7 never existed and HO-5 is no longer in use. Because of the length and complexity of the forms, only the main points and definitions are given in this chapter. In addition it should be noted that the aforementioned HO forms cannot be used for mobile homes. The ISO provides other forms for mobile homes.

To interpret Exhibit 8.4, look at the column under HO-1 Basic. In HO-1 Basic the dwelling must be owner occupied, and at least the minimum coverage of $15,000 must be purchased. Given that the average price of a new home is over $100,000, the $15,000 minimum coverage is far too low. As shown in the exhibit, under HO-1 Basic the **limited named perils** covered are fire, lightning, hail, aircraft, vehicles, smoke, vandalism and malicious mischief, theft, breakage of glass, explosion, riot, and civil commotion.

Limited named perils Usual basic perils covered by insurance.

Broad named perils Extended perils coverage.

*The list is more extensive under HO-2, thus it is termed **broad named perils**.* Coverage includes (1) fire and lightning; (2) windstorm and hail; (3) explosion; (4) riot and civil commotion; (5) damage by an aircraft; (6) damage by vehicles owned or operated by people not covered by the homeowner's policy; (7) damage from smoke; (8) vandalism or malicious mischief; (9) theft; (10) falling objects; (11) weight of ice, snow, or sleet; (12) collapse of a building or any part of a building; (13) accidental discharge or overflow of water or steam; (14) sudden and accidental tearing asunder, cracking, burning, or bulging; (15) freezing; (16) loss from artificially generated electrical current; and (17) volcanic eruption.[9]

Under HO-3 the minimum coverage is $20,000, and there is the term open perils. This refers to coverage for direct loss to property except for certain excluded items specified in the policy, such as no protection from loss of property due to wear and tear by animals kept by the owner.

Provision	HO-1 Basic Form	HO-2 Broad Form	HO-3 Special	HO-4 Renters	HO-6 Condominiums	HO-8 Older Homes
Owner Occupied	Yes	Yes	Yes	—	—	Yes
1-or 2-Family						
Minimum Limits	$15,000	$15,000	$20,000	$6,000 Personal Property	$6,000 Personal Property	$15,000
Perils Insured Against	Limited Named Perils	Broad Named Perils	*Open Perils* for Dwelling, Broad Named Personal Property	Broad Named Perils for Personal Property	Broad Named Perils	Same as HO-1
A—Dwelling	Amount Purchased	Amount Purchased	Amount Purchased	—	—	Amount Purchased
B—Other Structures	10% of *A*	10% of *A*	10% of *A*	Not insured	Not insured	10% of *A*
C—Unscheduled Personal Property	50% of *A*	50% of *A*	50% of *A*	Amount Purchased	Amount Purchased	50% of *A*
Unscheduled Personal Property at Secondary Residence	10% of *C* or $1,000 Whichever Greater	10% of *C* or $1,000 Whichever Greater	10% of *C* or $1,000 Whichever Greater	10% of *C* or $1,000 Whichever Greater	10% of *C* or $1,000 Whichever Greater	10% of *C* or $1,000 Whichever Greater
D—Additional Living Expenses	10% of *A*	20% of *A*	20% of *A*	20% of *C*	40% of *C*	10% of *A*
E—Comprehensive Personal Liability	$100,000	$100,000	$100,000	$100,000	$100,000	$100,000
F—Medical Payments to Others	$1,000 per Person	$1,000 per Person	$1,000 per Person	$1,000 per Person	$1,000 per Person	$1,000 per Person

*Note: This table describes the general provision of policies. Specific provisions vary by state and by insurer. Most people should buy an "all perils" or "broad named perils" policy (HO-2, HO-3, HO-4, HO-6) depending on type of residence or whether rented or owned that gives the broadest protection rather than a limited named perils policy.

EXHIBIT 8.4
Basic Coverages of Homeowners' Policies

Form HO-8 is the newest one and was developed to meet the needs of people moving back into older neighborhoods and remodeling homes. The market value of the older home may be $75,000, but its replacement cost may be $160,000. In other words if there is damage it may cost a lot more to repair or replace an older home than a newer one. Under the HO-8 coverage insureds can collect on a cost-to-repair basis (repair hardwood floors or replace a destroyed slate roof with shingles). The rates for HO-8 are higher than for HO-1, but it offers worthwhile protection.

In Exhibit 8.4 note the designations A for Dwelling, B for Other Structures, C for Unscheduled Personal Property, D for Additional Living Expenses, E for Comprehensive Personal Liability, and F for Medical Payments to Others. These are explained further as

- *"A"* refers to the residence premises.
- *B"* includes garages, greenhouses, tool sheds, and other structures separated from the dwelling by clear space or connected by a fence or utility line. Structures for business purposes are not included so that, for example, an

architect's office in a separate structure on the residence premises would not be covered by the homeowner's policy.

* *"C"* covers personal property owned or used by the insured anywhere in the world, such as clothes, appliances, luggage, and bicycles. Certain things are not covered to the amount that they may be worth, such as expensive computers, cameras, golf clubs, guns, silverware, antiques, musical instruments, stamp and coin collections or jewelry. These should be covered by a floater policy discussed earlier. There are some exceptions. For example, sound equipment in automobiles is usually not covered under a floater policy because it is covered by auto insurance. Individual homeowner policies should be checked to determine what is and what is not covered regarding personal property.
* *"D"* covers the increased cost of living that occurs when a residence is damaged and uninhabitable. Under "D" the insureds are permitted to maintain their standard of living which means they can rent a dwelling comparable to what was damaged.
* *"E"* provides basic comprehensive personal liability (CPL) coverage of about $100,000. What this means is that typical homeowner's policies provide $100,000 in liability coverage. CPL covers bodily injury and also property damage.

Beyond the basic coverage, endorsements can be added to cover business pursuits, personal injury, personal umbrella liability, or watercraft liability. For example, business people or physicians may add personal umbrella liability insurance in the millions of dollars to protect against catastrophic loss. Not everyone needs personal umbrella liability and some who need it have difficulty getting it, such as entertainers or athletes. Later in the chapter there is a section on umbrella policies.

* *"F"* provides medical payments to others. The basic amount of coverage is $1,000 of no-fault medical payments. A basic policy may also offer $250 of no-fault property damage coverage. Since these are low amounts, a homeowner may want to pay more for added protection.

Summary Checklist For Choosing a Homeowner's Policy *The overriding guiding principle is to choose a policy that meets your needs and keeps the cost to a minimum. Here is a checklist of main points to consider:*

* Purchase enough insurance to protect your home and personal possessions. Understand the difference between replacement cost and actual cash value coverage.
* Make an inventory (list and videotape) and valuation of contents (receipts) to help determine the type and amount of insurance that you need. Store the inventory in a safe-deposit box or other secure place.
* As stated earlier, the goal is to be appropriately insured, neither underinsured nor overinsured. Insurance agents will explain the options and the costs. Find out if additional coverage is needed.
* Check with your agent at least once a year to make sure your policy provides adequate coverage and inform your agent of any additions or major improvements to your home.
* Also check with your agent about discounts for fire and burglary safeguards or for hurricane shutters or tornado shelters.
* Another feature to inquire about is additional living expenses that pay the extra costs (motel and restaurant bills) if your home is damaged and you can no longer live there. To verify a claim, all receipts will have to be submitted.

Renter's Insurance

Form HO-4 is designed for renters. Renter's insurance covers the major perils, provides liability protection and additional living expenses if the dwelling is rendered uninhabitable by one of the covered perils. HO-4 protects the insured from losses to the contents of a dwelling rather than the loss of the dwelling itself. So in the case of an apartment building the owner should have insurance for the building, and the renter should have insurance to cover his or her possessions. An often overlooked part of HO-4 is the covering of personal property. This includes property that may or may not be in the dwelling such as a college student's books or bicycle stolen on campus.

Renter's insurance is relatively inexpensive yet fewer than half of renters in the U.S. have it. Annual premiums typically cost less than one car payment.[10] If you rent an apartment or a house, consider all of your possessions that could be lost or damaged from fire or smoke, vandalism, burglary or theft, natural disasters (earthquakes, hurricanes, tornadoes), water damage from a leaky roof or ruptured pipes, roof collapses, lightning or explosives.

The more possessions are worth, the more important it is to have renter's insurance. Exhibit 8.5 provides an inventory checklist of typically owned items. Use the checklist to determine how much coverage you need, keeping in mind that most policies have limits as to how much they will reimburse. For example, a typical policy will only cover $1,000 for jewelry and watches and $2,000 for guns.

Reducing Homeowner's and Renter's Insurance

Homeowners and renters have several ways to save money. Issuers may offer price breaks if you:

* raise the deductible.
* use the same insurer (*if you buy most or all of your insurance from the same company, they may offer the best deal; it also simplifies recordkeeping*).
* buy or rent new construction or construction of a certain type. For example, brick homes in the eastern United States are given a discount because they better withstand wind and hurricane damage than other types of construction. Frame homes in western states are given a discount because they withstand earthquake damage better. Reduced rates, as much as 15%, are often offered for newly constructed homes because they are in better shape than older homes.
* have a security system and smoke detectors.
* have a house or apartment near a fire hydrant or fire station.
* do not smoke (some insurers offer reduced rates to nonsmokers).
* are a senior citizen or if you belong to a group—such as an alumni, fraternal, or business organization—that offers group coverage.[11] Insurance companies' "senior" discount can begin as early as age 50.

CONSUMER ALERT

A house or apartment that is vacant for 30 consecutive days may be considered "abandoned." If you are going to be away for this length of time, your insurance company may not cover burglary, vandalism, or a fire. Have a housesitter or someone regularly check your property. Police will do this on request.

Some policies do not cover natural disasters such as earthquakes so you may need extra coverage. As a general rule do not assume coverage, ask.

EXHIBIT 8.5
Inventory Checklist

INVENTORY CHECKLIST

Item	Estimated Current Value	Estimated Replacement Value*
Appliances	______	______
Art/Collectibles	______	______
Bicycles	______	______
Books	______	______
Clothing	______	______
Computers	______	______
Exercise Equipment	______	______
Furniture	______	______
Guns	______	______
Jewelry	______	______
Silverware/China	______	______
Stereo/tv/phones	______	______
Tools	______	______
Miscellaneous	______	______

*Replacement value is the dollar amount needed to replace damaged items of like kind and quality, without deducting for depreciation. Policies specify the maximum dollar amount allowed to replace an item. For example, a stolen computer of a make and model that is no longer available, could be replaced with a new computer up to a stated maximum dollar amount. For the purposes of filling out the chart, replacement values are higher than current values.

As a final note, as a student you may be covered by your parents' insurance even if you maintain a separate residence so check with them before purchasing renter's insurance. Since both renters and homeowners can be sued, the next section discusses liability and lawsuits.

LIABILITY AND LAWSUITS

Thousands of people in the United States are sued and sue others every day. Sometimes lawsuits result from misunderstandings that can be resolved through alternatives such as mediation (dispute resolution) rather than going to court.

Liability lawsuits may arise from property damage, bodily injury, libel, or other damages caused by the insured. The insurance company agrees to pay for damages if they are awarded by the court up to the limited amount stated in the policy. Regarding home-related injuries, lawsuits can arise from a slip on an icy sidewalk at your house or someone falling off the balcony of your apartment. If an injury takes place on your property, you could be held liable for the injury if you failed to take reasonable precautions to avoid a foreseeable risk (called negligence). For example, in the icy sidewalk situation if you knew the sidewalk was slick and did nothing about it and did not warn an expected visitor, you might be liable.

A homeowner's policy with liability coverage should protect you. Medical payments coverage may apply also even if you were not at fault. If you are sued in a case such as this, your insurance company will probably provide a lawyer (at no charge) to protect your interests. If you are sued, contact your insurance company to find out about your coverage, to find out if the company will provide a lawyer, if necessary, and to give the company notice that a possible claim may be made.

If a lawyer is not provided, or if mediation is not an alternative, then you may need to hire a lawyer. This decision rests on the complexity of the case and the amount of money involved. Most colleges and universities provide free initial legal consultation to students, so this may be a first step if there is no insurer.

Insurance and Home Offices

As more people work from their homes full-time or part-time, there is a growing awareness of the special insurance needs that this may cause. For example, homeowner's or renter's policies should be checked to see how much equipment is covered. A standard amount is $2,500. This may be too low to cover several computers, faxes, printers, and copy machines. There are also liability issues. If a client is hurt while visiting a home office, the homeowner's or renter's policy may not cover the accident. A rider could be added on to an existing policy or a separate policy may be necessary. A standard separate at-home business policy costing approximately $300 a year would provide $1,000,000 in liability, $15,000 in coverage of contents, and twelve months of "reasonable" business interruption coverage, to cover the cost of renting a temporary office. A home worker who does not see clients may need only a rider to cover equipment.

LIABILITY AND AUTO INSURANCE

Objective 4: Identify the different types of automobile insurance.

In the United States the single, largest line of property and liability insurance is automobile insurance. With 124 million cars on the road, the possibilities of injuries, damage, and repairs are enormous. The Insurance Information Institute (www.iii.org/) reports that there are 35 million auto accidents a year. Fortunately half of these accidents are merely fender benders but even these invoke some costs, depending on the type of car and the type of accident.

The first part of this section discusses personal auto insurance and the latter part discusses uninsured motorists. Although the word "auto" or "car" will be used throughout, it also refers to pickups, recreational vehicles, sports utility vehicles, and minivans.

Personal auto policy
Legal liability arising out of the ownership or operation of a covered auto.

The **personal auto policy (PAP)** drafted by the Insurance Service Office (ISO) is used throughout the United States to cover legal liability arising out of the ownership or operation of a covered automobile. It includes insurance against physical damage losses to the vehicle. The coverage is divided into four parts:

* Part A: Liability coverage.
* Part B: Medical payments coverage.
* Part C: Uninsured and underinsured motorists coverage.
* Part D: Physical damage (collision and comprehensive) coverage.

These are described in Exhibit 8.6 and the following sections.

EXHIBIT 8.6
Parts of a Personal Auto Policy

Coverage	Persons Covered	Property Covered
A. Liability		
(1) Bodily injury*	Relatives living in insured's household injured in auto accident	N/A
(2) Property damage*	Relatives living in insured's household injured in auto accident	Autos and other property damaged by insured while driving
B. Medical		
(1) Medical payments	passengers injured	N/A
(2) PIP (in no-fault states)	passengers injured	N/A
C. Uninsured/ Underinsured Motorist	Anyone driving insured's auto with permission and insured's family members driving non-owned autos with permission	N/A
D. Physical Damage		
(1) Collision	Drivers of insured's auto with insured's permission	Insured's Auto
(2) Comprehensive	N/A	Insured's Auto and contents

Note: N/A means not applicable. *Must buy at least the state minimum. There are also optional forms of coverage such as roadside assistance and towing.

Part A: Liability Coverage

The most important part of PAP is the liability coverage because it protects a covered person against a lawsuit or claim arising out of the negligent operation of an automobile. The insurer pays any damages for bodily injury or property damage for which an insured is legally responsible because of an accident.

Single limit
Auto insurance pays up to a stated amount.

Split amount
Auto insurance coverage is split between three categories.

Typically coverage is **single limit**, which means up to a certain amount—for example $300,000—will be paid. Coverage can also be a **split amount**, for example, $200,000/$300,000/$100,000. In this case $200,000 is allocated for each person, $300,000 for each accident, and $100,000 for property damage. The insurer also covers legal costs.

A critical aspect of liability coverage is the definition of covered auto. Here are the four classes of vehicles considered to be covered autos:

1. *Any vehicle listed in the declaration.* The declaration is the first page of the PAP. It states the name of the insured, coverages provided, and amount of insurance.

2. *Any vehicle acquired by the insured during the policy period.* If you own one car and buy a second, the second will be insured for up to 30 days after purchase. Within that time the insurer has to be informed of the purchase, and a premium for the second car must be paid.

3. *A trailer owned by the insured* (a boat trailer attached to an auto involved in an accident).

4. *A temporary substitute auto* (a loaner car while the owned car is undergoing repairs).

Have you ever been told by someone that you cannot drive their car or they cannot give you a ride because of their insurance coverage? If so, they were referring to the concept of insured persons. Here are the four groups insured under the liability section of the PAP:

1. *The named insured and family members.* Typically this includes the husband, wife, and children. If children are away from home temporarily (attending college) they are still covered under the parents' policy.

2. *Any other person using the named insured's covered auto is insured provided that person can establish a reasonable belief that permission to use the covered auto exists.* So if someone borrows your car without your permission, that person is not covered.

3. *Any person or group legally responsible for the acts of a covered person while using a covered auto is insured.* If an employee drives his or her own car on business and injures someone, the employer is covered for lawsuits or claims.

4. *Any person or group legally responsible for the named insured's or family members' use of any automobile or trailer* (other than a covered auto or one owned by the person or group). For example, if an employee borrows a fellow employee's car and injures someone while on a work errand, such as mailing a package, then the employer is covered for any lawsuit or claim.

Part B: Medical Payment Coverage

Medical payment coverage pays for reasonable medical, dental, x-ray, surgical, and funeral expenses incurred by a covered person within three years from the date of an accident. Expenses are covered up to the policy limits such as $1,000 or $10,000.

Part C: Uninsured and Underinsured Motorists Coverage

Uninsured motorist coverage
Pays for bodily injury caused by an uninsured motorist or hit-and-run driver, or negligent driver with an insolvent insurer.

Many, but not all, states require automobile insurance therefore it is possible to be involved in an accident with an uninsured motorist. **Uninsured motorist coverage** pays for bodily injury (and in some states property damage) caused by an uninsured motorist, by a hit-and-run driver, or by a negligent driver whose insurance company is insolvent.

The insurer pays compensatory damages (medical bills, lost wages, permanent disfigurement) that an insured is entitled to receive from the owner or operator of an uninsured motor vehicle because of bodily injury caused by an accident. Whether the claim is paid or not and for how much depends on whether the uninsured motorist was legally liable, the stated maximum limit in the policy, and whether the people involved agree or not. The claim may be subject to arbitration.

Part D: Physical Damage Coverage (Collision and Comprehensive)

Coverage for damage to autos provides basic coverage for physical damage to, or theft of the insured's auto. Two optional forms of coverage are available: collision loss and comprehensive.

Collision
Violent striking of an auto by another object.

In a personal auto policy, **collision** means the violent striking of an automobile with another object, such as a tree, another car, or a building. The word "object" is defined quite broadly by PAP. Collision losses are paid regardless of who is at fault.

Comprehensive loss includes all physical damage losses except collision losses and other specified losses (a tree limb falls and damages your car, fire, theft, hail, riots, contact with a bird or animal, earthquakes).

Part D also covers an auto, pickup, van, or trailer not owned by or furnished or made available for the regular use of the named insured or family member, while it is in the custody of or is being operated by, the named insured or family member. Thus if a person occasionally drives a borrowed van, that person's physical damage insurance will cover the borrowed van. But if the van is regularly used, then Part D coverage does not apply.

Exclusions, Deductibles, and Other Provisions of PAP

Policies have deductibles and many exclusions. For example, if you carpool to work or drive your neighbor's child to school, these activities may be excluded from your coverage. Also two-wheeled vehicles, such as motorcycles and minibikes may not be covered.

Regarding deductibles, many policies have a deductible of $250 or $500 to cover collision or comprehensive losses. This is one of the reasons why many people don't bother filing a claim for a slight fender bender or a scratch on their car door. The higher the deductible, the lower the annual premium, so many people choose to have a higher deductible.

Other provisions include payments, bankruptcy, territory of coverage, and termination of policies. For example, if you rent a car in another country, PAP may not cover it, so you may need to purchase separate insurance.

If You Are in an Accident

If you are involved in an accident, follow these steps:

1. Stop your car and determine if anyone is hurt. If so, call the police or highway patrol; they will call the nearest medical unit.
2. Give help to the injured, but avoid moving anyone.
3. Provide the police with whatever information is required. Find out where you can obtain a copy of the police report. Your insurance company will need it.
4. Protect the scene of the accident. Protect yourself, others, and the cars from further damage. Highway pile-ups are not uncommon, one accident may lead to many others.
5. Make notes and record the details of the accident; write down the names and addresses of drivers and passengers, and license numbers. Obtain the names and addresses of witnesses, police, and emergency personnel. Even if there are no injuries or property damage, notify your insurance agent immediately.

If you are judged to be legally liable for an accident, you may be held responsible for property damage, hospital and medical payments, rehabilitative care, lost income, and the pain and suffering of any injured person. If the cost exceeds the amount of your liability insurance, you may have to pay the rest.[12]

State	Minimum Liability Limits	Compulsory Coverage	State	Minimum Liability Limits	Compulsory Coverage
Alabama	**20/40/10***	FR only	Montana	25/50/10	BI, PD
Alaska	50/100/25	BI, PD	Nebraska	25/50/25	BI, PD
Arizona	15/30/10	BI, PD	Nevada	15/30/10	BI, PD
Arkansas	25/50/15	BI, PD	New Hampshire	25/50/25	FR only
California	15/30/5	BI, PD	New Jersey	15/30/5	BI, PD, PIP, UM
Colorado	25/50/15	BI, PD, PIP	New Mexico	25/50/10	BI, PD
Connecticut	20/40/10	BI, PD, PIP, UM	New York	25/50/10	BI, PD, PIP, UM
Delaware	15/30/10	BI, PD, PIP	North Carolina	25/50/15	BI, PD
District of Columbia	25/50/10	BI, PD, UM	North Dakota	25/50/25	BI, PD, PIP, UM
Florida	10/20/10	BI, PIP	Ohio	12.5/25/7.5	BI, PD
Georgia	15/30/10	BI, PD	Oklahoma	10/20/10	BI, PD
Hawaii	25/40/10	BI, PD, PIP	Oregon	25/50/10	BI, PD, PIP, UM
Idaho	25/50/15	BI, PD	Pennsylvania	15/30/5	BI, PD, PIP
Illinois	24/40/15	BI, PD, UM	Rhode Island	25/50/25	BI, PD, UM
Indiana	25/50/10	BI, PD	South Carolina	15/30/5	BI, PD, UM
Iowa	20/40/15	20, 40, 15	South Dakota	25/50/25	BI, PD, UM
Kansas	25/50/10	BI, PD, PIP, UM	Tennessee	25/50/10	FR only
Kentucky	25/50/10	BI, PD, PIP	Texas	20/40/15	BI, PD
Louisiana	10/20/10	BI, PD	Utah	25/50/15	BI, PD, PIP
Maine	50/100/25	BI, PD, UM	Vermont	20/50/10	BI, PD, UM
Maryland	20/40/10	BI, PD, PIP, UM	Virginia	25/50/20	BI, PD, UM
Massachusetts	20/40/5	BI, PD, PIP, UM	Washington	25/50/10	BI, PD
Michigan	20/40/10	BI, PD, PIP	West Virginia	20/40/10	BI, PD, UM
Minnesota	30/60/10	BI, PD, PIP, UM	Wisconsin	25/50/10	UM
Mississippi	20/40/5	FR only	Wyoming	25/50/20	BI, PD
Missouri	25/50/10	BI, PD, UM			

Note: *FR* only is financial responsibility only; *BI* is bodily injury liability; *PD* is property damage liability; *PIP* is personal injury protection; and *UM* is uninsured or underinsured motorist.

*Automobile liability insurance may be quoted as these figures:

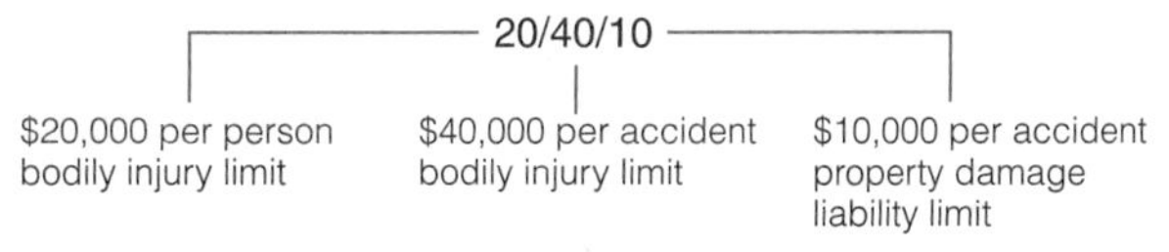

Source: Insurance Information Institute, www.iii.org. Reprinted with permssion.

EXHIBIT 8.7
State and District of Columbia Financial Responsibility Limits and Compulsary Auto Insurance Requirements

To File a Claim

If there has been an accident or your car is damaged by fire, a natural disaster (a hurricane or an earthquake), or vandalism, you will need to file a claim. Take these steps: Phone your insurance agent, ask your agent how to proceed, fill out the necessary forms, and keep records of expenses (repair bills) and paperwork.

Auto Insurance and the Law

Many states require automobile insurance but even when they do require insurance there are underinsured motorists. Because of this, in every state and in all the provinces of Canada, there are laws designed to solve the problem of the uncompensated victim of financially irresponsible automobile drivers. Automobile insurance financial responsibility limits and compulsory auto insurance requirements by state are given in Exhibit 8.7.

No-Fault Insurance

No-fault insurance
After an auto accident, each party collects from his or her insurer.

No-fault insurance provides that after an auto accident, each party collects from his or her own insurer regardless of fault. Thus the parties involved in the accident do not have to prove blame. *About half of the states and the District of Columbia offer no-fault insurance.* One of the reasons for this option is to reduce the number of frivolous lawsuits.

No-fault insurance benefits usually cover medical bills up to a limit, a stated percentage of an injured person's earnings, service expenses, funeral expenses, and survivors' benefits, but no payments are made for pain and suffering. Since no-fault laws vary by state, consumers should check what this insurance covers.

COST OF AUTO INSURANCE

Auto insurance is expensive. Costs range between a few hundred dollars to several thousand dollars per year. Insurers use the following factors to determine cost:

1. *Territory.* City drivers and dwellers in primarily urban areas generally pay higher rates than rural drivers. Certain states have considerably higher auto insurance rates than others. According to the *1998 Statistical Abstract of the United States*, the five most expensive places to purchase auto insurance are New Jersey (on average over $1,000), District of Columbia, Hawaii, New York and Connecticut. The least expensive are Nebraska, Wyoming, South Dakota, Idaho, and North Dakota (on average about $400). Multiply these figures by a four car family and you can see that auto insurance can be a high yearly expense.

2. *Age, sex, and marital status.* Drivers under the age of 20 have proportionately higher accident rates so their insurance rates are higher. From ages 20–25 men have more accidents than women, so men usually pay more for coverage. At ages 65 and up the accident rate increases, so insurance rates go up. Those over the age of 70 may have a difficult time obtaining insurance if they are applying for the first time or changing insurers. Regarding marital status, sometimes lower rates are offered to married persons than to single persons.

3. *Use of the automobile.* The greater the number of miles typically driven, the higher the rates.

4. *Driving record.* A poor driving record means higher rates or denied coverage.

5. *Number of cars insured.* A multicar discount may be offered.

6. *Good student discount.* Discounts may be offered for high grades usually a B average or above or the taking of an approved driver education course.

7. *Year, make, and model of the auto.* Cars are rated on age and susceptibility to damage, the cost of repair, and the rate at which they are stolen. For example, in recent years, Honda Accords and Toyota Corollas were more frequently stolen than Saturns. Insurance agents can tell you the estimated rate, or you can go online to find lists of expected insurance costs.

Ways to Lower Auto Insurance Costs

Ways to lower insurance costs include the following:

1. Comparison shopping—rates and coverage. Remember that the cheapest is not always the best. Make sure you purchase at least the minimum limits of Personal Injury Protection (PIP) and Property Damage Liability coverage, as required by law. In addition, depending on the state, you may need other coverage, such as Bodily Injury Liability.

2. Asking for higher deductibles.

3. Dropping collision and/or comprehensive coverage on older cars. Once the car is worth less than $2,000, you would be better off putting the money you save on premiums into a savings account toward a new car.

4. Buying a low profile car. Avoid cars that are easy to steal, desirable for joyriding, and expensive to repair. Luxury cars are often stolen for parts. For more information write to the Insurance Institute for Highway Safety, 1005 North Glebe Road, Arlington, VA 22202 and ask for the Highway Loss Data Chart.

5. Taking advantage of low mileage discounts.

6. Considering insurance cost when moving.

7. Asking agents about possible discounts, including seat belts, antilock brakes, driver history, and air bags.

8. Receiving a discount if all the family's cars are insured by the same company.

On a final: note a minor fender bender can increase premiums several hundred dollars a year, so one suggestion would be to pay for small accidents with your own cash and use insurance for major catastrophes. A DUI (driving under the influence) conviction can double or triple premiums.

Umbrella Policies 

Is basic home and auto insurance enough to cover most people's needs? Yes, but some experts would recommend more, especially if your income is high and you have a lot of assets to protect. An umbrella policy can provide liability coverage of $1 million or more at a cost of $200 to $400 a year.[13] It goes into effect after the underlying liability policy's coverage limits have been exceeded.

> "The cost of an umbrella policy continues to be a function of the size of jury awards, which have undoubtedly been on the rise," says Mr. Pinchak [vice president of underwriting for Prudential Property & Casualty in Newark, N.J.]. "But it's not as though the premiums have doubled or tripled either, and for anyone who can afford it, the policy offers tremendous peace of mind."[14]

Usually people purchase an umbrella policy—as the quotation indicates—to protect themselves against the possibility of a large jury award in a lawsuit. An umbrella policy can also protect the insured from events not usually covered in homeowner's or auto insurance policies such as slander (oral statements that damage a person's reputation) and libel (written or pictorial statements that damage a person's reputation).

ELECTRONIC RESOURCES

Comparing rates from companies used to take dozens of telephone calls, but now it is a mouse click away. For example, Rachel, age 30 with a 10-year old car, always thought she was paying too much for her car insurance, but she did not want to take the time to call around. Instead she went online and whittled her premium down from $880 a year to $480. Not only was the price lower (because of her age and good driving record), but she received more coverage.

Buying insurance online is still not as easy as buying a book online because, for example, a policy has to be specific to the driver, location, and type of car. So how does it work? A large site such as Quotesmith (**www.quotesmith.com**) offers instant quotes on auto, life, medical, dental, Medicare supplemental insurance, annuities, and other insurance from more than 300 underwriters. Other large sites are:

* **www.ecoverage.com** (Ecoverage) offering auto, homeowner, and marine insurance with an underwriting partner, their policies are for sale only on the Web.
* **www.insweb.com** (Insweb) offers comparison based upon leads from insurers and agents.
* **www.insuremarket.com** (Quicken Insurance) offers quotes from 21 carriers of auto, health, home, disability, small business, life, and other insurance. Promises instant, binding quotes.
* **www.ebix.com** (Ebix.com) offers reverse auctions in which insurers and agents bid for customers.

A comprehensive information site is maintained by the Insurance Information Institute at **www.iii.org/**. It explains how to cut costs and how to determine how much insurance is needed. It also provides links to insurance companies and service organizations.

The Insurance Regulatory Information Network (**www.irin.org**) provides general information on the industry as well and has links to most state departments of insurance. The federal government's Consumer Information Center Catalog website—with information on over 200 topics, including insurance—is **www.pueblo.gsa.gov.**

Most companies have their own websites that can be reached at **www.companyname.com**. Progressive Corporation and Geico Corporation are examples of companies that let customers buy policies online.

For coverage of a multitude of insurance topics, check out the Insurance Corner Index at **www.insurancecorner.com/myhome.htm.** For Insurance News Network's provision of Standard & Poor's Claims Paying Ability Reports check **www.insure.com/ratings/reports/**. For a wide variety of links targeted for consumers, agents, brokers, risk managers, and actuaries go to Insurance & Risk Management Central at **www.irmcentral.com/**

The Independent Insurance Agents of America provides a number of guides to insurance, including homeowner's and autos at **www.liaa.lix.com**. For general information about insurance, the National Insurance Consumer Helpline can be reached at 800-942-4242. For information about flood insurance call the National Flood Insurance Program at 800-427-4661.

1. Should homes be insured at (a) the amount it would take to rebuild them, or (b) the resale value of the homes?

   ■ The answer is (a), the amount it would take to rebuild them.

2. Does a homeowner's policy include flood insurance?

   ■ No. Over 90 percent of all U.S. disasters are flood related, but a homeowner's insurance policy does not cover flood damage; only Federal Flood Insurance does.

3. If a pipe bursts and there is water damage, will a homeowner's policy cover the damage?

   ■ Yes. The HO-3 covers accidental water damage from plumbing systems.

4. Do homeowner policies cover earthquake damage?

   ■ No. Typically a person in an earthquake-prone area will need additional coverage beyond a basic homeowner policy.

5. If someone slips on a neighbor's sidewalk, is the neighbor covered by the homeowner policy?

   ■ Usually, yes. The policy will pay for damages, legal and medical.

6. Do novice drivers have higher insurance rates because they are at greater risk for traffic accidents?

   ■ Yes. Beginning drivers often have automobile accidents and unfortunately, auto accidents are the number one cause of death for 15 to 20-year olds.

SUMMARY

Insurance is a financial arrangement in which individuals that are concerned about potential hazards pay premiums (make payments) to an insurance company which reimburses them in the case of loss or injury.

Risks associated with insurance include property risks, liability risks, and personal risks.

The connection between risk management and insurance is that most people rely heavily on insurance as the main way to handle risk.

Nearly all insurance is sold through agents who work on commission although there is a trend toward online insurance shopping that eliminates agents—the consumer deals directly with the company.

Individuals become insured when they sign a contract called a policy.

The insurance management process involves four steps: awareness, analysis, action, and evaluation.

Group plans are usually less expensive than individual plans. Property insurance pays for losses to houses, autos, boats, recreational vehicles, and other personal property due to theft, fire, vandalism, natural disasters, or other sources.

Liability insurance pays for losses from negligence of individuals resulting in bodily injury or property damage to others.

The most important feature of homeowner's insurance is the replacement-cost coverage (the amount it would cost to rebuild your home).

The guiding principle in selecting a homeowner's policy or a renter's policy is to choose one that meets your needs and keeps costs at a minimum.

In the United States the single largest line of property and liability insurance is automobile insurance.

The coverage of the personal auto policy (PAP) is divided into four parts: liability, medical, uninsured and underinsured motorist, and physical damage to autos.

Many, but not all, states require automobile insurance. About half of the states and the District of Columbia offer no-fault insurance.

Auto insurance costs between a few hundred dollars to several thousand dollars a year. Insurance companies consider a number of factors (make, model, and year of car, driving record geographical area), when setting the cost of insurance. Discounts are available.

At age 30 auto insurance rates reduce substantially.

Umbrella policies extend coverage beyond what is normally offered in homeowner's and auto insurance policies.

REVIEW QUESTIONS

(See Appendix A for answers to questions 1–10.

Identify the following statements as true or false

1. Actuarial risk is the risk an insurance undertaker takes in exchange for premiums.
2. The most important feature of homeowner's insurance is the replacement-cost coverage.
3. An insurance contract is called a policy.
4. The most important part of PAP is the liability coverage.
5. All states require auto insurance.

Pick the best answers to the following five questions.

6. ______________ is the main way people protect themselves and their belongings.
   a. Insurance
   b. Casualty safety
   c. Indenture
   d. Cost-plus margin
7. ____________(payments) are made by the policy holder to the insurer.
   a. Deductibles
   b. Annuities
   c. Premiums
   d. Indexes
8. ____________ is a claim on the assets of a business, organization, or individual.
   a. Incontestability
   b. Liability
   c. An exclusion
   d. Negligence
9. Which of the following is not one of the steps in the Insurance Management Process?
   a. Awareness
   b. Analysis
   c. Action
   d. Assumption
10. __________pays for losses to homes, autos, boats, recreational vehicles, and other personal property due to theft, fire, vandalism, natural disasters, or other causes:
   a. Property insurance
   b. Collision insurance
   c. Palimony
   d. Liability insurance

DISCUSSION QUESTIONS

1. Besides price what other factors should a person purchasing insurance consider? What are some of the ways that homeowner's or renter's insurance can be reduced? How can auto insurance be reduced?
2. The insurance management process has four steps including the second step which is analysis. Part of analysis is deciding on the right amount of coverage. Give an example of how someone could be underinsured and an example of how someone could be overinsured.
3. Using Exhibit 8.5, estimate how much money you would need to replace all your possessions in case of a fire in your campus residence, home, or apartment. What item would cost the most to replace? What item would represent the most sentimental or personal loss to you?
4. If there is a party at your apartment and someone falls off the balcony and breaks an arm, would you be liable? Would it matter or not if you had liability insurance? If you had liability insurance coverage up to $5,000 and the injured person is awarded $10,000 by the court, who would pay the extra $5,000?
5. Auto insurance varies widely. Using Exhibit 8.7, look up your state of residence and the minimum liability limits. What are they, and what does each category mean? In what states are the auto insurance rates the highest? the least? Why does so much variability exist?

DECISION-MAKING CASES

1. Anthony has been involved in an auto accident. The driver of the other car is named Briana. Anthony is at fault, and his car sustains $1,000 in damage as does Briana's. They live in a state that does not offer no-fault insurance. He has auto insurance with a $250 deductible. No one was hurt in the accident. How much will his company pay for the car repairs (his and hers), and how much will he pay given his deductible?
2. Samantha, age 19, is moving out of her dormitory to an apartment that she will share with three other students. She wonders whether she should get renter's insurance or not. Who should she talk to about it, and what factors should she consider before making a decision about purchasing insurance?
3. Adeline, a 25-year-old with a modest income, recently inherited $20,000 from her great uncle. She is going to buy a new car with the money and her parents told her to look into auto insurance rates besides the cost of the car. How should she go about obtaining and comparing insurance rate information?
4. Jose Franco is starting out his career after college graduation. Since he is no longer covered under his parents' policies, he needs to buy basic insurance in his own name or get it through his employer. As a minimum what kinds of insurance does he need to have?

GET ON LINE

Activities to try:

- Visit **www.quotesmith.com** and click on instant auto insurance quotes. Find the best rate for your car or a family car. Report on what you find.
- Go to **www.insuremarket.com** and click on insurance helpline, ask a specific question about homeowner's or renter's insurance, and report on the answer you receive.
- Visit **www.iii.org/** and click on auto. Scroll down the list of categories and pick one, such as "teen drivers." Report on what you find.

INFOTRAC

To access InfoTrac, follow the instructions given at the end of Chapter 1.

Exercise 1: This chapter introduced the topic of insurance. Using InfoTrac, go to the subject of "insurance." You will find encyclopedia excerpts, periodical references, subdivisions, and related topics. For an overview of the insurance industry click on "insurance companies." What are the trends in the industry? How has online access affected the insurance industry?

Exercise 2: The chapter also discussed the specifics of home and auto insurance. Using InfoTrac, go again to the subject of "insurance." Click on the subtopic of "insurance policies," then read up on the articles on *either* home or auto insurance. What do homeowner policies currently offer? If you select auto insurance, determine if you live in a no-fault-insurance state. Find an article discussing no-fault insurance. Is the article mostly positive or negative about no-fault policies?

chapter nine

health care, disability, and long-term care

Acquiring Appropriate Coverage at Minimal Cost

DID YOU KNOW?

* America spends 13.7 percent of its GNP, about $1 trillion, on health care each year.
* Most Americans have employer-based health insurance.

OBJECTIVES

After reading Chapter 9, you should be able to do the following:

1. Explain the rise in health care costs.
2. Identify three types of health care providers.
3. Describe the five main types of private health insurance policies.

CHAPTER OVERVIEW

As noted in the previous chapter, insurance is the main way people protect themselves and their assets from risks. Because of the potential severity of loss due to illness or injury, everyone needs some form of health insurance or a health care plan as a financial safeguard. Many of the insurance principles covered in Chapter 8 apply also to this one, such as finding out about a company's rating (A+, A-, etc.) before buying a policy.

The main purposes of this chapter are (1) to introduce the subjects of health care costs and coverage and (2) to explain health insurance plan options including disability income insurance and long-term-care insurance. Managed-care plans (HMOs, PPOs, and POSs) are discussed, as well as the insurance issues related to child-birth, children, marriage, and divorce. Whether your employer offers plans, or whether you purchase them on your own, you need to be aware of the range of costs and benefits. Five main types of private health insurance are covered, along with two rarer types. The government programs of Medicare and Medicaid are also discussed. The chapter concludes with future trends and an Electronic Resources section.

Objective 1:
Explain the rise in health care costs.

HEALTH CARE COSTS FOR INDIVIDUALS AND FAMILIES

The pie chart in Exhibit 9.1 shows changes in who is paying America's health care bill. According to the exhibit, the government is paying an increasing share, as is private health insurance. A lesser percentage is being paid directly by individuals out of their own pockets.

In the United States over $1 trillion a year is spent on health care, representing about 13.7 percent of the Gross National Product (GNP). On an individual basis this represents *over 15 percent of personal income.* On average the largest share goes for hospital care (about 42 percent) and physicians' services (22 percent).[1] Smaller amounts go to prescription drugs (10 percent), dental services (5 percent), nursing home care (9 percent), and the remainder to other health care services.

What does the average American family spend on health care? "American families with midrange incomes of $30,000–$40,000 a year spend an average of $1,500, or about *4.5 percent of their earnings,* on health insurance premiums not paid for by employers and other out-of-pocket costs," according to Consumers Union.[2]

Most Americans have employer-based private group health insurance. The general category of private group health insurance includes managed-care programs. A survey conducted by Mercer/Foster Higgins found that 85 percent of American workers with health insurance belong to some kind of managed-care plan, up from 52 percent four years ago.[3] Employers are moving away from offering traditional fee-for-service towards the use of more managed care offered by preferred provider organizations (PPOs), point-of-service organizations, (POSs), and health maintenance organizations (HMOs).

A side effect of the proliferation of employer-sponsored health plans is that "employees and their bosses are sparring as never before over what was once a private issue: personal health. Last year, 39 percent of all employers used some sort of financial incentive or penalty to encourage healthier behavior among their workers."[4] Companies' involvement in their employees' health may be relatively benign such as having on- site workout facilities, giving time off for exercise, and having arrangements with area gyms so that interested employees can get discounted rates. Or the involvement may become more intrusive as the following example shows.

> Rita DelRey, a legal secretary at a Portland, Ore., law firm, was given an ultimatum by her employer three years ago: Lose weight or lose free health coverage.

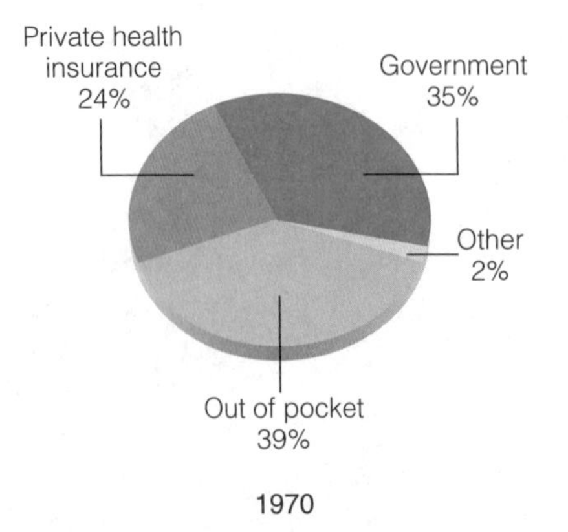

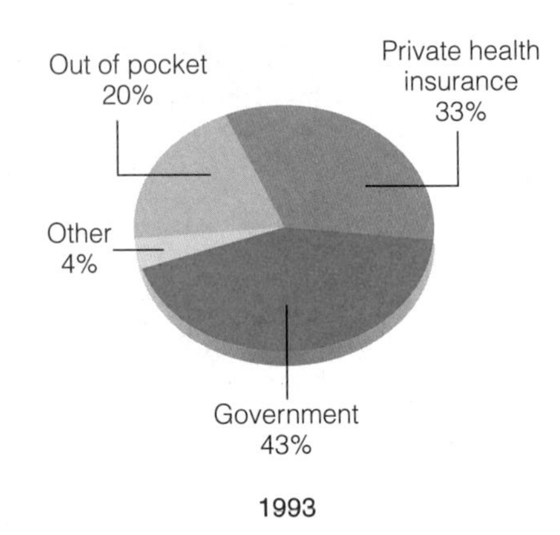

EXHIBIT 9.1
Changes in Who Pays for Health Care

Source: *Statistical Abstract of the United States 1994* (Washington, DC: U.S. Government Printing Office, 1995)

She had already dropped 33 pounds in the last two years, but that wasn't enough: She still weighs 12 pounds more than her employer, the law firm of Klarquist, Sparkman, Campbell, Leight & Whinston, considers healthy. So while her fitter cohorts receive health insurance at no charge, she must pay $56 a month for her coverage until she drops the weight.[5]

Rising Costs

What individuals and families pay for health care reflects the rising costs in the industry. Health care is rising at a rate higher than the overall rate of inflation. During a recent one-year period health care inflation was 3.6 percent—about half a percentage point higher than the Consumer Price Index and during a year of 2.3 percent inflation.[6]

Health care costs are rising for five main reasons:

* *Advances in medical technology and care, which prolong life.* Many of the new technologies are expensive and are used for longer periods of time.
* *A decline in lengthy hospital stays and more of an emphasis on outpatient services.* For example, someone may have back surgery and be sent home the same day to recover for three to six weeks. In the past that person would have been a patient in the hospital for a week. In the last ten years 700 hospitals have closed.[7]
* *Demographics.* People are living longer, and generally older people require more health services.
* *Increases in third-party payments.* Consumers with insurance policies rarely pay directly for medical care out of their own pockets. Because their bills go through a third party (an insurance company or the government), they do not realize what services cost and have little incentive to keep costs down. An average doctor's office visit may run $200 to $300.[8]
* *Rising costs in labor, prescriptions, supplies, and administration.*

To conclude this section and address the chapter's first objective: health care costs are taking an increasingly larger share of personal income. On average the highest percentage of expense goes to hospital care, followed by physicians' services, and there are five main reasons why costs have risen sharply.

OVERVIEW OF HEALTH CARE COVERAGE

Besides the large percentage (61%) of Americans with employer-based health insurance, 25 percent rely on Medicare [a federal health insurance program for the elderly and others who qualify] and Medicaid [a federal health insurance program for the poor], and 14–15 percent have no insurance.[9 & 10] Although these are across-the-board percentages, differences exist between states regarding the percentage of those who are not insured or who receive Medicaid.

Children, Health Plans, and Insurance

About 10 million children are uninsured in the U.S. As of this writing, *the federal government is exploring different ways to ensure better health care coverage for these children.* The situation is complicated because about three million of the uninsured children are eligible for Medicaid, but their parents (most of whom are workers who lack job-based coverage) have not signed them up.[11] Some parents are deterred by the complicated Medicaid rules (a 5-year old may be covered, but a 7-year old sibling may not be), others do not know that they qualify, and still others are too proud to seek help from a program linked to welfare. "Medicaid should not be perceived as welfare—it's also for families who are working and have children," says Sarah Shuptrine, an advocate with the Southern Institute on Children and Families in Columbia, South Carolina.[12]

Another trend in the last decade is that fewer children are covered by private health insurance. An important provider of health insurance to children is parents' employers, but in recent years employers have been shifting the cost of children's health insurance back to the parents. In addition one in four employees has no access to family coverage at any cost and many self-employed parents have no coverage.[13] The burden is heaviest on these parents because individually obtained insurance (not purchased through an employer or other group) for families can range from $375 to $800 a month.[14]

Health Care Plans or Insurance; Marriage and Divorce

Another aspect of health insurance coverage is the effect of marriage and divorce on insurance. If a couple is married and they each have insurance plans, they should study both plans and make appropriate decisions so they are not over insured. They should select the best coverage at the lowest price. This could mean choosing one plan over another or using a combination of plans.

Employer-based plans offered by companies have various rules about who is covered and under what conditions. For example, Jean and Max had been living together for ten years. They did not marry because she had a severe long-existing health condition that prevented her from working and required regular, costly treatments that were taken care of by a government program. If she married, she would no longer qualify for the government program because she would no longer be single and poor. His employer-sponsored plan would not cover her if they were married. Jean did not want to be a financial burden on Max, who had two children he was supporting from a previous marriage—especially if her condition worsened. When the company he worked for changed plans, the new plan allowed coverage for her health condition and they married immediately.

Under federal law if a couple divorces, medical coverage under a former spouse's medical plan can be continued for 36 months if the former spouse works for a company with 20 or more employees. If this is not the case or if 36 months is not deemed long enough, continuation of health insurance coverage may be part of the divorce settlement. Children are usually covered under the working parent's plan. Dependent and disabled children and dependent spouses will be discussed in regard to long-term care insurance near the end of the chapter. Since insurance laws and regulations vary by state, state insurance offices and local companies should be contacted to find out about typical coverage.

The Financial Pains of Childbirth

Childbirth coverage should be checked into before pregnancy occurs. The reality is that many pregnancies are unplanned, and insurance is checked into after the fact. Generally the way health plans work is that most doctors file a claim with the patient's insurer after a child is born.

> Most insurers pay obstetricians a single, "global fee" for prenatal care, delivery and a postpartum checkup, making the payment at some point after the baby is born. "It has been standard insurance practice forever," says Arthur Levin, a vice president for Prudential HealthCare.[15]

However, there are exceptions. An obstetric group in New York usually has the patients pay the $6,000 bill for a routine pregnancy in monthly installments that start after the first visit.[16]

A doctor who participates in a managed health plan has to accept the insurer's fee schedule. However, if you are in a managed-care plan that the obstetrician does not belong to or if your employer offers traditional indemnity coverage, you may wind up paying part of the doctor's bill. For example,

> Mrs. Goldin, 37 years old, of Brooklyn says her managed-care plan from Oxford Health Plans, Inc., Norwalk, Conn., will pay about 80% of maternity costs as part of its coverage for doctors that don't belong to its network. Mrs. Goldin...says she would rather pay upfront for doctors she trusts than pay virtually nothing for an in-network doctor, whose care might be compromised by the rules of managed care. "It's worth it," she says.[17]

Objective 2:
Identify three types of health care providers.

PROVIDERS: THREE MAIN CHOICES

There is no question that health coverage options have changed significantly over the last few decades.

> Health coverage used to be a no-brainer: You went to the doctor. Your insurance paid the bills. End of story. Today, as millions of Americans can attest, that story has dramatically changed. Instead of traditional health coverage, which pays medical bills regardless of what doctor they see, many workers now must choose among an alphabet soup of managed-care plans. Recent changes in Medicare have added new choices as well.[18]

The three main types of health care providers are private group health insurance (which includes managed care), individual health insurance, and government programs. Each of these will be described in the following sections beginning with managed care, the main type of health care that most employed Americans have.

Managed Care

Managed care
A prepaid plan that combines financing and delivery of health care.

Managed care refers to plans that combine the financing of health-related services and the delivery of health care. In managed care a list of health care providers (physicians, hospitals, and pharmacies) are identified by the plan. You do not pay a fee-for-service, but rather your health care is covered by a prepaid plan. More and more people are enrolled in managed-care health plans through their employers. Each type of managed care (HMOs, PPOs and POSs) has its own distinguishing features.

Health maintenance organizations (HMOs)
Type of managed care that provides all of one's health care for a fixed monthly fee.

Health Maintenance Organizations (HMOs) A **health maintenance organization**, or **HMO**, is an alternative to health insurance that has gained prominence in the past 20 years in the health care industry. In simple terms HMOs do not provide health insurance, rather, they provide health care. You pay a fixed premium in advance (once a month) in return for a wide range of health care services, including preventive care, hospitalization, and surgery. If you become an HMO subscriber, you must use the HMO's doctors and facilities. HMOs may provide services at more than one location, with a variety of doctors, assistants, and technicians. A subscriber selects a doctor that is designated as the primary-care physician. The primary-care physician oversees all medical treatments and referrals.

Once you join an HMO, you receive a contract, certificate, or member handbook.

This describes how your HMO works. Each document explains the services, benefits, exclusions and limitations of your coverage.

HMOs can be for profit or nonprofit. They usually operate in a geographic area. For example, *the oldest (1933) and the largest HMO is Kaiser Permanente* which covers California and 15 other states. To find out which HMOs serve your area, check with your employer or the Department of Insurance in your state or go online. If you are given a choice of HMOs, use the following guidelines for selection:

- Visit them and compare them as you would when you shop for other products.
- Ask about services (routine medical visits, wellness programs), physicians, and costs—including copayments and reduced-cost prescriptions. HMO copayments are usually less than traditional health insurance deductibles, the amount of expense a beneficiary must first incur before the policy begins payment for covered services—usually between $50 to $500 a year. A **copayment** means you pay a portion of your bill such as $5 or $10, each visit. Sometimes this copayment is waived if a person is ill or injured versus coming in for a routine checkup.
- Ask friends, coworkers, and neighbors.
- Find out the HMO's policy for treating pre-existing medical conditions.
- Find out when you can start receiving services.

Copayment
The portion of the bill that the patient pays.

Before an HMO member can see a specialist, the primary-care physician usually has to recommend the specialist. A limitation of an HMO is that the subscriber must use the HMO's network of physicians. Use of an "outside" specialist or "noncontracted" provider could cause the HMO to limit or deny benefits, making you personally liable for the bill. Another limitation is that someone who travels a great deal—or has dependents who live far away—may not find it convenient to use an HMO.

Preferred Provider Organizations (PPOs) **Preferred Provider Organizations (PPOs)** are similar to HMOs in that a group of physicians or hospitals provide services to employees through their employers, who get a group discount price. They are offered by Blue Cross and Blue Shield, hospital chains, and insurance companies. PPOs have been described as a looser form of managed care than HMOs. This is because they offer covered employees more choices (they do not have to use the personnel or facilities suggested by the PPO). Usually, though, there is a price to be paid for having more personal choice. For example, you will pay a higher deductible and more of the bill if you choose your own doctor or hospital rather than the ones suggested by the PPO. Also a fee may be charged for each use. A physician will charge less for a PPO enrollee than for someone who walks in off the street, but annual physical examinations may not be included.

Preferred provider organizations (PPOs)
Type of managed care that provides health care services to employees at a discount.

CONSUMER ALERT

With an HMO you are responsible for reading and understanding the contract. Also you are responsible for maintaining continuous coverage (switching often causes problems), and you are responsible for paying copayments and for paying medical bills if you select a specialist or other noncontract provider not authorized by your HMO. Some consumers disenchanted with their HMOs have switched to preferred provider organizations.

Point of service organization (POSs)
Type of managed care that combines traditional insurance with HMOs.

Point-of-Service Organizations or Products (POSs) The final type of managed care to be discussed is the **Point-of-Service Organizations or Products (POSs)**. It is becoming more commonplace for employers to offer POSs to employees. A POS combines traditional insurance coverage and HMO coverage. This product allows you to select which type of coverage you wish to have for certain illnesses or health care needs. POSs allow a patient to choose "in-network" providers or, for an extra fee, to go "out of network." Of the three types of managed care, this is the least selected choice, mostly because it has the highest deductible. With a POS you have two policies—an HMO and traditional coverage—and two carriers—an HMO and an insurance company—although you have only one premium to pay.

Further Discussion of Private Group Health Insurance and the Effect of Rate Changes on Personal Finance Group plans such as the managed care options just discussed are offered by private companies or large networked associations, such as Blue Cross and Blue Shield. Typically they provide basic coverage, major medical, and disability insurance. An advantage of most group plans is that you do not need to have a physical examination in order to get coverage. For many people with existing health problems this is the only kind of coverage they can get.

Although this book is written from a personal finance point of view, it should be noted that health insurance is a tremendous expense to employers. Any switch in policies can cost them millions of dollars. Because of the high cost some employers are requiring higher deductibles, laying off full-time workers with costly benefits and replacing them with temporary workers with few benefits, and increasing copayments.

For example, state employees in Florida in the first half of 1998 paid $26.02 a month for single coverage and $93.58 for family insurance, regardless of family size.[19] The governor proposed a plan (in response to the rising costs of health insurance to the state) to increase premiums 15 percent—$3.90 for singles and $14.04 for families—on July 1, 1998. He also proposed that copayments increase to $20 (previously they had been $10) for doctor's office visits in the state's preferred-provider network. If an employee used a doctor that was not in the network, the copayments would be $30. The employee's charge for emergency room treatments and for filling prescriptions would also rise in the plan.

The governor's plan, which went before the state legislature is an example of passing part of the employer's health insurance costs down to the employee, in this case to over 100,000 state employees. If employers do not do this, they have to find the money to pay for the rising costs from other sources, such as shifting money from other funds. From the employees' point of view, if pay raises are low during the year of increased premiums and copayments, their raises will be deeply cut into or even negated. Although you may be reading this and thinking "who cares about state workers in Florida," this example is given as an illustration of the interaction between rising health insurance costs and pay raises. Health insurance costs play a vital role in one's overall financial picture. It also illustrates that even if employees are covered by a group plan, they will still have to pay some of their own money. It is a misconception that employees with group plans get a free ride.

Individual Health Coverage

The second main category of health care providers is individual health coverage not obtained through employers or through the government. Blue Cross and Blue Shield and insurance companies offer insurance to individuals as well as to companies. Individual health coverage tends to be very costly. So costly that millions nationwide are locked out of traditional private health care systems. Consider the case of Ms. Gloria Brown:

> Ms. Brown isn't offered health insurance from her employer, Family Affair II, a hair salon in Suitland, Md. She has tirelessly approached private health insurers over the years to try to get coverage, but says she can't afford it on her income of $14,000 or so a year. Were it not for Medicaid, she'd have no health coverage at all. "I've found some good coverage, but [the private companies] want so much," says Ms. Brown.[20]

As Ms. Brown points out, the cost of privately obtained health insurance is very high. Consumers can reduce costs by shopping around for the best price for the coverage needed and watching for group insurance, which is usually less expensive. One way to lower the cost is to take a higher deductible. In many families one of the adults gets a job that includes health insurance so that everyone in the family is covered. Medicaid, the solution for Ms. Brown, will be discussed next.

Government Programs

The third type of health care provider is the government. The *two largest government-sponsored programs are Medicare and Medicaid.* Medicare and Medicaid were introduced in Chapter 7. *Medicare* is the federal health insurance program for people 65 or older people of any age with permanent kidney failure, and certain disabled people under 65. Medicare is administered by the Health Care Financing Administration. The Social Security Administration provides information about the program and handles enrollment. A person receiving Social Security or Railroad Retirement Board benefits when he or she turns 65 is enrolled in Medicare automatically and will receive a Medicare card in the mail. A person who is disabled will automatically get a Medicare card in the mail when he or she has been a disability beneficiary under Social Security or Railroad Retirement.

Medicare has both hospital (Part A) and medical (Part B) insurance. Both may have deductibles and copayments, so a person has to pay some of the expense out of his or her own pocket. Because hospitals are discharging people sooner, Medicare has picked up some of the costs of home visits.

Medicaid is a health care program for people with low income and limited assets. Some people qualify for both Medicare and Medicaid. In addition to hospital and medical expenses, Medicaid covers long-term care, so someone may exhaust the Medicare coverage and qualify for Medicaid. There are variances between states as to who receives Medicaid and how much they receive. For example, in one state a person earning $15,000 a year may receive Medicaid benefits that would cost more than $4,000 a year to purchase and a person earning the same amount of money residing in another state may receive no Medicaid benefits. For information regarding the laws and regulations in your state, contact the Department of Insurance.

Choosing a Plan

Individuals who are in a position of having to choose between several health care plans or health insurance options should ask themselves several questions:

1. What are my health care needs and those of my family? For example, if a child is planned, what are the health plan's guidelines for obstetrician selection and fee payment?
2. Do I regularly use preventive care? Will I or someone in my family need annual checkups, mammograms, immunizations, etc.?

3. Do I need specialists?
4. Is the policy renewable?
5. Do I mind filling out paperwork? Usually managed care plans require less paperwork.
6. Does the company have a good reputation? What is their rating?
7. Have I read the policy?

Once these questions are answered, a person can move on to the subject of reducing costs.

How to Save Money on Health Insurance

Money can be saved by:

* Comparison shopping, including going online to compare monthly premiums, deductibles, and copayments. Do not buy more than you need for the coverage level desired.
* Checking for preexisting condition exclusions. Preexisting conditions are usually health problems you went to see a physician about within the six months before the date the policy went into effect.
* Reviewing coverage needs over time. Do not pay for coverage you do not need. Find out what tests, immunizations, or physical examinations are covered. Also inquire about emergency room visit coverage and dental and vision examinations. If dental insurance is offered, does it pay for two routine checkups a year, but not root canal and more extensive work? Are regular eye examinations, glasses and contact lenses included? If so, are the premiums charged worth it, or would you be better off paying cash for dental or vision care?
* Evaluating the plan or policy after purchase. Is the billing process handled efficiently?

Objective 3:
Describe five types of private health insurance policies.

CHOICES IN PRIVATE HEALTH INSURANCE

The chapter now turns to a discussion of what is covered in typical private health insurance policies. Insurers sell five main types of private health insurance policies: (1) basic medical expense insurance, (2) major medical insurance, (3) disability income insurance, (4) Medicare supplement insurance, and (5) long-term-care insurance. Each of these will be discussed next.

CONSUMER ALERT

Because of the high costs involved (financial and healthwise), it is worthwhile to take the time to investigate the choices. Do not be pressured. Do not pay cash. Pay by check, money order or bank draft made payable to the insurance company—not to the agent or another individual. Get a receipt. Know with whom you are dealing. Check with your state insurance department if you have any questions or concerns.

Basic Medical Insurance

Basic medical insurance
Covers hospitalization and outpatient treatment.

Basic medical insurance covers hospitalization and most outpatient treatment. The hospital or the physician sends the bills directly to the insurer since in most cases there is not a deductible. So if the hospital bill is $3,000, the insurer—such as Blue Cross and Blue Shield—pays it although some insurers pay only up to a certain amount ($200 a day). For example, in maternity coverage the cost of a shared room may be covered, but not the expense of a private room. If a new mother does not want a roommate, she may be charged extra.

Basic medical expense insurance may include surgical contracts, which means that the insurer specifies a maximum amount of coverage for procedures. Again a reasonable amount is what is needed unless you have a preexisting condition or suspect that you will need more than average coverage.

Major Medical Insurance

Major medical insurance
Covers large medical expenses.

Major medical insurance as the name implies, covers potentially large medical expenses. It is useful in the event of catastrophic illnesses or injuries. Policies may include a high deductible ($500 and up), a participation provision (sharing of costs), and a high limit of liability ($50,000 and up).

Disability Income Insurance

Disability income insurance
Pays benefits to policyholders when they become incapable of working.

Millions of Americans have some type of disability. **Disability income insurance** is an insurance policy that pays benefits to a policyholder when that person becomes incapable of performing one or more occupation duties, either temporarily, on a long-term basis, or totally.[21] It replaces lost income because of the disability.

Policies typically pay 60 percent to 80 percent of one's paycheck. They are available through employers, or an individual can buy them personally. An option is a supplemental policy that individuals can purchase to take their coverage up to 80 percent if for example, their employer's policy goes only as high as 60 percent. Policies that cover 100 percent are not available.

Coverage may be short-term (weeks) or long-term (months or years). People do not use disability income insurance for less than a week or two because policies have **elimination periods**, a period of time that must elapse before insurance payments begin. Since it is linked to employment, disability income insurance usually ends when one is no longer employed (retires or quits).

Insurance policies state what constitutes a disability. The two main sources of disability insurance are Social Security and private disability insurance (either group usually through an employer or individual plans). Amounts vary by a number of factors, including the length and severity of the disability. For Social Security the disability

CONSUMER ALERT

Contracts cover varying amounts of hospital time such as 30 days, 60 days, 90 days, and so on. Usually a young person does not need coverage for more than 60 days. The idea is that the person will either be cured or sent home to recover in that time period. So, do not overinsure yourself (paying for coverage that is highly unlikely you will ever need).

has to be total—meaning the person cannot work—and disability coverage starts in the fifth month if the disability is expected to last a year or longer.

If a disability is due to illness or injury from a job, an employee may be eligible for **workers' compensation**. All fifty states have workers' compensation laws. State programs usually pay two-thirds of the worker's lost wages up to a maximum amount. Funds may also be provided for medical costs and for death benefits to the spouses and underage children in the case of deceased workers. Choice of physician may or may not be allowed.

Workers' compensation
Money paid if a disability is due to illness or injury from a job.

In addition a person under 65 whose earnings are lost or reduced because of certain disabilities may qualify for Supplemental Security Income (SSI). At the beginning of the chapter it was noted that Medicare is for the aged and others who qualify. The "others" are people under 65 who are entitled to receive Social Security disability benefits for a total of 24 months or who have severe kidney disease. The program is not based on income or financial need. For questions regarding SSI, your local office of Social Security should be contacted.

To summarize, a person with a disability may collect benefits from a variety of sources, especially if an illness or an injury occurred on the job. Dependents of someone with a disability, or who died as a result of work-related illnesses or injury, may also collect benefits from a number of sources.

If you are offered disability insurance through your employer, it is a good idea to take it even if it is expensive. To lower the premium

* buy the policy through the employer or another group since group coverage is usually cheaper than a policy obtained on your own.
* reduce coverage time. Give it up when you retire.
* settle for a lesser monthly benefit. If in all likelihood there will be other sources of income/benefits, lower the monthly benefit.
* lengthen the elimination period. If you have two to three months of sav
you may not need immediate support.
* accept a lower inflation adjustment (cost-of-living factor).
* accept a policy with a narrower definition of disability.

All of these factors should be considered in light of the type of life
work you do. If you travel to high-risk parts of the world or have a dan
ically demanding occupation, then disability insurance is more critica
a quiet desk job.

Medicare Supplement Insurance (Medigap Insuran

Because Medicare does not cover all medical expenses, servi
people purchase **Medicare supplement insurance** also know
Not everyone needs Medigap insurance. For example, if so
ings and investments, he or she may have enough mone
ered health care needs that arise beyond Medicare, or t
pay for many expenses.

Medicare supplement insurance or medigap
Covers medical expenses, services, and supplies not paid for by Medicare.

Medigap insurance is purchased from private insu
ment. It is regulated by federal and state law and m
supplement insurance. There are ten standard Me
J. Individuals choose which of these policies they
of becoming eligible for Medicare, the insure
existing conditions. If they wait longer than
coverage. A checklist with an example is g

EXHIBIT 9.2
Medicare Supplement (Medigap) Insurance Checklist

Comparison chart to assist when shopping for Medicare supplement insurance. Example and space for comparison company are given.

Company Name	Company 1	Company 2
Annual Premium: Self Spouse	 $2,045 $1,778	 $__________ $__________
How will you purchase the plan?	(Agent,) Mail, Online	Agent, Mail, Online
Is the company online or have a toll-free #?	(Yes) No 1-800-234-5678	Yes No
Is there a service office nearby?	(Yes) No	Yes No
What is the company's rating?	A. M. Best A–	
How long is the pre-existing conditions waiting period?	6 months	
Coverage	Basic Benefits (hospitalization and medical expenses)	

Long-Term-Care Insurance

...tics show that more than half of the U.S. population will need long-term care ... *Many people think that Medicare will pay their long-term care bills, but it cov-* ... *percent of all nursing home costs.*[20] For this reason long-term care insur- ... n increasingly popular option although no one knows for sure if he ... ate insurance companies offer individuals or group **long-term-** ... icies that provide benefits for a range of services not cov- ... e, or by Medicare or Medigap insurance. The decision ... including personal health and family health histo- ... rs. LTC is something individuals may never ... owner's insurance: It is there if they need it

... ypically policies cover extended stays in long- ... of assisted living in one's own home. Usually we ... being for the elderly, but a younger person may need ... y.

... y agents or through the mail or online. Some employers ... oyees. The policies usually pay a fixed dollar amount (an ... sts of the care up to a specified dollar amount for a specified ... nd of care. This is often referred to as the daily benefit amount. ... o $100 a day for a nursing home or in-home care is paid by long- ... ace.[22] An average one year stay in a nursing home can run between

...ngs,
... you lead and
...gerous or phys-
... than if you have

...ce)
...es, and supplies, many
...n as **Medigap insurance.**
...meone has substantial sav-
... to take care of any uncov-
...eir regular group policy may

...*ance companies*, not the govern-
...st be clearly identified as Medicare
...digap policies are labeled A through
... want. If they apply within six months
... has to accept them, regardless of pre-
...x months, they may be refused Medigap
...ven in Exhibit 9.2.

$35,000 and $100,000, depending on the area of the country. Generally costs are higher in cities.

According to the *New England Journal of Medicine*, nearly half of the people over 65 will spend time in a nursing home.[23] Medicare and Medigap insurance do not usually pay for the custodial care given in nursing homes. The only nursing home care that Medicare covers is skilled nursing care or skilled rehabilitative care in a Medicare-certified skilled nursing facility. This is one of the reasons people buy long-term-care insurance. It should be purchased before a crisis develops, but a lot of people put this decision off until their sixties or seventies for a number of reasons, including the following first-hand account:

> I've found exploring long-term health insurance about as pleasant as making pre-need burial arrangements. Both can be good ideas, but don't expect the sales presentations to be uplifting. Indeed, the fellow who sold me on the pre-need about 25 years ago had this parting thought as he stepped out the door: "Just remember, Mr. Whitsitt, I'll be the last to let you down."[24]

From the perspective of an older person making the decision to purchase LTC or not, Jane Bryant Quinn, a leading financial journalist, says

> If your income is modest—say, under $15,000—and your assets small, the answer is usually no. This coverage isn't cheap and may use up money needed to maintain your standard of living. If you enter a nursing home and run through your own resources, the Medicaid program will pick up the bill. . .. The insurance decision is harder for middle-income people—namely, most of us. You're balancing the risk of needing long-term care against the high price of coverage. You're asking whether you want to spend your savings on your own custodial-care expenses or leave the money to your kids.[25]

Another expert, Caralee Adams, agrees with Quinn's advice and adds

> For many older couples who may be well-off but not wealthy, long-term-care coverage can address the nursing home Catch-22: If you're wealthy, you can afford an extended stay. If you're poor, Medicaid will pick up most of the tab. If you're middle, Medicaid won't pay until you've spent most of the assets that you may have built up for your children.[26]

Anyone seeking LTC should get several proposals and not be pressured into making quick decisions. Before signing anything, current coverage should be checked as well as government coverage, including veterans' benefits. As in all insurance decisions, one should ask oneself (1) Do I need it? (2) Can I afford it? (3) What benefits does a particular policy offer? The costs, options, and potential risks should be weighed carefully. Exhibit 9.3 provides an example and a checklist for help in comparing long-term care policies.

Before making a decision, consider your income, what your employer offers, and your family's health history. If nearly everyone lives well into their nineties, then the likelihood is that you may need long-term health insurance. As the previous discussion explains, many people put off the decision to buy LTC, but the problem with that is if they wait too long they may be rejected for coverage because of a medical condition. A good time to start looking into LTC is between ages 55 and 60 although anyone with a health concern may want to consider it at a younger age.

Since LTC is relatively new, legislation is being passed to regulate it and to adjust taxes. Recently Congress created two tax breaks: Part of an LTC insurance premium may be deductible from tax returns, as a medical expense, and persons who receive long-term care can deduct that cost from their taxes.[27]

Long-term-care policies have their own vocabulary:

Assisted-living facility. This provides assistance to those who do not require institutional care, but could benefit from 24-hour assisted services, such as health care

Comparison chart to assist when shopping for long-term-policies.
Example and space for comparison company are given.

Company Name	Company 1		Company 2	
In addition to nursing home care, which of the following services are covered?				
Alzheimer's disease	(Yes)	No	Yes	No
Adult congregate living facility	(Yes)	No	Yes	No
Adult day care	(Yes)	No	Yes	No
Adult foster home	(Yes)	No	Yes	No
Assisted living	(Yes)	No	Yes	No
Home health care	(Yes)	No	Yes	No
Nursing service	(Yes)	No	Yes	No
Other	(Yes)	No	Yes	No
How much does the policy pay per day for:				
Adult day care	$100		$_______	
Adult congregate living facility	$100		$_______	
Adult foster home	$100		$_______	
Assisted living	$100		$_______	
Home health care	$100		$_______	
Nursing service	$100		$_______	
Other	varies		$_______	
What is the length of time of benefit?				
Nursing home care	5 years		_______	
Lower levels of care	3 years		_______	
How much does the basic policy cost per year?	$1,600		_______	
What is the age restriction on buying the policy?	any age, less expensive at ages 55–60			
How long before pre-existing conditions are covered?	6 months		_______	
Does the policy have inflation protection?	(Yes)	No	Yes	No
Are there any conditions that have to be met before benefits begin?	(Yes)	No	Yes	No
Does the policy have a premium waiver clause?	(Yes)	No	Yes	No
Is the policy guaranteed renewable?	(Yes)	No	Yes	No

Other questions to ask potential insurers: What happens if the person with long-term insurance dies before he or she uses any of its benefits? Is there a refund? Can the policy be cancelled at any time? What happens if cancelled? What is the waiting period? For example, can someone buy long-term care insurance at age 75 and start accessing benefits at 78?

EXHIBIT 9.3
Long-Term Care Insurance Policies Checklist

supervision, meals, housekeeping, and transportation. An individualized care plan is developed based on needs. Some involve expensive endowment (entrance) fees, others do not.

Continuing care retirement community. A group living arrangement with independent living units and a nursing facility; may be similar to assisted living facility.

Custodial care. Care for daily activities (bathing, dressing, eating) in the elder person's own home. Custodial care does not require medical training, but it does requires a physician's orders to be covered by insurance.

Home health care. This may include a wide variety of services including health care by nurses, social workers, and physical therapists; personal care; nutrition; housekeeping; and social and safety needs, such as transportation, companions, and a daily phone check. Insurance may cover part and the individual or his or her family pays the rest.

Hospice. A place caring for the terminally ill.

Intermediate care. Occasional nursing and care by a skilled medical person. Also must be ordered by a physician.

Skilled nursing care. Daily care by a skilled medical person, ordered by a physician.

An advantage of having a money supply or an LTC is that you can get into a better nursing home, if you can pay your own expenses for a year or so before going on Medicaid.[28] The next section explains this further, but first it should be noted that not all elderly people select these living/care options. Some live completely independently; others live with relatives and friends, or in senior living communities. An increasingly popular option is shared living arrangements (see the Electronic Resources section about this).

Long-Term-Care Insurance and Medicaid

To summarize, the main reasons people buy long-term care insurance are:

* to take care of themselves in old age.
* to preserve their assets (the money and investments they have built up over a lifetime) so that a spouse is provided for.
* to leave a legacy to children or grandchildren by avoiding a spenddown. A **spenddown** is the process by which medical and nursing-home care reduces a person's assets.[29] What it means is that a person cannot get assistance from Medicaid until nearly all their assets are depleted. One exception is that a person can keep a house if a dependent spouse or disabled or dependent children live in it. The person can also keep furniture, a car, a burial place, burial funds, and a small amount of cash.

Spenddown
Process by which medical and nursing home care reduces a person's assets.

Rules on what you can keep and spend vary by state. For example, in the late 1990s in 19 states—including Oregon, Colorado, Iowa, and Texas—you could not qualify for Medicaid nursing-home assistance if your income exceeded $1452 a month.

Rarer Types of Insurance

The previous discussion was about the main types of health insurance, concluding with the fifth type, long-term-care insurance. There are rarer types, and two of these will be discussed here: Accidental Death and Dismemberment and Specified Disease Policies.

Accidental death and dismemberment (AD&D)
Policies that provide benefits for accidental death, paralysis, loss of limb, speech, hearing, or eyesight.

Accidental Death and Dismemberment (AD&D). AD&D policies provide benefits for accidental death, paralysis, loss of limb, speech, hearing or eyesight. Usually the policies will not pay for accidents or death due to suicide, drug abuse, or mental illness. Employers may provide AD&D. It is expensive to buy on your own.

Specified disease policies Provide medical benefits for a specific disease such as cancer.

Specified Disease Policies. These provide medical benefits if you get a specified "dread" disease, such as cancer. They are not available in some states and the offer may come through the mail. The policies may duplicate coverage you already have. Usually individuals do not need specified disease policies unless the disease runs in their families and they do not have coverage for the disease already through their basic health policy.

Future Trends in Health Insurance

Many employers now allow their workers to make investment decisions about their retirement plans. Likewise, it has been suggested that in the future employees will have more say-so in how their health plans operate. For example, they may have more choices of health plans, deal directly with plan administrators, and opt to pay more pretax dollars for expanded benefits.[30] Another option may be "medical savings accounts" instead of traditional health insurance. Cornell University Professor Roger Battistella says that such accounts reward consumers for economizing, since unspent dollars can be applied to other uses, such as retirement savings.[31]

Another trend to watch for is changes in tax laws. For example, under a new law self-employed workers (consultants, writers) can deduct 60 percent of their premiums in 2000 and 2001, but this will rise to 70 percent in 2002 and 100 percent in 2003 and thereafter.[32] A final trend to note, as in all areas of personal finance, the Internet will be more heavily used to compare costs and options.

ELECTRONIC RESOURCES

As mentioned previously, shopping for insurance online has made it much easier to compare the prices of insurance policies than in the past. Quotesmith.com at **www.quotesmith.com** gives instant quotes on medical, disability, and dental insurance as does Quicken Insurance at **www.insuremarket.com**.

The federal government offers consumer catalogs on insurance topics at **pueblo.gsa.gov.**, and most insurance companies can be reached at **www.companyname.com**. Columbia/HCA Healthcare Corporation (the owner of many hospitals) has a website that covers insurance at **www.columbia.net**.

Regarding elder care, for home care or to find nursing homes or assisted-living facilities, call local Area Agency on Aging offices or phone the Eldercare Locator at 1-800-667-1116. "A Shopper's Guide to Long Term Care Insurance," published by the National Association of Insurance Commissioners, is available at **www.naic.org/**.

Elders who are interested in finding someone with whom to share housing can contact the National Shared Housing Resources Center operated by the AARP at **www.aarp.org/getans/consumer/homeshare/html**. The AARP has programs in more than 200 communities in the U.S.

For general insurance information try the Insurance News Network at **www.insure.com**. Check on the accreditation status of about 300 health plans by visiting the National Committee for Quality Assurances, the managed-care industry's leading accreditation group, at their website at **www.ncqa.org**. For an overview of trends in the insurance industry and copies of publications or reports on issues contact the Health Insurance Association of America at **www.com.hiaa.org**.

1. Are employers required to offer their employees a choice of HMO coverage?

   ■ An employer with at least 25 employees must offer employees an HMO choice.

2. What are the advantages of choosing an HMO?

   ■ With an HMO there are no large deductibles, and only small copayments. Also subscribers are not required to submit claim forms.

3. What are the disadvantages of choosing an HMO?

   ■ Subscribers must use the HMO's designated hospitals and doctors. Subscribers generally cannot use specialists without having a referral from their primary-care physician. Also if you travel a lot or have dependents that do not live close by, an HMO may not be convenient.

4. What is a point of service product?

   ■ This is a combination of traditional insurance coverage and HMO coverage, often with a high deductible. The product allows you to select the type of coverage you wish for certain illnesses or health care needs.

5. Does every older person need Medicare supplement insurance?

   ■ No, not everyone needs it. They may have other options, such as enough in savings to cover expenses not covered by Medicare, or they may belong to an HMO or a group policy with adequate coverage, or they may qualify for full Medicaid benefits (if income falls below a certain level).

6. What is the difference between Medicare and Medicaid?

   ■ Medicare is the federal health insurance program for people 65 or older or people of any age with permanent kidney failure, or for certain disabled people under age 65. Medicaid is a health care program for people with low income and limited assets.

7. Can someone have both Medicare and Medicaid?

   ■ Yes, some people qualify for both programs.

8. Can an insurance company refuse to insure a person?

   ■ Yes, a company has the right to refuse to sell coverage if a person is in failing health or if a person gives incorrect information on the application.

SUMMARY

Insurance is the main way people protect themselves and their assets from risks. Because of the potential severity of loss due to illness or injury, health insurance or health plans are necessary.

Eighty-five percent of American workers with health insurance belong to some kind of managed-care plan.

Health care costs are rising more rapidly than inflation. Five main reasons for the increase are advances in medical technology; a decline in lengthy hospital stays; changes in demographics; increases in third-party payments; and rising costs in labor, prescriptions, supplies, and administration.

On an individual basis, health care represents over 15 percent of personal income.

Sixty-one percent of Americans have employer-based health insurance, 25 percent rely on Medicare and Medicaid, and 14–15 percent have no insurance. Not everyone can get insurance and not all those that qualify for Medicaid apply for it. About 10 million American children are uninsured.

Health care insurance providers can be divided into three categories: private companies, independent companies, and the government. Most people use private companies.

The trend is towards managed care, defined as plans that combine the financing of health-related services and the delivery of health care. HMOs, PPOs, and POSs are types of managed-care plans.

HMO (health maintenance organizations) can be either for profit or nonprofit. In HMOs individuals usually are assigned to a primary-care physician who can recommend them to specialists.

PPOs (preferred provider organizations) offer a contract between local health care providers and employers to provide employees medical care at a discount. A wider range of physicians is usually offered than with HMOs.

POSs (point-of-service organizations or products) tend to have high deductibles and allow patients to choose "in-network" providers, or for an extra fee, "out-of-network" providers.

Five main types of health insurance coverage are basic medical expense, major medical, disability income, Medicare supplement, and long-term care.

Basic medical expense covers hospitalization, physician's services, and usually outpatient care.

Major medical covers the cost of catastrophic illnesses.

Disability income replaces income lost when the insured is unable to work for an extended period of time. Employers may offer disability insurance, or individuals can buy it or supplement insurance on their own. A person who is injured on the job may also receive workers' compensation or income from Social Security, depending on the disability.

Medicare supplement insurance (also known as Medigap insurance) "supplements" or fills in the "gaps" in Medicare benefits. For example, if a person retires early without continued health care insurance from the employer, he or she may need Medicare supplement insurance. Not every older person needs Medigap insurance.

Long-term care (LTC) covers many expenses, including nursing-home bills and in-home care for elder persons. Most people buy long-term care insurance to preserve the assets that a spouse may need to live on or to leave to children or grandchildren. Low-income people probably can not afford LTC and would be better off putting their money toward maintaining their current life needs.

Two rarer types of insurance are accidental death and dismemberment (AD&D) and specified disease policies.

Future trends in health insurance include more employee choice regarding plans and the investing of health plan dollars, medical savings accounts in which unspent dollars can be applied to other uses, and more use of the Internet to compare plans and options.

Recent changes in tax laws allow self-employed workers (consultants, writers) to deduct more of their health insurance costs.

More people are comparison shopping for health, disability, and long-term care insurance online.

REVIEW QUESTIONS

(see Appendix A for answers to questions 1-10)

Identify the following statements as true or false

1. Most Americans have employer-based health insurance.
2. Medicare is a federal insurance program for the poor.

3. Medigap insurance is offered by the government.
4. Typically young people need hospital insurance coverage for at least six months.
5. Group policies are usually less expensive than individually obtained policies.

Pick the best answers to the following five questions.

6. Health care costs are rising for all but which of the following reasons?
   a. Advances in medical technology
   b. Demography—people are living longer
   c. Increases in the number of first-party payments
   d. Rising costs in labor, prescriptions, supplies, and administration
7. Of the three main types of health insurance, which of the following is the most costly to individuals?
   a. Private group health insurance
   b. Individual health insurance
   c. Government programs
   d. Transfer payment insurance
8. Because _______________ does not cover all medical expenses, services and supplies, many people purchase Medigap insurance.
   a. Medicare
   b. Medicaid
   c. Health Aide
   d. Blue Cross/Blue Shield
9. The oldest and the largest HMO is
   a. Metropolitan USA.
   b. Kaiser Permanente.
   c. Nelson-Strozier.
   d. Prudential Plus.
10. People buy long-term care insurance for many reasons; which of the following is not one of them?
   a. For protection in old age
   b. To preserve their assets for a spouse
   c. To leave a legacy to children or grandchildren
   d. To make money on their investments

DISCUSSION QUESTIONS

1. Near the beginning of the chapter was the story of Rita DelRey, a legal secretary, who was told by her employer to lose weight. She did, but the law firm she worked for wanted her to lose more. According to the secretary, "I don't want to lose any more weight. I think I look good and I feel fine."[33] Do you think it is right for her employer to make her pay her own health insurance until she loses more weight? On what basis (for what reason) does her employer persist in this?

2. Health care costs are rising. What are the five main reasons for this? Which of these five do you think is the most significant in terms of driving up costs?

3. Managed-care plans sometimes pay for all childbirth costs and sometimes there are other options, such as partial payments, especially if a non network obstetrician is chosen. What do you think of these options? If you have a child or know someone who recently had a child, did the insurer pay all costs? Was the doctor paid after childbirth in one payment, or were there monthly payments?

4. Does disability insurance pay 100 percent of lost income? What percentage of lost income does workers' compensation typically pay? How does one qualify for disability insurance or workers' compensation?

5. Journalist Jane Bryant Quinn says she gets a lot of questions about long-term-care (LTC) insurance, mostly from baby boomers wondering whether their parents should buy it or not. If their parents' income is modest, should they buy LTC? If their parents have middle incomes, should they buy LTC? What factors besides income are important to consider before buying LTC? What is a good age for healthy individuals to start looking into LTCs?

DECISION-MAKING CASES

1. Kiera has been hired by a department store chain as a buyer, starting at $30,000 a year. The company offers a standard insurance package with a group health plan and disability insurance. Dental and vision insurance are extra, and the premiums are high. She figures that this first year most of her salary will go toward getting settled and paying off college loans. What should Kiera consider before purchasing extra insurance?
2. Crystal has stayed home to raise four-year-old twins and would like to stay at home until they are at least eight years old. Her husband has announced that he wants a divorce. He makes $50,000 a year. He has family health insurance coverage through his employer. Her divorce lawyer asked Crystal about her health insurance coverage. What should she tell her lawyer?
3. Gavin was injured on his job and laid up for a year. He has disability coverage. What percentage of his paycheck can he expect? Besides his disability insurance policy coverage, what other sources of income support may he be entitled to?
4. Justin's grandmother is 61 and knows she will receive Medicare at age 65. She has health problems, a medium income, and is thinking of quitting work at age 62. Will she need Medigap insurance? From whom can she purchase Medigap insurance? Should she consider getting long-term-care insurance?
5. Allyson is a healthy 23-year-old who works full time and has basic medical coverage. She received an offer for a special cancer insurance policy in the mail. She does not think she needs it, but would like to make sure. What should she do?

GET ON LINE

Activities to try:

- To learn more about managed care, go to www.aarp.org, the website of the American Association of Retired Persons. In the Quick Search box type *managed care*. Choose one of the following subtopics: Managed Care Complaints, Managed Care Costs, or Managed Care Benefits, and report on what you find under that subtopic.
- To obtain health, disability, or dental insurance quotes, go to www.insuremarket.com or www.quotesmith.com. If you select www.insuremarket.com, click on Quicken Insurance, then click on health insurance quotes and proceed with the type of insurance that you are interested in finding out more about. If you select www.quotesmith.com, scroll down to instant dental insurance quotes or click on instant quotes for families and individuals and report on what you find.

INFOTRAC

To access InfoTrac, follow the instructions given at the end of Chapter 1.

Exercise 1: A recent search of InfoTrac using the key words *health insurance* identified one Encyclopedia excerpt, four Reference Book excerpts, two Pamphlets, 451 Periodical references, 61 Subdivisions, and 30 Related subjects. One of the Subdivisions is Health Insurance Industry; click on that and find articles on HMOs. Report on what the articles say about the pros and cons of HMOs.

Exercise 2: Go again to *health insurance*; this time click on one of the following, select an article or reference, and report on what you find:

1. Uninsured Persons (health insurance).
2. Medicare Supplement Health Insurance (see Medigap).
3. Old Age, Survivors, Disability, and Health Insurance.

chapter ten

life insurance

Buying Smart
Revising Periodically

DID YOU KNOW?

* Seventy percent of American adults have life insurance.
* More than 25 percent of life insurance policies that are sold go uncollected.

OBJECTIVES

After reading Chapter 10, you should be able to do the following:

1. Describe the purpose of life insurance and how it works.
2. Determine who needs life insurance and if you need it.
3. Distinguish between term, whole life, universal life, and variable life insurance.
4. Discuss ways to find a secure company and reduce the cost of life insurance.

CHAPTER OVERVIEW

The chapter begins with a definition and explanation of life insurance, including an overview of the industry. Then the chapter explains the purpose life insurance serves and how it works. The application and underwriting processes are described. The chapter also addresses the questions of who needs life insurance and how to calculate how much is needed. Specific sections explore different types of life insurance and the insurance needs of college students, women, and children.

Since this is the last chapter on insurance, there is a summary insurance planning worksheet near the end of the chapter. On the worksheet you can list the different types of insurance you own (health, disability, auto, life), the companies you have insurance with, the estimated annual cost, and estimated worth along with a column to check plans for future insurance purchases. Near the end of the chapter is an Electronic Resources section and Summary.

LIFE INSURANCE DEFINED

Life insurance
Contract between an insurer and policyholder specifying a sum to be paid to a beneficiary on the insured's death.

Face amount
Stated sum of money on a policy.

Beneficiary
Recipient of policy proceeds if insured dies.

Life insurance is a contract between an insurer and a policyholder that specifies that a sum be paid to a beneficiary on the insured's death. A person joins a risk-sharing group (an insurance company) by purchasing a contract (a policy). A standard contract specifies that upon the death of a person whose life is insured, a stated sum of money (the policy's **face amount**; i.e., $250,000 or $1 million) is paid to the person(s) or organization(s) designated in the policy as the **beneficiary**.

An insured person can name both a *primary beneficiary* and *contingent beneficiaries*. The primary beneficiary (usually a spouse) will receive the entire death benefit if he or she survives the person who dies. If the primary beneficiary does not live longer than the insured person, then the money goes to the contingent beneficiaries (usually children). There are special situations that might affect the order of beneficiaries, such as the case of several family members losing their lives in a shared disaster (a car accident or plane crash). If this is the case, state laws regarding simultaneous deaths go into effect, and the policy will be checked for survivor clauses. For this reason and others it is important to take the time to think about who beneficiaries should be and to update policies whenever there is a change in family status (marriage, divorce, adoption, childbirth).

Life insurance is a key part of financial planning especially if there are dependents. But how much is needed, and what type to buy, is less clear. Since life insurance payments can cost over $500 a year, it is worth taking the time to explore the options.

National Overview

In the United States the average amount of life insurance held by households has steadily increased. In 1970 the average insured household had $25,500 in life insurance; now it is about $125,000. Most new policies are sold to individuals between the ages of 25 and 44. *Seventy percent of American adults have life insurance.*[1]

PURPOSE OF LIFE INSURANCE

Objective 1:
Describe the purpose of life insurance and how it works.

When death strikes, life insurance provides security and protection for survivors. *Most people buy life insurance to protect those who depend on them from financial loss.*

Enough life insurance should be purchased to cover dependents' living expenses for a number of years. Dependents are usually spouses and children, but they may be other relatives, partners, friends, businesses, or organizations. For example, an alumna of a university can name the university as a beneficiary of her life insurance policy.

In addition it is not unusual for a person to have more than one policy. Consider the following:

> For Larry Wigand, the easiest decision to make about his life insurance was to get a new policy. True, he already had one, but that was before the three kids came along. "I just wanted to make sure that my family would be taken care of in my absence," says Mr. Wigand, a construction contractor in Fort Wayne, Ind. "I want them to enjoy life like I was still around to provide for them."[2]

Cash value
Accumulated savings that provide a refund to the owner of the policy if the contract is terminated prior to death.

Some people also like to have a life insurance policy that has a savings feature (called the cash-value element or more simply **cash value**) that provides a refund to the owner of the policy if the contract is terminated prior to death. In this way money can be gained from a life insurance policy without anyone dying. This savings feature can be used for retirement or for other purposes, such as paying off mortgages or other debts. Examples of cash value life insurance policies are whole life, universal, and variable. These will be described later in the chapter.

HOW DOES LIFE INSURANCE WORK?

As with other types of insurance, the concept of life insurance is based on the pooling of risk. Life insurance is essentially a method by which large groups of individuals equalize the burden of financial loss from death by pooling their funds and distributing them through insurers to the beneficiaries of those who die.

Insurers charge premiums which are fees paid to the companies for insurance protection. The companies take the premiums and invest them, making profits. They base premium rates on their calculation of mortality rates across millions of people. Thus mortality rates form the basis of life insurance charges. The company has to make more money from premiums and investments than they pay out in death benefits, the amount of money paid to beneficiaries when a policyholder dies. A death benefit is the face value of the policy less any unpaid policy loans or other claims against the policy.

The National Association of Insurance Commissioners (NAIC) approved a mortality table that is used in the industry (see Exhibit 10.1). The figures in this exhibit show the life expectancy of a male or female at a certain age. For example, a 25-year-old female's life expectancy is 52.34 years meaning 52.34 is the average number of years all females alive at age 25 will still live. Based on this table, a 25-year-old female, on average, is expected to live to age 77.34. Insurers base their projections of benefit payouts on their studies of such factors as age, gender, heredity, and health habits. Also, they base their rates on analyses of these factors. Over the twentieth century, life expectancy rose dramatically, so one would assume in the twenty-first century that people will continue to live longer and life expectancy tables will be adjusted accordingly.

Once it is determined that life insurance is needed, then the search for the best company and policy begins. The search includes comparing prices and services. Besides searching locally, national companies give life insurance price quotes over the telephone or the Internet (see the Electronic Resources section). Insurance agency quotation services will send a variety of proposals from the highest-rated, lowest cost companies available. The services receive a commission if you buy a policy from them, but you are not obligated to do so. Before receiving information or signing anything make sure you know the costs, if any, involved.

After a company and a policy are decided upon an application is filled out. Application forms can be obtained from employers, insurance agents, over the Internet or the telephone, or they may show up unsolicited in the mail.

CONSUMER ALERT

Since the main purpose of insurance is protection, one must be cautious about using it for savings. Other investments can provide a better return.

Commissioners Standard Ordinary Mortality Table

Age	Male Mortality Rate per 1,000	Male Expectancy, Years	Female Mortality Rate per 1,000	Female Expectancy, Years	Age	Male Mortality Rate per 1,000	Male Expectancy, Years	Female Mortality Rate per 1,000	Female Expectancy, Years
0	4.18	70.83	0.89	75.83	50	6.71	25.36	4.96	29.53
1	1.07	70.13	0.87	75.04	51	7.30	24.52	5.31	28.67
2	0.99	69.20	0.81	74.11	52	7.96	23.70	5.70	27.82
3	0.98	68.27	0.79	73.17	53	8.71	22.89	6.15	26.98
4	0.95	67.34	0.77	72.23	54	9.56	22.08	6.61	26.14
5	0.90	66.40	0.76	71.28	55	10.47	21.29	7.09	25.31
6	0.85	65.46	0.73	70.34	56	11.46	20.51	7.57	24.49
7	0.80	64.52	0.72	69.39	57	12.49	19.74	8.03	23.67
8	0.76	63.57	0.70	68.44	58	13.59	18.99	8.47	22.86
9	0.74	62.62	0.69	67.48	59	14.77	18.24	8.94	22.05
10	0.73	61.66	0.68	66.53	60	16.08	17.51	9.47	21.25
11	0.77	60.71	0.69	65.58	61	17.54	16.79	10.13	20.44
12	0.85	59.75	0.72	64.62	62	19.19	16.08	10.96	19.65
13	0.99	58.80	0.75	63.67	63	21.06	15.38	12.02	18.86
14	1.15	57.86	0.80	62.71	64	23.14	14.70	13.25	18.08
15	1.33	56.93	0.85	61.76	65	25.42	14.04	14.59	17.32
16	1.51	56.00	0.90	60.82	66	27.85	13.39	16.00	16.57
17	1.67	55.09	0.95	59.87	67	30.44	12.76	17.43	15.83
18	1.78	54.18	0.98	58.93	68	33.19	12.14	18.84	15.10
19	1.86	53.27	1.02	57.98	69	36.17	11.54	20.36	14.38
20	1.90	52.37	1.05	57.04	70	39.51	10.96	22.11	13.67
21	1.91	51.47	1.07	56.10	71	43.30	10.39	24.23	12.97
22	1.89	50.57	1.09	55.16	72	47.65	9.84	26.87	12.28
23	1.86	49.66	1.11	54.22	73	52.64	9.30	30.11	11.60
24	1.82	48.75	1.14	53.28	74	58.19	8.79	33.93	10.95
25	1.77	47.84	1.16	**52.34**	75	64.19	8.31	38.24	10.32
26	1.73	46.93	1.19	51.40	76	70.53	7.84	42.96	9.71
27	1.71	46.01	1.22	50.46	77	77.12	7.40	48.04	9.12
28	1.70	45.09	1.26	49.52	78	83.90	6.97	53.45	8.55
29	1.71	44.16	1.30	48.59	79	91.05	6.57	59.35	8.01
30	1.73	43.24	1.35	47.65	80	98.84	6.18	65.99	7.48
31	1.78	42.31	1.40	46.71	81	107.48	5.80	73.60	6.98
32	1.83	41.38	1.45	45.78	82	117.25	5.44	82.40	6.49
33	1.91	40.46	1.50	44.84	83	128.26	5.09	92.53	6.03
34	2.00	39.54	1.58	43.91	84	140.25	4.77	103.81	5.59
35	2.11	38.61	1.65	42.98	85	152.95	4.46	116.10	5.18
36	2.24	37.69	1.76	42.05	86	166.09	4.18	129.29	4.80
37	2.40	36.78	1.89	41.12	87	179.55	3.91	143.32	4.43
38	2.58	35.87	2.04	40.20	88	193.27	3.66	158.18	4.09
39	2.79	34.96	2.22	39.28	89	207.29	3.41	173.94	3.77
40	3.02	34.05	2.42	38.36	90	221.77	3.18	190.75	3.45
41	3.29	33.16	2.64	37.46	91	236.98	2.94	208.87	3.15
42	3.56	32.26	2.87	36.55	92	253.45	2.70	228.81	2.85
43	3.87	31.38	3.09	35.66	93	272.11	2.44	251.51	2.55
44	4.19	30.50	3.32	34.77	94	295.90	2.17	279.31	2.24
45	4.55	29.62	3.56	33.88	95	329.96	1.87	317.32	1.91
46	4.92	28.76	3.80	33.00	96	384.55	1.54	375.74	1.56
47	5.32	27.90	4.05	32.12	97	480.20	1.20	474.97	1.21
48	5.74	27.04	4.33	31.25	98	657.98	0.84	655.85	0.84
49	6.21	26.20	4.63	30.39	99	1,000.00	0.50	1,000.00	0.50

Source: American Council of Life Insurance.

EXHIBIT 10.1 Commissioners Standard Ordinary Mortality Table. Reprinted with permission.

CONSUMER ALERT

When filling out the application, be truthful. Any omissions or errors on the application can negate its acceptance or the ability of the survivors to receive payment.

Preferred risks
People whom insurance companies prefer to insure.

Most insurers have restrictions on whom they will cover. Certain policies are sold only to those whom the insurance industry calls **preferred risks**. To qualify for preferred rates (reduced premiums), generally a person must be in excellent overall health. Also the individual does not participate in any hazardous activities, and there must be no history of drug or alcohol abuse. For example, the insurer may prefer not to insure people who in the last three years have engaged in flying as a pilot or crew member, parachuting or hang gliding, skin or scuba diving, or auto, motorcycle or boat racing. Accuquote (www.accuquote.com) estimates that *about 60% of people qualify for the regular "preferred" rate.*

Preferred-plus (or "superpreferred") risk is an even more stringent policy, meaning the person, for example, does not use tobacco in any form. These criteria are based on actuarial tables that show that people who are most likely to live the longest do not smoke, drink excessively, use drugs, or engage in hazardous activities. Standard risk refers to persons with minor health impairments such as high cholesterol levels or 50 or more pounds overweight. Substandard risk refers to people with more major health concerns and hazardous lifestyles.

The Underwriting Process

Underwriting
Insurance company process of evaluating a policy application.

Underwriter
Person who conducts a policy application review.

Once the application is submitted and your physical exam is completed (if necessary), the insurance company evaluates your case. This evaluation process is called **underwriting**. The person conducting the review is called an **underwriter**. The underwriter may ask for additional information, such as a copy of medical records from doctors, employment verification, and so forth. Once the decision has been made to issue a policy, a rate is set. An initial rate quote may be obtained from the Internet in four minutes, but the whole underwriting process takes approximately 2–8 weeks, depending on the amount of information needed.

Lapsed Policies and Grace Periods

Lapsed policy
Termination of a policy because of non-payment.

Grace period
Time allowed for overdue payment without penalty or the policy lapsing.

Once the company accepts the application and the policy is issued, prompt payment is crucial for its continuation. A **lapsed policy** means that it has been terminated due to nonpayment. Usually lapsed policies can be reinstated if they have not been turned into cash. To reinstate the policy, lapsed payments with interest (a penalty charge) will have to be paid.

State laws generally require that multiyear and cash-value policies include a **grace period**, a period of time (usually 30 days) during which an overdue payment can be paid without a penalty or the policy lapsing. In other words the insurance is still in force even though premiums have not been paid for 30 days.

How often do people lapse their policies? About half of the cash value policies are dropped within seven years.[3] Usually policies are dropped because people can not keep up the payments. Of course this is good for insurers because they have collected thousands of dollars in premiums without having to pay out a cent. If payments are not kept up, the policy is dropped and the purchaser loses everything that was paid in. The Consumer Federation of America estimates that about $6 billion each year is wasted this way.

Objective 2:
Determine who needs life insurance and if you need it.

LIFE INSURANCE AND FAMILIES

"Most American families need life insurance and most have it," according to Leonard Sloan, financial columnist of *The New York Times*.[4] Life insurance benefits, depending on the type, can help families to pay final expenses; pay for the education of children; pay off a mortgage, leaving a debt-free home; provide an adequate income for the family when the primary or secondary breadwinner dies; and provide a retirement income for the surviving spouse.

Determining Need

The need for life insurance depends on one's own financial circumstances and that of dependents. Life insurance should be considered if any of the following situations applies:

* You have a dependent spouse.
* You have dependent children. Parents should acquire life insurance when a child is born or adopted.
* You have an aging or disabled relative who depends on you for support.
* You own a business. Business partners can be named as beneficiaries to pay off any debts the business has incurred. In Charles Dickens' story *A Christmas Carol* published in 1843, Scrooge's wealth increased considerably when his business partner Jacob Marley, died, as noted in this passage:

> Scrooge and he were partners for I don't know how many years. Scrooge was his sole executor, his sole administrator, his sole assign, his sole residuary legatee, his sole friend, and sole mourner. And even Scrooge was not so dreadfully cut up by the sad event, but that he was an excellent man of business on the very day of the funeral, and solemnised it with an undoubted bargain.[5]

* Your savings and retirement pension are not enough to support a spouse against cost-of-living increases.
* You want to leave a large amount of money to an organization or a cause (a sorority, church, the American Heart Association).

As mentioned earlier in addition to these reasons there are other reasons people buy life insurance, including to accumulate savings through cash value (although as noted in the consumer alert, the return is low compared to other forms of investing), to take advantage of the tax-deferred status of most cash value policies, and to pay estate taxes. The last chapter in the book, on estate planning will discuss this last reason further. Basically anyone with a large estate should consult tax experts before making a life insurance purchase or changing an existing policy.

College Students and Life Insurance

The previous list should give you some ideas about whether you need life insurance or not. You should ask your parents if they have a life insurance policy on you already. In general most college students do not need life insurance unless they have dependents.

Unfortunately college students, especially those near graduation, are often targeted for high-pressured life insurance sales tactics. The sales talks can be scary, threatening, and guilt-ridden ("you don't want to be a burden to your loved ones"). Insurance salespeople especially push cash value policies because of the high commissions that companies pay them.

> Commissions on cash value life insurance range from 50 to 100 percent of your first year's premium. An insurance salesperson, therefore, can typically make ten times more money (yes, you read that right) selling you a cash value policy than he can selling you term insurance.[6]

According to an insurance expert:

> Life insurance salespeople near college campuses suggest three potential needs to college students: (1) a general need for life insurance protection, (2) a general need for savings, and (3) a need to obtain the future right to purchase life insurance. Under careful examination, these arguments have little validity for most students.[7]

If you are graduating, you will probably be better off waiting to get a group policy from your employer. If money is tight, health insurance should take precedence over life insurance.

Women and Life Insurance

More than half of all American married women and nearly all single adult women younger than retirement age work outside the home. Single-parent women need insurance, especially if they are the main or sole support of their children. Working women tend to be underinsured, but if their income is critical to their family's welfare they need life insurance.

Do homemakers need life insurance? The answer to this depends on family size, the age of children, and how much money each adult contributes to the family income. If insurance dollars are limited, the first line of defense (coverage) should be the main breadwinner. Next is the secondary breadwinner. Last is the non-wage earning adult. However, since homemakers make a valuable contribution to the management of the home and care of the family, the question is "If she dies prematurely, what financial hardship would befall the family? What replacement services would be needed to cover the contribution she makes?"

Children and Life Insurance

Do children need life insurance? The short answer is "no." Nevertheless life insurance for children is sold even on infants. Twenty-five percent of all cash-value policies sold are on children.[8] Some people buy life insurance for their children because of the forced savings feature offered in cash value policies, but parents could earn a higher rate of return from many other types of savings and investments.

Because children do not contribute income to the family (an exception may be a child celebrity such as an actor or athlete whose income is the main support of the family), there is less reason to insure them than their parents. An insurance expert sums it up this way:

> Only after the wage earner and spouse have been insured adequately should life insurance on children be considered. It is generally believed that life insurance

should fulfill economic needs. If such needs are not present, the purchase of life insurance cannot be justified logically.[9]

Insuring People Other Than Family or Business Partners

Most people take a life insurance policy out on themselves or someone financially dependent on them. But there are circumstances in which individuals may take a policy out on someone besides their children, spouse, or business partner. Life insurance policies specify that the purchaser of the policy must have an insurable interest in the other party. This usually means a financial interest or an affectionate bond. Thus people can not take policies out on others whom they do not know personally or are not financially connected to (the President of the United States, a sports figure, or an entertainer).

How Much Is Needed?

Once it is decided that life insurance is needed, the next question is "how much?" Basically a person wants enough money to cover dependents' immediate cash needs and living expenses. Savings and investments and any unusual anticipated expenses should be considered in estimating how much life insurance is needed. To make it easier to figure, an old rule of thumb was that you needed 5 to 7 times your annual salary, depending on your lifestyle, number of dependents, and dependents' other sources of income. So if a person who was the main support of the family made $20,000 a year, he or she should have $100,000 to $140,000 face value in a life insurance policy.

Given the 5-to-7 times formula, consider the following case. Don (the main support), age 50, is married to Linda and they have two teenagers. Early in their marriage he made $8,000 a year; now, 25 years later, he makes $90,000 a year. His term life insurance policy, which he bought years ago, is for $150,000 and costs him $500 a year. Given his current income and the needs of his family (living expenses, a mortgage, and college educations coming up), he is severely underinsured. How much life insurance should he have? According to the 5-to-7 times rule, Don should have $450,000–$630,000. This range, some experts would say, is too low. For example, using the Life Insurance Calculator available from Accuquote at www.accuquote.com, it is estimated that Don would need $689,632 of life insurance given his income level and desire to have enough insurance to take care of his children through college.

Exhibit 10.2 shows another case, this time an estimate of what a 25-year-old single person, Jessica—who earns $30,000 a year and has a two-year-old child—needs. Jessica would like 20 years of coverage, long enough to fund her child's college education. The calculator indicates that Jessica needs $438,583 of life insurance. Notice that the calculator takes into account two key factors: inflation and interest.

As shown by Don's and Jessica's examples, age and lifestage come into play when determining the amount of face value. Both Don and Jessica have dependent children, but a person whose children are grown and earning their own living, whose mortgage is paid, who has investments, and whose spouse has a good-paying job and a retirement plan, needs less life insurance than someone with small children, a nonworking spouse, no investments, and a large mortgage.

What Does Life Insurance Cost?

Many people pay more per year for life insurance than for any other kind of insurance so it is important to consider the factors affecting cost. In the previous example Don was paying $500 a year, but costs vary depending on many factors:

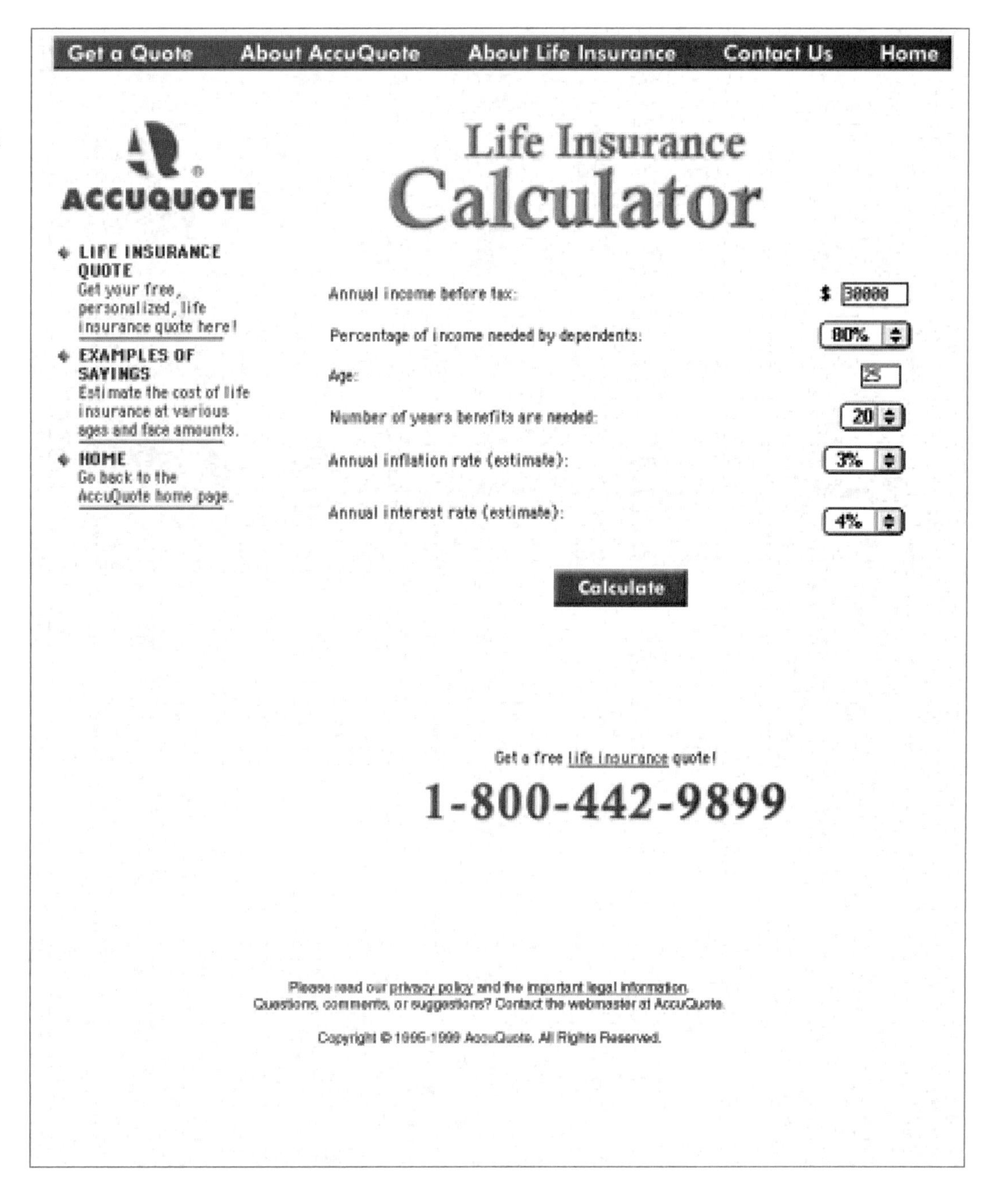

EXHIBIT 10.2
Jessica's Life Insurance Needs. Estimated using one online calculator.

This calculator is provided to you courtesy of www.accuquote.com, the nation's leading life insurance quoting and brokerage firm.

- The face value
- The type of insurance
- The company you buy it from
- The commission the agent gets
- How long the company thinks you will live
- The premiums (the amount you pay and for how long)
- Fees
- Whether your employer provides free or reduced rate life insurance.

As a general rule the least expensive insurance is a group policy. Most people obtain these from their employer. You may have to pay part or all of the premiums or your employer may pay the whole premium.

EXHIBIT 10.2
Continued

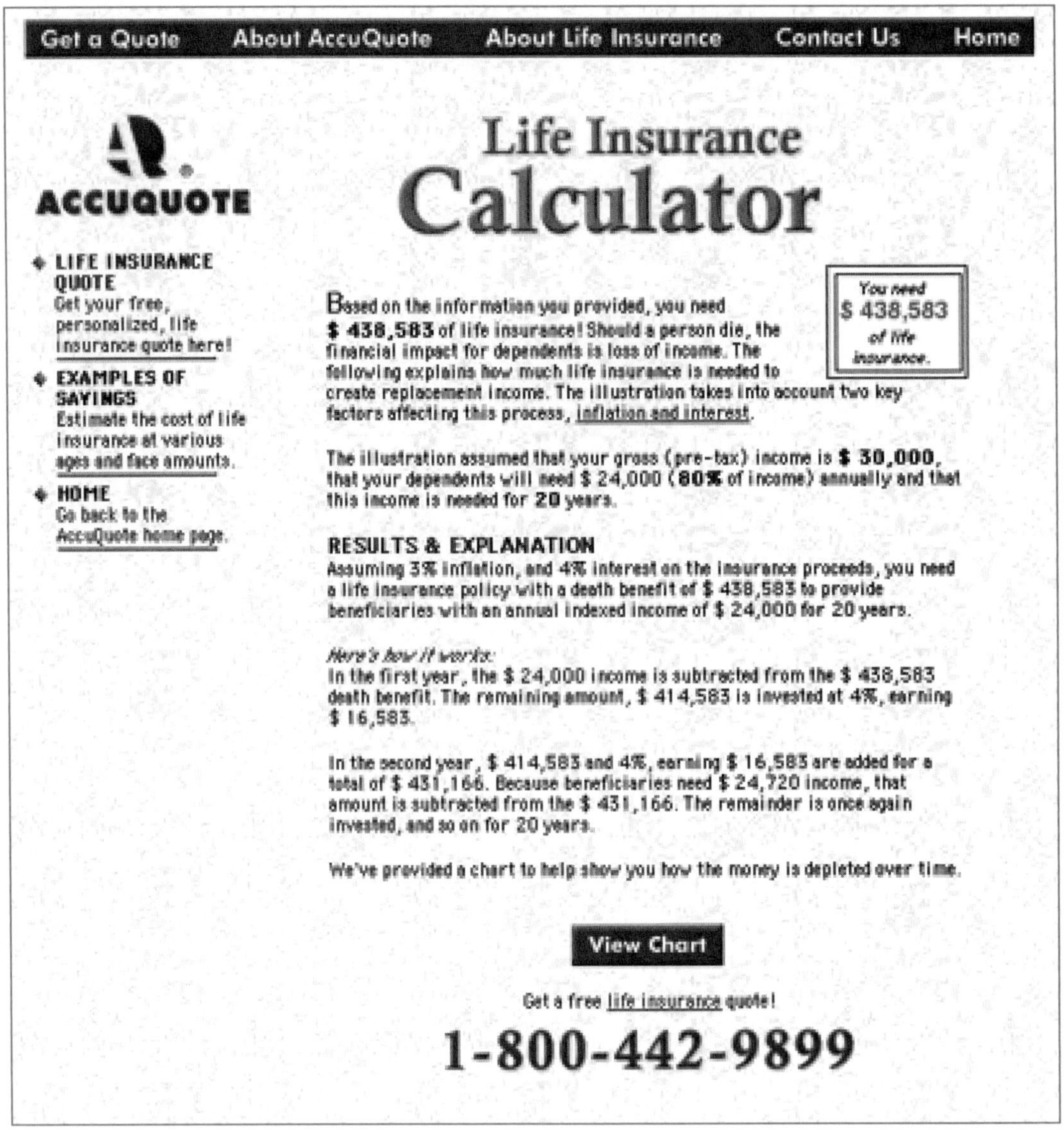

Get a Quote | About AccuQuote | About Life Insurance | Contact Us | Home

ACCUQUOTE

◆ **LIFE INSURANCE QUOTE**
Get your free, personalized, life insurance quote here!

◆ **EXAMPLES OF SAVINGS**
Estimate the cost of life insurance at various ages and face amounts.

◆ **HOME**
Go back to the AccuQuote home page.

Life Insurance Calculator

You need $ 438,583 of life insurance.

Based on the information you provided, you need **$ 438,583** of life insurance! Should a person die, the financial impact for dependents is loss of income. The following explains how much life insurance is needed to create replacement income. The illustration takes into account two key factors affecting this process, inflation and interest.

The illustration assumed that your gross (pre-tax) income is **$ 30,000**, that your dependents will need $ 24,000 (**80%** of income) annually and that this income is needed for **20** years.

RESULTS & EXPLANATION
Assuming 3% inflation, and 4% interest on the insurance proceeds, you need a life insurance policy with a death benefit of $ 438,583 to provide beneficiaries with an annual indexed income of $ 24,000 for 20 years.

Here's how it works:
In the first year, the $ 24,000 income is subtracted from the $ 438,583 death benefit. The remaining amount, $ 414,583 is invested at 4%, earning $ 16,583.

In the second year, $ 414,583 and 4%, earning $ 16,583 are added for a total of $ 431,166. Because beneficiaries need $ 24,720 income, that amount is subtracted from the $ 431,166. The remainder is once again invested, and so on for 20 years.

We've provided a chart to help show you how the money is depleted over time.

View Chart

Get a free life insurance quote!

1-800-442-9899

How Is the Beneficiary Paid?

When the insured dies, the beneficiary may:

* receive a lump-sum settlement of the face value (98 out of 100 policyholders choose this).
* receive proceeds over a given period.
* leave the money with the insurer temporarily and draw interest on it.
* use it to purchase an annuity guaranteeing regular payments for life.

CONSUMER ALERT

Since widows often receive a lump sum settlement ($100,000 or $200,000), they are targets of con artists and fraudulent investors. These disreputable people read the obituary columns in newspapers in search of potential victims. Widows and their families should protect their money by dealing with advisors with whom they are well acquainted and by educating themselves on financial matters.

EXHIBIT 10.2
Continued

Get a Quote | About AccuQuote | About Life Insurance | Contact Us | Home

ACCUQUOTE

- **LIFE INSURANCE QUOTE**
  Get your free, personalized, life insurance quote here!
- **EXAMPLES OF SAVINGS**
  Estimate the cost of life insurance at various ages and face amounts.
- **HOME**
  Go back to the AccuQuote home page.

Life Insurance Calculator

In the following chart, each year of income has been indexed to keep pace with 3% inflation. Inflation reduces the purchasing power of money. The $ 24,000 you earned today will not buy you the same amount of goods and services next year. That's why $ 24,720 is needed in the second year.

Additionally, interest at 4% was added to the remaining money each year. As you can see, over the course of 35 years, the original $ 438,583 has been depleted. Nothing is left.

Year	Age	Investment capital	Income at 3% index	Capital remaining	Interest at 4%
1	25	$ 438,583	$ 24,000	$ 414,583	$ 16,583
2	26	$ 431,166	$ 24,720	$ 406,446	$ 16,258
3	27	$ 422,704	$ 25,462	$ 397,243	$ 15,890
4	28	$ 413,132	$ 26,225	$ 386,907	$ 15,476
5	29	$ 402,383	$ 27,012	$ 375,371	$ 15,015
6	30	$ 390,386	$ 27,823	$ 362,563	$ 14,503
7	31	$ 377,066	$ 28,657	$ 348,408	$ 13,936
8	32	$ 362,345	$ 29,517	$ 332,828	$ 13,313
9	33	$ 346,141	$ 30,402	$ 315,738	$ 12,630
10	34	$ 328,368	$ 31,315	$ 297,053	$ 11,882
11	35	$ 308,936	$ 32,254	$ 276,682	$ 11,067
12	36	$ 287,749	$ 33,222	$ 254,527	$ 10,181
13	37	$ 264,708	$ 34,218	$ 230,490	$ 9,220
14	38	$ 239,710	$ 35,245	$ 204,465	$ 8,179
15	39	$ 212,643	$ 36,302	$ 176,341	$ 7,054
16	40	$ 183,395	$ 37,391	$ 146,004	$ 5,840
17	41	$ 151,844	$ 38,513	$ 113,331	$ 4,533
18	42	$ 117,864	$ 39,668	$ 78,196	$ 3,128
19	43	$ 81,324	$ 40,858	$ 40,465	$ 1,619
20	44	$ 42,084	$ 42,084		

Get a free life insurance quote!

1-800-442-9899

Annuity
Contract in which the insurer promises the insured a series of periodic payments.

An **annuity** (conventional or fixed) is a contract in which the insurer promises the insured a series of periodic payments. *It is the opposite of life insurance because an annuity pays as long as the person lives.* Life insurance pays when someone dies unless there is a cash value option. In an annuity someone near retirement age makes a single payment or multiple payments (called premiums, discussed earlier) to build an annuity fund in the hope that he or she will live long enough to get more back from the insurer than was paid in.

Variable annuities
Investments that fluctuate with those of underlying securities.

There are also **variable annuities,** whose value fluctuates with that of an underlying securities portfolio. This contrasts with a conventional or fixed annuity, whose rate of return is constant and therefore vulnerable to inflation effects. Variable annuities are tax-deferred, and the return to the investor may be in the form of a periodic payment that varies with the market value of the portfolio or a fixed minimum payment with add-ons based on the portfolio's appreciation.

Usually annuities are associated with older people, but any adult can have an annuity. *What annuities and life insurance have in common is that they are both purchased from life insurance companies.* More discussion of annuities will take place in the retirement chapter.

Objective 3:
Distinguish between term, whole life, universal life, and variable life insurance.

FOUR BASIC TYPES OF LIFE INSURANCE

The four basic types of life insurance are term, whole life, universal life, and variable life. The latter three (whole, universal, and variable) fall under a category called permanent life insurance, also called cash value insurance. In permanent life, each type of policy offers a different way to pay and to invest the cash-value account. The four types will be discussed in the following sections followed by a discussion of more obscure types.

Term

Term life insurance
Policies that offer pure protection, no savings feature, over a specified period of time.

Term life insurance policies are often called "pure protection" or "pure insurance" because they offer protection for only a specified number of years. It is, by definition, temporary insurance. Protection expires at the end of the period (unless it is renewed), and there is no cash value remaining. The only way to collect is to die during the term. If death occurs, beneficiaries collect the face amount—for example $100,000—of the policy, free of income tax. To conclude, there is no savings element in term insurance, and the time period is limited. Because of these features term insurance is usually the least expensive of the four basic types.

Term insurance is contingent on regular payments. If you stop making payments, the insurance stops. If you die during a period when payments have not been made, then the beneficiaries do not get the life insurance money. Coverage stops at the end of the term, but it can be renewed if there is a guaranteed **renewable clause**. So if you buy term, you want a guaranteed renewable clause. When you renew, because you are older, the cost increases.

Renewable
A clause that allows for renewing the policy.

Convertibility
A provision that allows the conversion of a term policy to whole life.

An important provision in term policies is **convertibility**. This provision allows the insured to convert the policy to a whole life policy providing the same death benefit without proof of insurability (no medical examination).

Term policies can be for a single year, five years, long term (10, 15, 20, 25, or 30 years) or multiyear. In multiyear term policies the most common types are level and decreasing. **Level-premium term insurance** (or **guaranteed-premium**) has a constant face value all the years of the policy, but the premium increases (costs more) each time you renew your policy.

Level-premium term or guaranteed-premium
Policy with constant face values but premiums increase with each renewal.

Decreasing term
Policy has decreasing face value but premium remains the same.

Decreasing term insurance has decreasing face value, but the premium remains the same. The reason someone may want to have decreasing term is to protect a big-ticket item, such as paying off a home mortgage. Once the mortgage is paid off, the person may feel comfortable with less life insurance.

To summarize, not all term insurance is alike. For example, renewable policies, as the name suggests, can be renewed annually, but each time they are renewed, the premium goes up. A level-premium policy means that the payments are fixed for five or ten years, so although initially premiums can be high, they should in time level out (given inflation). For example, a renewable term policy could start out with premiums of $432 a year and reach $920 by the tenth year. In contrast, a level premium at $513

CONSUMER ALERT

Most level-premium term policies contain guarantees of level premiums, however, some policies do not. Without a guarantee the insurance company can raise premiums. Be sure you understand the terms of any insurance policy under consideration.

a year for ten years would be cheaper in the long run. The level-premium policy would cost $1500 less than the renewable term policy.[10]

To illustrate these concepts, comparisons of level term vs. decreasing term and level-term premiums vs. renewable term premiums are given in Exhibit 10.3.

Whole Life

Whole life insurance
Policy that combines insurance with a savings feature.

Whole life insurance, as the name implies, runs for the insured's whole life; it does not end after a specified number of years, and it *has a cash value*. It is a form of permanent insurance designed to remain in effect one's whole life. Generally the premi-

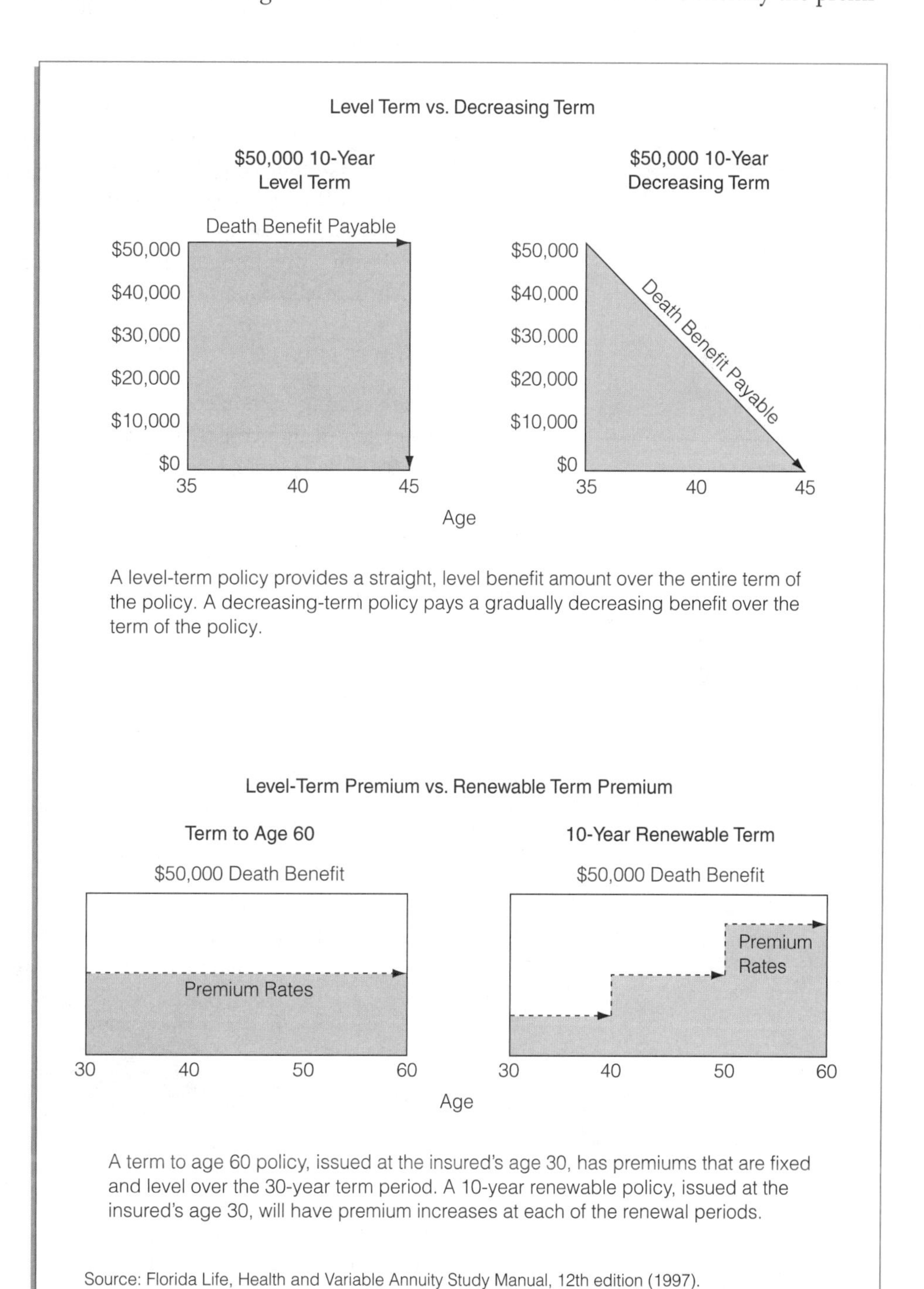

EXHIBIT 10.3
Policies Compared in Two Aspects

ums remain the same throughout the life of the insured. During the early years of the policy, premiums are much higher than for term life insurance. The cash value is paid when the contract matures or is surrendered. It accumulates on a tax-deferred basis. One of the curiosities of the insurance industry is that if you live long enough (age 100), you get all the money back from a cash-value life insurance policy. Exhibit 10.4 shows how cash value builds up.

Anyone signing up for whole life should inquire about the interest rate paid on the contract (usually it is low compared to interest paid in other forms of savings and investing). Here is how one expert figured the difference:

> Peter LoBello, a senior vice president at financial planners Strategic Capital Corp. in Midland Park, NJ, calculated what a 30-year-old male would earn from buying a whole-life policy and compared it with returns of buying a term policy and investing the difference in premiums, which are usually higher for permanent policies. For a $250,000 whole-life policy, Mr. LoBello chose one from a major insurer with a premium of $2,450 a year. After 30 years, Mr. LoBello figured, the cash value on the policy would be more than $193,000 with a death benefit of over $408,000. The same man would be able to find a 30-year term policy for $250,000 with a premium of just $435 a year. Using historical returns on mutual funds, Mr. LoBello found that the difference between the term and whole-life policies, if invested in the typical growth and income fund, would have grown to more than $731,000.[11]

If you are dealing with agents, they may promote the whole life or another type of permanent life insurance policy because they get a higher commission. Financial planners or fee-only insurance advisors who are not selling insurance may give more objective advice than agents who represent one company. Agents, on the other hand, can personalize the insurance experience by helping complete policy applications; delivering and explaining policy contracts; collecting and reporting premiums; changing beneficiary designations, if requested; assisting with policy loans; and servicing health and death claims.

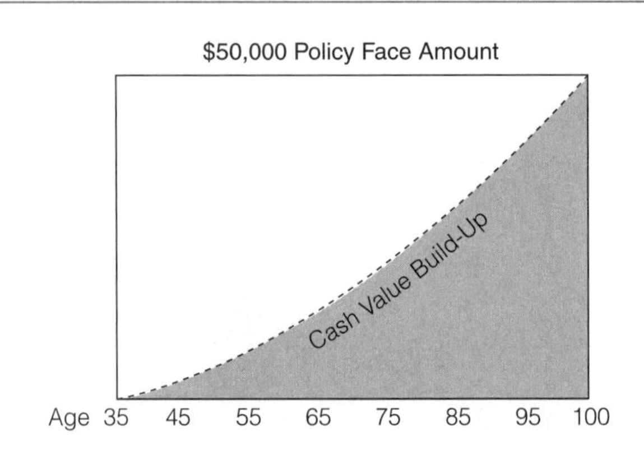

At the insured's age 100, the cash value of a whole life policy equals the face amount and will be paid to the insured as a "living benefit," if he or she is still living.

Source: Florida Life, Health and Variable Annuity Study Manual, 12th edition (1997). Tallahassee, FL: Dept. of Insurance, p.42.

EXHIBIT 10.4
Whole Life Insurance

Universal Life

Universal life insurance
Flexible type of policy that combines term life insurance with a tax-deferred savings account.

Universal life insurance is a fairly new option that combines term insurance protection with a tax-deferred savings account allows you, after an initial payment, to pay premiums at any time and in any amount, and to modify the face amount. It is considered a flexible life insurance plan since the policies are interest-sensitive and permit owners to adjust the death benefit and/or premium payments to fit their situation. However, investment choices are more limited than they are for variable life insurance (to be covered in the next section).

Also some universal policies allow you to skip payments. If a payment is skipped, the administrative and death benefit costs are deducted from the policy's cash value. Insurers send a yearly statement that shows how much interest the policy is earning and how much cash value it has.

Variable Life

Variable life insurance
Policy that allows for flexibility in types of investments.

Variable life insurance offers another life insurance option by providing the flexibility of investing in more than one type of account (stocks, bonds, money market portfolios), as well as the ability to transfer between the accounts. Since these types of accounts go up and down, the death benefit will vary, depending on the performance of the market in which the funds are invested. So more risk is incurred, but more gain is possible. Any loans or withdrawals may reduce cash values or death benefits.

Policy Choice: Term Versus Permanent

Policy choice essentially comes down to your preference for purchasing insurance with or without the savings feature (cash value). Term life insurance is pure insurance. The other types discussed fall under the category of permanent insurance which means they have a cash value associated with them. They allow some money back or have a fund from which you can borrow, but you pay for this feature in higher premiums. Generally experts agree that:

> If you're uncertain about whether to go for a term policy or a permanent policy, go for the term insurance. By cashing in a permanent policy after only a couple of years, you will probably face surrender charges as well as the risk of not being able to get a permanent policy later in life because you are older or have an illness. "You really should stick with the basics," says Mr. Bellmer of Deerfield Financial Advisors. "You're buying the insurance for the death benefit."[12]

Another aspect to consider in making a choice between term and permanent is time. If it is needed for less than ten years term is usually the best choice.

Other Types of Life Insurance

Term and the three types described are undoubtedly the ones most often chosen. There are lesser-known types that offer a variety of coverage: survivorship or 2nd-to-die, credit life insurance, endowment insurance, family income policies, and family policies.

Survivorship or 2nd-to-die
Policy that pays a death benefit at the later death of two insured individuals.

Survivorship or 2nd-to-die This type of coverage is offered as whole life or universal life that pays a death benefit at the later death of two insured individuals, usually husband and wife. Since the mid-1980s this has been a popular choice of wealthy individuals as a method of discounting their future estate tax liabilities. Most wealthy individuals try to arrange their estates to delay payment of any estate taxes until the second insured's death. In so doing, an individual is looking out for the welfare of his or her spouse and passing on estate-tax problems on to heirs who may be in a better position (healthwise and financially) to deal with them.

Credit life insurance
Policy used to repay debts.

Credit life insurance This is used to repay a personal debt (an auto loan or home mortgage) if the borrower dies before doing so. The idea behind it was that one should not leave debts as a burden to one's family. Usually these policies are not the best buy, and an individual would be better off buying less expensive decreasing term insurance.

Endowment insurance
Policy offers protection for a limited time with a savings feature.

Endowment insurance This offers protection for a specified period of time—for 10, 20, or 30 years or up to age 65. Funds can be accumulated with the face value paid at the maturity date in either installments or a lump sum. Premiums are relatively high because the full face amount is normally paid sooner than for whole life insurance. Some people like to get endowments to meet future known expenses, such as college educations for their children or their own retirement fund.

Family income policies
Combine whole life with decreasing term insurance typically for the main breadwinner.

Family income policies These combine whole life with decreasing-term insurance for additional temporary protection. Typically the main breadwinner is covered. For example, a person may have a $200,000 base policy with a 15-year family-income provision. If the insured dies within that time, his beneficiaries would receive $200,000 in benefits plus one percent of the face value, or $2,000 a month for fifteen years. If the insured lives beyond the 15 years after the start of the policy, the decreasing-term portion of the policy that provided the monthly income expires; the $200,000 in the whole life portion remains intact.

Family policies
Combine whole life with decreasing terms for all family members.

Family policies also combine whole life with decreasing-term insurance, but the coverage is typically for all members of the family (husband, wife, children) until each reaches a certain age. The highest amount of coverage is on the main breadwinner, less on a nonworking or lesser employed spouse, and the least amount on the children.

Clauses

The previous discussion of policies is fairly straightforward, but clauses (parts of policies) add to the number of decisions to be made. The usual clauses include named beneficiaries, a description of the type of settlement (lump sum, timed payments), and an explanation of the premium schedule. Payments may be due monthly, quarterly, semi-annually, or annually. Sometimes discounts are offered for annual payments because this saves the insurer billing expenses.

There is also a description of what types of death are covered and which ones are not. For example, two of the more interesting clauses that you may have heard about are:

Double- or triple-indemnity clause
A clause that doubles or triples the amount of insurance proceeds if the insured dies in certain types of accidents.

* Accidental death clause: This guarantees an additional amount of insurance, usually equal to the face amount of the policy, if the policyholder dies as a result of an accident. It is often referred to as **double- or triple-indemnity clause** because the effect of this clause is to double or triple the amount of insurance coverage. Not all accidents are covered. Normal exclusions include death resulting from airplane disasters, suicide, riots, or the commission of a felony (serious crime).
* Suicide clause: This clause limits the insurer's liability in the event of death by suicide within two years of the policy's issuance to the amount the policyholder already paid in premiums. So someone cannot take out a $1 million policy, then kill himself or herself the next day, and expect the family to collect.

Objective 4:
Discuss ways to find a secure company and reduce costs.

FINDING A FINANCIALLY SECURE INSURER

What if you paid into a company for years and then found out it went bankrupt or the agent skipped town with your money? What if you receive a life insurance policy offer in the mail from a company or group that you have never heard of? Since there are thousands of companies selling life insurance policies, and even though the industry is regulated, there are a few bad apples. How can you separate the good from the bad? Insurance company reputations can be checked through the Better Business Bureau, state departments of insurance, or publications (available at libraries) of independent rating agencies such A.M. Best, Moody's, Investors Services, or Standard and Poor's, Dunn & Phelps, and Weiss Research, as noted in Exhibit 8.3 in chapter 8. As a review, an A.M. Best rating of A++ or A+ means superior and a Standard & Poor's rating of AA+ means excellent. As an illutration of how the rating systems work, the American General Life and Accident Insurance Company of Nashville, Tennessee was given a rating of A++ by A.M. Best for financial condition and operating performance and an AA+ from Standard and Poor's for their ability to pay claims.

Besides the ratings by independent companies, good word of mouth is also valuable. Does the company have a good reputation for customer service and fair prices?

Cutting Costs

Several ways of cutting costs have already been suggested, but here they are summarized along with a few additional points:

Open-enrollment periods
Times when insurance companies or employers offer the opportunity to add on to existing policies.

* Buy a group plan which is usually cheaper than an individually obtained plan. Take advantage of **open-enrollment periods** that are offered by your insurance company or employer. During open enrollment you can increase the face value of an already held policy up to a certain amount, usually at a reasonable cost.
* Buy term instead of permanent life.
* Do not buy life insurance if you do not need it (no dependents).
* Comparison shop for competitive rates and features.
* Buy one $50,000 policy instead of five $10,000 policies.
* Do not smoke, drink excessively, or take drugs. Take care of your health.
* Add a rider (also called an **endorsement or floater**) to an existing policy. A rider is an amendment or addition to the basic insurance policy that can expand coverage to accommodate specified needs. *It is usually less expensive to add to an old policy instead of getting a new policy.* With a rider you could, for example, increase an existing policy from $50,000 to $100,000 Read the following for a further explanation:

CONSUMER ALERT

If life insurance is from a group employer-based plan and the employee leaves or retires, then the life insurance policy may be terminated. Before making any job changes, any possible effects on insurance coverage should be checked.

Replacing an old policy with a new one rarely makes sense, even if additional coverage is needed. The reason: New policies carry enormously steep upfront costs; the first year's entire premium can go to fees and commissions. If the policyholder has aged significantly since taking out the original policy, a new one likely involves steeper premiums. A better alternative is usually to keep the original policy and buy a new one only for the desired amount of additional coverage.[13]

Time to Revise

Policies should be reviewed every few years or more often if certain events warrant a more immediate reexamination such as:

* a marriage or divorce. Before marriage, prospective spouses should discuss each other's insurance policies to determine what needs to be changed regarding amounts and beneficiaries. In a divorce settlement life insurance should not be overlooked in terms of the amount, especially if there is a cash value and in regard to the naming of beneficiaries.
* an adoption or the birth of a child.
* a significant change in your health or your spouse's.
* buying a new house or refinancing an existing one.
* a promotion, an inheritance, winning the lottery (any large amount of money).
* retirement.
* children have left the nest.

The main question to ask yourself when purchasing or revising life insurance is "What provisions for the welfare of my family after my death do I need to make?" Life insurance is primarily a protection for loved ones. To illustrate the first point regarding marriage and divorce, read the following from a nationally syndicated newspaper column:

> Dear Ann Landers: After I read the letter from the widow who discovered she was not the recipient of her husband's insurance benefits, I phoned my husband's life insurance carrier. Sure enough, his ex-wife was still listed as the beneficiary. I then called his stock and savings plans manager and learned that his ex-wife was the beneficiary there, as well. This possibility never crossed my mind, and when I mentioned this to my husband, he said he hadn't thought about it, either.[15]

CONSUMER ALERT

Insurance agents may discourage policyholders from adding on to existing policies because they make bigger commissions when people sign up for new ones. Trying to persuade a policyholder to switch policies in a transaction that would do little or nothing more than generate commissions for the insurance agent is called **churning**.

Several companies who have been accused of churning have gone before federal appeals courts in multimillion dollar class-action suits brought on by policyholders. To rectify problems caused by churning, the industry is working on programs under which salaried underwriters help consumers complete their insurance applications to ensure they understand what they're buying, what it costs, and how they're paying for it.[14]

Churning
Practice of encouraging insureds to switch policies often to generate commissions.

Unclaimed Life Insurance Benefits and Unclaimed Property

More than 25 percent of life insurance policies that are sold go uncollected, according to David Folsom, an unclaimed-money expert.[16] Companies tend to file policies by policy number rather than Social Security numbers which makes it difficult for heirs to know about the policies. Another reason much of the money goes unclaimed is that individuals forget about old policies or neglect to leave a record of them. To rectify this situation, a search should be made of the deceased relative's canceled checks for the names of insurance companies and agents.

Unclaimed life insurance benefits are part of a general category called unclaimed property including stocks, retirement benefits, pensions, bank accounts, union benefits, and money from real estate. To find out if you have inherited property begin your search in your state's Unclaimed Property Division, which is usually located within the Department of Revenue or Treasury. If deceased relatives lived in other states those state's Unclaimed Property Divisions should be contacted also.

Insurance Planning Worksheet

The present chapter marks the end of a three-part series on insurance. The amount of information given in the insurance chapters can seem overwhelming. Part of the problem is that frequently insurance information is wrapped in almost incomprehensible jargon.[16] Two of the goals of the insurance chapters were (1) to help you decipher the code and (2) to prevent you from underinsuring or overinsuring yourself.

To further help you sort through the information and to apply it to your financial plan, Exhibit 10.5 provides an example of an insurance planning worksheet that was

EXHIBIT 10.5
Donna's Insurance

Personal Characteristics
Age.......22
Marital Status.......single
Annual salary.......$24,000
Number of dependents.......0
Number of automobiles.......1
Rents apartment

Type of Insurance	Company* or Gov.	Annual Cost	Estimated Worth	Plan to Purchase Within 5 years
Auto	Geico	$740	$30,000	
Disability				maybe
Health (HMO)	Capital Health Plan	$1200	unlimited	
Life				no
Long-Term Care				no
Medigap				no
Personal Liability (with Auto Ins.)	Geico	included with Geico, more will come with renter's insurance		
Renter's				yes
Other				maybe

*If blank below, Donna does not have this type of insurance.

filled out by Donna, a recent college graduate working in a government agency. Her annual salary is $24,000. Donna's health and disability insurance are provided by her employer, but she pays for her own auto insurance. She has no dependents so she does not have life insurance. Donna moved around while in college, but now has settled into her own apartment that she plans to stay in for at least two years, so she is considering getting renter's insurance. Exhibit 10.6 provides a worksheet for you to summarize your current insurance coverage and check off any types of insurance you plan to purchase in the next five years.

ELECTRONIC RESOURCES

As mentioned in previous chapters, most insurance companies have their own websites, the federal government has a catalog with information on insurance at **www.pueblo.gsa.gov**; and Quicken Insurance at **www.insuremarket.com** and Quotesmith.Com at **www.quotesmith.com** provide instant quotes on all kinds of insurance including life insurance. Other services are Insurance Quote Services at **www.iquote.com** and Accuquote at **www.accuquote.com**. Compared to other forms of insurance, the unique aspect of life insurance is the type of lifestyle questions that are asked, such as tobacco use or "Have you been a pilot, copilot, or crew member of an aircraft in the last 3 years?" When you use quoting services, your answers are sent confidentially to participating insurers. Within minutes you should receive price quotes from several companies.

The Independent Insurance Agents of America provides consumer information at **www.liaa.lix.com/**. The Insurance News Network provides Standard & Poor's

Personal Characteristics
Age.................................... ______________
Marital Status....................... ______________
Annual salary....................... ______________
(current or on graduation)
Number of dependents.......... ______________
Number of automobiles......... ______________
Own home or rent?................ ______________

Type of Insurance	Company or Gov.	Annual Cost	Estimated Worth	Plan to Purchase Within 5 years
Auto				
Disability				
Health				
Life				
Long-Term Care				
Medigap				
Personal Liability				
Renter's or Homeowner's				
Other				

EXHIBIT 10.6
Your Insurance Worksheet

Claims Paying Ability Reports at **www.insure.com/ratings/reports/**. LifeNet covers life insurance at **lifenet.com/index/html**. The American Council of Life Insurance in cooperation with the U.S. Office of Consumer Affairs and Consumer Information Center, gives tips about buying life insurance at **www.gsa.gov/staff/pa/cic/acli/index.htm**.

1. Does everyone need life insurance?

   ■ No, not everyone does. Usually life insurance is for those with dependents.

2. Which is less expensive—term or whole life?

   ■ Term insurance is less expensive.

3. Why do agents prefer selling cash value life insurance policies rather than term insurance?

   ■ Agents selling cash-value life insurance policies make more in commissions than if they sell term insurance policies.

4. Why is term insurance considered to be the only type of pure insurance?

   ■ Because term insurance is purely insurance with no add-ons. For example, term insurance does not have a savings feature.

5. How do life insurance needs change over the lifecyle?

   ■ Life insurance needs increase through midlife if one has more dependents and responsibilities and then it decreases as one approaches retirement age.

6. What happens at age 100 regarding life insurance?

   ■ If individuals live this long, they will get all the money in their cash-value life insurance policies from the insurer.

7. What is an annuity?

   ■ An annuity provides a series of steady level payments for life. It is a type of investment vehicle sold through life insurance companies.

8. What is a variable annuity?

   ■ Variable annuities allow you to place your money in a separate account, much like a mutual fund. An important benefit of a variable annuity is the ability to defer taxes when your investment is accumulating interest.

9. Can a life insurance policy be canceled by the company because of the insured's poor health?

   ■ No. Policies are never canceled because of a change in health. However, the insured has the right to cancel a policy at any time.

SUMMARY

Life insurance is a contract between an insurer and a policyholder specifying a sum to a beneficiary upon the insurer's death.

The main purpose of life insurance is to provide security and protection to someone from financial loss caused by the insured's death.

Seventy percent of American adults have some form of life insurance. Most American families need life insurance especially if they have dependent children. The main breadwinner should be covered first, the secondary breadwinner next.

Single working women with dependents tend to be underinsured. Children usually do not need life insurance.

As a general rule the least expensive insurance is a group policy. A group insurance plan insures a large pool of people under a single policy without requiring medical examinations.

To estimate how much life insurance is needed, individuals could multiply their income by five to seven times, use a general website calculator, visit insurance company websites, or use any of these methods as a basis and adjust them for special considerations, such as unusual anticipated expenses or the value of other investments and savings.

An individual wanting life insurance applies for a policy. Once the application is accepted and a physical exam (may be optional) has taken place, the underwriting process begins. The evaluation of an application is called underwriting and the person doing the review is called an underwriter.

Churning means that agents persuade policyholders to switch policies often so that the agents can generate higher commissions.

Usually beneficiaries receive a lump sum (the face value of the policy).

An annuity is the opposite of life insurance because it pays while you live. Annuities (conventional or fixed) and variable annuities are purchased from life insurance companies, usually by older people who use them for retirement income.

The four basic types of life insurance are term, whole life, universal life, and variable life. The latter three fall under the category of permanent life insurance that has a savings (or cash value) feature. Term insurance does not offer saving; it is considered "pure" insurance.

If term life is selected, the policy should be checked to see that it offers guaranteed renewability. This means the policy will be automatically renewed without a physical examination.

Life insurance policies have clauses naming beneficiaries and explaining the terms of settlement and premiums. In addition there are accidental death and suicide clauses.

Policies should be reviewed every two years to determine if coverage should increase, decrease, or remain the same, and to make sure the beneficiaries are still the ones desired. Policies should be reviewed more often if there are significant life changes.

College students are often approached by insurers. Usually they do not need life insurance.

Policies are rated by independent companies. Rating lists are available in libraries and on the Internet.

It has been estimated that more than 25 percent of life insurance benefits are unclaimed.

REVIEW QUESTIONS

(See Appendix A for answers to questions 1–10)

Identify the following statements as true or false

1. Seventy percent of American adults have life insurance.
2. Cash value and face value are the same thing.
3. Term insurance is usually more expensive than whole life insurance.
4. College students need life insurance more than they need any other kind of insurance.
5. The main purpose of life insurance is to provide security and protection for loved ones.

Pick the best answers to the following five questions.

6. A standard contract specifies that upon the death of a person whose life is insured, a stated sum of money (face amount) is paid to the person(s) or organization(s) designated in the policy as the ______________.

   a. premium
   b. adjuciary
   c. beneficiary
   d. annuitant

7. Using the 5-to-7 times rule, how much life insurance does a 45 year old father of two school-age children need? His annual income is $30,000.

   a. $14,000-$16,000
   b. $150,000-$210,000
   c. $250,000-$275,000
   d. $300,000-$310,000

8. An ____________ is a contract in which the insurer promises the insured a series of periodic payments.

   a. annuity
   b. appraisal
   c. affidavit
   d. actuary statement

9. Which of the following is not one of the four types of life insurance?

   a. Term
   b. Whole life
   c. Universal life
   d. Indenture

10. Trying to persuade a policyholder to switch policies in a transaction that would do little or nothing more than generate commissions for the insurance agent is called ____________.

    a. chaffering
    b. reversing
    c. churning
    d. factoring

DISCUSSION QUESTIONS

1. Life insurance planning takes into account a person's financial goals, assets, and future income and expenses. If a single working father has a three-year old daughter, does he need life insurance? Does the daughter need life insurance? What future expenses may the father want to provide for?
2. How does the concept of pooling work in regard to life insurance? How do insurers make money?
3. What might be gained by buying term insurance and investing the difference versus buying whole life insurance?
4. A company has been given an A.M. Best rating of "A+." Does this mean the company is superior and would probably be okay? Why or why not?
5. Changes in life events may indicate the need to revise current life insurance coverage. Besides marriage and divorce, what other life changes may indicate the need to reexamine life insurance policy coverage?

DECISION-MAKING CASES

1. Brenda is approached by a life insurance agent. She is a college student with no dependents and no large debts. Does she need life insurance? Explain your answer.
2. Using Exhibit 10.2 as a guide and the Internet website www.accuquote.com calculate how much life insurance a married couple, Allen and Erika, each 28 years old, should have. They each make $30,000 a year (so their combined income is $60,000). They have one son, Jared, age 5, whom they want to go to college. Allen and Erika have a small savings fund but no other investments. They do not plan to have more children.
3. Emma's grandfather just died. Her grandfather left Bertha, his wife and Emma's grand-

mother, a life insurance policy with a lump sum of $200,000. This is the most money Bertha has ever had. Since Emma's grandfather's obituary appeared in the newspaper, Bertha has been deluged with sales calls. What advice should Emma give her grandmother?

4. Mattie is a stay-at-home mom. A life insurance agent has called her, suggesting that she get a policy for herself. Her husband says "no way" since he is already covered by a policy and their money is tight. What are the arguments for and against life insurance for homemakers?
5. Tyler, age 27, has a small term life insurance policy, but now that he has married and his wife is pregnant he wants more insurance. Should he get another policy or add on to the one he already has? Which choice would probably be less expensive? If he looks into a new company, what websites can he use to get rate quotes.

GET ON LINE

Activities to try:

- Visit www.accuquote.com/ and click on the Life Insurance Needs Calculator to get a form that collects the information needed. Fill in the information requested and find out how much life insurance you need, using the calculator.
- Visit www.insuremarket.com and quotesmith.com to get price quotes on a term $100,000 life insurance policy. Type in your own information at both sites and compare the price quotes that you receive.
- Visit www.insuremarket.com and click on Annuities and then click on step one, "What is an Annuity?" Then click on step 2, "Do I need one?" and you will see four questions. Answer the four questions and see what response you receive.

INFOTRAC

To access InfoTrac, Follow the instructions at the end of Chapter 1.

Exercise 1: Type in *life insurance* for the key word search and you will find reference book excerpts, periodical references, 16 subdivisions, and 19 related subjects. You will also see articles on the life insurance industry and on specific companies. Click on Life Insurance Industry articles and scroll down the list of articles until you find one of interest; report on what you find. Although many of the articles cover the types of general information given in this chapter, you will also find current controversial issues within the life insurance industry such as the requiring of HIV-1 tests before issuing policies.

Exercise 2: This time, in the subject guide type in *annuities* for the key word search. This should bring up periodical references and related subjects. Under the periodical references you will find articles pro and con the purchasing of annuities.

(a) Select a point of view (pro or con) and report on the issues involved regarding annuities.

(b) You will also find the subtopic of variable annuities. Many of these articles will discuss the tax advantages of variable annuities; report on what these tax advantages are.

part three
investing

chapter eleven
fundamentals of investing

Getting Started Building Wealth

DID YOU KNOW?

* In 1637, in the Netherlands, a single tulip bulb was worth the cost of the best house in Amsterdam.
* One hundred shares of AT&T purchased in 1983 at a cost of $6,140 is now worth about $74,562, representing a 17.7 percent annual return.

OBJECTIVES

After reading Chapter 11, you should be able to do the following:

1. Define investment and explain why people invest.
2. Know when to start investing.
3. Describe the four steps in investing.
4. Explain the different types of investing.
5. Discuss various aspects of investing and investment strategies.

CHAPTER AND PART III OVERVIEW

As the tulip story in the "Did You Know?" shows, investments can rise and fall. Currently, most tulips bulbs sell for less than $1.00, but in 1637 their rarity and high demand drove up the price. In more recent times the value of stocks overall have soared. AT&T, one of the most widely held stocks, has ridden the wave along with Intel, Oracle, Microsoft and a host of others. How do you know how to pick a winner? And how do you get started? This chapter introduces the investment portion of the book that addresses these questions. The chapter's coverage is based on the assumption that readers have thought about their overall financial plan and are now ready to learn how to manage and enhance wealth through investing.

The present chapter first explores the concept of investing and why people invest. Next, knowing when to start investing is discussed followed by the four steps in the investing process. This is followed by a description of the ten main types of investments. The last section is devoted to investment strategies and other aspects of investing. As in all chapters there will be an Electronic Resources section near the end of the chapter.

The underlying principle guiding all the chapters in Part III is that to build wealth, one needs an investment plan and a good return on money. Since investments stretch your income further, these four investment chapters are critical to your financial success.

Objective 1:
Define investment and explain why people invest.

INVESTMENT DEFINED

Investment
Commitment of funds to achievement of long-term goals.

Investment is the commitment of funds (capital) to the achievement of long-term goals or objectives. The field of investments involves the study of the investment process. Investments serve a vital role in financial planning because they concern the management of wealth. They are the key *to long-term financial success.* Planning, information, and time are the investor's chief allies; fear of the unknown and risk are the main enemies.

The Purpose of Investment

Individuals invest to make money. Investments help money grow so that there is more to spend, save, and reinvest. Investors hope to enhance their future consumption options by increasing their wealth while protecting it from inflation, taxes, and other factors. To be more specific, through investing individuals hope to: achieve financial goals; increase current income; gain wealth and financial security; and have funds for homes, children's college educations, and retirement.

Investing Is Becoming More Important

Investing has always been an important part of building wealth and security, but in recent years it has become more critical because:

- of rising prices (inflation). For example, it is estimated that a cup of coffee that costs a dollar today will cost $2.00 in 2016. Likewise, the costs of cars, homes, and college educations will more than double by then.
- in many jobs salary raises are barely keeping pace with inflation. To get ahead, people need to invest. Also downsizing, early retirements, and job instability have given people more reasons to invest.
- people are living longer, their money has to last longer.
- of worries that Social Security will not exist by the time it will be needed.
- of the turn toward self-directed retirement plans. In a self-directed investment plan, the individual guides how retirement funds are invested. Over half of all retirement plans are self-directed. While employees in the past did not have much say about how their retirement funds were invested, current and future employees are being asked how they want their retirement funds invested.

Objective 2:
Know when to start investing.

PREPARATIONS FOR INVESTING

Before you start investing, the following tasks should be completed:

- Your life is stabilized. You are living in a place you intend to stay for awhile, you have a job, your moving expenses are paid for, your car runs, you have insurance, student loans are paid off or under control, etc. Sometimes students get anxious to start investing, but they should wait until their lives have some semblance of order.

* Part of this order is having a net worth estimate (what you own, what you owe) and a budget so you know how much you take in each month, how much you spend, and how much you have left to invest. If you have $50 a month left over then you could invest in a mutual fund, for example.
* Next, you should have a regular savings plan and money set aside for emergencies. The idea here is that you need to be a saver before you become an investor. Emergency funds can be stored in a number of ways including bank money market accounts, certificate of deposits (CDs), or money market mutual funds.
* Further, credit cards should be paid off or under control. For many people this is the hardest part. The reason that this is important is that it does not make financial sense to pay 18% on credit cards and make only 5% on investments. Interest paid on credit cards is money down the drain that can rarely be recouped through investments.
* Lastly, you should be enrolled in your employer's sponsored retirement plan. These plans have tax advantages and most employers match at least part of your contribution. Actually this is a form of investment, but it is so automatic (out of payroll deduction) that people tend not to think about it—but they should because retirement plans are the biggest investment most people have.

Objective 3:
Describe the four steps in investing.

FOUR STEPS IN INVESTING

Once you are ready to invest, a guiding principle is that the earlier you start the greater the return. Therefore you are encouraged to begin investing as soon as you can. Exhibit 11.1 shows the benefits of starting early.

Exhibit 11.2 shows the four steps in the investment process: setting goals and developing an investment attitude, assessing risk and return, selecting the right investments and allocating assets, and managing investments.

Setting Goals/Developing an Investment Attitude

One of the most important factors affecting goals is current income, but as a student you are in a different situation. Most students should base their investment strategy on anticipated income after graduation. Because students are moving toward new goals, college is a good time to think about what you want to achieve. For you to succeed, your goals should be flexible, realistic, specific, desirable, and evaluated. For example, a goal of turning a $28,000 yearly income into a million dollars in two years fails the reality test. A more realistic goal would be to save up a nest egg of $1,000 to invest.

Setting goals is one thing, fulfilling them is another, and this is where attitude and other factors come into play. As one example, will you rely solely on your employer's retirement package to reach the goal of a secure retirement, or do you think you should have more investments on top of that? Life and job circumstances as well as aspirations affect goals and the steps you will take to reach them.

> "You need an attitude that's sort of like following a sports team," says Hersh Shefrin, a finance professor at Santa Clara University in California. "There are some dry years and some good years, but people stick with their team. It's that personal identification that's key to commitment.[1]"

EXHIBIT 11.1
Start Investing Early

How to build $250,000 for retirement
The earlier you start, the easier it is

If you start at age...	You'd have to invest this much each month to reach that $250,000 goal
25	$45/month
35	$121/month
45	$349/month
55	$1,257/month

See footnote below

The Benefits of Starting Early

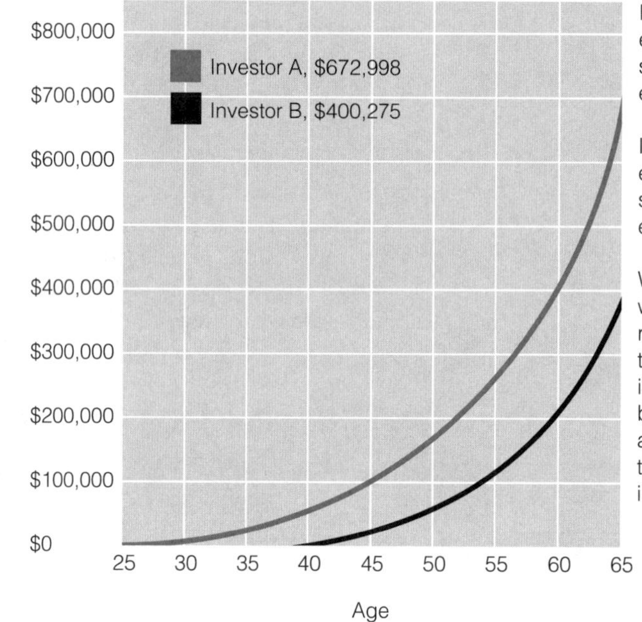

Investor A invests $2,000 each year for 10 years starting at age 25 and ending at age 35.

Investor B invests $2,000 each year for 30 years starting at age 35 and ending at age 65.

Which investor ends up with more money for retirement? Remarkably, the investor who only invested for ten years—but invested early—achieved higher returns than the investor who invested for 30 years!

Note: Projected returns based on the value of an investment with a 10% fixed rate of return.

Source: "The Power of Equities." (1997). Oakbrook Terrace, IL: Van Kampen American

The following questions are helpful in setting goals and developing an investment frame of mind:

1. What do I really want?
2. How much money will it take to get there?
3. How much money do I have (or will I have)?
4. How long will it take to get the money?
5. How much risk can I take?
6. Am I willing to give up things I want now to live better in the future?
7. Given my circumstances, what can I realistically expect?

Regarding point #1, if there are many goals you want to achieve or things you want to own or activities you want to do, how would you prioritize these? What do you want most of all?

EXHIBIT 11.2
Steps in the Investing Process

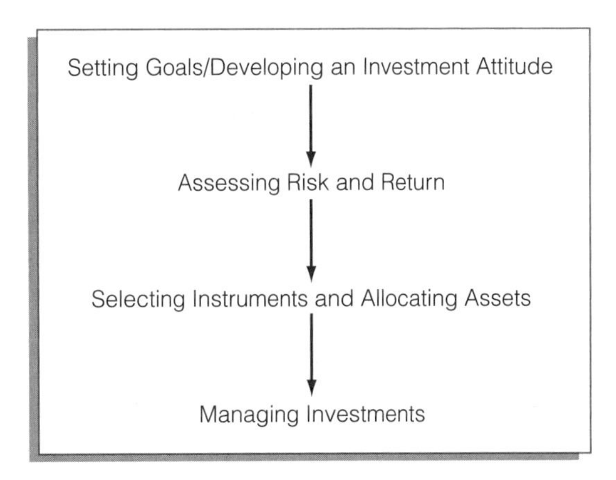

To practice goal setting, the goals of Colleen, a 23-year-old working on her masters degree, are given in Exhibit 11.3, and there are spaces for you to fill in with your goals. In the exhibit, goals are divided by time; short-term (0 to 3 years), midterm (3 to 7 years), and long-term (7+ years).

Today's college students are goal oriented according to a study commissioned by Northwestern Mutual Life Insurance Company. In their poll of 2,000 freshmen, 81 percent said owning a home is very important to them, 74 percent said life insurance is important and 46 percent said individual retirement accounts (IRAs) and pensions are important. Only thirty-percent said stocks, bonds, and mutual funds are very important for their security.[2]

Assessing Risk and Return

Once goals are set, the next step is figuring out risk and return. This evaluation revolves around three questions: How long do you plan to invest? What amount of expected return are you aiming for? How do you feel about risk?

The first question refers to investment horizon. How long do you need to save or invest to get something that you want? Is retirement 25 years away or 40 years away?

Financial Goals	Colleen's Goals	Your Goals
Short-Term—0 to 3 years		
Goal 1:	begin full time job	______
Goal 2:	increase savings	______
Goal 3:	repay student loans	______
Midterm—3 to 7 years		
Goal 1:	pay off last student loan	______
Goal 2:	buy new car	______
Goal 3:	invest in mutual fund	______
Long-Term—7+ years		
Goal 1:	buy a house	______
Goal 2:	travel	______
Goal 3:	buy new furniture	______

*Colleen is a 23-year-old graduate student, single, who owes $23,000 in student loans. She is self-supporting and will pay these off herself.

EXHIBIT 11.3
Listing Financial Goals by Time

As an example, long-time sports enthusiasts, Kimberly and her husband Jason are saving $300 a month so they can go to the next Olympics two years away. At this rate they will have $7,200 (plus about 2% interest if they put it in a savings account). That should cover their flights and a week's stay plus tickets to the events.

The second question revolves around anticipated future returns referred to as **expected returns**. For example, a bank may offer a certificate of deposit (CD) with an expected return of 6.5 percent if $1,000 is left in for one year. Kimberly and Jason could take the $3,600 they have saved at the end of a year and put it into a CD for one year and make a higher return than if they left it in a savings account.

Expected returns
Anticipated future returns.

Current income is money received from an investment. For example, the rent collected by the owner of an apartment building provides current income for the owner. Interest on savings accounts or stock dividends also provides current income.

Current income
Money received from an investment.

Capital gains refer to income that results from an increase in the value of an investment. Conversely, **capital losses** refer to lost income that results from a decrease in the value of an investment. Investors interested in the long term often favor foregoing current income for future capital gains. Capital gains and losses are realized only when an investment is sold. Capital-gains taxes are taxes on profits from the sale of capital assets.

Capital gains
Income that results from an increase in the value of an investment

Capital losses
Lost income resulting from a decrease in an investment's value.

The length of time an investment is held affects the tax rate. Currently if a stock is sold within 12 months of purchase, there is a short term capital gains tax. If it is held longer, the capital gains tax is reduced. Capital gains taxes make people hesitate before quickly buying and selling stocks. They encourage investors to leave their money in the market. The exception to this general rule is that stocks and bonds held in retirement accounts (since they are not accessed until retirement) may be bought and sold without capital gains taxes at each purchase and sale. Financial advisors or stock brokers should be consulted regarding capital gains tax assessments and other taxation policies that affect investments.

How do capital gains and losses relate to return? The **return** is the total income from an investment based on current income plus capital gains or minus capital losses.

Return
Total income from an investment.

The best measure of return is **total return,** which is the annual return on an investment including appreciation and dividends or interest.[3] It tells how well an investment performed in a year. For example, the total return on an investment that pays $200 in interest or dividends and increases in value by $100 a year is $300.

Total return
Annual return on an investment, including appreciation and dividends or interest.

Yield is return on a capital investment. Most commonly the word "yield" is used to refer to bonds. **Current yield** is the annual interest on a bond, divided by market price. It is the actual rate of return.

Yield
Return on a capital investment, typically bonds.

> For example, a 10% (coupon rate) bond with a face (or par) value of $1000 is bought at a market price of $800. The annual income from the bond is $100. But since only $800 was paid for the bond, the current yield is $100 divided by $800, or 12½%.[4]

Historically, stocks have been the most powerful hedge against inflation. Exhibit 11.4 shows how stocks have fared compared to inflation and other forms of investment over a seventy year period. Unfortunately past performance is an indicator, but not a guarantee of future performance.

Regarding question #3 at the beginning of this section, the subject of risk has come up many times in this book, but it is an especially important concept when it comes to investing. Risk refers to the chance of loss and perhaps a significant loss. How you handle risk greatly influences the investment decisions you will make. A government savings bond is virtually risk-free because it is backed by the federal government, whereas a stock in a corporation is riskier because no one can guarantee the

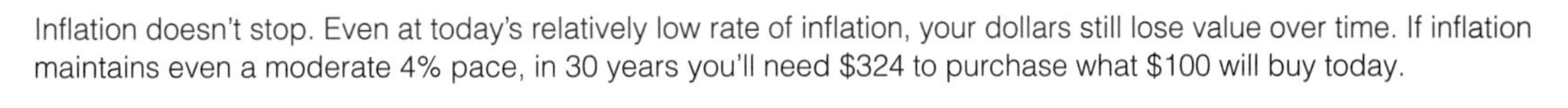
Inflation doesn't stop. Even at today's relatively low rate of inflation, your dollars still lose value over time. If inflation maintains even a moderate 4% pace, in 30 years you'll need $324 to purchase what $100 will buy today.

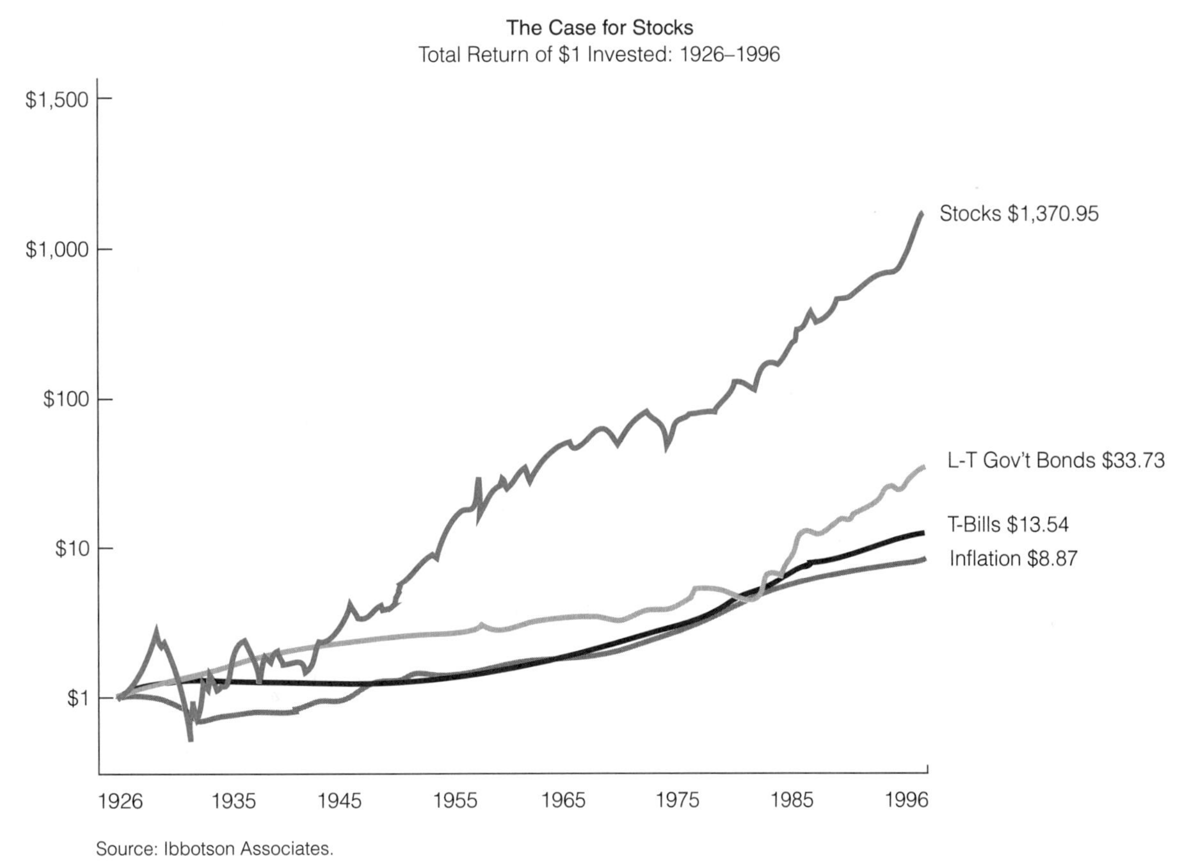

The S&P 500 is an unmanaged statistical composite measuring the performance of 500 stocks from 83 industrial groups. U.S. Treasury bonds are considered long-term investments and are subject to price fluctuations. The value of long-term bonds declines as interest rates rise. U.S. Treasury bills are short-term obligations of the U.S. Government which are purchased at a discount and mature at face value. Both the principal and yield of investment securities, however, will fluctuate with changes in market conditions. Inflation is measured by the Consumer Price Index, which is not an investment. This chart is for illustrative purposes only; it is not meant to forecast, imply, or guarantee the future performance of any particular investment vehicle. Common stocks involve greater risks than government bonds, as they are more volatile and have a greater potential for loss of principal.

EXHIBIT 11.4
Common Stocks Versus Inflation

future success of that company. Usually the greater the risk, the higher the potential yield. The question is "How much risk can you stomach?" Students who have experienced only savings accounts (that limp along at 2%, but are secure) have to think about this.

Financial planners and stockbrokers take into account the degree of risk that investors are comfortable taking and also the different levels of risk, depending on the type of investment. One way planners assess risk-taking personality is to find out about an investor's past investment behavior. Another way to assess it is by giving the investor a risk-tolerance quiz. Take the quiz shown in Exhibit 11.5 to determine if you are a conservative, moderate, or aggressive risk taker.

Such quizzes as the one in Exhibit 11.5 are useful self-assessment tools, although a variety of psychological factors may come into play (some people may think they are bigger risk takers than they really are), and reaction to risk may change over time.

Investment risk
Uncertainty over an investment's actual return.

Personality aside, the amounts of risk and return involved in investments themselves are of paramount importance in successful investing. **Investment risk** refers to uncertainty over what an investment's actual return may be over a period of time. The greater the uncertainty, the greater the risk, but also the greater the chance for a higher return.

Exhibit 11.6 shows an investment ladder that reveals the tradeoffs between risk and return. For example, savings bonds at the bottom of the ladder are one of the safest investments you can make, whereas futures contracts at the top are among the riskiest. Also note that bonds are generally safer than stocks, although within categories of investments, risk and return vary also. For example, municipal bonds (at Step #2) are safer than riskier junk bonds (at Step #4). Sometimes risk is low because there is a stated rate of return, such as in a certificate of deposit, and sometimes risk is high because there is no guaranteed rate of return, such as in a stock.

Selecting Investments and Allocating Assets

So far it has been said that investment selection should be based on your goals and an assessment of risk and return. Other factors include price, availability, desirability, and potential for growth. The reason "desirability" is on the list is that you can make money being a landlord or owning a convenience store, but you may not like the problems associated with owning rental property or running a store. Investing in mutual funds involves less personal time and energy.

Asset allocation
Target mix of different types of investments.

Eventually what investors seek is a comfortable, workable level of assets. **Asset allocation** is a target mix of stocks, bonds, and cash investments, such as Treasury bills, CDs, and money market funds. Exhibit 11.7 shows an asset allocation model that balances risk with returns. It is another way of showing some of the same concepts covered in Exhibit 11.6. The difference is that in Exhibit 11.7 you can place an X where you stand. Money market and short-term bonds are at the bottom because they usually involve low risk and low return and emerging markets are at the top because they involve the greatest risk but also the potential of the greatest return.

Exhibit 11.8 shows two other ways to visualize asset allocation. The simplest way is shown by the series of circles at the top of the exhibit. The circles take your investment type (as determined in the quiz in Exhibit 11.5—conservative, moderate, aggressive) and assigns it a suggested mix of securities (bonds, stocks, and cash management instruments). A more complex illustration of diversified portfolios is shown beneath the circle illustration. Investors should **diversify** their investments which means to put money into different forms of investing, such as stocks, bonds, mutual funds, and real estate. Diversification is an investment strategy that is employed to balance risk by ensuring that "not all your eggs are put into one basket."

Diversify
Putting money into different types of investments.

Portfolio
Combined holdings to increase diversification and to reduce risk.

A **portfolio** is created whereby the assets held by an investor are evaluated and managed as a group. A balance is sought by combining investments and securities in such a way as to achieve the highest potential return given the levels of risk involved. For example, investors may wonder if they have too many stocks or too few or if they have the right kind. According to Exhibit 11.8 a conservative investor may have 15 percent of his or her portfolio in blue chip stocks (IBM, General Electric), 10 percent in a mutual fund and so on. A moderate investor would have a smaller percentage in blue-chip stocks. In portfolio management the investors figure out what they are missing or what they need more of and fill in the gap with the next purchase or by selling a category they have too much of. A financial planner or account executive would help you arrive at this balance (the right combination of holdings) or you can try it on your own. Sources of self-directed help are given in the interactive websites described in the Electronic Resources section.

What's Your Risk Tolerance?

Circle the letter that corresponds to your answer.

1. Just 60 days after you put money into an investment, its price falls 20%. Assuming none of the fundamentals have changed, what would you do?
   a. Sell to avoid further worry and try something else
   b. Do nothing and wait for the investment to come back
   c. Buy more. It was a good investment before; now it's a cheap investment too

2. Now look at the previous question in another way. Your investment fell 20%, but it's part of a portfolio being used to meet investment goals with three different time horizons.

2A. What would you do if the goal were five years away?
   a. Sell
   b. Do nothing
   c. Buy more

2B. What would you do if the goal were 15 years away?
   a. Sell
   b. Do nothing
   c. Buy more

2C. What would you do if the goal were 30 years away?
   a. Sell
   b. Do nothing
   c. Buy more

3. The price of your retirement investment jumps 25% a month after you buy it. Again, the fundamentals haven't changed. After you finish gloating, what do you do?
   a. Sell it and lock in your gains
   b. Stay put and hope for more gain
   c. Buy more; it could go higher

4. You're investing for retirement, which is 15 years away. Which would you rather do?
   a. Invest in a money-market fund or guaranteed investment contract, giving up the possibility of major gains, but virtually assuring the safety of your principal
   b. Invest in a 50–50 mix of bond funds and stock funds, in hopes of getting some growth, but also giving yourself some protection in the form of steady income
   c. Invest in aggressive growth mutual funds whose value will probably fluctuate significantly during the year, but have the potential for impressive gains over five or 10 years

5. You just won a big prize! But which one? It's up to you.
   a. $2,000 in cash
   b. A 50% chance to win $5,000
   c. A 20% chance to win $15,000

6. A good investment opportunity just came along. But you have to borrow money to get in. Would you take out a loan?
   a. Definitely not
   b. Perhaps
   c. Yes

7. Your company is selling stock to its employees. In three years, management plans to take the company public. Until then, you won't be able to sell your shares and you will get no dividends. But your investment could multiply as much as 10 times when the company goes public. How much money would you invest?
   a. None
   b. Two months' salary
   c. Four months' salary

Scoring Your Risk Tolerance

To score the quiz, ad up the number of answers you gave in each category a–c, then multiply to find your score

(a) answers ________ x 1= ________ points
(b) answers ________ x 2= ________ points
(c) answers ________ x 3= ________ points

YOUR SCORE: ________ points

If you scored...	You may be a:
9–14 points	Conservative investor
15–21 points	Moderate investor
22–27 points	Aggressive investor

Source: Scudder Kemper Retirement Services

EXHIBIT 11.5
Risk-Tolerance Quiz
Reprinted by permission.

Once established, portfolios should be managed and revised. *Active portfolio management involves buying and selling.* A more passive approach to a portfolio requires revisions from a tax standpoint and the maintenance of the desired risk level. More on how to manage assets will be covered in the next section.

EXHIBIT 11.6
Investment Ladder: Tradeoffs Between Risk and Return

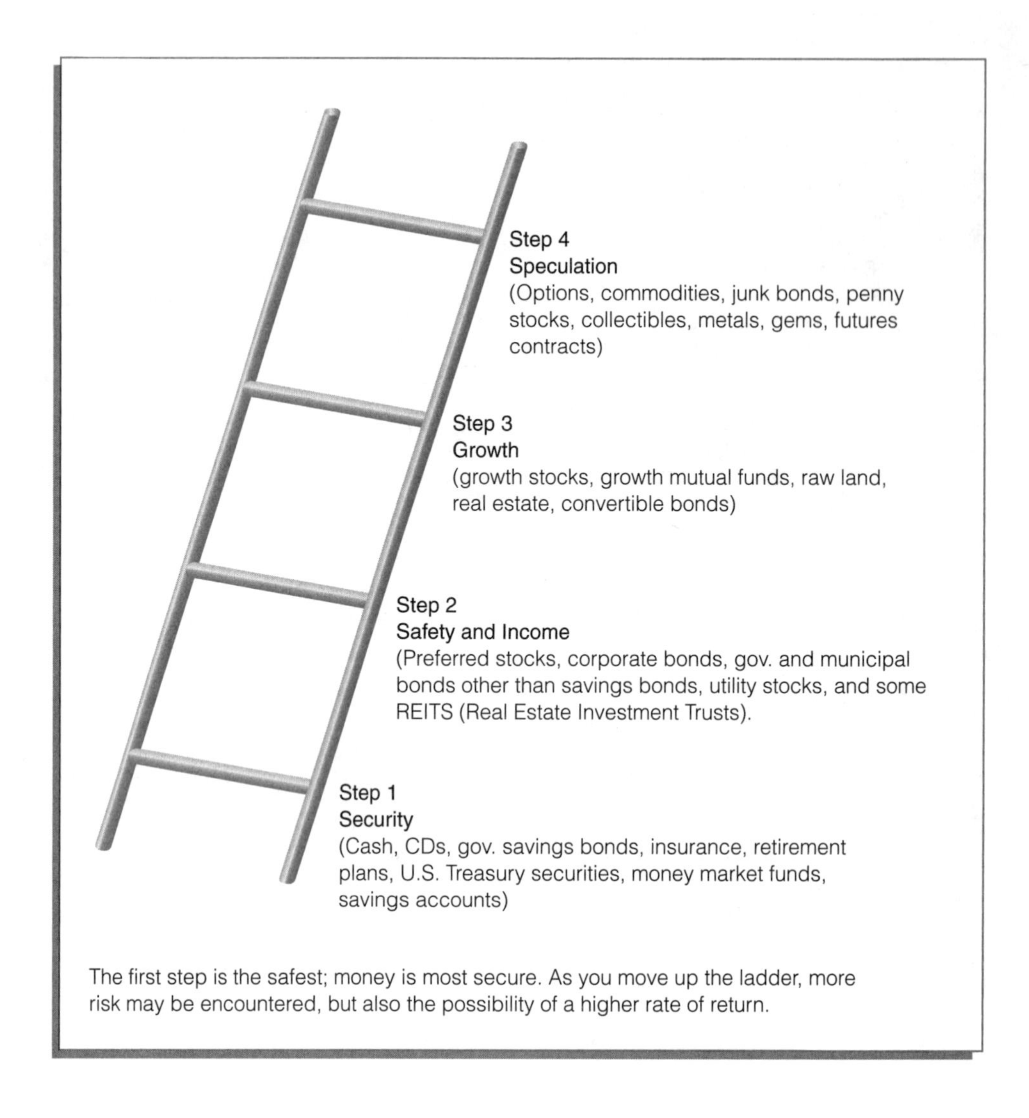

Step 4
Speculation
(Options, commodities, junk bonds, penny stocks, collectibles, metals, gems, futures contracts)
Step 3
Growth
(growth stocks, growth mutual funds, raw land, real estate, convertible bonds)
Step 2
Safety and Income
(Preferred stocks, corporate bonds, gov. and municipal bonds other than savings bonds, utility stocks, and some REITS (Real Estate Investment Trusts).
Step 1
Security
(Cash, CDs, gov. savings bonds, insurance, retirement plans, U.S. Treasury securities, money market funds, savings accounts)
The first step is the safest; money is most secure. As you move up the ladder, more risk may be encountered, but also the possibility of a higher rate of return.

Managing Investments

Managing investments revolves around three questions:

1. *How much* should be invested?

2. *How long* should investments be held? Short- and long-term strategies include

* Buy and hold. This is a long-term strategy involving keeping investments for many years.

"The key word is conviction," says Donald MacGregor, a senior research associate at the Decision Science Research Institute in Eugene, Ore. "Buy and hold is a tough discipline. You need to have a strong sense of where you're going and how you're going to get there. Often, that doesn't happen. People just get into the market, and they start messing around."[5]

* More activity. In short-term strategies the performance of individual investments often leads to changes. For example, a stock or mutual fund that loses money or remains stagnant over a period of time should be sold and replaced with an investment with more potential.

3. *Who* should be involved in investment decisions and processing buy and sell orders? There are about 80,000 registered representatives (stockbrokers) employed in the United States to choose from.[6] They can be contacted in person,

EXHIBIT 11.7
Asset Allocation Model

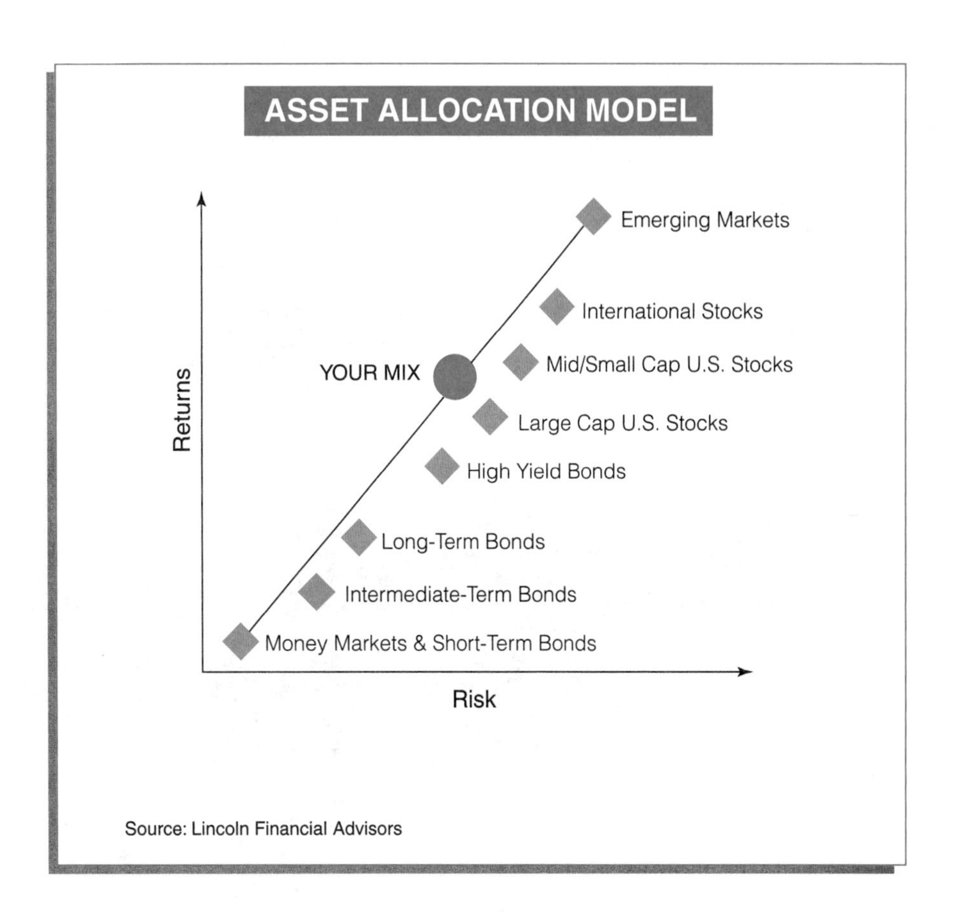

by fax, telephone, or over the Internet. So besides selecting investments, one has to select the broker, firm, or institution to use for buying and selling.

Objective 4:
Explain the different types of investing.

TYPES OF INVESTMENTS

Investments include a wide range of options that can be grouped into the following ten main types of investments.

Savings.

The goal of savings is to accumulate funds in a risk-free, conservative manner based on *interest* (the amount earned on the amount placed in the investment) over time. Chapter 4 on "Managing Cash and Savings" outlined the various types of savings accounts. Since savings accounts pay such a low rate of interest, they are often not listed as investments although they are. The next type is what is more commonly thought of when people use the word "investments."

Stocks, bonds, mutual funds, and money market instruments.

Stocks
Ownership in a corporation represented by shares.

Stocks. **Stocks** represent ownership in such corporations as Intel, Pepsi-Cola, and IBM. So if you own 10 shares of Intel stock you actually own part of that company. Ten shares may not seem like a lot, but someone who bought it at $90 a share in the 90s found that this $900 investment turned into $4,000 in a little over a year. If that person leaves the $4,000 in and it grows at the historic market increase of 11 percent, that could turn into $32,000 in 20 years time, all from a $900 investment left in and not touched.

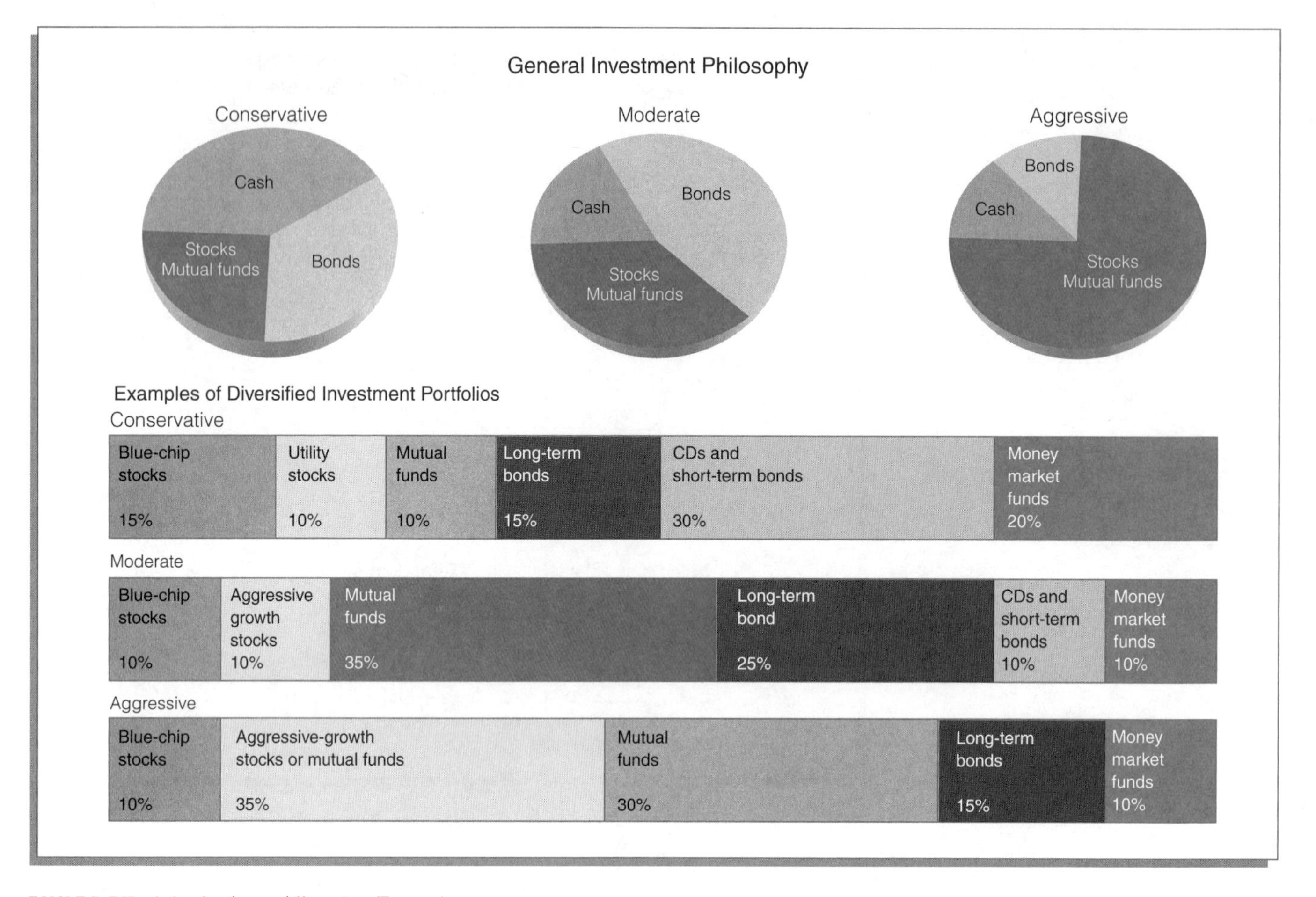

EXHIBIT 11.8 Asset Allocation Examples

Bonds
Investments involving lending money to organizations.

Bonds. **Bonds** are investments in which you lend money to some organization (corporation or governmental) for a period of time. The issuer of the bond is obligated to pay you, the bondholder, a specified sum of money, usually at specific intervals, and to repay the principal amount of the loan at maturity. Corporate bonds come from a company. Government bonds are written pledges of a government or a municipality to repay a specified sum of money along with interest.

Maturity dates on bonds range from one month to many years, interest may be paid every six months. Interest or **dividends** are distributions of money that a corporation or government pays to bondholders or stockholders. Holding the bonds until they mature or holding onto the stocks and collecting the dividends along the way is one way to make money on this type of investment.

Dividends
Distributions of money from a corporation or government to investors.

Mutual funds. Mutual funds are groups of stocks, bonds, or other securities managed by an investment company. Their key advantages are the diversification they offer, a low minimum amount to get started, and professional management. Fees are charged for the management services.

Mutual funds
Groups of stocks, bonds, or other securities managed by an investment company.

Money market. *Money market instruments* mean that you lend money to an organization such as a bank (certificates of deposit) or the government. They offer security and liquidity.

Real estate

If you have a rental property, it serves as a source of income and a hedge against inflation. If you own land or own and live in a house, this is also a hedge against inflation. As a general rule real estate increases in value and you will make a profit, but there are no guarantees. Location and the timing of transactions are important. Real estate is less liquid, less readily turned into cash, than most other forms of investments. Here is one way to look at a home you may own and live in:

> Once you pay off the mortgage, however, you do avoid paying rent. "The implicit income from your house is a part of your real income in retirement," argues Robert Soldofsky, a retired finance professor who now works as an investment advisor in Iowa City, Iowa. He compares owning your own home with owning a bond.[7]

Social Security

Those close to retirement or already in retirement will receive a lifetime of Social Security checks. For those further away a lifetime of checks may or may not happen depending on the future of Social Security.

A number of alternatives are being considered to keep it going fully funded or partially funded. These include issuing partial checks, or delaying the receiving of checks to a later age (as mentioned in earlier chapters), or reducing the percentage that non-working spouses receive based on their working spouse's contribution.

While this book was being written, a national debate was going on regarding whether or not to allow individual decision making about how one's Social Security account was invested. Traditionally the government has made these decisions.

> In theory, Americans overwhelmingly endorse making personal investment accounts part of Social Security. But feelings change when pollsters point out the risks. In a recent Associated Press poll, 80 percent approved of the accounts, but only 46 percent said they would risk their own money in them.[8]

Since Social Security is in transition it is difficult to know how to figure how much it will be worth to individuals by the time they retire. The best advice is to keep abreast of changes and consult with financial advisors and with the Social Security Administration regarding one's specific case.

Company pensions

Like Social Security, company pensions can provide a monthly check at retirement. Social Security has built-in inflation protection, whereas a pension may not, it can be a set number of dollars per month without change. With a fixed pension, as your retirement progresses, your pension has less buying power. Thus pension money that does not go for immediate living expenses should be invested with growth in mind. This is easier said than done. Consider the following:

> You go to the investment seminar and decide to take charge of your future. You take the plunge and move your savings into stocks—a growth fund. But inside, doubt drills away. What if the market dives?[9]

This quote brings us to an interesting point. Should a person panic? When it comes to pension funds, people are particularly prone to panic attacks. Exhibit 11.9 discusses when to panic (or not).

Large holdings

This type of investment is less common, but what it means is that executives and family owners of companies may have a lot of money tied up in stock in that company (the family owners of large amounts of WalMart; Michael Eisner of Disney; Bill Gates of Microsoft).

EXHIBIT 11.9
When to Panic (or not)

Given all you know from this book, you should not panic.
But some people do for the following reasons:

- the market drops.
- their particular stock or mutual fund tanks.
- people around them are panicking.
- there is bad economic news, and media sources fan the flames.

None of these are cause for panic, because

- history shows us that the market will rebound.
- hopefully, you bought the stock or mutual fund because it is fundamentally sound; remind yourself why you bought it in the first place.
- you should not overreact to bad news; put it in perspective; a lawsuit of $1 million against a huge corporation in which you own stock is a drop in the bucket; or if the Dow is at 10,000 and drops 100 points, that is only 1 percent.
- even if there is cause for genuine alarm, you should know what to do; ride out the storm, buy more, sell, etc.

Your own business

Owning a business is risky, but it can be thought of as an investment since owning a stake in your own business is similar to owning stock in a large corporation. If a business is successful, it can be a great investment.

Anticipated inheritances

Naturally things can go wrong. A will can change or final health expenses and taxes can take away any money that was left in an estate, but it can work out. Here is one way to estimate the possibility of an inheritance:

> "If your parents are in their early 60s and they have a modest estate, I wouldn't include an inheritance as part of your financial plan," says Minneapolis financial planner Ross Levin. "But if they're in their late 80s and they've shared their estate plan with you, you may want to include it."[10]

Precious metals and collectibles

Some people have gold, silver, art, antiques, stamps, collectibles, and jewelry that serve as a hedge against inflation. For most people though, the value of their collection is overestimated. Consider the following:

> Eventually, we realize that our baseball cards, Beanie Babies and comic books are worth a lot more to us than they are on the open market. Or, as Mr. Levin puts it, "I wouldn't want to be betting on my Adam-12 lunchbox to provide my retirement income."[11]

It is important to know your subject (stamps, coins, antiques etc.) and to buy from reputable dealers. Chapter 14 will discuss these types of investments.

Annuities:

As described in Chapter 10, these are contracts bought from life insurance companies in which the company promises the insured a series of periodic payments. If the company is secure, an annuity pays as long as a person lives.

Transaction Costs Involved in Investing

Transaction costs
Charges for buying and selling securities.

A share of stock can cost anywhere from 50 cents to $200. Besides the cost of the investment itself sometimes there are **transaction costs,** which are the charges for buying and selling securities. Since transaction costs vary widely, investors are wise to comparison shop for costs and services. For example, a full-service broker may charge $96 to buy 100 shares of a stock at $36 a share. The cost to the investor is $3,600 for the shares, plus the $96 transaction fee, for a total of $3,696. In order to recoup this cost, the 100 shares of stock will have to rise in value above $3696 (not counting capital gains, the commission charged for the sale, and taxes).

A full-service broker as the name implies gives advice and works with you on a regular basis to build an investment portfolio. Buying online, transaction costs can range from anywhere from $5 a trade to $29.95 a trade depending on which brokerage firm is used, how much is invested, and what services the firm provides. Some online firms require $2,000 upfront to open an account.

Since transaction costs, capital gains, and taxes affect the actual cost of a stock or mutual fund, investors hesitate before selling. The most realistic approach is to try to keep investment costs (fees) to a minimum, but still get the services that you need. As a guide to saving money in selecting a broker, investment advisor, or company consider what they offer, what their investment style or philosophy is, how accessible they are; and what they charge.

Objective 5:
Discuss various aspects of investing and investment strategies.

INVESTING STRATEGIES AND TERMS

There are many investing strategies beyond the ones already discussed in the chapter such as diversification and asset allocation. The following section addresses several more aspects of risk, explains more investment terms, and describes further on the strategies already introduced.

Risk and Investment

Building on the previous discussion of the investment ladder, asset allocation, and risk-taking, there are two kinds of investment risk: the risk of losing money by being too aggressive, and the risk of losing buying power by being too conservative.

Most people are aware of the first risk, but are less aware of the second. An example of being too conservative would be putting large amounts of money in savings accounts that pay such a low rate of interest that the invested money does not keep up with inflation especially considering taxes. For investments to grow, they should have the potential to grow beyond the inflation rate. In the long run you should seek higher rates of return than savings accounts can provide.

Tradeoffs Between Risk and Return

Realized return
Actual return on investments.

Investors buy, hold, and sell financial assets to earn returns on them. There is a difference between the expected return (the anticipated future return) and the **realized return** (the actual return). Although stocks on the average have outperformed bonds and savings accounts, an individual stock may perform poorly, and there have been times (when the stock market has dropped) when bonds and savings accounts have had better actual returns.

Investors want to maximize their return. If they try to minimize risk, they lower their expected return. If someone wanted to be virtually risk-free he or she should put the money into savings accounts, U.S. savings bonds, and Treasury bills.

Tax-Shelters

Tax-free or tax-exempt
No taxes.

Tax-sheltered
No taxes are charged on profits until a later date.

If an investment is **tax-free** or **tax exempt** the investor keeps the entire return free from tax liabilities. If an investment is **tax-sheltered,** there are no taxes due on the profits until one or more years in the future. This allows the profits to grow untaxed. Eventually taxes will be paid, but this may be delayed until taxes are lower (when a person retires and income is lower). Investors in high tax brackets seek out tax-free or tax-sheltered investments such as tax-free bonds so they can legally avoid or reduce tax liabilities.

Dollar-Cost Averaging

Dollar-cost averaging
Systematic investing of equal sums of money at regular intervals regardless of price fluctuations in an investment.

Dollar-cost averaging is the systematic investing of equal sums of money at regular intervals, regardless of fluctuations in the price of an investment. An example would be putting $50 each month into a mutual fund or a single stock such as Coca-Cola, over several years. In so doing, the investor buys whether the mutual fund or stock price is rising or falling. When the price of the mutual fund or Coca-Cola is down, the investor's $50 goes further and more shares are bought, conversely, when the price soars, the $50 monthly investment buys fewer shares.

The underlying philosophy is that, given the history of the stock market, in the long run stocks and mutual funds (invested in stocks) will increase in value; therefore systematically investing will lead to increased revenue for the investor. This approach requires discipline and staying the long course.

Dividend reinvestment plans (DRIPs)
Automatic reinvestment of shareholder dividends to buy additional stock.

Direct investment plans
Investors buy stock directly from a corporation.

Dollar-cost averaging is usually arranged by a stockbroker to invest regularly a specified amount ($50 to $100 every month) in a monthly investment plan (MIP). Another way dollar-cost averaging can work is through **dividend reinvestment plans (DRIPs)**. DRIPs eliminate the need for a broker because under a DRIP, dividends are automatically reinvested in additional shares purchased directly from a company. More than 1,000 companies—such as McDonald's, Exxon-Mobil, and Proctor & Gamble—offer DRIPs. They are a good idea for the long-term investor because shares can be bought in a consistent, inexpensive manner. To find out which companies offer dividend reinvestment plans or **direct investment plans** (in which investors initially purchase stock directly from a corporation without using a broker), contact a company directly and ask, check your library for the Standard & Poor's Directory of Dividend Reinvestment Plans, use the sources in the Electronic Resources section at the end of the chapter, or contact any of the following:

* DRIP Investor, 7412 Calumet Ave., Suite 200, Hammond, IN 46324-2692 (800-233-5922).
* The Moneypaper, 1010 Mamaroneck Ave., Mamaroneck, NY 10543 (800-388-9993).
* Direct Stock Purchase Clearinghouse Hotline (800-774-4117)

Employee stock ownership programs or stock options
Program in which employees can buy stock in the company they work for at a reduced price or with matching funds from employers.

Another form of dollar-cost averaging plans is **employee stock ownership programs** (also known as **stock options**) offered to employees as a fringe benefit. These permit employees to buy shares in the company usually at a reduced rate because the employer makes a matching contribution to the purchase. If the employee purchases $50 worth of stock a month, the employer may add $25 so that the employee ends up with $75 worth of stock each month. For example, Taylor, a 24-year-old taking a Family Financial Analysis class, found out that she had over $400 in stock options from the company who owned the Red Lobster restaurant where she had been waitressing part-time for several years. She remembered signing up for the stock option, but never really understood what it meant or what it was worth until she took the class.

In addition some employers give shares of stock as a bonus at holiday or raise time. Given what you have learned so far about investing, if an employer offered you $100 in cash or $100 in company stock (assuming that it is a solid company) which would you choose? Also if you are comparing job offers and one offers a stock option plan or bonus plan and the other does not, you should take this into account when making a final decision. The only real downsides to stock options are (1) if the company goes under or loses worth significantly or (2) if an employee holds his or her own company's stock exclusively (thus lacking a diversified portfolio).

Investment Formulas and Experts

Investment decisions are both an art and a science grounded in experience. If there was a single, proven formula everyone would follow it. Unfortunately there is no perfect formula just as there is no perfect way to lose weight that works for everyone. A universal investment formula does not exist because everyone's goals and resources are different and market conditions are always changing. In the next chapter numerical measures of stocks will be given and strategies discussed further, but the point is that after all is said and done, there is a lot of trial and error in investing. The asset allocation models given in the exhibits are examples of recommended ways to split up investments. They do not fit everybody.

Along with this caveat about being wary of one-way-only investment formulas, investors should be skeptical of anyone or any group that says they have all the answers. For example, in 1998 an investment group known as the Beardstown Ladies, authors of *The Beardstown Ladies' Common-Sense Investment Guide*, admitted that they had incorrectly stated their return on investment. Instead of the 23.4 percent return they originally reported, an outside audit of their books revealed that they actually had a 9.1 percent annual rate of return on investment. This compared unfavorably with a Standard & Poor's Average during the same ten years of 14.6 percent. According to financial columnist Jane Bryant Quinn, "In truth, it's hard for professionals to beat the market year after year, let alone a start-up neighborhood group."[12]

Rebalancing and Life Cycle Investing

Rebalancing
Adjusting asset allocation.

Rebalancing refers to readjusting your asset allocation. It means that you reset targets for what percentage you want held in stocks, bonds, and cash investments, such as money market funds or Treasury bills. At one age you may want 60% in stocks, 30% in bonds, and 10% in other. Given your age or career stage, level of risk tolerance, time horizon and tax exposure, you may want to rebalance or change this mix so that you have more growth or more income type investments (see Exhibit 11.10).

Generally *the younger you are, the more risk you can take* because you have the time to build wealth and have an income that is increasing. A 25-year-old with a high risk tolerance personality may want 80% in growth stocks, whereas a 60 year old may want only 10% in growth stocks. Exhibit 11.11 shows the portfolios of two people, Mary age 25 (with 80% in growth stocks and 20% in bonds) and Frank age 46 (with 32% in growth stocks, 40% in bonds, and 28% in stable, guaranteed investments). In the Exhibit, the term **small cap** refers to small capitalization stocks or mutual funds holding such stocks. These stocks have a market capitalization of $500 million or less.

Small cap stocks
Companies with market capitalization of $500 million or less.

Small capitalization stocks represent companies that are less well established, but in many cases faster-growing than mid-cap stocks (from $500 million to $3–$5

EXHIBIT 11.10
Life Cycle Investing.
Reprinted with permission.

Life Cycle Investing: This table shows how an investor's profile may change over time.

	Early Career	Middle Career	Late Career	Early Retirement	Late Retirement
Risk Tolerance	High	High	Moderate	Moderate	Low
Return Needs	Growth	Growth	Growth	Growth/Income	Income
Time Horizon	Long	Long	Long	Short/Long	Short/Long
Tax Exposure	Lower	Higher	Higher	Lower	Lower

Source: American Association of individual Investors, www.aaii.org.

billion) or large cap stocks ($1 billion or more). (Ranges vary somewhat and may overlap, depending on the funds or indexer defining them). Since they are less established, small cap stocks are usually more volatile than blue chips.[13]

Once people become accustomed to high returns on stocks, sometimes they have a hard time switching to less risky (less exciting) investments, especially when they are nearing retirement.

> Scaling back your portfolio's risk level may be the prudent thing to do. But many investors won't like the idea. 'The specter of greed rears its ugly head,' Mr. Balasa (a financial planner in Schaumburg, Ill.) says.[14]

Starting to Invest

The most immediate concern for the majority of students reading this book is how to get started. Once you begin working full-time, you can take advantage of stock options and retirement plans offered to you and sign up through payroll deduction to have a regular amount put into savings bonds, stocks, mutual funds, or other forms of investing. If you enroll in an automatic investment program into a mutual-fund company, allowing it to remove $50 a month electronically from your bank account (or your paycheck) for investment, many companies will waive their requirements for a minimum deposit, which typically would be anywhere from $1,000 to $3500. Examples of mutual-fund companies that offer $50-a-month plans are T. Rowe Price (800-638-5660), Invesco (800-525-8085), and Strong (800-368-1030).

Youth is an advantage. The best advice is to start as soon as you can no matter how small the amount.

Deciding to Sell

Buying is only half of investing, selling is the other half. How do you know when to sell? Here are three guidelines: sell when you reach your goal (or need the money), cut losses quickly, and decide if what you own is a good investment today. In other words do not get so tangled up in the price you bought it at or the transaction costs or taxes involved that you cannot make a decision to sell. Ask yourself if that Gillette stock or that property you bought is still a sound investment today. The general advice is to buy and hold, but not stranglehold.

Time to Invest

Investing takes time, but if you set investing as a priority it is worth the time. If your investments are doing well, it is a tremendous psychological boost as well as an eco-

EXHIBIT 11.11
Case Studies

Case Studies

To illustrate how investment mix should change over time, let's consider the portfolios of two individuals of varying ages: Mary, 25; and Frank, 46. Let's also assume that both individuals have access to a retirement plan specialist who is trained to conduct a risk tolerance questionnaire and portfolio optimization.

Mary: 25-years-old...
40 years to retirement

At age 25, Mary's biggest investment advantage is her youth, which translates into a long time until retirement. Based on Mary's responses to a risk tolerance questionnaire, she discovered that she was an aggressive investor.

Mary's Portfolio

Overall portfolio historical mean return: 13.28%
Standard deviation or risk factor: 12.62%

Investment/ Objective Category	Class	Portfolio Weighting
Growth/stocks		80%
	International	32%
	Domestic small cap	18%
	Domestic mid cap	15%
	Domestic large cap	15%
Income/bonds		20%
	Non U.S. Government	10%
	U.S. Treasury bonds	10%
Stability/ guaranteed investments		0%

Frank: age 46...
19 years to retirement

Frank is a typical baby boomer with some awesome responsibilities. He has kids in college and takes care of an elderly parent. He must also invest for his own retirement.

Because of his age and financial responsibilities, Frank's risk tolerance questionnaire identifies him as a "moderate" risk-taker.

The chart (right) illustrates Frank's optimized portfolio. As you can see, his portfolio has less in growth investments (32%) than Mary's (80%). His portfolio is more heavily weighed in less-risky income investments (40%) and stability investments (28%).

Frank's Portfolio

Overall portfolio historical mean return: 10.98%
Standard deviation or risk factor: 6.61%

Investment/ Objective Category	Class	Portfolio Weighting
Growth/stocks		32%
	International stocks	6%
	Domestic small cap	11%
	Domestic mid cap	5%
	Domestic large cap	10%
Income/bonds		40%
	Non U.S. Government	25%
	U.S. Treasury bonds	15%
Stability/ guaranteed investments		28%
	Fixed annuities	28%

Source: Investment Digest, VALIC, Fall 1996, Vol. 10, No. 3. (1-800-633-8960)
2919 Allen Parkway, Suite L6-01, Houston, TX 77019

nomic one. Growing investments make it easier to swallow a mediocre raise. Even if your salary is barely keeping up with inflation, your investments should be beating inflation.

If your investments are not beating inflation, see it as a chance to rebalance your portfolio. Investing does not have to take a lot of time especially if you accept it as part of life and find reliable financial help and advice.

OTHER ASPECTS OF INVESTING

A few last aspects of investing (global investing, women and investing, and investment clubs) will be discussed next.

Global Investing

No longer is it sufficient to think of investing as a domestic concept. While foreign investments have been available for a number of years, until recently most investors did not buy and sell in international markets.

Foreign markets have grown rapidly. The Japanese market, for example, is one of the three largest stock markets in the world. Other Asian and South American markets are emerging. Sometimes these markets are doing better, sometimes they are doing worse, than the United States markets. So there is risk involved in global investing just as there is in domestic markets. There are other potential downsides including a lack of knowledge about foreign countries' currencies and companies and potential volatility. This was made readily apparent in 1998 when problems in the Asian economy not only affected Asian markets, but created a short downturn in the United States markets.

Another aspect of global investing is the recognition that many famous U.S.-based companies—such as Colgate-Palmolive and Ford—derive a large percentage of their revenues from abroad. For example, eighty percent of Coca Cola's revenues are from abroad. In summary, global investing through U.S. companies with foreign sales or through foreign-based companies offers investors another option to consider when diversifying their portfolios.

Women and Investing

Historically women investors have been ignored by financial services although this is changing in recognition of the amount of money women are earning and investing and of their influence on household financial decisions. Research shows that women investors want the same things as men investors: performance, information on which to base decisions, and a plan to guide their investments in order to meet goals. But there is reason for concern because three research studies indicated that college women know less about investing and plan to invest less than men students even when the women have higher GPAs and when majors are considered.[15-17]

This may be because many women find that investing becomes more of a priority as they age and gain life experience. For example, Patricia Larkin, age 36, who is a senior portfolio manager with Dreyfus Corporation said that, "When I first started working at 21, I cared more about shopping than saving . . . but it only took a few years for me to understand how big the benefits of starting young are.[18] Regardless of their age or current marital status, women should invest with the assumption that at some point they will be solely responsible for their financial well-being. Statistics bear this out:

- Immediately after women go through divorce, they experience a "26 percent decrease in their standard of living, while that of their ex-husband's increases by 34 percent. This means women have less money to spend on housing, transportation, and other expenses."[19]
- The National Center for Women and Retirement Research reports that a widow is four times more likely, and a single or divorced women is five times more likely, to live in poverty than a married woman.[20]

* CoreStates Bank in Philadelphia estimates that, because women receive less retirement income, a 35-year old woman must save $50,000 more than a man the same age to retire comfortably. Due to average life expectancy, her money needs to last seven more years than his.[21]
* Women in general have lower incomes than men and typically receive fewer retirement benefits. For example, less than half—45 percent—of women age 25 to 64 participate in a pension plan.[22]
* "Women also have shorter careers, spending 14.7 percent of their working years raising children or taking care of sick relatives. That means women have less time to get raises or contribute to retirement plans."[23]

To summarize, "the thing women need to take into consideration is that between the divorce rate and the death rate, they need to know what is going on with their finances," said Celia Hayhoe, a University of Kentucky professor of family studies and a certified financial planner. Further, she said, "All women should plan their lives like they would be single at some time. They need to plan so they will know what to do."[24]

Investment Clubs

Most people invest as individuals, but some people belong to investment clubs where members pool their money and invest in stocks such as the Beardstown Ladies described earlier. Investment clubs have existed in the United States for over a century. The oldest known club began in 1898 in Texas.

> Most of the early clubs were built around a combination of like social interests and a desire to try one's wings in the blue skies of investment. There was little, if any, attempt to establish uniform principles for guidance or education of members. The modern investment club movement began in Detroit in 1940 because circumstances dictated a serious attitude toward investment results.[25]

Investment clubs
Groups that form partnerships to invest in stocks and to learn about investing.

Today the clubs are world-wide. **Investment clubs** have on average 17 members that invest a set amount each month (usually between $25-$50 per person) into a partnership. Then they invest the money as a group. Typically the clubs meet once a month to educate themselves about investing, share stock market reports, and review their portfolios.

To find an existing club ask colleagues or friends or call local brokerage firms who usually know about area clubs. Or, you can start a club yourself with friends, relatives, or co-workers.

ADVICE FOR BEGINNERS

Imbedded in the chapter were a number of investment tips, but here is a summary of key points:

* Know your personality and ability to handle risk.
* Investigate before you invest, know what you are doing and where you are going.
* Hope for the best, but expect ups and downs.
* Stay the course.
* Get in there, you can't beat the market if you are not in the market.
* Start small, learn with each investment.
* Diversify.
* At first, focus on accumulating but when the time comes do not be afraid to sell.

* Either start with a professional broker or financial advisor or turn to one when you are in doubt or need another level of advice.

ELECTRONIC RESOURCES

Since investing is one of the most popular subjects on the Web, there are countless sites that deal with investments. This section will highlight some of the main ones. For an overview of investment basics try the American Association of Individual Investors at **www.aaii.org**.

For a list of popular investment websites see Exhibit 11.12. Exhibit 11.13 shows a UPS stock tipsheet from the **www.thomsoninvest.net** website.

For companies that sell shares directly to the public contact the Securities Transfer Association at **www.netstockdirect.com**. For investment club information try the National Association of Investors Corporation (NAIC) at **www.better-investing.org**.

Most brokerage houses have websites with investment information. An example of a company offering free investment advice for women is Prudential Securities at **www.prudential.com**. For overall advice and data the New York Stock Exchange can be contacted at **www.nyse.com**.

Examples of sources of free online financial services and information from print sources are given in Exhibit 11.14.

EXHIBIT 11.12
Popular Investment websites

Site	URL	Brief Content Summary
Money Central	moneycentral.com	news, commentary
Quicken.com	www.quicken.com	data, tools*, easy to use
CNNfn	www.cnnfn.com	news, analysis
Morningstar.com	www.morningstar.com	asset allocation, research
CBS MarketWatch	cbsmarketwatch.com	news, analysis
Yahoo!Finance	quote.yahoo.com	comprehensive, links
The Street.com	www.thestreet.com	news, commentary, data
Silicon Investor	www.siliconinvestor.com	chat, data, tools
The Motley Fool	www.fool.com	chat, humor, advice
Bloomberg	www.bloomberg.com	news, data
Wall Street City	www.wallstreetcity.com	historical data, tools
Multex Investor	www.multexinvestor.com	over 300,000 reports
StockPoint	www.stockpoint.com	data, links
Lycos	investing.lycos.com	reports, updates
Clearstation	www.clearstation.com	3-prong analysis
Briefing.com	www.briefing.com	commentary, analysis
INVESTools	www.investools.com	clearinghouse, resources
Thomson Investors	www.thomsoninvest.net	stocks, general info.
Raging Bull	www.ragingbull.com	message boards

*Under content summary, when the word "tools" is used, it can mean an interactive activity or a means to conduct research about a stock, mutual fund, etc.

Note: AOL Finance Channel, a top investment website, is not on the list because you need to belong to AOL to get in. The other sites should have free access for at least initial inspection, but this could change at any time. The trend is to open up websites to the public rather than making them exclusive.

EXHIBIT 11.13
A Sample Stock Tipsheet

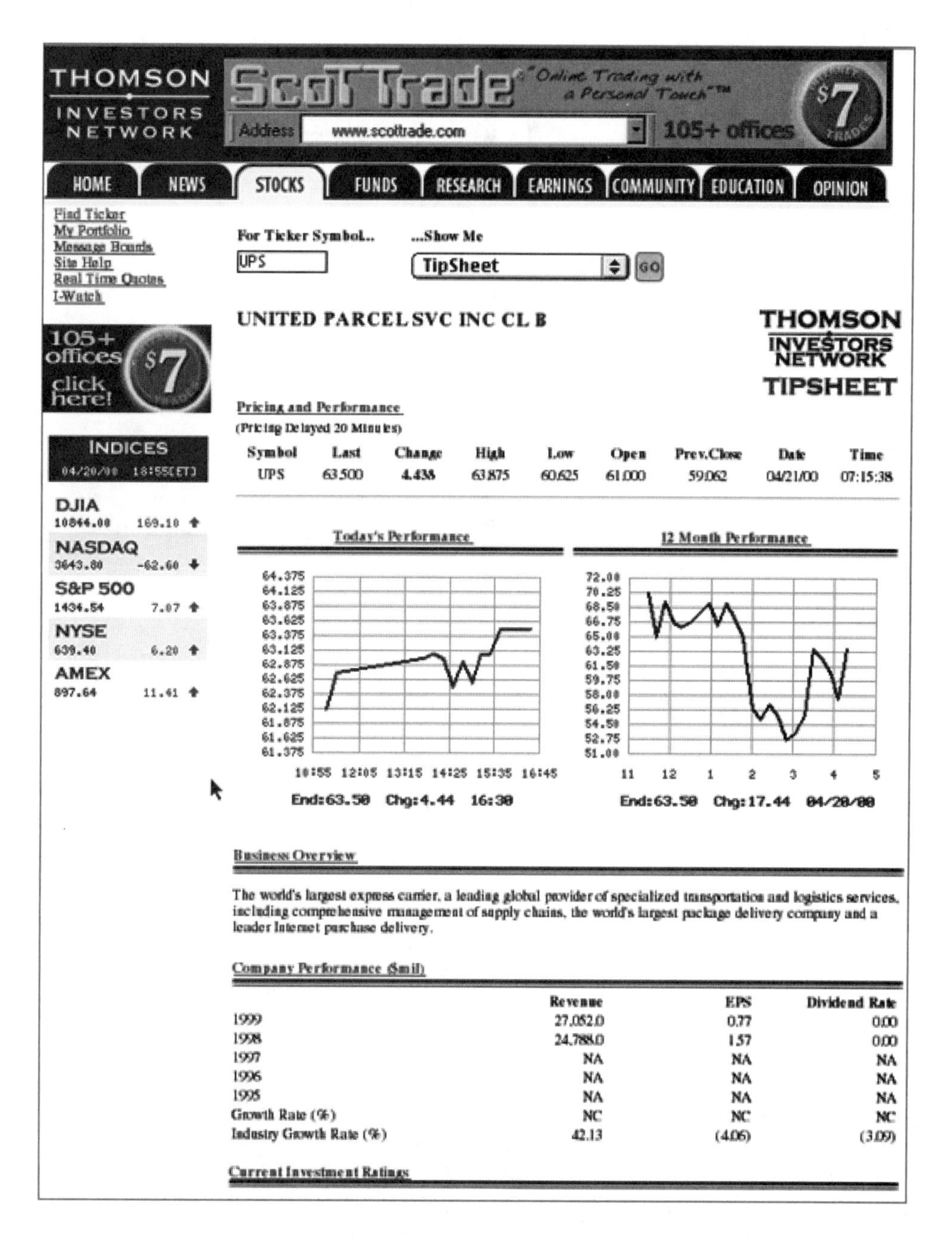
THOMSON INVESTORS NETWORK

Scottrade "Online Trading with a Personal Touch"™ $7 TRADES

Address www.scottrade.com 105+ offices

HOME | NEWS | STOCKS | FUNDS | RESEARCH | EARNINGS | COMMUNITY | EDUCATION | OPINION

Find Ticker
My Portfolio
Message Boards
Site Help
Real Time Quotes
I-Watch

105+ offices $7 click here!

For Ticker Symbol.. UPS ...Show Me TipSheet GO

UNITED PARCEL SVC INC CL B

THOMSON INVESTORS NETWORK TIPSHEET

Pricing and Performance
(Pricing Delayed 20 Minutes)

Symbol	Last	Change	High	Low	Open	Prev.Close	Date	Time
UPS	63.500	4.438	63.875	60.625	61.000	59.062	04/21/00	07:15:38

INDICES 04/20/00 18:55(ET)

Index	Value	Change
DJIA	10844.00	169.10
NASDAQ	3643.80	-62.60
S&P 500	1434.54	7.07
NYSE	639.40	6.20
AMEX	897.64	11.41

Today's Performance

End:63.50 Chg:4.44 16:30

12 Month Performance

End:63.50 Chg:17.44 04/20/00

Business Overview

The world's largest express carrier, a leading global provider of specialized transportation and logistics services, including comprehensive management of supply chains, the world's largest package delivery company and a leader Internet purchase delivery.

Company Performance ($mil)

	Revenue	EPS	Dividend Rate
1999	27,052.0	0.77	0.00
1998	24,788.0	1.57	0.00
1997	NA	NA	NA
1996	NA	NA	NA
1995	NA	NA	NA
Growth Rate (%)	NC	NC	NC
Industry Growth Rate (%)	42.13	(4.06)	(3.09)

Current Investment Ratings

FAQs

1. What is risk, and how does it affect return?

   ■ Risk involves the ups and downs of returns. The greater the uncertainty, the greater the risk, the chance of loss. A return consists of current income, plus capital gains due to growth, minus losses incurred. Both risks and returns are uncertain.

2. Why is diversification important?

   ■ Diversification reduces risk by spreading it over a variety of investments. The idea is that if one type of investment goes down, another may go up. Diversification can be spread across different kinds of industries (aerospace versus consumer goods like toothpaste and soaps) or asset categories such as stocks, fixed income and money market investments.

EXHIBIT 11.14
Publication-based Investment websites

Newspapers	URL
■ The Wall Street journal—published daily M-F	info.wsj.com
■ USA Today—published daily M-F	www.usatoday.com
■ The New York Times—published daily	ww.nytimes.com
Business Periodicals	
■ Barron's—published weekly	ads.barrons.com
■ Business Week—published weekly	www.businessweek.com
■ Fortune—published twice a month	www.pathfinder.com/fortune/index.html
■ Forbes—published twice a month	www.forbes.com
■ Newsweek—published weekly	www.newsweek.com
■ U.S. News & World Report—published weekly	www.usnews.com
■ Fast Company—bimonthly	www.fastcompany.com
Personal Financial Publications	
■ Consumer Reports—published monthly	www.consumerreports.org
■ Kiplinger's Personal Finance Magazine—published monthly	www.kiplinger.com
■ Money—published weekly	www.money.com

Note: A master list of business publications is available at www.excite.com/news/business.

3. What is asset allocation?

■ Asset allocation is the process or practice of dividing assets among different investment categories, such as stocks, bonds, mutual funds, and cash. As individuals age and their investments grow, they usually change their asset allocation. If they are nearing retirement, they may change to more financially secure investments.

4. How does a person know he or she is on the right track?

■ The right track should feel right. An effective investment portfolio has a balance of risk and return based on the investor's personality and goals. The main strategy is to get started and not agonize over small deviations in an overall asset allocation plan or the market. Learning to invest is a lifelong process.

SUMMARY

Investors should know key investment terms, the ten main types of investments and the four steps in the investment process, as well as the principles of risk and return.

People invest to make money. Investment plans are based on goals. But since goals change as well as markets, a fresh look at investments is needed periodically to determine if an investment should be bought, increased, or sold.

Asset allocation refers to a target mix of stocks, bonds, and other investments.

Assess risk, get started as soon as possible, and find help. Individuals can become more knowledgeable about investing through a variety of means including reading, online sources, and by advice from professionals. However there is not a single formula or expert who has all the answers. Investment plans should be individualized to a person's resources, goals, personality, and lifestage or career stage.

Investment strategies include diversification, tax shelters, dollar-cost averaging, dividend reinvestment plans, direct investment plans, and stock options.

Global investing offers investors another avenue to consider when diversifying their portfolios.

Historically women have been largely ignored by financial services. This is being remedied as financial services become more aware of women's earning and investing power.

Investment clubs have members that learn about investing and invest in stocks as a group.

REVIEW QUESTIONS

(See Appendix A for answers to questions 1-10)

Identify the following statements as true or false.

1. Bonds represent ownership in a company.
2. Savings accounts and Social Security are types of investment.
3. One of the reasons people need to invest is that since they are living longer, their money has to last longer.
4. In dollar-cost averaging the investor invests money only when a stock price is low.
5. It is illegal for a broker to charge transaction costs.

Pick the best answers to the following five questions.

6. ____________ is the commitment of funds (capital) to the achievement of long term goals or objectives.
   a. Investment
   b. Escrow
   c. Time-sharing
   d. Pegging
7. Investing as a means of building wealth and security has become more critical in recent years for all of the following reasons except:
   a. rising prices.
   b. worries about the future of Social Security.
   c. the trend towards self-directed retirement plans.
   d. salary raises are always higher than inflation.
8. ____________ represent ownership in corporations such as Intel, Pepsi Cola, and IBM.
   a. Bonds
   b. Stocks
   c. Money market instruments
   d. Pensions
9. ____________ refer to income that results from an increase in the value of an investment.
   a. Capital annuities
   b. Capital gains
   c. Capital losses
   d. Capital risks
10. ____________ is a target mix of stocks, bonds, and cash investments.
    a. Dollar cost averaging
    b. A tax shelter
    c. Asset allocation
    d. A direct investment plan

DISCUSSION QUESTIONS

1. Stanley and Danko, authors of *The Millionaire Next Door*, say, "It's much easier in America to earn a lot than it is to accumulate wealth."[26] Why do you suppose they say this? Do you agree or disagree with their statement? Explain your answer.
2. According to a quote in the chapter, Americans overwhelmingly endorse making personal investment accounts part of Social Security. But feelings change when pollsters point out the risks. If you were given the choice of having some say in how your Social Security account is invested versus letting the government decide, how would you react? Would you feel comfortable making investment decisions about your Social Security money?
3. Using the investment ladder, which of the following would be the most risk-free: Treasury bills, savings accounts, or stocks? Discuss the reasoning behind your selection.

4. Why would an investor select investments at the top of the investment ladder (Step #4)? How does the age of the investor figure into the selection of different levels of investment risk?
5. Women and men investors both want financial growth and security, but historically women have been ignored by financial services. What are some of the reasons why women should be especially concerned about investing for their future?

DECISION-MAKING CASES

1. Cynthia Lee is 23 years old, unmarried, has no children, and has a job paying $26,000 a year that provides health insurance. She has about $1,500 set aside in savings and she does not owe anything except her monthly car payment. She wants to start investing. Can she begin with $25 or $50 a month? What is one way she can get started?
2. James Soburn is 42 years old, married, with one child, and is a computer programmer. He has a mutual fund and 200 shares of Wendy's and Disney. He has joined a top-rated technology company that offers an employee stock ownership program (stock option) in which he can buy $100 of the company's stock a month and the company will match it with $50 worth of stock. What should James do?
3. Rochelle works at a government job and has heard from a friend about an investment club. She is thinking about joining, but she does not know anything about clubs like this. What can you tell her about investment clubs?
4. Kenneth wants to avoid paying a broker by buying stock directly from a company. How can he find out which companies offer direct stock purchasing?
5. Evan and Vita are nearing retirement age and their broker, Josh, has suggested that they need to rebalance their portfolio. Right now they have 70% in growth stocks and Josh is suggesting a reduction in this category. Why?

GET ON LINE

Three activities to try:

- Go to www.Thomsoninvest.net, click on Thomson Investors Network, and then enter Ticker UPS and hit go. You should find a tipsheet similar to the one in Exhibit 11.13.
- Try another stock, see what it is currently selling for and its highs and lows during the day you check.
- Go to SmartMoney.com and try their Asset Allocator. It shows you how to tailor your investments based on your particular needs. To access, enter your personal information and financial goals, then watch as the allocator calculates your optimum asset mix. SmartMoney.com is the Web site for SmartMoney magazine, The Wall Street Journal's personal finance magazine.
- Go to one of the publication websites in Exhibit 11.14 and see what they have to offer. Select one stock or mutual fund and investigate what the publication has to say about it.

INFOTRAC

To access InfoTrac follow the instructions at the end of Chapter 1.

Exercise 1: This chapter introduced the topic of investments. Using InfoTrac, run a search on one of the following subjects related to investments: dollar cost averaging, risk, asset allocation, or diversification. For the topic you have selected, answer some or all of the following questions. What is it and how is it part of the general topic of investing? What are its advantages and disadvantages?

Exercise 2: InfoTrac has a number of resources on investing. In recent years, global investing has become more popular. How many articles are there on this subject and what is the general advice given? Do the articles suggest that now is a good time to invest internationally? Why or why not? Are any areas over the world suggested more than others as likely to bring strong returns?

chapter twelve

stocks

Choosing Well
Staying the Course

DID YOU KNOW?

* If you had invested $1,000 in Yahoo! at the end of 1996 and sold it at the end of 1997, it would have been worth $6,110 (a 511% gain).
* Conversely, $1,000 invested in Boston Chicken would have been worth $179 at the end of the year (an 82.1% loss).

OBJECTIVES

After reading Chapter 12, you should be able to do the following:

1. Differentiate between common stocks and preferred stocks.
2. Locate and evaluate information on stocks.
3. Use numerical measures and investment theories to evaluate stocks.
4. Know how to buy and sell stocks.

CHAPTER OVERVIEW

Stocks are defined as ownership in corporations such as Intel, Pepsi Cola, and Dow Chemical. Over the last ten years the percentage of Americans who own common stocks has doubled. In this chapter common stocks will be differentiated from preferred stocks, and different methods of evaluating stocks will be covered.

Stocks are important to study because more and more you will be asked to make choices regarding how many and what types of stocks you want in your retirement account and, more immediately, you will want to know how to make your money grow to take care of future needs. Since you can lose money as well as make it in the stock market, the more you know about stocks the better off you will be.

As examples of making money, if you invested $1,000 in Yahoo! stock at the end of 1996, you could have sold it for $6,110 twelve months later. This represents a 511 percent leap. During the same time period, another rising stock, Dell Computer had a 216 percent return.[1] Bonds or savings accounts cannot touch these levels of returns. But if you invested $1,000 in Boston Chicken (now known as Boston Market) during the same time period, it would have ended up being worth $179.[2] This represents a plummet of 82.1 percent.

How do you pick a stock like Yahoo! or Dell Computer and avoid a stock like Boston Chicken? We would all be rich if we could answer that question. Of the 40 million Americans who own stock, most are not rich. They own less than $10,000 worth of stock, and most of their stocks are in their 401(k) or 403(b) retirement plans. The most popular stock in America is not Procter & Gamble or Microsoft. It is the company the investor works for, since many companies offer stock options wherein the employee is given the option to purchase shares at or below the market price with the possibility of an employer match as discussed in the previous chapter. The largest American firms with majority ownership held by employees are Publix Super Markets, with 131,000 employees, and United Airlines, with 77,900 employees.

Since stocks can make or lose money, this chapter provides guidelines for separating the chaff from the wheat. Your judgment about the future profitability of the company, the state of the domestic and the global economy, the company's competitors, the possibility of mergers, along with chance and other factors, all play a part. Near the end of the chapter is an advice section on how to choose the first stock to purchase. Sources of information on which to base your decisions will be given throughout the chapter, which concludes with an Electronic Resources section.

COMMON STOCK

Objective 1: Differentiate between common stocks and preferred stocks.

Common stock
Shares or units of ownership in a public corporation.

Common stocks are a popular form of investment owned by millions of individuals and financial institutions. **Common stock** refers to shares or units of ownership in a public corporation. As the name implies, *this is the most typical kind of stock.*

Owners of common stock usually are entitled to vote on the selection of members of the company's board of directors to receive dividends on their holdings if dividends are declared and to attend the company's annual meeting.

Proxy
Written statement assigning shareholder votes to another person who votes in place of shareholder.

Typically each share of stock held entitles the holder to one vote. A small stockholder will not have enough voting power to sway a large corporation, but collectively small investors have a voice. Corporations hold annual stockholder meetings (usually lasting one hour) to which stockholders are invited, regardless of their number of shares. Since most small stockholders cannot afford to attend these meetings which are usually at or near company headquarters, they can vote by **proxy** (a written statement assigning their votes to another person who will vote for them). Information about annual meetings and other details about a company can be obtained by contacting a company's investor relations department by phone, mail, e-mail, or the Internet.

Issuers of Common Stock

All corporations have shareholders, but not all stocks are *publicly traded.* A *private corporation* is not traded openly in the market. In this book we are interested in discussing *public corporations*, whose stock is available to the public, meaning that it is openly bought and sold in the markets. For example, Walgreen and American Telephone & Telegraph (AT&T) are traded on the New York Stock Exchange.

Publicly traded stocks can be purchased from companies directly (as explained in the previous chapter), through brokerage firms, through payroll deduction, and from financial institutions, such as banks and insurance companies. Shares of stock can also be purchased through a procedure known as a public offering. In a public offering a company—such as World Wrestling Federation Entertainment, Inc. or Simpson Manufacturing—working with an underwriter, offers the public a certain number of shares of its stock at a certain price. Usually offerings are advertised in newspapers or announced on financial news programs.

The Value of Common Stock

The value of common stock increases in four ways: from the dollar appreciation of stock value, stock splits, mergers, and income from dividends. Each of these will be described in the following sections.

Dollar Appreciation of Stock Value. Stockholders purchase a stock and hold it for a period of time hoping the stock will increase in value. When the stockholder sells the stock, he or she makes a profit from the difference between the purchasing and the selling price (less the transaction costs and capital gains taxes discussed in the previous chapter).

Stock split
When a corporation declares that stocks will split.

Stock Splits. Investors can increase their holdings through a stock split. In a **stock split**, shares of the stock owned by existing stockholders are divided into a larger number of shares. For example, you could own 100 shares of XYZ stock, and if the corporation declares a 2-for-1 split, you will then have 200 shares, a 4-for-1 split yields 400 shares.

Usually corporations split stocks because more people will buy their stock at $50 a share than at $100 a share. The corporations are looking for an ideal price that is attractive to investors; a high price is unwieldy for the company as well as for the investor. Psychologically it feels better for the investor to own 200 shares of stock in a company than 100. Typically a stock will go up in value after a split. If, after the split, shares initially sell for $38, often what happens within a matter of hours or days is the price will rise to $39 or $40, although there are no guarantees that this will happen. Usually investors do well in a split whether they are previous owners whose number of shares increases or whether they buy in immediately at the new low price.

Mergers. When two or more companies join together (merge), their stocks will be affected, often dramatically. Even news of a merger will affect stock prices of the companies immediately involved and of companies in the same industry. For example, in 1998 the merger of Citicorp and Travelers created Citigroup, a leading world financial services provider. When the merger appeared imminent, shares of Citicorp rose $37.62 and Travelers gained $11.31 in one day. Mergers do not always cause both stocks to go up; sometimes one or both stocks go down. The one thing that can be counted on is that a merger will affect the prices of the stocks involved in one way or another and stockholders should be ready to respond by holding in anticipation of further growth, selling, or buying more shares in the existing company or newly formed company.

Dividends. As defined in Chapter 11, *dividends* are distributions of earnings that a corporation pays to stockholders. Dividends are paid out of the company's profits and the company's board of directors decides what the dividend will be. Not all companies pay dividends. Dividends can be

1. *Cash dividends*, which are usually issued quarterly (four times a year). Some large corporations which have had an extremely good year will also issue a year-end dividend. For example, on November 2, 1999, Allnc Capital (symbol AC) was selling at 30 1/8 per share and was offering a yearly dividend of 8%.
2. *Company products, property, or discounts.* For example, Wrigley & Co. sends

packs of gum to their shareholders. As another example, shareholders of Sara Lee stock receive a catalog offering a 15% discount on Coach products, such as purses and briefcases. Coach is one of Sara Lee's businesses.

2. *Stock dividends,* which are new shares of stock issued to existing stockholders. For example, if a company declares a 10 percent stock dividend, each stockholder would receive one-tenth of a share of stock for each share owned. A stockholder with 200 shares of XYZ stock would receive 20 additional shares. Whether or not dividends are offered is an important consideration when deciding whether to pick one stock over another.

Dividend-paying stocks can be a significant part of an individual investor's portfolio because the income they pay out tends to cushion the ride during market downturns. Knowing that those quarterly dividends are coming in can give an investor the courage to stay the course when things get rough, rather than selling when stock prices are depressed. Retirees often count on dividends to pay for a healthy part of their living expenses.

At the same time, though, there are some important caveats that investors should keep in mind. Among other things, focusing too much on dividends could lead to a portfolio that is overly concentrated in a few industries, such as utilities and banking, that tend to be dominated by dividend-paying stocks. It also could lead an investor to overlook promising new companies simply because they don't pay quarterly dividends.[3]

Common Stock versus Preferred Stock

Common stocks may or may not pay dividends, whereas preferred stocks pay dividends before common stockholders are paid. Another difference is that if the corporation is liquidated, the claims of secured and unsecured creditors and owners of bonds and preferred stock take precedence over the claims of owners of common stock. In this sense preferred stock is safer than common stock. However, as discussed in the previous chapter with more risk there is a greater chance for return, usually common stock has more potential for appreciation. Many companies such as KMart and Dupont offer both common and preferred stock.

Common stock certificates are documentation of a shareholder's ownership of a company. They are engraved on heavy paper with intricate designs to deter forgery. Since so much trading is done electronically these days, most people do not have the actual paper certificates. This makes practical sense because many stocks (especially with the advent of day trading) are bought and sold within a few days or a few hours.

CONSUMER ALERT

Day trading, an occupation involving rapidly buying and selling stocks, has received a lot of publicity. According to an article in *Business Week* (November 1, 1999, p. 214), a group of securities regulators found that 70 percent of day traders "will not only lose, but will almost certainly lose everything they invest...Only 11.5% of the accounts reviewed evidenced the ability to conduct profitable short-term trading. "The main way people are making money on day trading is by offering training, software, and subscriptions. Anyone considering day trading should do so with caution.

It would not be feasible to have paper certificates mailed out and returned in this short time. So, for the most part, documentation of stock ownership is stored on computer and sent to investors periodically in a statement; but if a stockholder wants the actual certificates, they can be obtained and usually they are stored in a shareholder's safe-deposit box or in a stockbrokerage firm's safe.

PREFERRED STOCK

Preferred stock
Shares that pay dividends and have precedence over common stock.

Par value
Face value of a security.

Preferred stock dividends are stated as an amount of money for each share of preferred stock or a certain percentage known as the **par value**. The par value, the face value of a security, is printed on the stock certificate. For example, if the par value for a preferred stock issue is $100 and the dividend rate is 5 percent, the dollar amount of the dividend is $5 ($100 x 5% = $5).

The dividends and the security of knowing that the corporation must pay preferred shareholders before paying common stockholders make preferred stocks attractive, especially (as previously mentioned), to retired people. The "preferreds come first" factor is an important feature if the company is experiencing financial difficulties. This does not mean that preferred stockholders will always be paid if the company fails; it does mean that preferred stockholders will be paid before common stockholders. *Preferred stock does not include voting rights.*

The four main types of preferred stocks (cumulative, participating preferred, adjustable-rate preferred, and convertible preferred) are discussed next.

1. Most preferred stock is *cumulative*; if dividends are passed (not paid for any reason), they accumulate and must be paid before common dividends. In other words dividends will be paid first to holders of preferred stock and second to holders of common stock.

2. *Participating preferred* stock, a rare type, entitles its holders to share in profits above and beyond the declared dividend, along with common shareholders, as distinguished from *nonparticipating preferred*, which is limited to the stipulated dividend. *Most preferred stock is nonparticipating meaning that the company pays only the stipulated dividends.*

3. *Adjustable-rate preferred (also called floating rate or variable rate preferred)* stock pays a dividend that is adjustable, usually quarterly, based on changes in the Treasury bill rate or other money market rates.

4. *Convertible preferred* stock is exchangeable for a given number of common shares. This gives the investor a chance to make a change from preferred to common. From the issuer's standpoint the convertible feature is a sweetener, a way to attract investors.[4]

Callable preferred stock
Clause that allows for the corporation to recall the stock.

Preferred stock can also be categorized as a **callable preferred stock,** which means that the stock in that corporation may be exchanged for a specified amount of money. In other words, the corporation has an option through a clause in the stock certificate, that allows them to recall the preferred stock issue. The company may want to do this and substitute a new preferred stock that pays a lower dividend or they may want to offer only common stock. Typically, callable preferred stock pays a higher dividend than "plain" preferred stock. Callable preferred stock is another way that corporations keep their options open so that they can move with the market.

The way to tell if a stock is common or preferred is to ask an account executive (stockbroker) or refer to the newspaper financial pages. A preferred stock has a "pf" written after the name of the company.

Objective 2:
Locate and evaluate information on stocks.

STOCK REPORTS IN THE NEWSPAPER FINANCIAL SECTION

An evaluation of which stocks to buy and sell begins with a search for information. Several Web sites on the Internet and most large daily newspapers give prices of stocks listed on NASDAQ, the New York Stock Exchange, the American Stock Exchange, and other exchanges. Stocks are listed alphabetically in newspapers. Find the exchange that has the name of the company you are interested in, and then read across the line to find the closing price (60 would mean it closed at $60 a share) and net change (+1/4 would mean that a single share of stock went up 25 cents) during the previous day (see Exhibit 12.1 for an example of a *Wall Street Journal* listing of WalMart). It should be noted that near the end of the year 2000, stock price quotes will change to cents, so instead of a stock price listed at 18 1/4 per share the new listing will be 18.25.

Since a lot of people do not want to wait until the next day to find out how their stock is doing, they can find updated quotes on the Internet or by watching televised financial news reports. Internet Web sites for stock tracking are given in the Electronic Resources section at the end of the chapter.

Stock Advisory Services

Besides newspapers, magazines, the Internet, and television, another source of financial information is stock advisory services. Some of these charge for a service, such as a newsletter or report; others are available for free at the library. Most public libraries or university libraries have Value Line, Moody's Investors Service, and Standard & Poor's reports, or you can pay to have these reports sent directly to your home by surface mail or through e-mail or Internet access. These services provide another way for you to check out a stock, what the corporation sells, and the corporation's financial past and predicted future. So, before investing in a stock, you may want to look it up in one of these sources. Exhibit 12.2 shows 2 pages of a Standard & Poor's report for WalMart.

Stock Classifications

An investor, in order to diversify, may want stocks that fall into different categories. Stocks are generally divided into six categories:

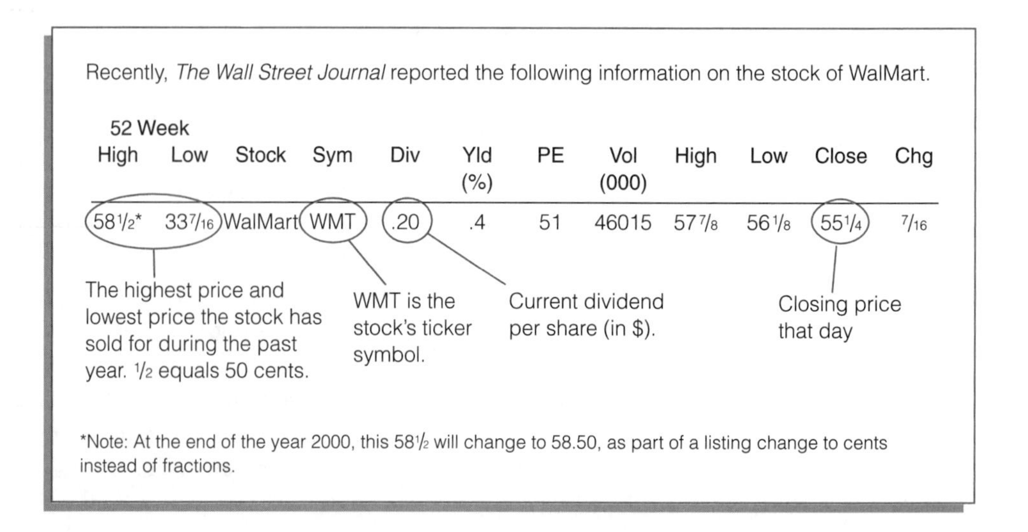

EXHIBIT 12.1
Reading a Stock Price Quotation

STANDARD
&POOR'S
STOCK REPORTS

Wal-Mart Stores

NYSE Symbol **WMT**

In S&P 500

11-MAR-00

Industry:
Retail (General Merchandise)

Summary: Wal-Mart is the largest retailer in North America, operating a chain of discount department stores, wholesale clubs and combination discount stores and supermarkets.

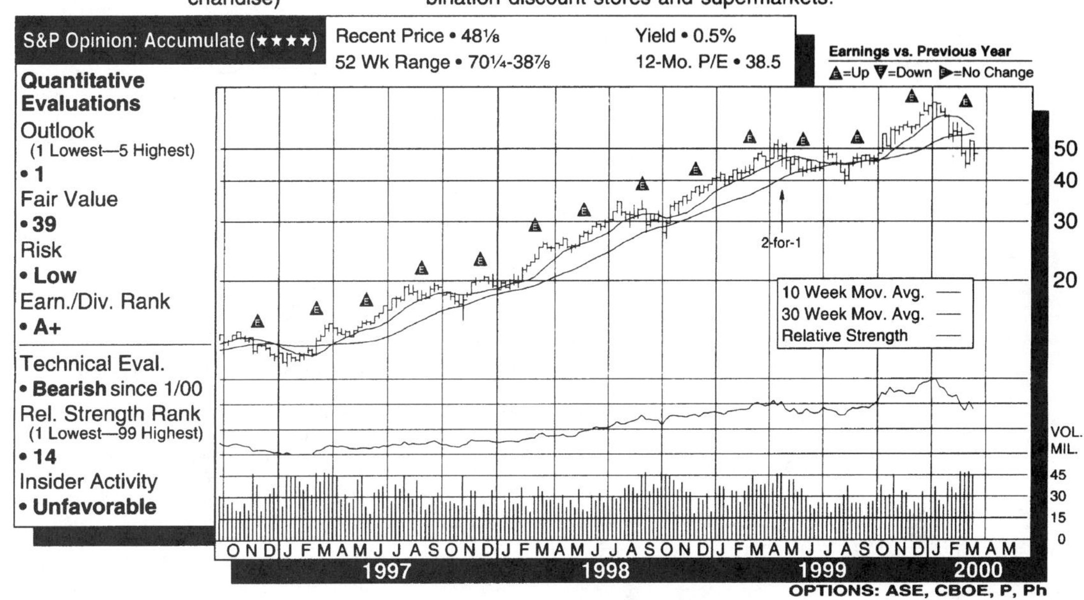

Overview - 09-MAR-00

Sales in FY 01 (Jan.) should increase about 15%, with same-store sales up about 5%. Discount stores and supercenters represent about 68% of total revenues. WMT has reduced prices on a broad assortment of items by 10% leading to strong market share gains. As a result of its efficient inventory management systems, gross margins should widen. Expense ratios should decline with leverage from a growing store base, and operating income should increase by about 16%. Interest expense should decline slightly. WMT supercenters are the key to WMT's domestic growth. Supercenters, which carry both groceries and general merchandise, are typically larger than WMT's traditional discount stores. The company operates more than 720 of these units, generating over $60 million per unit annually, and is expected to add about 165 units in FY 01, mostly replacing smaller discount units. International expansion also enhances long-term growth prospects. WMT is repurchasing $2 billion of its common stock over time.

Valuation - 09-MAR-00

The shares are down some 20% year-to-date, reflecting concerns that higher interest rates will dampen consumer spending. We are maintaining our accumulate recommendation for this premier discount chain. Fourth quarter earnings were on target, up 26%. Same store sales rose 7.0%. Operating profit increased 14.3% at Wal-Mart stores, rose 7.8% at Sam's Clubs and jumped 91.5% in the international division. We expect the company to continue to benefit from a retail spending shift toward sellers of lower- priced products. A joint venture has been created with Accel Partners to expand the development of the company's Internet site -- Wal-Mart.com. The shares of this well managed company will continue to receive a premium valuation, reflecting increased levels of free cash flow.

Key Stock Statistics

S&P EPS Est. 2001	1.45	Tang. Bk. Value/Share	4.75
P/E on S&P Est. 2001	33.2	Beta	1.14
Dividend Rate/Share	0.24	Shareholders	257,200
Shs. outstg. (M)	4453.7	Market cap. (B)	$214.3
Avg. daily vol. (M)	12.814	Inst. holdings	36%

Value of $10,000 invested 5 years ago: $ 46,805

Fiscal Year Ending Jan. 31

	2000	1999	1998	1997	1996	1995
Revenues (Million $)						
1Q	34,717	29,819	25,409	22,772	20,440	17,690
2Q	38,470	33,521	28,386	25,587	22,723	19,942
3Q	40,432	33,509	28,777	25,644	22,913	20,418
4Q	51,868	40,785	35,386	30,856	27,550	24,448
Yr.	166,809	137,634	117,958	104,859	93,627	82,494
Earnings Per Share ($)						
1Q	**0.25**	**0.18**	0.14	0.13	0.12	0.11
2Q	**0.28**	**0.23**	0.17	0.15	0.14	0.13
3Q	**0.29**	**0.23**	0.17	0.15	0.14	0.13
4Q	**0.43**	**0.35**	**0.28**	0.24	0.20	0.23
Yr.	**1.25**	**0.99**	**0.78**	**0.67**	**0.59**	**0.58**

Next earnings report expected: mid May

Dividend Data (Dividends have been paid since 1973.)

Amount ($)	Date Decl.	Ex-Div. Date	Stock of Record	Payment Date
0.050	Jun. 03	Jun. 16	Jun. 18	Jul. 12 '99
0.050	Aug. 12	Sep. 15	Sep. 17	Oct. 12 '99
0.050	Nov. 11	Dec. 15	Dec. 17	Jan. 10 '00
0.060	Mar. 02	Mar. 15	Mar. 17	Apr. 10 '00

A Division of The McGraw-Hill Companies

EXHIBIT 12.2 Standard & Poor's report for Wal-Mart in 2000.

STANDARD
&POOR'S
STOCK REPORTS

Wal-Mart Stores

SELL HOLD BUY

WALL STREET CONSENSUS

10-MAR-00

Analysts' Recommendations

Stock Prices

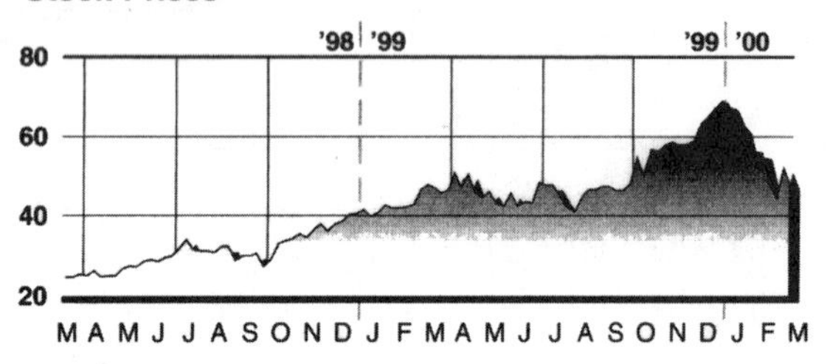

Analysts' Opinions

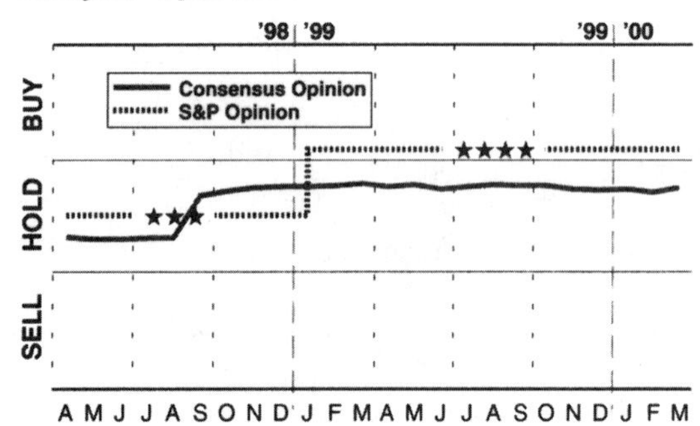

Number of Analysts Following Stock

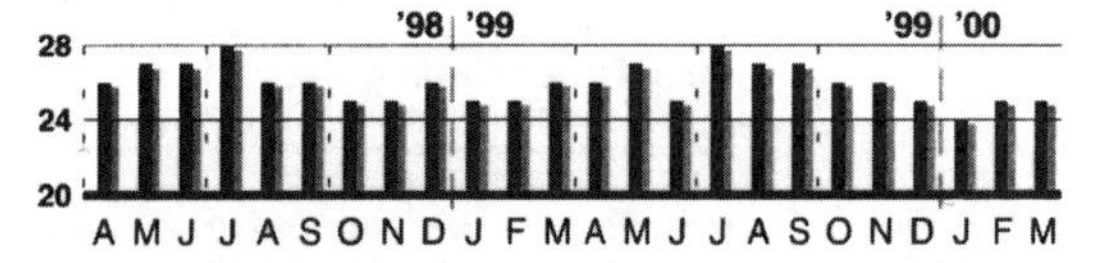

Analysts' Opinion

	No. of Ratings	% of Total	1 Mo. Prior	3 Mo. Prior	Nat'l	Reg'l	Non-broker
Buy	8	31	7	7	6	0	2
Buy/Hold	14	54	15	14	6	5	2
Hold	3	12	3	4	1	0	0
Weak Hold	0	0	0	0	0	0	0
Sell	0	0	0	0	0	0	0
No Opinion	1	4	1	0	1	0	0
Total	26	100	26	25	14	5	4

Analysts' Consensus Opinion

The consensus opinion reflects the average buy/hold/sell recommendation of Wall Street analysts. It is well-known, however, that analysts tend to be overly bullish. To make the consensus opinion more meaningful, it has been adjusted to reduce this positive bias. First, a stock's average recommendation is computed. Then it is compared to the recommendations on all other stocks. Only companies that score high relative to all other companies merit a consensus opinion of "Buy" in the graph at left. The graph is also important because research has shown that a rising consensus opinion is a favorable indicator of near-term stock performance; a declining trend is a negative signal.

Standard & Poor's STARS

(Stock Appreciation Ranking System)

★★★★

★★★★★	Buy
★★★★	Accumulate
★★★	Hold
★★	Avoid
★	Sell

Standard & Poor's STARS ranking is our own analyst's evaluation of the short-term (six to 12 month) appreciation potential of a stock. Five-Star stocks are expected to appreciate in price and outperform the market.

Analysts' Earnings Estimate

Annual Earnings Per Share

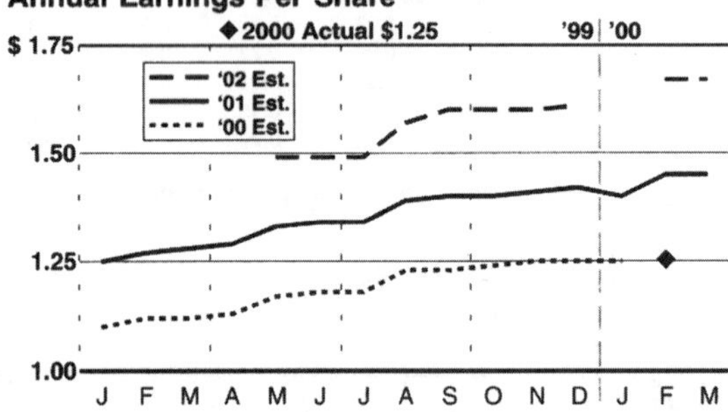

Current Analysts' Consensus Estimates

Fiscal years	Avg.	High	Low	S&P Est.	No. of Est.	Estimated P-E Ratio	Estimated S&P 500 P-E Ratio
2001	1.45	1.48	1.43	—	21	33.2	21.4
2002	1.67	1.75	1.62	—	15	28.8	—
1Q'01	0.28	0.29	0.28		15		
1Q'00	0.25 Actual						

A company's earnings outlook plays a major part in any investment decision. S&P organizes the earnings estimates of over 2,300 Wall Street analysts, and provides you with their consensus of earnings over the next two years. The graph to the left shows you how these estimates have trended over the past 15 months.

EXHIBIT 12.2 continued

Blue chip stocks
Stocks of nationally known companies with quality reputations and long records of profit growth, and which usually pay dividends.

Blue chip stocks Stocks of nationally known companies with long records of profit growth, in many cases dividend payments, and reputations for quality management, products, and services earn the designation of blue-chip stocks.[5] Examples are AT&T, IBM, McDonald's, General Motors, Campbell's Soup, Bethlehem Steel, and General Electric. These stocks are usually high priced, and they are less volatile than start-up companies. The term "blue chip" comes from the blue chips used in poker; the blue chips are the most valuable ones.[6]

Cyclical stocks
React quickly to changes in the economy.

Cyclical stock When a stock follows the business cycle—meaning it rises and falls with advances and declines in the economy—it is a cyclical stock. Examples are automakers, chemical companies, and homebuilders.

> Cyclical stocks are favorites of market timers..."Figuring out when to buy cyclicals is always tricky, and it's even more difficult right now," says Fred Taylor, chief investment officer at U.S. Trust. "You need to worry about where they are in their own cycle, where you are in the business cycle, and where you are in the stock-market cycle." Cyclicals are companies whose fates are linked directly to the health of the economy.[7]

Defensive stocks
Provide consistent dividends and remain stable regardless of changes in the economy.

Defensive stock A stock classified as **defensive** gives consistent dividends and remains stable even with advances and declines in the economy. If the market is weak, they tend to hold their value better than most stocks because they tend to be from companies that offer basic, everyday goods. Examples are food and drug companies, such as Winn Dixie and Bristol-Myers.

Growth stocks
Earn profits above average for all firms and are expected to prosper.

Growth stock Corporate issues earning profits above the average of all the firms in the economy, but with no assurance of success, are termed growth stocks. They usually do not pay dividends. Typically growth stocks are relatively young companies with expanding product lines, global expansion, new manufacturing facilities and outlets for products. Examples are Dell, Intel, Cisco Systems, and Microsoft.

Income stocks
Pay higher than average dividends.

Income stock The higher-than-average dividends paid by income stocks make them attractive to investors (retired persons, others on fixed incomes) seeking income and not necessarily a significantly higher selling price. The emphasis is on current dividends rather than the company's future prospects. Dividends received are taxable in the year gained unless the stock is held in a tax-deferred account, such as an IRA or a Keogh plan. Generally these are stocks from telephone, electric, gas, and other utilities; banks; insurance companies; and real estate investment trusts. Examples are Southern Company and Pacific Gas & Electric.

Speculative stocks
Risky, usually newer stocks invested in with the anticipation of gain and the realization of possible loss.

Speculative stock Investors who are not adverse to taking a chance, for example, with a biotech company offering a new wonder drug or a start-up Internet company, often deal in speculative stocks. Since these stocks are far riskier than average, be very careful before buying; but the payoff can be big if they grow (Microsoft and Intel were once speculative stocks).

Over time a company's stock can change categories, as the examples of Microsoft and Intel show. Also an individual stock may fit more than one category. For example, a pharmaceutical stock may fall into the blue chip and the defensive categories.

Stock Classifications by Industry

Another way to diversify holdings is to purchase stocks in a variety of industries. Thus if one industry is having financial trouble, holdings in another industry may offset these losses. The following list shows the classifications used by S&P 500 with examples in each industry:

1. Aerospace & Defense: Boeing, Northrup Grumman
2. Automotive: Ford Motor, General Motors
3. Banks: Chase Manhattan, First Union
4. Chemicals: Dupont, Monsanto
5. Conglomerates: General Electric, TRW
6. Consumer Products: Gillette, Nike
7. Containers & Packaging: Owens-Illinois, Avery Dennison
8. Discount & Fashion Retailing: Gap, TJX
9. Electrical & Electronics: Honeywell, Intel
10. Food: Hershey Foods, Campbell's Soup
11. Fuel: Texaco, Exxon Mobil
12. Health care: Schering-Plough, Pfizer
13. Housing & Real Estate: Owens Corning, Sherwin-Williams
14. Leisure Time Industries: Disney, Mattel
15. Manufacturing: Applied Materials, Caterpillar
16. Metals & Mining: Bethlehem Steel, Reynolds Metals
17. Nonbank Financial: Aetna, Fannie Mae
18. Office Equipment & Computers: Compaq, Oracle
19. Paper & Forest Products: Kimberly-Clark, Mead
20. Publishing & Broadcasting: CBS, Dow Jones
21. Service Industries: Crane, Waste Management
22. Telecommunications: Tellabs, Lucent Technologies
23. Transportation: Southwest Airlines, Union Pacific
24. Utilities: Duke Energy, Pacific Gas & Electric

Another example of a classification system is the one used by the Dow Jones U.S. Industry Group. Regardless of the classification system used, the point is that investors should strive to diversify by investing in different companies in different industries. As mentioned previously, a balanced portfolio will guard against a collapse or downturn in one area of the market.

STOCK MARKET INDICATORS

Dow Jones Industrial Average (DJIA)
Market index of the movement of 30 leading stocks.

The ups and the downs of the market can be observed in reports of various market indicators (indices). The oldest market measure, the **Dow Jones Industrial Average (DJIA)** nicknamed the Dow, was created by Dow Jones cofounder Charles H. Dow. In recent years it has been computed from 30 leading industrial stocks whose composition changes over time to reflect changes in the economy. The list of the original 12 DJIA stocks from October 1, 1928, and the current list are given in Exhibit 12.3. *The only stock on both the original and the current list is General Electric.*

On November 1, 1999, four stocks representing the "new economy" were added (Microsoft, Intel, Home Depot, and SBC Communications), replacing Chevron,

EXHIBIT 12.3
Dow Jones Industrial (DJIA) Stocks

Dow Jones Stocks

October 1,1928 Original 12 stocks
American Cotton Oil
American Sugar Refining Co.
American Tobacco
Chicago Gas
Distilling & Cattle Feeding Co.
General Electric
Laclede Gas Light Co.
National Lead
North American Co.
Tennessee Coal, Iron & Railroad Co.
U.S. Leather
U.S. Rubber Co.

2000 Today's 30 stocks
Alcoa
Allied Signal
American Express
AT&T
Boeing
Caterpillar
Citigroup
Coca-Cola
Dupont
Eastman Kodak
Exxon Mobil
General Electric
General Motors
Hewlett-Packard
Home Depot
IBM
Intel
International Paper
Johnson & Johnson
McDonald's
Merck
Microsoft
Minn. Mining (3M)
J.P. Morgan
Phillip Morris
Procter & Gamble
SBC Communications
United Technologies
Wal-Mart Stores
Walt Disney

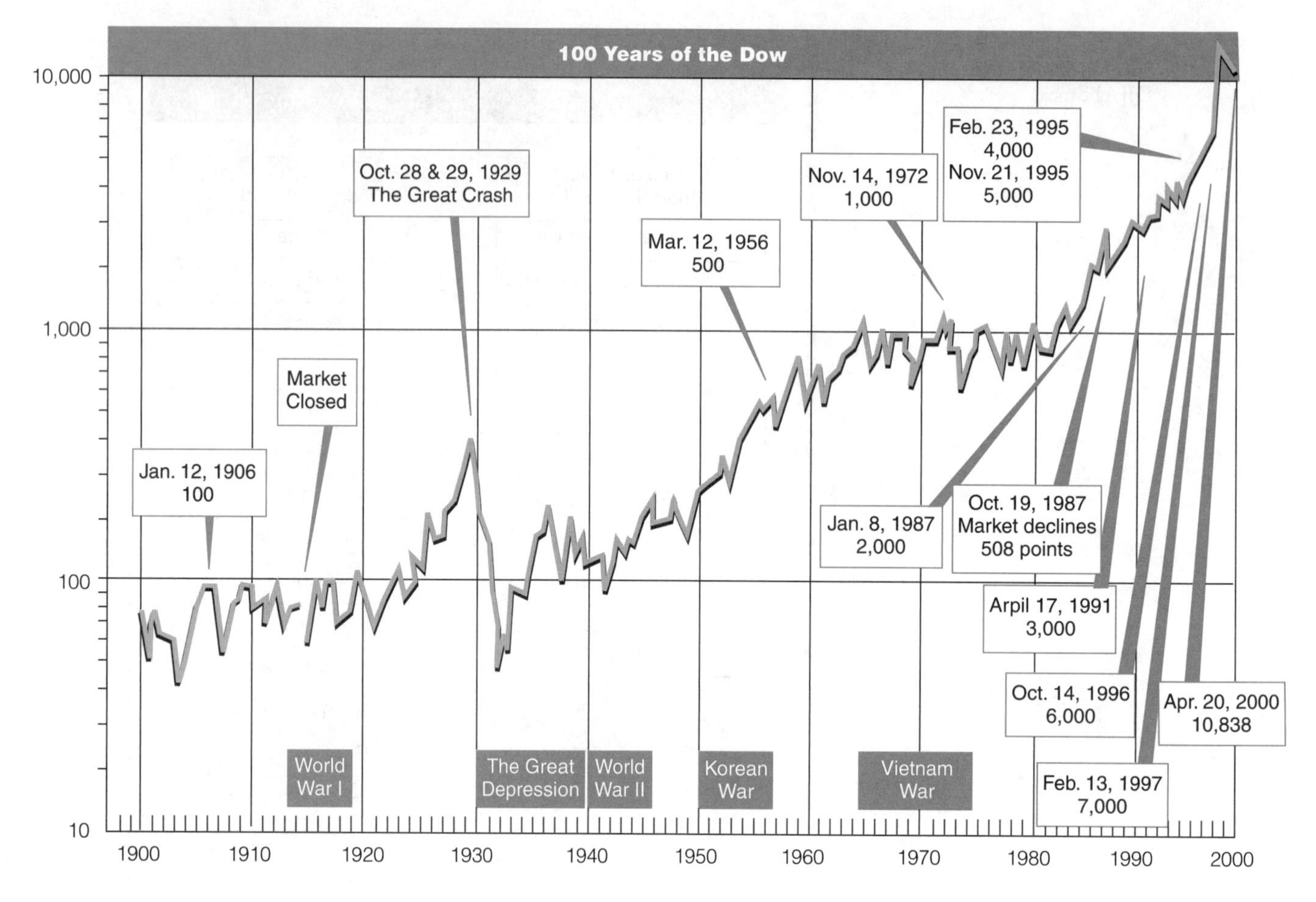

EXHIBIT 12.4
One Hundred Years of Dow Movement

Goodyear Tire and Rubber, Sears, Roebuck & Co., and Union Carbide. This was a unique occurrence because Microsoft (the nation's largest company measured by stock market value) and Intel are traded on NASDAQ. Previously all 30 companies had been listed on the New York Stock Exchange. Adding Microsoft and Intel to the Dow reflects a broader shift toward newer, higher technology ways of doing business.

Exhibit 12.4 shows the movement of the Dow over the last 100 years. Stocks were at first considered very risky. Even Mr. Dow recommended to readers of the *Ladies' Home Journal* to buy bonds instead of stocks.[8] Dow Jones & Company (DJ) also computes and publishes a transportation average and a public utility average.

Standard & Poor's Index (S & P 500)
Market index of the movement of 500 stocks.

Another leading market indicator is the **Standard & Poor's 500 Composite Index (S&P 500)** which, as the name implies, covers the market activity of 500 stocks. *It was introduced in 1957.* The S&P 500 represents a huge universe, accounting for 70 percent of the market capitalization of all U.S. stocks.[9]

NASDAQ Index
Market index of fast moving technology and financial services stocks.

The **NASDAQ (National Association of Security Dealers Automated Quotations) Index** monitors stocks traded over the counter as well as for many New York Stock Exchange listed securities. It is an especially good indicator of technology and fast-moving financial service stocks since the NASDAQ Index is heavily weighted toward computer, biotechnology, telecommunications, and start-up companies.

To give an example of how the rises and falls of these indices are reported, here was the news on March 26, 1998: "Standard & Poor's 500-stock index fell 1.13, or 0.10%, to close at 1100.80...The NASDAQ Composite Index improved 4.03, or 0.22%, to close at 1828.54, extending its record."[10] Exhibit 12.5 provides an update.

EXHIBIT 12.5
Movement for Major Indexes on a Single Day

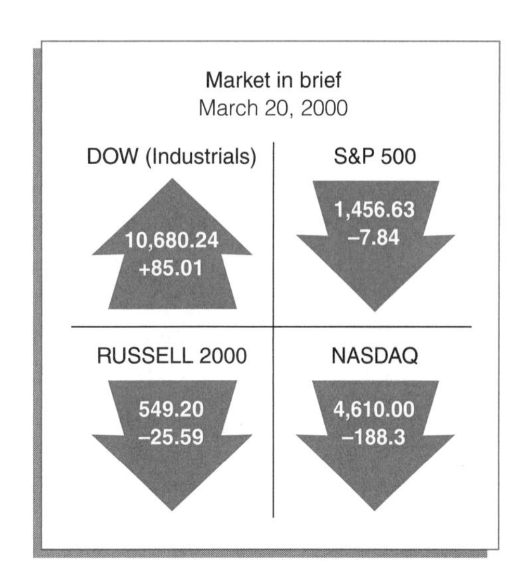

Russell 2000 small stock index
Measure of the stock price performance of small companies.

Wilshire 5000 equity index
Broadest of all market indexes, the market value of all U.S. equities on the main stock exchanges.

Indexes other than those already covered are the Russell Indexes of which the most well-known is the Russell 2000 small stock index, the AMEX Major Market Index, Value-Line Composite Average, and the Wilshire 5000 Equity Index. Of these the ones often quoted on the news are the **Russell 2000 small stock index** which measures the stock prices performance of small, usually young companies and the **Wilshire 5000 Equity Index.** The Wilshire is the broadest of all the market indexes. This index represents the value of all U.S. headquartered equities on the NYSE, AMEX, and the NASDAQ markets. Equities are another word for stocks.

All of these indexes have their strengths and weaknesses because they represent different segments of the market. The DJIA is criticized because it follows only 30 stocks; however, even though it represents so few companies, they are well-diversified industry leaders. So, for the most part, the DJIA parallels the movements of broader market-value-weighted indices such as the S&P 500 and the NYSE indexes. The NASDAQ Composite is usually more volatile than the Dow. *The purpose of the indexes is to give investors a sense of where stocks stand.* They also serve as indicators of the health of the general economy.

Bull and Bear Markets

Bull market
The market is up.

Bear market
The market is down.

A **bull market** *means the market is up*; investors are optimistic about the economy and are buying stocks. The 1990s experienced ups and downs in the market, but for most of the decade it was a bull market. Conversely, a **bear market** *means the market is down*; investors are pessimistic about the economy, and they are not buying stocks or they are selling stocks. Overall, bull markets continue for longer periods of time than bear markets.[11]

As an individual investor you make decisions to buy and sell based on many of the ideas already presented in this chapter (gathering information, following the ups and downs of the market, and diversifying). The next section describes ways of measuring a corporation numerically and from efficient market theories standpoints.

Objective 3:
Use numerical measures and investment theories to evaluate stocks.

FOUR NUMERICAL MEASURES

As a guiding principle, the better a company does and the higher its profits, the more stockholders can make. To help determine how well a company's stock is performing, an investor may use four numerical measures: book value, earnings per share, price-earnings ratio, and the beta coefficient.

Book Value

Book value
Numerical measure of a company derived by deducting liabilities from a corporation's assets and dividing the remainder by the number of outstanding shares of common stock.

The **book value** of a stock, the net asset value of a company reported in financial publications, is determined by deducting all liabilities from the corporation's assets and dividing the remainder by the number of outstanding shares of common stock. The way book value is calculated is shown in this example:

$$\frac{\begin{array}{c}\text{Assets} - \text{Liabilities} \\ \$10{,}000{,}000 - \$5{,}000{,}000\end{array}}{100{,}000 \text{ shares of common stock}} = \$50 \text{ per share}$$

Asset and liability figures can be obtained from the company's annual report, stock advisory services, or stockbrokers.

When a stock's market value is lower than, or the same as, the book value, some investors think they have found a sure thing. For example, if the book value is $50 for a share, and the stock market price of a share is $40, an investor may find this stock worth purchasing. Unfortunately, it is not as easy as that because the figures may be off, and there are so many other factors to consider in stock selection such as company management and new product lines. Generally, using book value as a guide works best on companies that are directly involved in natural resources such as oil (Texaco, Exxon Mobil) or forestry companies.

Earnings per Share

Earnings per share
Numerical measure of the portion of a company's profit allocated to each outstanding share of common stock.

Earnings per share is the portion of a company's profit allocated to each outstanding share of common stock.[12] For example, a corporation that earned $100 million last year and has 100 million shares outstanding would report earnings of $1 per share. This figure is calculated after paying taxes and preferred shareholders and bondholders. Earnings per share is highly regarded as an important statistic in evaluating a stock. An increase in earnings per share is a healthy sign for investors.

Price-Earnings (P/E or PE) Ratio

Price-earnings ratio
Numerical measure of the relationship between the price of one share of stock and the annual earnings of the company.

The price-earnings (P/E or PE) ratio is one of the better known and most critical numerical measurements. *It refers to the relationship between the price of one share of stock and the annual earnings of the company.* The PE ratio, which expresses the value of a stock in terms of company earnings rather than selling price, is given in newspaper financial listings, as shown in Exhibit 12.1. If you want to figure it yourself, here is the formula.

$$\text{P/E ratio} = \frac{\text{Price of stock}}{\text{Earnings of the stock}}$$

For XYZ stock:

$$\text{P/E ratio} = \frac{\$20}{\$1} = 20$$

The PE ratio may be reported as *trailing*, meaning based on reported earnings from the latest year, or *forward*, meaning an analyst's forecast of next year's earnings.

In the XYZ stock example, the stock sells for $20 a share, and last year's earnings per share was $1, resulting in a trailing PE of 20. If the same stock has a projected PE of $2, it will have a forward PE of 10.

The PE ratio provides investors an idea about how a company is doing. *PE ratios typically range between 5 and 20.* Those 20 and over are typically young, fast-growing companies. They are *riskier* than those with lower PE ratios. Low PE ratios are common in older, established companies such as blue chips. Usually, low PE ratio stocks have higher yields than higher PE ratio stocks, which often pay no dividends at all.[13] However, this is a matter of opinion. Some investors avoid high PE ratios especially those over 25, because higher PE ratios may indicate volatility. Other investors pay no attention to PE ratios especially if they invest in new technology companies that have very high PE ratios.

Another way to evaluate PE ratios for a particular stock is to look at them over a three-year range. Let's say XYZ's PE ratio has ranged between 12 to 25 over the last three years and now it is at 15, this may potentially be a good investment.

Beta Coefficient

Beta coefficient
Numerical measure of a stock's volatility.

The **beta coefficient** (also known as **beta**) measures the stock's volatility. It estimates the risk of a stock based on how much its market price changes compared to changes in the overall stock market. Conservative investors who want to preserve their capital (and are less concerned with high returns) should look for stocks with low betas; a high-risk taker may be drawn to high betas.

Here is how it is figured. The S&P 500 Stock Index and the NYSE have beta coefficients of 1. Note that in Exhibit 12.2 on the right side of the first page, the beta for WalMart is 1.14. Since most stocks have betas between 0.5 and 2.0, 1.14 is within that range and is very close to the movement of the general market.

A beta of 1.75 would be more volatile; if the market goes up 10 percent, it is estimated that the value of the stock would rise to 17.5; but if the market fell 10 percent, the value of the stock would decrease 17.5 percent. A beta closer to 1.00 would be less volatile.

A beta of less than one is supposed to move less in price than the market in general. A risk-averse investor may prefer lower beta stocks so that if the market falls, the predicted loss will be less. On the flip side, a low beta stock would do less well in a bull market than a high beta stock.

THREE INVESTMENT THEORIES

As mentioned earlier, numerical measures are a way to evaluate stocks. Other ways are based on three investment theories: efficient market theory, fundamental theory, and technical theory.

Efficient Market Theory

Efficient market theory or random walk theory
Assumes that it is impossible to beat the market, that it is futile to forecast market movements.

Of the three this is the most talked about theory. **Efficient market theory (also called Random Walk Theory)** is based on the assumption that stock price movements are random (hence the nickname Random Walk Theory). It posits that an individual stock investor cannot outperform the average for the stock market over a period of time. This theory assumes that the market is efficient, and buyers and sellers have considered all the information available on individual stocks. Believers in this theory are skeptical of the next two theories.

Fundamental Theory

Fundamental theory
Assumes that a stock's real value is determined by a company's future earnings.

Fundamental theory assumes that a stock's real value is determined by the company's future earnings. Future earnings are compared to present earnings; if future earnings are estimated to be substantially higher, it is assumed that the stock will increase in value. Investors consider not only future earning potential but also the type of industry the company is in (whether it is in a high growth area like technology), the strength of the individual company, new product development (a pharmaceutical company coming out with a new drug), and the general state of the economy.

Technical Theory

Technical theory
Assumes that trends and patterns can be found in the overall market.

Technical theory is not based on the strengths of the individual company nor expected earnings, but on trends and patterns found in the market as a whole. Technical analysts use charts or computer programs to identify and project price movement (number of shares bought and sold) and other market averages. They are not as concerned with the financial position of an individual company as fundamental analysts are.

Objective 4:
Know how to buy and sell stocks.

BUYING AND SELLING STOCKS

Up to this point the chapter has focused on how to get information about stocks, conduct numerical measures, and use investment theories. Now we turn to the mechanics of buying and selling stocks.

After deciding what stock you want, you contact an account executive in person, by telephone, or over the Internet. He or she in turn buys the stock from a primary or secondary market.

Primary market
The market for purchasing new issues of securities.

Account executives (stockbrokers)
Employees of brokerage firms who advise and handle orders for clients and perform other financial services.

* A **primary market** involves purchasing new issues of securities through an investment bank or other representative. The **account executive** (more typically known as a *stockbroker*) acts as an agent for an investment banking firm and then the money paid for the stock goes to the corporation that issued the stock. The term "account executive" is really more accurate than "stockbroker" because they do more than handle stocks. A full-service account executive buys and sells stocks, bonds, and other securities; manages money market funds; and provides financial planning including tax guidance regarding investments. Thousands of company stocks are sold through the primary market.

Secondary market
Previously owned shares of stock that are resold on this market.

* In a **secondary market** existing stock shares are sold between investors through an account executive. *After stocks are sold in the primary market, they can be sold repeatedly in the secondary market.* This means that previously owned shares of, for example, Disney are also sold through the secondary market. If you sell your shares, someone else can buy them on the secondary market. If no one is selling shares, there are none to buy, and you have to wait until shares come on the market. Unavailability can occur when there are small or highly sought-after corporations. For example, if you want to purchase shares in a small regional bank listed on the NYSE, there may be none on the market on the day you want to make a purchase. You will have to wait until some shares come on the market.

Securities Exchanges

Securities exchange
The marketplace where member brokers represent investors who want to buy and sell.

The **securities exchange** is the marketplace where member brokers represent investors who want to buy and sell. Most securities of national companies are sold through the NYSE (the New York Stock Exchange, also known as the Big Board) or NASDAQ (the National Association of Securities Dealers Automated Quotations system).

The NYSE and NASDAQ are the two largest stock exchanges in the U.S. The New York Stock Exchange is, as the name implies, located in New York City. NASDAQ, owned and operated by the National Association of Securities Dealers, is a computerized system that provides brokers with price quotations for securities traded over the counter as well as for many NYSE-listed securities. Recently there has been discussion of merging the boards of the NASDAQ Stock Market's parent and of the American Stock Exchange.[14] In addition to these large exchanges, there are smaller regional exchanges in Chicago, Boston, San Francisco, and Philadelphia, usually offering stocks from regional corporations. Also there are 142 foreign exchanges.

Most of the NYSE's 1,366 members (or seats) represent brokerage firms that charge commissions on trades made by their representatives for their investors. What this means is that when you contact brokers, they place an order with their firm's NYSE member or members. There are also specialists who buy or sell particular stocks on the floor of the exchange.

Each of the stock exchanges has listing requirements regarding minimum standards of:

- annual earnings of the corporation before taxes.
- net tangible assets.
- market value.
- number of common shares offered to the public.
- number of stockholders owning 100 or more shares of stock.

Over-the-Counter (OTC) Market

Not all stocks are listed on a securities exchange. In fact the vast majority of U.S. stocks (many of them very small, beginning stocks) are not traded on an exchange at all. Many companies—even a large one like Apple Computer Inc.—prefer being traded over the counter (OTC). In OTC a network of account executives (brokers) buy and sell the securities of corporations. Securities are traded through a specialist. OTC trades take place over the telephone or the Internet.

CONSUMER ALERT

As discussed in the insurance chapters, there is a practice called churning which is to be avoided. In investments the term churning refers to the illegal practice (under the Securities and Exchange Commission) of excessive buying and selling of securities to generate commissions. Also as discussed in the previous chapter, most full-service brokerage firms charge between $20–$100 for a transaction; usually if you trade over the Internet or through discount brokers this transaction fee will be less. In addition be wary of cold calls (unsolicited) from brokers whom you do not know.

Since there are so many OTC companies, only the most actively traded issues are listed in *The Wall Street Journal* under the NASDAQ National Market Issues. Most OTC companies pay no dividends because they are small, start-up companies that need to put their earnings back into the business.

Steps in a Typical Transaction

Most investors contact their account executive or brokerage service and place an order to buy or sell a stock. Once the account executive receives the order, he or she initiates the following process:

1. Relays the order electronically to a representative at the stock exchange.
2. A clerk of the firm signals the transaction to a partner on the stock exchange floor.
3. The partner finds what the specific stock is trading for and trades with another stock exchange member who is buying or selling.
4. The partner signals back to the clerk once the transaction is made. The purchase or sale is recorded.
5. A confirmation is relayed back to your account executive, who notifies you of its outcome.

THE LANGUAGE OF BUYING AND SELLING

Market order
Request to buy or sell a stock at the best price available currently.

Limit order
Request to buy or sell a stock at a specified price.

Stop order
A type of limit order in which a request is made to sell a stock if it falls below a specified price.

You can make a **market order**, which is a request to buy or sell a stock for the best buy or sell price available in the current market value. The way you would say this is, "Buy or sell 100 shares of XYZ for me, at the best price you can get."

You can also make a **limit order** which is a request to buy or sell a stock at a specified price or better. You would say, "Trade for me only if and when you can do better than $55" (or whatever other price you set).

If the price only goes up to $54 the account executive will not buy the stock. If you think the price may fall rapidly, you may also decide to sell and place a **stop order** by saying, "If the stock goes as low as $50 (or any other price you name), sell my shares."

To summarize, you could place an order to buy 100 shares of XYZ at the going rate (market order), or you could say you will buy 100 shares when the price hits or falls under a set price, such as $75 a share. You could also place an order to sell at a certain price. With a limit order, orders by other investors may take precedence over your limit order. The account executive will confirm the price you bought or sold the stock for and your monthly statement from the brokerage firm will list the price.

Discretionary order
Lets the account executive decide when to buy a stock or bond.

You can also let the account executive decide when it is best to make a transaction. This is called a **discretionary order**. It is generally preferred that account executives and clients work together to build an investment strategy and work as a team in making buy and sell decisions.

Round lots
Transactions of 100 shares or multiples of 100 shares or $1,000 or $5,000 par value on bonds.

Odd lots
Securities traded for fewer than 100 shares.

Transactions are called **round lots** if 100 shares or multiples of 100 shares are traded or $1,000 or $5,000 par value for bonds. **Odd lots** refer to fewer than 100 shares. This distinction is important to make because some brokerage firms charge a lower transaction fee for round lots than odd lots because odd lots have to be combined into round lots before a trade can be made.

Securities Regulations

Because of past abusive and fraudulent practices, securities regulations were developed. Securities regulations can be divided into two broad categories:

State regulation Most states have laws that provide for registration of securities, licensing of brokers and securities salespeople, and prosecution of any person fraudulently selling stocks and bonds.

Federal regulation In 1934 the U.S. government created through the Securities Exchange Act *the Securities and Exchange Commission (SEC). This agency enforces federal regulations whose primary purpose is to protect investors by regulating the activities of stock traders.* The two main objectives of the SEC are to see that investors are fully informed about securities being offered for sale and to prevent misrepresentations, deceit, and other fraud in securities transactions.[15]

The SEC administers many federal acts. A recent, highly publicized one was the Insider Trading Act of 1988 that made brokerage firms responsible for reporting transactions based on inside information to the SEC. Failure to report such violations of the law can result in fines up to $1 million.

In 1997 the SEC conducted 350 insider trading investigations.[16] If people buy or sell company securities while they have material information not available to the public, it is considered trading illegally. Material means it is important enough for a reasonable investor to take into account in buying, selling, or holding decisions.

Long-Term Investment Strategies: A Review

As a personal investor you can be a long- or short-term strategist or a combination of both. In the previous chapter the long-term strategies of buy-and-hold, dollar cost averaging, direct investment plans, and dividend reinvestment plans were explained. The *buy-and-hold* strategy means that you select well and hold on to your stocks. This allows time for the stock to increase in value and minimizes the costs of paying commission fees for buying and selling.

Dollar-cost averaging refers to a series of equal dollar investments made at regular intervals. For example, a person could invest $100 a month in a stock over a three-year period. With this strategy, when the stock is selling at a low price, the investor gets more shares; when the stock is selling at a high price, the investor gets fewer shares. The advantage of this strategy is in the consistency of purchase regardless of the ups and downs of the market.

Direct investment plans allow investors to initially purchase stock directly from a corporation without going through a broker. *Dividend reinvestment plans (DRIPs)* also eliminate the need for a broker because dividends are automatically reinvested in additional shares purchased directly from the company. The main advantage of both of these plans is that commissions are avoided by dealing directly with the corporation.

Short-Term Strategies

The following short-term strategies (buying on margin, selling short, and trading options) are more risky and require more careful study than the long-term strategies. They are not for all investors and especially not recommended for beginners. Although the word "Stock" is used because this is the most usual application, these terms can also be applied to buying and selling commodities such as grains, metals, and foods.

Margin
Amount an investor borrows when buying a stock.

Buying Stock on Margin. In this method an investor borrows part of the money needed to buy a stock. The Federal Reserve Board sets the margin requirement; currently it is 50 percent. This means an investor can borrow up to half of the stock purchase price. Obviously this involves risk.

If the stock increases in value substantially the investor can pay off the loan and come out ahead. If the stock drops to approximately one-half of the original price, the investor will receive a margin call from a brokerage firm. After the call the investor must pledge more cash or securities to serve as collateral for the loan. If the investor does not have the money or securities, then the stock will be sold and the proceeds used to help pay for the loan. There are charges for these services and transactions.

Selling short
A technique of selling a securities or commodities contract that was borrowed.

Selling Short. This refers to selling a stock that has been borrowed from a stock brokerage firm and needs to be replaced at a later date. If individuals sell short today, that means they know that the firm must buy or cover their transaction at a later date. People sell short for two reasons: to take advantage of an anticipated decline in price, and/or to protect a profit in a long position. This latter reason is connected with the motive of protecting a capital gain in shares that are already owned. In other words the short sale may be used to delay realization of the gain until it qualifies as a tax-favorable long-term gain rather than a short-term gain taxable as ordinary income. Persons interested in the tax angles to selling short should talk with their account executive.

For example, someone could decide that IBM is overpriced because of increased competition from other makers of computers. As a result that person could decide to get rid of IBM stock before the price drops lower. He or she could sell short 200 shares of IBM stock. If the stockbroker goes through with the sale and the price drops they will come out ahead. However, if the stock rises in value, the person will lose money on a sell-short order. To make money, you must be accurate in your estimation that the stock will decrease in value. In reality in the case of IBM, you would have to base this estimation on more than an observation about the rise and fall and competition in the computer market because IBM is involved in much more than computers and they are involved in global markets. Observations about the U.S. market are not sufficient to make a judgment. Brokerage firms handle sell-short orders and charge the usual commissions.

Option
Securities transactions agreements to have the option to buy or sell stocks or other types of investments.

Trading Options. An **option** is a contract that conveys to its holder the right, but not the obligation, to buy or sell stocks (usually for 100 shares), foreign currency, interest rates, or futures (corn, crude oil). Options are traded on the NYSE and other exchanges. As in stock trading, option trading is regulated by the Securities and Exchange Commission (SEC).

A stockbroker registered to trade stocks can also trade stock options. A commodity broker can trade options on futures. There are commission and tax considerations involved in trading options. Prices for exchange-traded options are published in large circulation newspapers.

You *can* make money. If, for example, you have the right to buy a stock at $40, and it is selling at $44, you could exercise your call option, pay $40 a share, then turn around and sell the shares at $44. The higher the stock price rises, the higher your profits. You can lose money if the stock drops. If you owned the stock and the price fell $20 a share you would lose the $20 a share. With *options you do not actually own the stock; you have only an option to buy that stock.* If the stock drops in price, you would lose the cost of that option. The profit is in trading the rights to your option for more money than you paid.

There are two types of options, calls and puts.

Call
Gives the holder the right to purchase.

Put
Gives the holder the right to sell.

1. A **call** is an option contract that gives the holder the right to *purchase* stocks or futures contracts at a fixed price until the expiration date (3 months, 6 months) and to sell (short) known as writing a call. If holders do not exercise the option before the agreed-upon date, they can lose money.

2. A **put** is an option contract that gives holders the right to *sell* a specified num-

ber of shares of a security at a particular price within a specified time period. They can sell short by writing a put, which means selling the right to sell the stock or futures contract until the expiration date. The **strike price** is the agreed upon selling price. The seller must be committed to sell if there is a buyer. This commitment enables the seller to receive money from the buyer, called an **option premium**.

Strike price
Agreed upon price.

Option premium
Amount of money the seller receives from the buyer.

Options are desirable because they can provide flexibility, leverage, and the ability to participate in price movements without committing large amounts of funds needed to buy stock outright. The risks lie in speculating about the rise and the fall prices and, if you are a buyer, you pay a nonrefundable amount in return for time to decide whether or not to conclude a deal.

One of the main differences between stocks and options is that stocks can be held indefinitely whereas options have a limited life, each has an expiration date. Another difference is that with stocks you get a certificate if you so desire evidencing ownership. With options you do not get a certificate; rather, options are indicated on statements from brokerage firms. There are a number of different types of options. **Covered options** limit the amount of risk because they are backed by shares, whereas other option choices involve substantial risks. Options are open to individuals and institutional investors around the world. Obscure markets and commodities are involved, as well as blue chip stocks, in every industry.

Covered options
Contract backed by shares of underlying option.

ADVICE ON PICKING YOUR FIRST STOCK

Sometimes the basics of picking a stock are lost among all the details. Here is a short guide to getting started.

* Select two or three stocks to study, follow their progress.
* Get familiar with their ups and downs and with their management philosophies. Focus on stocks with good records and potential.
* Check how the stocks are rated by services such as Value Line and Standard & Poor's.

Advice on Buying Your First Stock

When you get ready to purchase, you may want to start with a full-service firm, where you can get individualized help and develop an investment plan. Many full-service firms are also offering less-expensive Web-based trading services so the differences among brokerages (such as cyberfirms like E*Trade), discount firms (such as Fidelity Brokerage), and full service firms (such as Solomon Smith Barney, Morgan Stanley Dean Witter, and Merrill Lynch) are blurring. So even if you start with a full-service firm, you may switch to their Web-based services to save money on commissions. If you feel confident, alternatives include online trading or contacting a company directly or through the websites given in the next section to find out if they offer direct investment plans.

ELECTRONIC RESOURCES

At the end of Chapter 11, there were lists of general investment Web sites and publication-based websites. This section focuses on information available on the Internet about stocks.

The New York Stock Exchange's Web site is **www.nyse.com** and NASDAQ's is

www.nasdaq-amex.com. The NASDAQ Web site offers charts showing the performance of certain stocks from the last six months to five years and links to the company's home page as well as the SEC's EDGAR Database.

If you are interested in technical stocks (computers, software, communications, and semiconductors), try the Silicon Investor site at **www.techstocks.com**. This site offers quotes, company profiles, news stories and a news group.

A Web site called "Dogs of the Dow" lists weekly investment information on the top Dow stocks with the highest dividend yields, at **www.dogsofthedow.com**.

For a list of companies that allow direct enrollment into their stock purchase programs try netSTOCK Direct at **www.netstockdirect.com**.

For information on options trading, contact the Chicago Board of Trade at **www.cbot.com/**, or the London International Financial Futures and Options Exchange at **www.liffe.com**, or *The Wall Street Journal Interactive Edition* at **http://wsj.com.**

For information about the IPOs (initial public offerings) industry contact **www.capmarkets.com,** the Capital Markets Financial Center site.

For information on stocks and making better investment decisions try the SEC's Web site at **www.sec.gov**.

The boom in online trading has revolutionized the way stocks, bonds, and mutual funds can be bought and sold and also the way investors can track investment information. In 1996 1.5 million online trading accounts were opened and it is predicted this will hit 10 million in 2001.[17] The appeal of personal computer (PC) trading is that it is quick and inexpensive, yet there are still transaction fees, so be sure to check these out. For example, e.Schwab (on the list following) charges a flat $29.95 for a transaction of up to 1,000 shares versus spending $55 per 100 shares for in-person help with a transaction.[18] Ameritrade advertises on television that they charge $8 per trade. Here is a partial list of brokers who offer PC trading:

- American Express InvestDirect: **www.american express.com/direct.**
- Ameritrade: **www.ameritrade.com.**
- Accutrade: **www.accutrade.com.**
- eBroker: **www.ebroker.com.**
- Charles Schwab: **www.schwab.com.**
- E*Trade Securities: **www.etrade.com.**

CONSUMER ALERT

There are stock chat rooms on the Web that offer "insider" rumors, tips, and gossip. Federal investigators are looking hard at these chat rooms because some investors have been duped. For example, in May 1997

> thousands of E-mail messages were sent to investment forums touting Comparator Systems, a tiny fingerprint-identification company. As excited investors traded, the stock jumped from pennies a share to nearly $2 in four days before collapsing back to 57 cents. It now trades at around 1 cent per share on the "pink sheets," a thinly traded part of the over-the-counter market.[19]

The National Association of Securities Dealers offers a toll-free phone service, 800-289-9999, where investors can find out if someone is a legitimate broker. This is important because e-mailers can use fake names, and sometimes the hyped companies do not even exist.

* PC Financial Network: **www.pcfn.com.**
* Quick & Reilly: **www.quick.reilly.com.**

FAQs

1. How do stock prices go up and down?

   ■ Stock prices are shaped by such forces as changes in companies and industries, and by national and international economies.

2. How can someone become a better investor?

   ■ A smart investor keeps up with information about American and international business and news on existing, emerging, and merging companies.

3. How can investors tell if they have good brokers or financial advisors?

   ■ A good broker or financial advisor answers questions (no matter how basic) and keeps investors' goals in mind when making recommendations. Investors should expect a yearly report on how their investments are performing compared to previous years.

4. What is the Dow Jones Industrial Average (DJIA)?

   ■ The DJIA is computed by averaging the day's prices on 30 key companies. The Dow is recorded in points, not in dollars and cents. Usually a rise in the Dow indicates a strong economy, whereas a sustained drop in the Dow indicates an economic downturn.

5. What is the NYSE Composite Index?

   ■ The composite index is an average of the price changes of all common stocks listed and traded on the New York Stock Exchange; thus it is a broader measure of the market than the DJIA. Other major market indicators include the NASDAQ Index, AMEX Index, Value-Line, Russell Indexes, Wilshire 5000, and Standard and Poor's Indexes.

6. What does the Securities and Exchange Commission (SEC) do?

   ■ The SEC is the investor's advocate. They watch out for fraud and deception and provide investor education.

SUMMARY

Stock ownership is up. In the past ten years, the percentage of Americans who own common stock has doubled. Much of the stock is being held in 401(k) or 403(b) retirement accounts.

Common stocks refer to units of ownership in a public corporation.

Preferred stocks pay dividends. If a corporation is liquidated, owners of bonds and preferred stocks take precedence over the claims of owners of common stock.

Stocks can be classified as blue chip, cyclical, defensive, growth, income, or speculative and also by type of industry.

A bull market means the market is up.

A bear market means the market is down.

The price-earnings (P/E or PE) ratio is one of the better known and most critical numerical measures of a stock's value. Others are beta coefficient, earnings per share, and book value. The PE ratio is the relationship between the price of one share of stock and the annual earnings of a company. High PE ratios may indicate riskier stocks; low PE ratios may indicate more established, stable stocks.

Numerical measures are one way to evaluate stock value, another way is to follow one or more investment theories: efficient market theory, fundamental theory, and technical theory.

The largest stock exchanges are the NYSE and NASDAQ.

There are a number of stock indexes including the Dow Jones Industrial Average, Standard & Poor's Index, NASDAQ Index, Russell 2000 Small Stock Index, Wilshire 5000 Equity Index, the AMEX Major Market Index, and Value-Line Composite Index.

The Securities and Exchange Commission (SEC) enforces federal regulations whose primary purpose is to protect investors.

Long-term investment strategies include buy-and-hold, dollar-cost averaging, direct investment plans and dividend reinvestment plans.

Short-term strategies include buying on margin, selling short, and trading options. Generally short-term strategies involve more risk than long-term strategies.

The Internet is a useful tool for financial information, but misinformation exists. In particular, investors should be wary of stock information exchanged in chat rooms.

REVIEW QUESTIONS

(See Appendix A for answers to questions 1–10)

Identify the following statements as true or false.

1. Generally common stocks are less secure than preferred stocks. T
2. The only stock in the original Dow 12 and the current DJIA is Chevron. F
3. All common stocks pay dividends. F
4. A growth stock pays higher-than average dividends. F
5. The beta coefficient measures a stock's stability. High-risk takers are drawn to ~~low~~ high betas.

Pick the best answers to the following five questions.

6. The term ________ stands for stocks of the largest, most consistently profitable companies.
   a. cyclical
   b. blue chip
   c. defensive
   d. Dow 100
7. The Standard & Poor's 500 Composite Index (S&P 500)
   a. was introduced in 1957.
   b. represents a huge universe, accounting for 70 percent of the market capitalization of all U.S. stocks.
   c. is a leading market indicator.
   d. All of the above are true.
8. The __________ Index is an especially good indicator of technology and fast-moving financial service stocks.
   a. American Stock Exchange
   b. Dow Jones Industrial Average
   c. NASDAQ
   d. Russell 2000
9. If you think the price of a stock may fall rapidly, you may decide to sell at a set price by placing a
   a. stop order.
   b. limit order.
   c. market order.
   d. discretionary order.
10. In 1934 the U.S. government created the ________________ to enforce federal regulations whose primary purpose is to protect investors by regulating the activities of stock traders.

a. Securities and Exchange Commission (SEC)
b. E-Commerce Commission
c. Truth in Lending Act
d. Federal Reserve

DISCUSSION QUESTIONS

1. What are the differences between common and preferred stocks? What are the advantages of each?
2. What are the six stock classifications? Which classification attracts you the most? Why?
3. Four numerical measures of a corporation relevant to stock value were given. Which one of the four seems the most useful to you, and why?
4. What are some simple guidelines to follow when picking a first stock? Why are trading options and other short-term strategies not for beginners?
5. Why do we have the Securities and Exchange Commission? What is insider trading?

DECISION-MAKING CASES

1. Jorge's parents never invested in the stock market because they thought it was too risky. Instead, they put their money in certificates of deposit in banks and purchased real estate. Jorge is hesitant to invest in the market, but would like to try. How can he get started? What sources of information should he consult?
2. Paul, age 27, a finance major in college now employed full-time, is fearless when it comes to investing in the stock market. He would like to find the types of stocks that will bring the most return. He is debt-free and has $3,000 he can afford to lose. Which classification(s) should he consider?
3. Melinda has inherited $10,000. After paying off a few bills, she would like to invest the rest in the stock market. She wants to diversify, so she is considering buying round lots in several industries. If you were in her shoes, which industries would you select, and why?
4. Ethan, age 35, heard a hot tip about a new, emerging stock. His account executive says she is not sure this is such a good idea because the stock has no history on which to base a judgment. Ethan still wants to pursue it. How can he evaluate the stock using the Internet?
5. Mahogany is interested in using numerical measures to determine which stock to buy. She has found an oil stock with the following: a book value of $100 a share (versus a stock market price of $90 per share), an earnings per share that has been rising, a P/E ratio of 10, and a beta coefficient of 1.05. Do these measures indicate that this may be a promising stock to invest in?

GET ON LINE

To see what brokers have to offer, visit the website of a brokerage firm that offers online trading, such as those listed previously. Most offer a trading demonstration. See if you can obtain current price information, company news, a chart of how the stock has fared in recent months, and information about how much the firm would charge per trade. Note that commissions can vary from customer to customer even within the same company.

As a final comment, there are many legitimate sources of financial information on the Web. A central website for stock quotes and the ups and downs of the Dow, S&P 500, and NASDAQ is www.cnbc.com. However, whenever money is concerned the potential for fraud exists, as the Consumer Alert on page 300 explained.

INFOTRAC

To access InfoTrac follow the instructions at the end of Chapter 1.

Exercise 1: Type in the subject word "stocks." You'll see hundreds of periodical references and dozens of subdivisions. Click on Blue Chip stocks and review one of the articles selected. What does the article say about blue chip stocks?

Exercise 2: Type in the subject word "stocks." Click on Preferred stocks and see what two articles have to say about preferred stocks as an investment choice. Compare their advice, report on what you find.

chapter thirteen

bonds

Amid Market Worries, Bonds Offer Security.

DID YOU KNOW?

* You can buy bonds online.
* About 55 million Americans own savings bonds worth $185 billion.

OBJECTIVES

After reading Chapter 13, you should be able to do the following:

1. Describe the role of bonds in an investment plan.
2. Explain corporate bonds and describe the various types.
3. Explain government bonds and describe the various types.
4. Understand the importance of bond ratings and other aspects of evaluating bonds.

CHAPTER OVERVIEW

A basic principle stressed throughout the book is the need to evaluate potential investments in terms of your own goals. Bonds are no exception. After reading this chapter, you should be able to decide if bonds fit into your investment plan.

The subject of bonds was introduced in the Chapter 4 discussion of savings bonds. This chapter goes more in-depth by covering all types of bonds. The first part of the chapter explains the advantages of owning bonds. Then the discussion turns to key terms, calculations to determine yield, and an explanation of the factors involved in investing in bonds.

Since bonds may be obtained from corporations or from government, most of the rest of the discussion will be grouped first by corporate bonds and then by government bonds. In addition other aspects of bonds (bond ratings, junk bonds, how to read the bond section of the newspaper) are given. A personal check list is provided to use in evaluating a bond purchase, and advice for beginning bond purchasers is given. As in all chapters, the end section will provide a description of Electronic Resources.

BONDS AND BONDHOLDERS

Bonds are investments in which money is lent to an organization for a period of time. In other words bonds are loans.

Governmental units or corporations issue bonds to borrow money to use for expansion, construction, and other purposes. Usually in return for the loan, investors called **bondholders** receive interest payments twice a year; at the end of the bond term, they get their principal paid back.

Bondholders
Owners of bonds.

Bonds can be purchased directly from corporations or from the government or through a third party, such as a bank or a broker. They can also be bought individually or as part of a mutual fund. Bondholders can be individuals, or they can be institutions, such as banks, corporations, or mutual funds.

Objective 1:
Describe the role of bonds in an investment plan.

THE ROLE OF BONDS IN AN INVESTMENT PLAN

Stocks representing ownership in companies have historically brought the greatest return of all types of investments. So who needs bonds? The answer is, just about everyone.

> "Even a young person who has many years to retirement should put some money in bonds," says Danuta Zierlonka, a financial consultant at Smith Barney. "If you put all your money in stocks, even blue chips, you could lose part of your principal in a major downturn. Holding part of your money in bonds will help off-set the drop in equities".[1]

Bonds play an important role in an investment plan, along with stocks and cash (including savings and money market funds). The key to success is finding the best combination of stocks, bonds, and cash to fit your time horizon (years to retirement or whatever else you are saving for) and your risk tolerance. Bonds earn higher interest than cash in a savings account, for example, but they are less stable. Their value moves up and down as interest rates change, as shown in Exhibit 13.1. But compared to the stock market, the bond market is more stable. Even so, it does fluctuate. For example, in 1994 bonds turned in their worst performance in decades, but this was followed in 1995 with one of the best showings in modern times.[2] From July to December 1997 the Dow Jones Industrial Average produced a four percent return while the 30-year U.S. Treasury bond (informally called Treasuries) posted a total return of 15 percent. The reason that bonds did so well during this time period is that interest rates dropped. When the stock market takes a downturn, people wish they had more bonds.

Investing in Bonds: How Much and Why?

How much should be invested in bonds? As suggested earlier, the answer depends on a person's goals and risk tolerance and the competitive rates and expected growth of other types of investments.

CONSUMER ALERT

High rates of return (15%) cannot always be counted on. Generally speaking, when interest rates drop, bond prices rise. The reverse is also true: when interest rates surge, bond prices drop, as shown in Exhibit 13.1. So bonds are not without risk, even U.S. Treasuries are not immune from interest rate risk.

EXHIBIT 13.1
The Relationship Between Interest Rates and Bond Prices

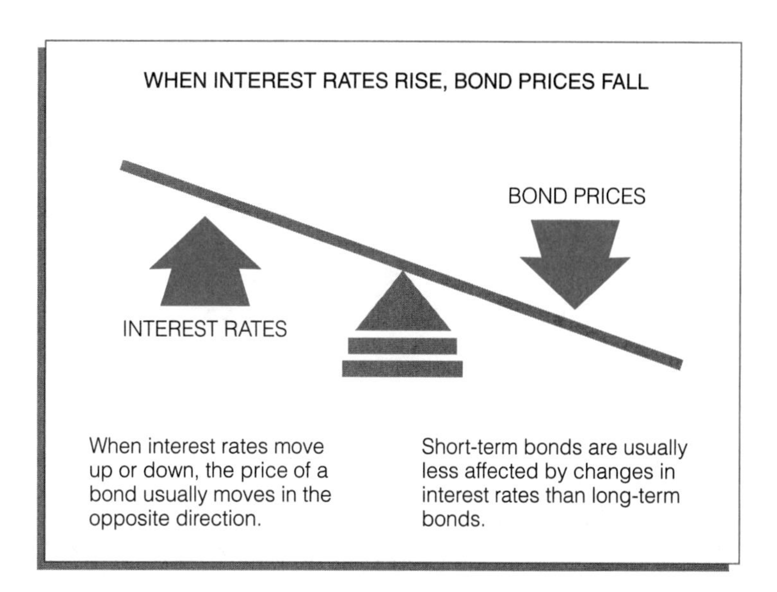

Some bonds are riskier than others. Individuals who are very risk-averse may be drawn to Treasury bonds, which are considered one of the safest investments around because they are backed by the U.S. government. Other individuals may be drawn to a variety of different types of bonds including short- and long-term corporate bonds, if their goal is to diversify their holdings.

To summarize, one of the main *advantages of bonds is that they can serve as an attractive means to diversify a portfolio.* Because they respond differently than stocks, bonds can provide a balance during rocky economic times. Bonds are known for sustaining price swings in the overall economy, but they do not as a rule beat inflation as well as stocks do. *Since 1925 stocks have outpaced inflation by seven percentage points a year, while bonds have beaten inflation by two percentage points.*[3] This provides an overall context for weighing the potential consequences of investing in stocks (equities) versus bonds, but note:

> Stock market drops tend to be abrupt and violent, quite a problem for anyone retiring during a market storm. Yes, smart long-term investing is built on an equity base. But it makes sense to diversify into other assets—especially bonds.[4]

Rate of Return

The issuer of the bond promises the bondholder (1) *a fixed rate of return* (the coupon rate), (2) over *a specified period of time* or at specified intervals, and (3) *repayment of the face or par value* at maturity. The **coupon rate** is the interest rate, expressed as an annual percentage of face value, on a debt security that the issuer promises to pay to the holder until maturity. For example, a bond with a 5 percent coupon rate will pay $5 per $100 of the face amount per year, usually in installments paid every six months. The coupon rate cannot change over the maturity of the bond.

Coupon rate
Interest rate expressed as an annual percentage of face value.

The par value refers to what a bond will be worth at maturity. It is also known as the face or stated value of a security such as $1,000. The par value is different from the **market value** (price), which is determined by such considerations as net asset value, yield, and investors' expectations about future earnings. For example, a fixed-rate bond's market value changes so that the yield on the bond will match the current market interest on similar bonds.

Market value
Price determined by net asset value, yield, and investor's expectations of future earnings.

Determining Market Value

The following calculations are used to determine market value. Usually the individual investor does not do these calculations; the broker or the institution, such as a governmental unit from which the bondholder purchased the bond, performs the calculations and can tell bondholders what they want to know. The formula is given here so that students can see how brokers or institutions arrive at the market value or price.

Let MV = P = Market value or Price
r_c = Coupon rate
PV = Par value
r_m = Market rate of interest on similar bonds
M = Remaining years to maturity, assume interest is paid annually

If M = 3, then

$$MV \text{ or } P = \frac{r_c \times PV}{(1 + r_m)^1} \quad \frac{r_c \times PV}{(1 + r_m)^2} \quad \frac{r_c \times PV}{(1 + r_m)^3} \quad \frac{PV}{(1 + r_m)^3}$$

Note that the terms in the various numerators are constant. So as r^m changes (either up or down), MV or P will change in the opposite direction.

Yield and Yield Calculations

As defined in Chapter 11, yield is the rate of return earned by a bondholder who holds a bond for a stated period of time. A yield of 7.15% is shown in the advertisement for debentures (unsecured bonds) from Flowers Industries, Inc. in Exhibit 13.2. What this means is that if the bond was issued with a coupon rate of 7.15% and had a par value of $1,000, the investor would receive $71.50 per year in interest. If the bond was then held to maturity, its yield-to-maturity would also be 7.15%. If sold prior to maturity its yield would depend on the interest and capital gain (loss) the investor received. The $71.50 was calculated by multiplying .0715 x $1,000.

Current yield
Rate of return determined by dividing a bond's annual income amount by its current market value.

Current yield and yield-to-maturity are two methods used to measure the yield on a bond investment. **Current yield** is determined by dividing an investment's annual income amount by its current market value. For bonds the current yield can be shown by the following formula:

$$\text{Current yield} = \frac{\text{Interest amount}}{\text{Market value}}$$

Assuming in the Flowers example that the current market value is $960, then

$$\text{Current yield} = \frac{\$71.50}{\$960}$$
$$= .0744 \text{ or } 7.4 \text{ percent}$$

To interpret this, the higher the current yield, the better, so that a current yield of 7.4 percent is better than a current yield of 5 percent.

Maturity or redemption date
When the borrower must repay the maturity value of a bond.

The **maturity date (or redemption date)** is the date when the borrower must repay the maturity value of a bond. In Exhibit 13.2 the bond is due to mature in 2028. At this time the issuer will return the bondholder's principal and final interest payment. For the promised full rate of return, a person should not sell a bond before it

EXHIBIT 13.2
An Advertisment for Debentures

reaches maturity. If an emergency forces an individual to sell when the bond's price is down, he or she is stuck with the loss. If held to maturity, the bondholders cash in the bond and get their money back and decide what to do next.

Short-term bonds mature in one to five years. They are relatively safe and do a reasonable job of keeping up with or slightly surpassing inflation, if the inflation rate is fairly steady. Intermediate bonds mature in five to ten years. Many mature in ten or more years.

Yield-to-maturity
Calculation that determines the rate of return on a long-term investment such as a bond.

Current yield is an important guide in choosing bonds. To go even further in determining all the ramifications of choosing a particular bond over another bond or type of investment, one could approximate the **yield-to-maturity (YTM)**. This can be done by using computer programs, bond tables, or calculators or by using the formula that follows. The YTM takes into account the purchase price of the bond, redemption value, time to maturity, coupon yield, and time between interest payments. To calculate the YTM, the following formula can be used:

$$\text{YTM} = \frac{\underset{\text{amount}}{\text{Interest}} + \frac{\text{Face value} - \text{Market value}}{\text{Number of periods}}}{\frac{\text{Market value} + \text{Face value}}{2}}$$

To show how this works, here is an example. On January 1, 2000, you could have purchased at the current market price of $900 a corporate bond with a $1,000 face value issued by Acme Company. The bond pays 6 percent annual interest, its maturity date is 2015 and you hold the bond for 15 years from 2000 to 2015. The yield-to-maturity is 7 percent, calculated as follows:

$$\text{YTM} = \frac{\$60 + \frac{\$1{,}000 - \$900}{15}}{\frac{\$900 + \$1{,}000}{2}}$$

$$= \frac{\$66.67}{950}$$

$$= 0.0702 \text{ or } 7 \text{ percent}$$

In this example the purchase price ($900) was below the face value ($1,000), and the yield-to-maturity (YTM) is greater than the stated interest rate (7% versus 6%). This is a positive outcome since you would make more money than you thought, given the stated interest rate. Comparing YTM to stated interest rate, then, is an important calculation to make before making a bond purchase. All other things being equal (risk, maturity date), the higher the YTM, the better.

The YTM is essentially the interest rate that would make the market value or current market price equal to the discounted value of all future cash payments promised by the bond. On a three year, 5 percent coupon rate, $1,000 par value bond currently selling at $950, the calculations would be as follows:

$$\$950 = \frac{50}{(1 + \text{YTM})} \quad \frac{50}{(1 + \text{YTM})^2} \quad \frac{50}{(1 + \text{YTM})^3} \quad \frac{1000}{(1 + \text{YTM})^3}$$

Liquidity

Cautious investors like bonds because they usually offer better yields than money market funds and less risk than with stocks so many investors regard bonds as a safe compromise between yield and risk. However, *one of the drawbacks of bonds is that they are less liquid than money market funds or stocks. Usually you buy a bond with the intention of holding it until it matures* whether it is 30 days or 20 years.

Coupon Rate (interest rate)

In the example given in Exhibit 13.2 the corporate bond from Flowers Industries, Inc. offered 7.15% interest at the same time that money market funds were offering about 3%. Also bonds provide expected cash flows (as in the Flowers Industries, Inc. example of $71.50 a year) that are known in advance, unlike stocks where returns are unpredictable. An interest-bearing bond provides periodic interest payments and the maturity value when the bond matures unless the issuer defaults.

As mentioned earlier, anyone interested in investing in bonds should be aware of the relationship between interest rates and bond prices (as shown in Exhibit 13.1). As interest rates rise, the price of outstanding bonds decreases. They become less attractive than newly issued bonds with higher rates of interest. If interest rates drop, the market price of existing bonds increases because their higher coupon rates make them more valuable. Short-term bonds are usually less affected by changes in interest rates than long-term bonds.

Bonds with maturity dates far in the future are more difficult to predict investmentwise than short-term bonds because it is difficult to predict interest rate changes over a long period of time. The issuers of the Flowers Industries, Inc. bonds shown in Exhibit 13.2 set an interest rate of 7.15% for debentures due in 2028 based on their best assessment of current market conditions and indicators. They are taking a chance by projecting this far in advance. Investors have to weigh how they think a 7.15% coupon interest rate will fare in the future given the vagaries of the market.

Besides high coupon rates, investors seek other factors, such as security, length of time the bond will be held, purchase price, fees, tax considerations, and how often interest payments are made. Individuals on fixed incomes—especially retired people—find receiving interest payments from bonds on a regular basis to be a reliable way to supplement their retirement income from other sources such as Social Security and employer-sponsored pension plans. In the Flowers Industries, Inc. example $71.50 a year may not seem like much, but if a bondholder has 20 bonds at $1,000 totaling $20,000, then he or she will receive $1,430 a year in interest payments. Note that the Flowers Industries, Inc.'s bonds in Exhibit 13.2 pay interest twice a year, as do most bonds.

Objective 2:
Explain corporate bonds and describe the various types.

DEFINITION AND EXPLANATION OF CORPORATE BONDS

Corporate bonds come from companies, such as the Flowers Industries, Inc., a manufacturer of bread and other bakery products. They represent a corporation's pledge to repay a specified amount of money with interest within a specified period of time. When the bond matures (reaches the *maturity date*), the bondholder receives the face dollar amount on the bond. A typical face value is $1,000, but they can be $10,000 and up.

Bondholders are essentially holding an IOU from the corporation, but they do not have corporate ownership privileges, 25 stockholders do (the right to vote). Corporations sell bonds to support ongoing business activities, expansion, and as a tax-reduction measure. Bonds differ from stocks in that bonds must be repaid. In the event of a company's bankruptcy, bondholders will be paid before stockholders. So *a major advantage of bonds is the lower risk that they offer, plus contractual periodic interest payments.*

As mentioned earlier, bonds are especially attractive to investors on fixed incomes such as retired people, but bonds may also be useful to people with salary or income gaps such as teachers and professors on nine-month contracts. They may seek bonds that make interest payments in the summer to help tide them over until their paychecks start again. If, for example, a teacher had a $10,000 bond that pays 8% interest a year, he or she could receive $400 in January and $400 in July.

Types of Corporate Bonds

Debenture
Debt obligation backed by the borrower and documented by an indenture, a formal agreement.

The most common type of corporate bond is a ***debenture****,* an unsecured bond, as described earlier in the Flowers Industries Inc. example in Exhibit 13.2. A **debenture** is a debt obligation backed by the earning power of the borrower and documented by an agreement called an indenture, a formal agreement between an issuer of bonds and the bondholder.

There are also secured bonds, subordinated debentures, and convertible bonds. Each of these will be described next.

Secured bonds
Bonds backed by collateral, mortgages, or other liens.

Secured Bonds. **Secured bonds** are backed by the pledge of collateral (assets), a mortgage, or other lien. Since so many are backed by mortgages, secured bonds are also known as mortgage bonds. Since they are secured, you pay for this added security by receiving lower interest rates on mortgage bonds than on debentures (unsecured bonds).

Subordinated debentures
Unsecured bonds that give bondholders a claim secondary to that of designated bondholders.

Subordinated Debentures. Unsecured bonds that give bondholders a claim secondary to that of their designated bondholders are called **subordinated debentures.** Because of the increased risk, subordinated debentures usually receive higher interest rates than debenture holders.

Convertible bonds
Bonds that can be exchanged for stock.

Convertible Bonds. When bonds are capable of being exchanged for stock shares instead of getting cash back, they are **convertible bonds** This conversion feature attracts potential buyers if they think the corporation is growing and that the price of its stock will rise. The buyer gets the security of a bond but with the option to convert it to a common stock at a given rate of exchange. Because of this feature, issuers may pay lower interest rates on convertible bonds.

Callable bonds
Type that allows corporations to call in or buy back outstanding bonds.

Callable Bonds. Given changes in companies and the economy, most corporate bonds are **callable bonds** meaning a corporation can call in or buy back outstanding bonds before their maturity date. *Because investors whose bonds are called must then reinvest usually at less favorable rates, issuers must offer call premiums, which increase the repurchase value above the normal face value, in order to compensate for this reinvestment risk.*

The bond issuer and full-service brokers with whom you work will alert you when a bond is called. Also *The Wall Street Journal* publishes "Redemption Notices" every Tuesday.

How Investors Purchase Corporate Bonds

Bonds are bought much like stocks through primary markets (banks, account executives, companies) or secondary markets (from other investors) through the exchanges discussed in Chapter 11. In the secondary market, investors are looking for previously issued bonds that are being offered at what they deem to be good prices.

The usual trading amount is $1,000, although occasionally $100 or $500 bonds called baby bonds are issued. As stated earlier, maturity dates are short term (1–5 years, often financials), intermediate term (5–10 years, commonly banks), or long-term (10–20 years, usually utilities or industrials).

If you purchase a stock from an account executive, you will pay a commission—for example, $25–$50 for a $1,000 bond—and you will pay a commission when you sell. When you place an order to buy, the account executive enters the order on his or her computer and sends it to a Wall Street broker-dealer. The Wall Street broker-dealer buys a bond from one investor and then sells it to another, profiting on the **spread**, or difference in buying and selling prices. Alternative methods to this include buying online direct or placing an order through a broker OTC, meaning over the counter, not through an exchange. Trading of corporate bonds listed on the New York Stock Exchange is done through an electronic trading system that enables brokers to provide clients with nearly instantaneous trading execution.[5]

Spread
Difference between buying and selling prices.

Reasons for Purchasing Corporate Bonds

The reasons why investors purchase corporate bonds have been alluded to previously, but to summarize and explain further, investors buy corporate bonds for the following reasons:

1. *To provide interest income.* There are three methods used to determine interest income, depending on whether the holder has registered bonds, coupon bonds, or zero-coupon bonds.

Registered bonds
Bonds recorded in name of bondholder.

Coupon bonds
Bonds issued with detachable coupons that are presented to a paying agent or issuer for payments.

Zero-coupon bonds
Bonds sold at a deep discount.

* A **registered bond** is recorded in the name of the holder on the books of the issuer and can be transferred to another owner only when endorsed by the registered owner.
* A **coupon bond** is a bond issued with detachable coupons that must be presented to a paying agent or the issuer for semiannual interest payments. These are bearer bonds, so whoever presents the coupon is entitled to the interest.[6]
* **Zero-coupon bonds** are sold at a *deep discount* of their face values. These can be more volatile than other types of bonds, but they are popular investments in IRAs or Keogh accounts because they are tax-sheltered. Zero-coupon bondholders do not collect interest; instead the value of the bond increases to its full value when it matures. So, for the most part, zero-coupon bonds are held for a long time. The potential disadvantage to this strategy is that holding investments for a long time can leave you stuck with a low-paying investment. Although this section is on corporations, zero-coupon bonds are available from governments as well.

2. *To provide potential capital gains.* The price of a bond can fluctuate until its maturity date. If it goes up in value, it is said to be selling at a premium. If it goes down in value, it is said to be selling at a discount. Prices of bonds can be checked through account executives or listings in your local newspaper, *The Wall Street Journal*, or *Barron's*, or over the Internet.

3. *For the bond repayment at maturity.* The usual scenario is that bonds are paid their face value at maturity (5 years, 10 years). You can keep a bond until maturity or sell it earlier to another investor or back to the company, not necessarily for face value.

Tax Consequences of Corporate Bonds

Capital gains and interest from corporate bonds are taxable. Investors consider tax consequences before deciding which bonds to purchase and, if subject to taxation, when to cash them in. Account executives or bond issuers can explain to potential bondholders the possible tax consequences of their choices. The next section covers government bonds. Taxation policy is different on government bonds than on corporate bonds.

Objective 3:
Explain government bonds and describe the various types.

GOVERNMENT BONDS

Government bonds are written pledges of a government or municipality to repay a specified sum of money along with interest. Bondholders are essentially holding IOUs from the government.

Bonds can be obtained directly from federal, state, or local governments in a variety of ways including online or through stockbrokers, employers, or banks. Federal government-issue bonds will be discussed first, followed by state and local government-issue bonds.

Types of Government Bonds

Government bonds fall under the following types:

* *U.S. treasury bills (T-bills), notes, and bonds.*

* *Agency bonds* (the most popular and well known of these are the bonds of mortgage associations with nicknames, such as Fannie Mae or Ginnie Mae).
* *Municipal bonds* (referred to as munis, issued by governmental units smaller than the federal government (city, county, municipalities) to finance new construction, schools, roads, etc.). In addition U.S. investors can buy international government bonds.

Federal

Introduction to Treasury Bills, Notes, and Bonds. Because they are backed by the federal government, these instruments are considered by financial experts to be very secure investments. The government issues bonds to support government activities and to finance the national debt.

Treasury bills, notes, and bonds can be purchased from the 12 Federal Reserve Banks or one of their branches. The banks' public information telephone number is 213-624-7398. To start the process, one would fill out a **tender form**, which is an application to purchase U.S. Treasury securities (see Exhibit 13.3).

Tender form
Application form to purchase U.S. Treasury securities.

No commission is charged on bonds if they are purchased directly from the government. Savings bonds can be purchased at work or from banks and from the Internet. An Internet site, called the Savings Bond Connection, is the latest step to make savings bonds easier to buy (24 hours a day, seven days a week) and a more attractive investment for Americans. About 55 million Americans own savings bonds worth $185 billion, according to the Treasury Department.

Prices can be checked by contacting local banks, Federal Reserve banks (through their Web site; see the Electronic Resources section for this as well as the Savings Bond Connection Web site), or from lists in financial newspapers or the financial section of your local newspaper.

Treasury bills (T-bills)
U.S. securities sold in minimum units of $1,000 that mature in one year or less.

Treasury Bills (T-bills) **Treasury bills**, informally called **T-bills**, *are sold in minimum units of $1,000.* They can be issued in amounts up to $1 million. They are the largest component of what is called the money market, which is the market for short-term debt securities. T-bills are valuable investments for individuals, and they also serve as a major tool of the Federal Reserve in controlling the nation's money supply. Their maturity dates range from 13 weeks to 52 weeks.

T-bills are sold at a discounted rate, and the actual amount paid is less than the maturity value. For example, a $10,000 T-bill may have a stated interest rate of 4 percent. To determine the discounted amount ($400), multiply the maturity value by the interest rate ($10,000 x 4% = $400). To determine the purchase price ($9,600), subtract the discount amount from the maturity value ($10,000 - $400 = $9,600). In actuality the investor will get more than 4 percent, probably 4.2 percent, so that he or she would receive $420 interest on a $9,600 investment. This represents a 4.2 percent return when calculated on an annual basis.

Treasury bills can be bought directly from the government or through a broker. Safety is maximum but the return may be low.

Treasury notes
U.S. securities that mature in one to 10 years and range in denomination from $1,000 to $1 million or more.

Treasury Notes. **Treasury notes** are intermediate Securities meaning they mature in one to ten years. They range in denomination from $1,000 to $1 million or more. Usually interest rates are slightly higher than for T-bills, and the interest is paid every six months.

Treasury bonds
U.S. securities that are long-term debt instruments ranging from 10 to 30 years issued in minimum denominations of $1,000.

Treasury Bonds. **Treasury bonds** usually pay higher interest rates than T-bills or notes because they have a longer maturation period—over 10 years, usually 30 years. The exception to this is that during infrequent periods when the yield curve inverts

EXHIBIT 13.3
A Tender Form

COMPLETING YOUR TREASURY DIRECT TENDER

As the first step in purchasing Treasury securities either from a Federal Reserve Bank or Branch or from the Bureau of the Public Debt, an investor must complete and submit a tender form (shown below). The tender form has two copies, one of which is to be retained by the purchaser.

PD F 5176-3
Department of the Treasury
Bureau of the Public Debt
(Revised May 1995)

OMB NO.1535-0069
Expires 09/30/95

TREASURY DIRECT®

52-WEEK TREASURY BILL TENDER

TENDER INFORMATION — AMOUNT OF TENDER: $ ______

BID TYPE (Check One) ☐ NONCOMPETITIVE ☐ COMPETITIVE AT . %

TREASURY DIRECT ACCOUNT NUMBER

INVESTOR INFORMATION

ACCOUNT NAME

ADDRESS

CITY STATE ZIP CODE

TAXPAYER IDENTIFICATION NUMBER

1ST NAMED OWNER — SOCIAL SECURITY NUMBER OR EMPLOYER IDENTIFICATION NUMBER

TELEPHONE NUMBERS (FOR NEW ACCOUNT ONLY)

WORK ( ) - HOME ( ) -

PAYMENT ATTACHED — TOTAL PAYMENT: $ ______

CASH (01): $ ______ CHECKS (02/03): $ ______

SECURITIES (05/06): $ ______ $ ______

OTHER (07): $ ______ CHECKS ARE DEPOSITED IMMEDIATELY

DIRECT DEPOSIT INFORMATION (FOR NEW ACCOUNT ONLY)

ROUTING NUMBER

FINANCIAL INSTITUTION NAME

NAME ON ACCOUNT

ACCOUNT NUMBER

ACCOUNT TYPE (Check One) ☐ CHECKING ☐ SAVINGS

AUTOMATIC REINVESTMENT

1 2

Circle the number of times you want this 52-week Bill purchase automatically reinvested noncompetitively after your initial investment.

AUTHORIZATION

I submit this tender pursuant to the provisions of Department of the Treasury Circular, Public Debt Series No. 2-86 (31 CFR PART 357), and 1-93 (31 CFR PART 356) and the applicable offering announcement.

Under penalties of perjury, I certify that the number shown on this form is my correct taxpayer identification number and that I am not subject to backup withholding because (1) I have not been notified that I am subject to backup withholding as a result of a failure to report all interest or dividends, or (2) the Internal Revenue Service has notified me that I am no longer subject to backup withholding. I further certify that all other information provided on this form is true, correct and complete.

SIGNATURE DATE

SEE INSTRUCTIONS FOR PRIVACY ACT AND PAPERWORK REDUCTION ACT NOTICE

FOR DEPARTMENT USE

TENDER NUMBER
912794

CUSIP

ISSUE DATE

RECEIVED BY

DATE RECEIVED

EXT REG ☐
FOREIGN ☐
BACKUP ☐
REVIEW ☐

CLASS ☐

NUMBERS

or at least becomes very flat, Treasury bonds may return a lower interest rate than T-bills or notes.

Because they have such a long maturation period, Treasury bonds are also referred to as long bonds. The going interest rate at Treasury auctions and the price of long-term bonds in the secondary market are used as benchmarks of investor attitudes toward the general economy.[7]

Treasury bonds are sold in units of $1,000 and, like notes, pay interest every six months. U.S. Treasury notes and bonds can be bought through brokers or directly from any Federal Reserve Bank. Treasury bonds are not to be confused with savings bonds, a type of government bond introduced as a beginning savings investment in

Chapter 4 and explained further in this chapter.

If you bought a Treasury bond and wanted to find out how much it was worth at a certain point in time, you could call whomever you purchased it from or the Bureau of Public Debt or a service company that evaluates stocks and bonds for individuals for a fee. The government Internet sites are given in the Electronic Resources section.

Agency Bonds. Government agencies besides the Treasury Department offer bonds and securities. These include Fannie Mae (formally known as the Federal National Mortgage Association, FNMA) and Ginnie Mae (formally known as the Government National Mortgage Association, GNMA). They offer a marginally higher risk and higher interest than Treasury bonds and slightly higher interest rates than Treasury issues.[8]

A usual minimum denomination is $1,000 to $25,000 and up. Maturities can range from 30 days to 40 years. They can be bought directly from the government or from banks or account executives. Since Fannie Maes and Ginnie Maes are real estate investments as well as examples of government bonds, they will be discussed further in the next chapter.

State and Local Government Securities

Municipal bonds (munis)
Bonds issued by city, state, or local governments for improvements.

Almost four million Americans earned municipal bond income in 1995. Nearly half had incomes under $50,000.[9] So **municipal bonds (munis)** are a popular form of investment, but they are not necessarily just for the wealthy. They have the advantage of being usually exempt from state and federal income taxes. Because of this munis are particularly attractive for people in high income tax brackets. Munis are issued by cities or state or local governments to support such activities as building roads, bridges, and schools. There are more than one million munis to choose from, and the number continues to grow.

Munis are general obligation bonds backed by the full faith and credit of the issuer, so they are usually safe, but occasionally they default. To avoid the possibility of default, check the bonds' rating and purchase insured municipal bonds. The starting price for munis is usually $5,000, and brokers often prefer a multiple of this as a minimum investment. Occasionally, mini-munis are available for $1,000. They are designed for investors who want tax-exempt munis, but do not want to invest large sums of money. Maturation dates for munis vary widely from 30 days to 30 years. Munis offer tax-exempt interest and lower interest rates than comparably rated corporate bonds and U.S. Government Securities.[9]

Tax Consequences of Government Bonds

As mentioned in the preceding section, investors, especially those in high tax brackets—are attracted to government bonds because many are tax-exempt or tax-deferred. Most federal government-issued bonds are exempt from state and local taxes. For example, a bondholder would owe federal tax on the interest that Treasuries pay but not state or local tax when the bondholder invests with aftertax money. However, if the Treasuries are in a tax-deferred retirement account the withdrawals would be taxed at the state level, too. Many municipal bonds (munis) are tax-free because they are exempt from local, state, and federal taxes. The best advice is to ask about tax consequences before buying or selling bonds.

Objective 4:
Understand the importance of bond ratings and other aspects of evaluating bonds.

BOND RATINGS

Investors can check the strength of a given corporate or municipal bond's rating from Standard & Poor's Corporation or Moody's Investors Service, Inc. Their ratings are available in the library or on the Internet (see the Electronic Resources section for Web sites).

The highest ratings are Aas in Moody's and AAA in Standard & Poor's. These ratings indicate that the issuer's capacity to pay interest and to repay principal is extremely strong. There is little or no default risk. **Default** refers to the failure of a debtor to make on-time payments of interest and principal as they come due. Bonds of the highest quality are, of course, safest and therefore most desirable.

Default
Failure of debtor to make payments of interest and principal when due.

Ratings of Aa in Moody's and AA in Standard & Poor's indicate a very strong capacity to pay interest and repay principal. "A" ratings from both rating services indicate a strong capacity to pay interest and repay principal although these bonds are somewhat susceptible to adverse changes in financial or economic conditions.

Bonds rated BBB or above are considered **investment grade,** meaning the risk of default is fairly low. A Baa rating in Moody's or a BBB in Standard & Poor's indicates adequate capacity to pay interest and repay principal and more susceptibility to adverse changes in economic and financial conditions.

Investment grade
Bonds rated BBB or above.

When bond ratings drop further, caution should be taken. Bonds rated Ba in Moody's or BB in Standard & Poor's are considered speculative, meaning that their futures cannot be assured. Bonds rated C or D have poor standings. Institutions that invest other people's money are not, under most state laws, allowed to buy low-rated bonds.

Usually U.S. Treasury securities are not rated because they are virtually risk-free. The exception to this rule is long-term municipal bonds that are rated similarly to corporate bonds. Shorter-term municipal bonds, which mature in three years or less are rated by Standard & Poor's as

- SP - 1: Very strong or strong.
- SP - 2: Satisfactory.
- SP - 3: Speculative.

Junk Bonds and Junk Bond Funds

Junk bonds (also referred to as high-yield bonds) have low ratings and are considered below investment grade. The traditional description of junk bonds is as follows:

Junk bonds or High yield bonds
Bonds with ratings of BB or lower, volatile.

> **Junk bonds** are bonds with a speculative credit rating of BB or lower by Standard & Poor's and Moody's rating systems. Junk bonds are issued by companies without long track records of sales and earnings, or by those with questionable credit strength. Since junk bonds are more volatile and pay higher yields than investment grade bonds, many risk-oriented investors specialize in trading them.[10]

In short, junk bonds carry greater risk than other types of bonds. Beginning investors, especially, are warned to be careful. In the late 1990s junk bonds received a lot of press because in the first half of 1998, $64 billion in junk debt came into the market vs. $23 billion for the same period in 1997. Demand was intense due to the robust economy and low default rates, just 2 percent compared to 10 percent in 1989.[11] In essence junk bonds gained new respect and became more desirable than they were a decade earlier. Given their volatility though, how does an individual investor make decisions about junk bonds?

- *If the economy remains robust,* then perhaps junk bonds are a choice. As an added measure of protection, "the best move is to steer clear of individual issues and stick to funds."[12] Bond funds are collections of funds that are professionally managed. They are a type of mutual fund (to be discussed further in the next chapter). Look for high yield bond funds with solid, long-term records from companies that you know.

> ... right now, you're probably better off sticking with funds that "plod along with consistently good records," says Merrill Lynch global high-yield strategist Martin Fridson. That could help shield you from disappointment later on. Says Fridson: "Slow and steady wins the race."[13]

Many well-known companies such as Viacom, Revlon, Sprint, and Kmart—have offered high-yield bonds.

* *If the economy slows greatly and default rates rise*, investors may avoid junk bonds and instead turn to government bonds, such as Treasuries. "That may sound boring. But if you're going to lose sleep, reserve your angst for the stock portion of your portfolio, not bonds."[14]

Reading the Bond Section of the Newspaper

The Wall Street Journal, *Barron's*, and larger local newspapers publish information on corporate and government bond prices. For corporate bonds prices are given as a percentage of the face value, which is usually $1,000. To find the actual price to be paid, multiply the face value ($1,000) by the newspaper quote. For example, a price quoted as 90 means a selling price of $1,000 × 90% = $900. Information is also given about current yield, volume, the closing price, and net change. For an example of a newspaper listing of corporate bond prices see Exhibit 13.4.

Corporate bond prices are quoted in increments of points and eight fractions of a point based on a par value of $1,000. Thus

1/8 = $1.25
1/4 = $2.50
3/8 = $3.75
1/2 = $5.00
5/8 = $6.25
3/4 = $7.50
7/8 = $8.75

So a bond quoted at 76 1/2 would be selling for $765, and one quoted at 100 3/8 would be selling for $1,003.75. For government bonds there are listings of bond yields, tax-exempt bonds, and international bonds. International government bonds' rates are given in *The Wall Street Journal* or are available over the Internet.

Bonds as Part of an Investment Strategy

Investment strategy
Plan to allocate assets among diverse investments.

Bonds can play an important part in an overall **investment strategy**, which is a plan to allocate assets among such choices as stocks, mutual funds, and real estate. This chapter provided many ways to compare bonds. A check list to aid in comparisons is given in Exhibit 13.5. Information to fill in the blanks can be obtained from newspapers, brokers, rating services, financial planners, issuers of bonds, and the Internet.

As shown in the exhibit, one of the factors in a bond purchase is the general state of the economy (if the economy slows considerably and default rates for the corporate issuers rise, junk bond yields will drop) and the financial strengths and weaknesses of the bond source. In a spiraling down economy, top-rated corporate and government bonds are attractive. Another factor is the investor's age. As one ages, the usual investment strategy is to decrease the percentage of stocks (although not eliminating them entirely) and to increase the percentage of bonds. For example, a financial planner may recommend 80% in stocks and 20% in bonds for those in their forties or younger and 60% in stocks and 40% in bonds for those in their fifties. By the sixties a suggested mix may be 30% in stocks, 65% in bonds, and 5% in money market funds. Your invest-

EXHIBIT 13.4
Newspaper Listing of Corporate Bonds

NEW YORK EXCHANGE BONDS

Quotations as of 4 p.m. Eastern Time
Thursday, May 18, 2000

Volume $10,582,000

	Domestic Thu.	Domestic Wed.	All Issues Thu.	All Issues Wed.
Issues Traded	165	172	174	181
Advances	61	56	66	56
Declines	78	78	81	86
Unchanged	26	38	27	39
New highs	0	1	0	1
New lows	25	21	25	21

SALES SINCE JANUARY 1
(000 omitted)

2000	1999	1998
$987,141	$1,345,120	$1,508,216

Dow Jones Bond Averages

1999 High	1999 Low	2000 High	2000 Low		2000 Close	2000 Chg.	2000 %Yld	1999 Close	1999 Chg.
106.88	96.80	96.51	93.23	20 Bonds	93.36	+0.13	8.32	103.05	+0.07
104.72	94.96	95.09	90.69	10 Utilities	90.83	−0.10	8.17	100.86	−0.07
109.44	98.31	99.27	95.53	10 Industrials	95.89	+0.36	8.47	105.25	+0.23

CORPORATION BONDS
Volume, $9,499,000

Bonds	Cur Yld.	Vol.	Close	Net Chg.
AES Cp 8s8	8.4	10	94¾	+ 1¾
AMR 9s16	8.9	30	101⅝	+ ¾
ATT 5⅛01	5.2	35	98	...
ATT 7⅛02	7.2	61	99½	+ ⅛
ATT 6¾04	7.0	18	96⅝	...
ATT 5⅝04	6.1	45	92⅞	+ ⅛
ATT 7s05	7.2	73	96¾	+ ⅞
ATT 7½06	7.6	65	98⅜	− ¼
ATT 7¾07	7.8	60	99¾	+ ¼
ATT 6s09	7.0	188	85½	− ¾
ATT 8⅛22	8.3	46	97¾	− ¼
ATT 8⅛24	8.4	55	97⅜	− ⅝
ATT 8.35s25	8.5	7	98¼	− ¼
ATT 6½29	8.2	68	79⅛	− ⅛
ATT 8⅝31	8.8	572	98⅝	− ⅜
AForP 5s30	8.8	16	56⅝	...
Amresco 10s03	18.4	18	54¼	− ¼
Amresco 10s04	18.9	84	53	+ ⅛
AscCp 8.15s09	8.3	45	98⅛	− 1
ARch 10⅞05	9.3	10	117⅜	− 2⅝
BkrHgh zr08	...	5	78	− ¾
BkOne 7¼04	7.5	10	96⅝	− 2⅞
BayView 9s07	13.2	60	68¼	...
Bellso 6⅜28	8.1	125	79⅛	+ ⅜
BellsoT 7s05	7.2	110	97⅛	+ ¼
BellsoT 5⅞09	6.9	436	85⅝	− 1¾
BellsoT 8¼32	8.4	15	98	− ⅜
BellsoT 7⅞32	8.4	7	93¾	+ ¼
BellsoT 7½33	8.4	105	89	+ ¼
BellsoT 6¾33	8.4	191	80⅝	− ½
BellsoT 7⅝35	8.4	30	90⅝	+ ½
BethSt 8.45s05	8.9	56	95⅜	+ 1⅜
Bluegrn 8¼12	cv	35	67	+ 1½
Bordn 8⅜16	11.5	1	73	...
BosCelts 6s38	11.3	16	53¼	− ⅝
BoydGm 9¼03	9.5	5	97⅞	+ 1⅞
ChaseM 7½03	7.5	100	100¼	+ ⅛
ChaseM 6⅛08	7.0	35	88	...
CPoV 7¼12	8.0	10	90⅝	...
ChiqBr 10s09	12.3	10	81¼	+ 1
ChryF 6⅝00	6.6	15	99⅝	+ 1/16
vjClardg 11¾02f	...	22	56⅛	+ ¾
ClrkOil 9½04	11.2	35	85	− 1⅝
Coastl 10¼04	9.7	35	106	+ ¾
CoeurDA 7¼05	cv	5	47½	...
Coeur 6⅜04	cv	104	47⅞	− ¼
CmclFd 7.95s06	8.7	150	91¾	− 3¼
Consec 8⅛03	11.4	282	71⅛	− 1⅞
Conseco 10¼02	15.4	289	66½	− 2½
ConPort 11s06	16.6	53	66⅛	− 1⅞
DVI 9⅞04	11.9	15	83	− 5
Dole 7s03	7.5	20	93¼	− ⅝
DukeEn 6⅞23	8.3	36	83¼	− ½
DukeEn 6¾25	8.3	10	81	− 1⅞
DukeEn 7½25	8.3	17	90⅝	− 1⅜
Exxon 6s05	6.5	50	93	...
FedDS 8⅛02	8.1	20	100	...
Florsh 12¾02	14.0	50	91	+ 1½
FordCr 6⅜08	7.1	40	89⅝	− ⅛
GEICap 7⅞06	7.6	5	104	− ¼
GMA 6⅝02	6.9	1	96½	− ½
GMA 8½03	8.4	15	101	− ½
GMA 5⅞03	6.2	4	95	− ⅛
GMA dc6s11	7.2	15	83	...
GMA zr12	...	20	348⅛	− 1⅞
GMA zr15	...	19	288	+ 1
GenesisH 9¾05f	...	211	14	...
Hallwd 10s05	10.5	12	94⅞	+ 1½
HlthcrR 6.55s02	cv	15	89¾	...
Hlthso 9½01	9.5	125	100	...
Hilton 5s06	cv	14	77	...
Hollngr 9¼07	9.4	25	98	+ ⅛
Hollngr 9¼06	9.4	50	98	...
IBM 6⅜00	6.4	10	99 27/32	...
IBM 7¼02	7.3	171	99½	− ½
IBM 6.45s07	6.9	5	93	− 1
IBM 7½13	7.7	65	97¼	− ¾
IPap dc5⅛12	7.2	10	71⅜	− ⅜
vjKCS En 8⅞08f	...	110	51¾	+ ¼
KaufB 9⅜03	9.4	24	99⅞	+ ⅞
KaufB 7¾04	8.5	105	91¾	+ ¾
KaufB 9⅝06	10.0	10	96	+ ⅛
KentE 4½04	cv	140	86½	− ¾
KerrM 7½14	cv	72	93½	+ ⅝
Kolmrg 8¾09	cv	87	99	− ⅛
LTC Pr 8¼01	cv	15	80	+ 2½
Lilly 8⅛01	8.1	10	100⅛	− ⅞
Loews 3⅛07	cv	85	90⅞	+ ⅞
LslsLt 7s04	7.3	7	95½	− ½
LglsLt 9s22	8.9	7	101⅝	− ⅛
LglsLt 8.2s23	8.8	20	93¾	− ¾
Lucent 6.9s01	6.9	25	99½	...
Lucent 7¼06	7.4	10	98½	− 1⅞
Lucent 6½28	7.8	25	83	− 1
MBNA 8.28s26	9.9	4	84	− 2
MSC Sf 7⅞04	cv	53	90	− ½
MarO 7s02	7.2	10	97¾	+ ¼
Mascotch 03	cv	10	75½	− ½
MPac 4¼05	5.1	21	82⅝	+ ⅜
MPac 4¾20f	...	61	62	...
MPac 5s45f	...	1	50⅝	− ½
Motrla zr13	...	8	102	− 2
NRurU 7.3s06	7.8	5	94¼	− ¾
NStl 8⅜06	9.0	25	93½	− 1½
NJBTl 7¼11	7.8	15	92½	+ ⅛
NYTel 5⅝03	6.1	10	92⅞	...
NYTel 4⅝04	5.1	27	90½	+ ⅜
NYTel 6⅛10	7.0	1	88	− ½
NYTel 7⅜11	7.9	10	93⅝	− 1⅛
NYTel 6.70s23	8.1	15	82¼	+ ⅝
NYTel 7s25	8.3	10	84¼	− 1⅛
OcciP 10⅛01	9.9	6	102⅝	− ⅞
OcciP 11⅛19	11.1	11	100	− 1
OreStl 11s03	14.6	185	75½	+ 1½
ParkerD 5½04	cv	10	77	− 1
PepBoys zr11	...	24	54	+ ¼
PhilEl 6½03	6.8	125	95½	− ¾
PhilPt 6⅜09	7.4	25	86	...
PhilPt 8.49s23	8.5	10	100	− ¾
PhilPt 7.92s23	8.6	1	92	+ 1⅞
PhilPt 7.2s23	8.6	5	83½	− ⅜
PhilIP 7⅛28	8.6	40	83	− 1
Polaroid 11½06	11.2	80	102⅞	+ ⅝
PSvEG 6⅛02	6.4	117	96⅜	+ ¼
PSEG 9⅛05	8.7	2	105	+ ½
Quanx 6.88s07	cv	101	83	− 3
Rallys 9⅞00	10.4	206	94⅝	− ⅛
RalsP 9¼09	8.6	20	107	+ ¼
RalsP 7¾15	8.2	10	94¼	− 3¾
RelGrp 9s00	10.1	50	89½	− ½
ReynTob 7⅝03	9.7	8	78½	+ ¾
ReynTob 8¾04	11.1	60	78½	+ ½
ReynTob 8¾05	11.4	60	76½	+ 1½
RobMyr 63	cv	20	95	+ 1
Safwy 10s01	9.7	25	102¾	...
Safwy 9.65s04	9.2	20	105¼	− ¼
Safwy 9⅞07	9.1	6	109	+ 3¼
Simula 8s04	cv	78	50½	+ 1½
Sizeler 8s03	cv	20	89	...
Sprint 6⅞28	8.1	10	85	− 4¼
StdCmcl 07	cv	3	47⅝	− ¾
StdPac 8½07	9.4	11	90	...
StoneC 10¾02A	10.7	256	100⅛	− ⅛
StoneC 10¾02O	10.5	106	102	− 1¼
StoneC 11½04	11.0	74	104½	+ ¾
StoneCn 6¾07	cv	10	82	+ 1
TVA 6⅛03	6.4	402	96	...
TVA 5⅜08	6.3	26	85¼	+ ⅛
TVA 8¼34	8.3	15	99	+ ¾
TVA 7¼43	8.0	46	90⅝	− ⅝
TVA 6⅞43	7.8	40	88⅛	+ ⅛
Tenet 8⅝07	9.1	10	94½	+ 1
TmeWar 8.11s06	8.2	12	99	− ⅞
TmeWar 8.18s07	8.3	15	99	+ ¾
TmeWE 7¼08	7.8	110	92⅝	− ⅝
TmeWar 9⅛13	8.9	4	103	...
TolEd 8s03	8.0	2	100	+ 1⅜
US Timb 9⅝07	11.1	3	87	+ 2
UtdAir 10.67s04	10.1	30	106	− ¾
WsteM 4s02	cv	100	89½	+ ½
Webb 9¾03	9.9	75	98⅞	+ ⅞
Webb 9¾08	11.6	50	84	− 1
WebbDel 9⅝09	11.4	126	82	− 1⅛
Weirton 11⅜04	11.2	146	102	...
Weirton 10¾05	10.7	10	100⅜	+ ¼

FOREIGN BONDS
Volume, $1,012,000

Bonds	Cur Yld.	Vol.	Close	Net Chg.
CGDina 8s04f	cv	16	38	+ 1
Inco cv04	cv	8	91	...
Inco 7¾16	cv	30	85⅞	+ ⅝
SeaCnt 12½04A	17.5	640	71⅝	− ⅝
SeaCnt 12½04B	17.3	50	72¼	− ½
SeaCnt 10½03	13.0	5	81	+ 1
SeaCnt 9½03	12.3	57	77	− ¾
TelArg 11⅞04	11.6	250	102¾	+ ½
TrnMarMx 03	...	10	81¼	+ ¼

EXPLANATORY NOTES
(For New York and American Bonds)
Yield is Current yield.
cv-Convertible bond. cf-Certificates. cld-Called. dc-Deep discount. ec-European currency units. f-Dealt in flat. il-Italian lire. kd-Danish kroner. m-Matured bonds, negotiability impaired by maturity. na-No accrual. r-Registered. rp-Reduced principal. st, sd-Stamped. t-Floating rate. wd-When distributed. ww-With warrants. x-Ex interest. xw-Without warrants. zr-Zero coupon.
vj-In bankruptcy or receivership or being reorganized under the Bankruptcy Act, or securities assumed by such companies.

AMEX BONDS

Volume $81,000

SALES SINCE JANUARY 1

2000	1999	1998
$49,599,000	$71,400,000	$132,202,000

	Thu.	Wed.	Tue.	Mon.
Issues Traded	6	2	7	4
Advances	1	0	1	2
Declines	4	0	5	2
Unchanged	1	2	1	0
New highs	0	0	0	0
New lows	0	0	1	0

Bonds	Cur Yld.	Vol.	Close	Net Chg.
AltLiv 5¼02	cv	1	46	− 2
DRS T 9s03	cv	2	105	− 2
FriedeGld 4½04	cv	34	48	− 2½
ThmoInst 4s05	cv	20	84½	+ ½
TrnsLux 7½06	cv	4	79	− 2½
TWA 11⅜06	38.1	20	29⅞	...

NASDAQ

Convertible Debentures
Thursday, May 18, 2000

Issue	Vol.	Close	Net Chg.
DrgEmp 7¾14	25	20	...
DuraPh 3½02	134	83	+ ½
Jacobsn 6¾11	20	73	+ 5½
Metamor 2.94s04	6	84	+ 10
OHM 8s06	96	85¼	+ ¼
PhyCor 4½03	40	23	...

Source: The Wall Street Journal

ment mix or asset allocation should be checked once a year to see if changes need to be made.

As mentioned at the beginning of the chapter, over the long run, stocks have historically outperformed bonds and U.S. Treasuries. In order to diversify (spread the risk) and lessen their tax burden, however, most investors seek a balance among stocks, bonds, and other investments. With careful study on your own, or with the aid of finance professionals, you should be able to achieve this balance.

Laddering

Laddering
Buying issues of varying maturities so that bonds expire at different intervals.

A final concept in bond purchase and management is laddering. **Laddering** means buying issues of varying maturities so that bonds will expire at different intervals providing cash when you need it. Financial experts recommend laddering as a strategy to use when you are saving for a goal—the down payment on a house or retirement—or for investment income, or if there is a portion of your portfolio that you are not ready to commit to stocks.

EXHIBIT 13.5
Bond Evaluation Checklist

Bond Evaluation Checklist

The following questions will help you begin an evaluation of a potential bond purchase.

Section I: Information about Issuer

1. What is the issuer's name? ______
2. Where is the issuer located? ______
3. Check if this is a government bond ___ or a corporate bond. ___
4. Briefly describe what the issuer does, makes, produces.

______
______
______

Section II: Bond Factors

1. What is the face value of the bond? ______
2. What is the maturity date? ______
3. What is the interest rate? ______
4. How often are interest payments made? ______
5. Is the bond rated? yes ___no ___. If yes, what is the rating? ___ By what rating system? ___
6. When was the bond issued? ______
7. Is the bond callable? yes ___ no ___. If yes, when is the call date? ______
8. Is the bond tax-exempt? yes ___ no ___
9. Is the bond secured? yes ___ no ___

Section III: Other Factors

1. Describe any other information that you found that is pertinent to the bond evaluation.

______
______
______

2. Think about the state of the general economy. Are interest rates up? ___ or down? ___. Is the stock market up? ___ or in a downward spiral? ___
3. How does a bond fit into your overall financial plan? (i.e., does it help diversify your investments?, does it add balance? does it provide income? does it provide tax benefits?)

______
______
______
______

ELECTRONIC RESOURCES

As mentioned earlier, corporate and municipal bonds (munis) are rated by two main investment service companies, Moody's and Standard & Poor's. You can obtain the ratings by contacting Moody's at **www.moodys.com/** or Standard & Poor's at **www.ratings.standardpoor.com.**

Currently there is a surge in e-bond (e stands for electronic) trading because more bonds are being offered online and more investors are finding out that this is an option. In 1998 about five percent of U.S. government bonds and five percent of other types of bonds were traded electronically. It is predicted that the percentage will rise to 55 to 60 percent for U.S. Government bonds and 40 percent of all other bonds in 2001. In other words more investors will be buying directly from the issuers of the bonds, eliminating the middleman.

To actually buy online, here are some sites and addresses:

E-Bond Site*	**Address**
Bondagent	**www.bondagent.com**
Charles Schwab	**www.schwab.com**
E*Trade	**www.etrade.com**
Merrill Lynch	**www.mlol.ml.com** or **AskMerrill.com**
Morgan Stanley Dean Witter	**www.online.msdw.com**
Munidirect	**www.munidirect.com**
Tradebonds	**www.tradebonds.com**
U.S. Treasury	**www.publicdebt.treas.gov**

*Note that prices, bond selection, minimums to open an account, and commissions vary by site. Regarding selection, E*Trade, for example, offers corporate, agency, U.S. Treasury, municipal, and zero-coupon bonds from several dealers. Purchasing online usually saves you money on commissions.

For government securities a system called Treasury Direct allows you to own and deposit interest and principal electronically in your bank account. If you use Treasury Direct, you can buy Treasuries without paying a sales commission and also reinvest Treasury bills automatically up to two years. For information about Treasury Direct and other news about U.S. government notes and bonds, and to purchase Treasuries, contact the U.S. Treasury at their website listed above.

To find the current value of a savings bond you can call whomever sold it to you and ask, or you can go on the Internet to the Bureau of Public Debt at **www.savings-bonds.gov**, where you will find a bond value calculator.

To find out more about buying U.S. savings bonds online, go to the Savings Bond Connection at **www.savingsbonds.gov** and link to a page to conduct a transaction. At this site the traditional Series EE bonds, which sell for half their face value, are available in denominations ranging from $50 to $1,000. Series I bonds, which carry a lower interest rate but are adjusted to reflect inflation, are sold in $50 to $500 denominations. There is information as well on HH/H bonds and investing for kids.

For information on the whole topic of investing in bonds, check **www.bondsonline.com**, which lets you see prices as part of a database. This site provides information only; it does not sell bonds. Another source for information and prices for over 1,000 munis and corporate bonds is the Bond Market Association's Web site at **www.investinginbonds.com**.

Another source of information is the Fixed Income Home Page, which provides up-to-the-minute prices and news for the stock market and Treasury bonds. They can be reached at **www.fixedincome.com**.

FAQs

1. What is a bond?

   ■ A bond is an "IOU," a promise to repay, a loan or debt instrument.

2. Why do corporations or government units issue bonds?

   ■ They want large sums of money. A company may need the money for new equipment or a new factory. A city may need money for a new school or sewer system.

3. How can a beginner start to add bonds to his or her investment portfolio?

   ■ A good way to begin is to start with U.S. Savings Bonds or other highly secure government bonds or highly rated corporate or municipal bonds.

4. Are municipal bonds safer than corporate bonds?

   ■ Municipal bonds issued by states, cities, or other agencies may or may not be as safe as corporate bonds. Finding out a corporate bond's rating—for example, through Moody's Bond Ratings—will help you assess risk levels. In Moody's, an Aaa is the highest quality rating, followed by Aa.

5. What are U.S. bonds?

   ■ U.S. bonds are issued by the Treasury Department and other government agencies and are considered very safe.

6. How much does a corporate bond cost?

   ■ Usually corporate bonds are issued in units of $1,000.

7. Who buys bonds?

   ■ The usual buyers of bonds are individuals, banks, and mutual funds—generally people and institutions who are risk-averse (want safety) and are trying to balance or diversify their investment portfolio.

8. How can an investor maximize bonds to achieve greater investment security?

   ■ Start by buying short-term bonds and replace them with intermediate-terms bonds when you have more money to allocate to bonds. Eventually, try to have several intermediate- and long-term bonds that mature every year or two, thus providing safety plus access to money and, in most cases, twice yearly payments. This timing of bond maturity is called the laddering technique.

SUMMARY

Bonds are loans to corporations or governments. When you buy a bond, you are a creditor of that company or government unit.

Interest from corporate bonds is taxable by state and federal governments.

Usually when interest rates move up, the price of bonds move down. When interest rates drop, the prices of bonds go up. Short-term bonds are usually less affected by changes in interest rates than long-term bonds.

In evaluating a specific bond, individuals should consider several factors, including the purchase price, the coupon rate, the bond rating, taxes and commissions, the maturity date, and the yield.

Corporate bonds are issued by companies to raise funds for their growth. Because these bonds carry a greater risk than government securities, they may offer higher rates of interest and may be unsecured (debentures).

Investors purchase corporate bonds for three main reasons: to provide interest income, to provide potential capital gains, and for the bond replacement at maturity.

Zero-coupon bonds are sold at a deep discount—or a fraction of their par value—by corporations and governments. They have several potential drawbacks, including taxes on the interest accrued and a higher degree of price volatility.

Convertible bonds offer the option of acquiring stock instead of getting cash back. In other words they can be converted to stocks. If it is a growth company, this can be an attractive feature.

Some bonds have a call provision (they are said to be callable), meaning that they can be paid off before their due date.

Municipal bonds (munis) are debts issued by states, counties, and local government agencies to finance projects, including the building of roads, schools, and

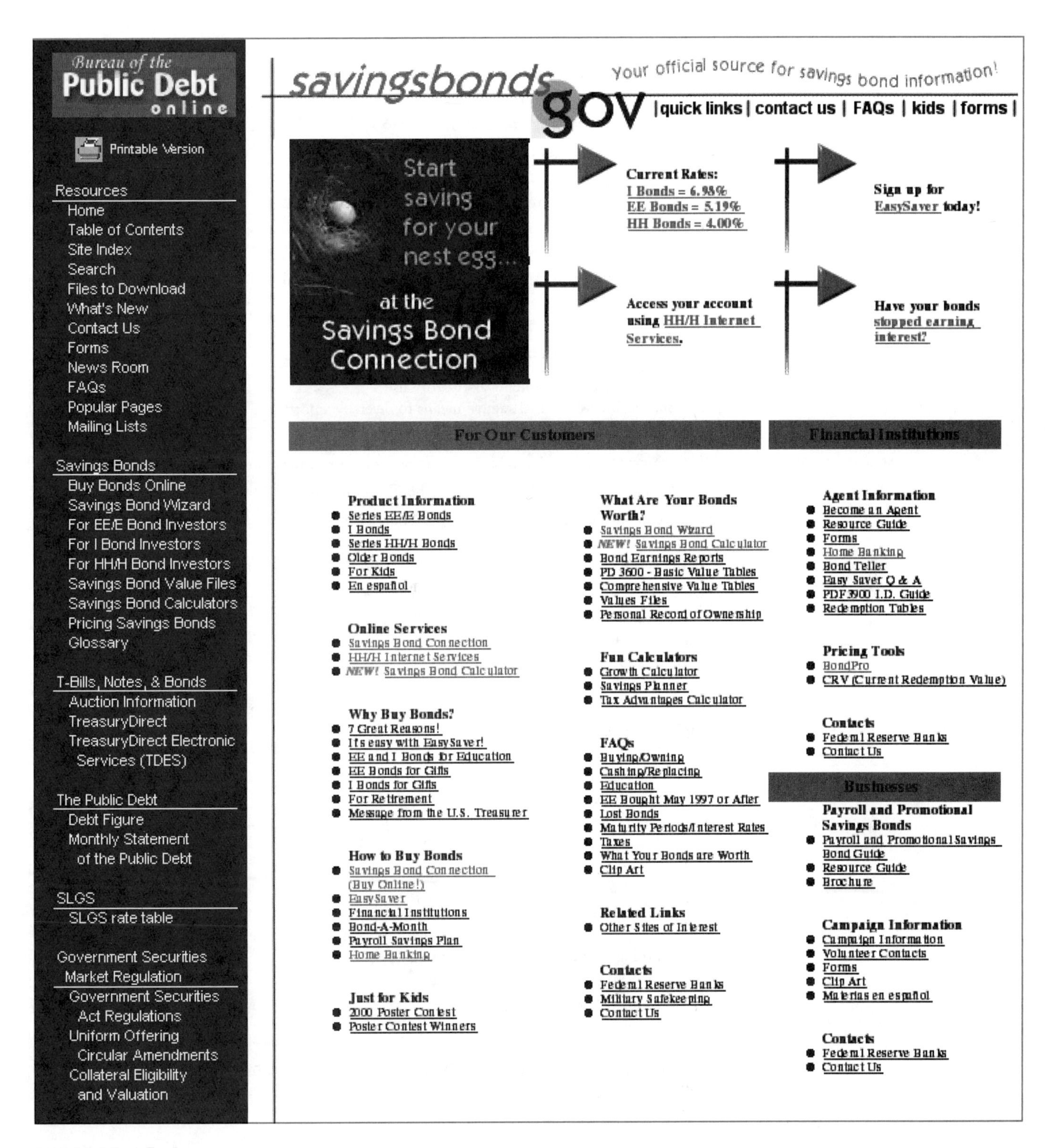

EXHIBIT 13.6
U.S. Savings Bond Information Online

bridges. Interest earned on most municipal bonds is exempt from state and federal income taxes.

Corporate and municipal bonds are rated by independent rating services, such as Moody's and Standard & Poor's. High-quality bonds are considered relatively safe investments.

Junk bonds—also known as high-yield bonds—have low ratings and are not considered investment grade. They incur more risk than other types of bonds.

Government bonds, issued by the U.S. Treasury, are not rated because their interest and principal payments are backed by the full faith and credit of the United States Government. They are considered some of the safest investments that exist.

U. S. Treasury bills are available for $1,000 and up. They mature in one year or less. Treasury notes mature in 1 to 10 years and Treasury bonds up to 30 years.

More people are buying bonds online.

To avoid sales commissions on federal government bonds, buy Treasuries through the U.S. Treasury.

Laddering is an investment strategy in which issues of varying maturity are bought so that bonds mature at different time intervals.

REVIEW QUESTIONS

(See Appendix A for answers to questions 1–10.)

Identify the following statements as true or false

1. Bonds can serve as an attractive means to diversify a portfolio.
2. Since 1925 bonds have outpaced stocks.
3. Owning a corporate bond means that you own part of the company.
4. The maturity date is the date when a bond is purchased.
5. A chief advantage of top-rated corporate bonds is the security they offer plus payments usually twice a year.

Pick the best answers to the following five questions.

6. ____________ cost a minimum of $1[illegible],000 each.
   a. Treasury bonds
   b. Series EE bonds
   c. Treasury bills
   d. Series HH bonds
7. ________ refers to the investment strategy of buying securities that mature at different times.
   a. Laddering
   b. Billeting
   c. Remargining
   d. Staggering
8. Which of the following is <u>not</u> a reason for purchasing a corporate bond?
   a. To provide interest income
   b. To provide potential capital gains
   c. For bond repayment at maturity
   d. For a quick return on money
9. __________ are issued by governmental units smaller than the federal government to finance such things as new schools and roads.
   a. Agency bonds
   b. Municipal bonds (munis)
   c. Treasury bills
   d. Fixed bonds
10. To start the process of buying a U.S. Treasury security, one would fill out a ____________.
    a. tender form
    b. 1040A tax form
    c. limited partnership form
    d. market form

DISCUSSION QUESTIONS

1. After reading this chapter and considering the state of the current economy and your personal financial goals, if you had $1,000 today to invest in a bond or a stock, which would you select? Explain your answer.
2. What characteristics of bonds make them attractive to cautious investors? Of all the bond types discussed, which bonds are the most secure? What are the major drawbacks of bonds, compared to money market funds and stocks?

3. Why would someone choose government bonds over corporate bonds? What are the tax considerations involved in choosing corporate or government bonds?
4. Why should someone be cautious before investing in junk bonds? How could he or she investigate the rating of a bond?
5. Brittany's grandparents in their late sixties have about 65 percent of their investments in bonds, 30 percent in stocks, and 5 percent in money market funds. They also have fully paid off their house. Given the stronger financial history of stocks, she wonders why they have so much invested in bonds instead. Why would retired people on fixed incomes prefer bonds?

DECISION-MAKING CASES

1. Amanda's grandmother gave her $2,000 for a graduation present. She spent $1,000 paying bills and buying clothes and wants to invest the remaining $1,000 in a corporate bond that matures in three years. It is rated Aaa by Moody's, and it pays an interest rate of 8 percent. How much money can she expect to be paid in interest each year?
2. The Acme Corporation needs to raise money for expansion overseas, so they have issued a debenture. Each debenture has a face value of $1,000, a coupon rate of 0.1 (10 percent), and a term of 5 years. Calculate the value of this debenture if Miguel invests $10,000 in it and keeps the debenture until it matures. Remember to figure yearly interest payments as well as the amount of the debenture.
3. Marian can not decide between putting a bonus check ($5,000) into a short-term or a long-term corporate bond. If interest rates are predicted to fluctuate wildly in the next several years, on the basis of Exhibit 13.1 and the discussion in the chapter, would she be better off with a short-term bond or a long-term bond?
4. Ashley has received a letter from XYZ Corporation wanting to call back the bonds it issued to her several years ago that were not due to mature until the year 2020. The corporation is offering her a higher interest rate necessary to compensate for the reinvestment risk. What factors may influence how she responds to their offer?
5. Tristan has a high need for security, so he is more drawn to federal government-secured bonds than to munis or corporate bonds. He has $3,000 saved up. Which of the following could he purchase with $3,000: a Treasury bill, note, and/or savings bond? How does he purchase the one(s) he has selected?

GET ON LINE

Here are two activities to try.

1. To check out savings bonds, go to www.savingsbonds.gov. Exhibit 13.6 shows how much information is available about government bonds. If you actually want to buy a bond, you will need to provide your name, Social Security number, and mailing address, just as you do when you purchase bonds from banks or other financial institutions. Bonds bought online must be purchased by credit card; as of the writing of this book, only MasterCard and Visa were accepted. After orders are processed, the bonds are mailed to buyers.
2. Compare prices of munis and corporate bonds by going to www.investinginbonds.com and locating the previous day's price on a specific bond, such as a Microsoft bond, an IBM bond, or a Dade County bond. Then go to www.bondsonline.com and check their price for the same bond.

INFOTRAC

To access InfoTrac follow the instructions at the end of Chapter 1.

Exercise 1: Type in the subject word "bonds." You will find over one hundred periodical references (articles) and several subdivisions. Select an article under the subdivision "corporate bonds" and report on what it says. Is this a general article or are specific corporate bonds recommended?

Exercise 2: Type in the subject word "bonds." Click on the subdivision "municipal bond." Select two articles on municipal bonds and compare what they say. Do they recommend municipal bonds as good investments? Why or why not?.

chapter fourteen

mutual funds, real estate, and other investments

Diversifying
Spreading the Risk

DID YOU KNOW?

* More money is invested in mutual funds than is deposited in banks.
* Over half of American households own mutual funds.

OBJECTIVES

After reading Chapter 14, you should be able to do the following:

1. Define mutual funds and describe their characteristics.
2. Classify mutual funds by investment objectives and evaluate for your investment purposes.
3. Describe how and why mutual funds are bought and sold.
4. Identify and evaluate types of real estate investments.
5. Analyze the risks and rewards of investing in precious metals, gems, and collectibles.

CHAPTER OVERVIEW

This chapter marks the culmination of the investment section of the book. It covers mutual funds, real estate, and a variety of other investments, including precious metals, gems, and collectibles.

Mutual funds are the success story of investing. They have taken off at an incredible rate. Currently over half of American households own mutual funds. Why are they so popular? The main reasons are that they are easier for people to handle than direct ownership of stocks and bonds, and in many cases they are the only choice available in employer-sponsored retirement programs. More of the advantages and also the disadvantages of mutual funds and other types of investing will be covered in this chapter. As in all chapters, there is an Electronic Resources section before the Summary.

Objective 1:
Define mutual funds and describe their characteristics.

MUTUAL FUNDS DEFINED AND EXPLAINED

Mutual funds are a form of investment operated by companies that raise money from shareholders and invest it in stocks, bonds, currencies, futures, options, or money market securities. The best way to differentiate mutual funds from other forms of investing is to think of them as *groups (or pools) of investments.* When you invest in a mutual fund, you invest in several companies, thereby spreading the risk.

Although their popularity has soared in recent years, pooled funds have been around since the 19th century. The version that we are familiar with dates from 1924, the Massachusetts Investment Trust started in Boston. Currently there are more mutual funds than there are stocks on the New York Stock Exchange. This is apparent in *The Wall Street Journal* where the lists of mutual funds are longer than the lists of stocks.

In April 1998 mutual funds crossed the $5 trillion mark. This sum rivals the deposits Americans had in banks which was $4.5 trillion at the end of 1997.[1] Currently the world's two biggest mutual funds, with over $90 billion each, are the Vanguard 500 Index Fund and the Magellan Fund.

How a Mutual Fund Works

Mutual funds are organizations that raise funds by selling shares to the public. In turn these funds are used to purchase groups of securities consistent with the fund's objective (see Exhibit 14.1). What happens is that you, as an investor, own shares in the mutual fund company, which has diversified holdings in such companies as Microsoft, Ford, Citigroup, Texaco, and UPS. Although diversity is the key, some funds specialize in areas of the market, such as technology funds, but what the company does within that category to achieve diversity is to hold a variety of securities representing different types of technology, such as telecommunications, software, hardware, Internet, and retail stocks.

Diversity, then, is one of the key advantages of a mutual fund. By owning one share of a mutual fund, an investor automatically has a more diverse holding than if he or she held one share of Coca-Cola stock. So, for a beginning investor, mutual funds make a lot of sense. The whole concept of mutual funds rests on pooled diversification, similar to the concept of pooling discussed in the insurance chapters. Individuals pool their resources for the collective benefit of all who contribute. The benefits of diversification are stressed repeatedly in the investment chapters in this book. It can mean the difference between making decent money and failing to meet your investment goals.[2]

Other advantages are that *mutual funds require small minimum investments (usually $1,000) and they are professionally managed.* In other words professionals buy and sell the stocks within the mutual fund and thus decide on their mix. You, as an investor, pay a fee for the professional management of the fund.

To describe further how mutual funds work, a mutual fund company seeks profit from (1) income from interest and dividend payments received from the fund's securities, and (2) increases in the value of the fund based on the rising value of the underlying securities. If all goes well (the underlying securities increase in value), the mutual fund investor receives income as dividends and also shares in the profits when the shares are sold.

Before further describing the advantages, the negative aspects or costs involved in mutual funds should be discussed. These include (1) the risk factors involved that vary considerably between funds, (2) the lack of personal control, and (3) the fees, hidden and overt.

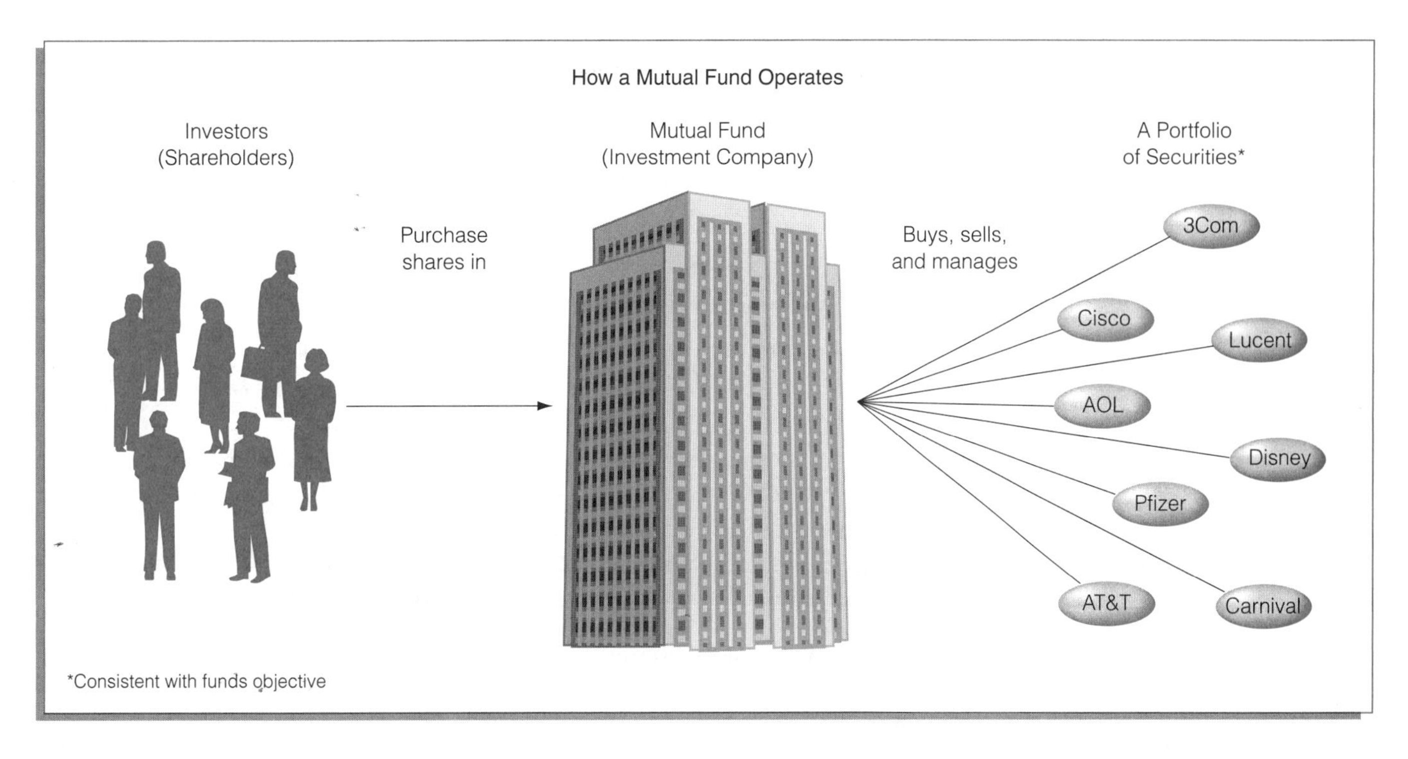

EXHIBIT 14.1
How a Mutual Fund Operates

Regarding risk, some funds perform better than others, and some are more speculative than others. The lack of personal control refers to the fact that you are turning your money over to professional fund managers, who vary in skill, foresight, and experience. They may buy securities that you do not care for or at a time or in amounts that you would not do personally if you were buying your own securities.

Fees. All sorts of fees are involved in mutual funds, and there is a fee war taking place. Consumers are becoming more savvy about finding no-load, low-load, or minimal annual fees. Traditionally the charge for professional management services was usually between 0.25 percent and 2 percent of the fund's total assets per year. Funds sold by full-service firms that charge fees, called sales commissions or "loads," are also referred to as **load funds**. These fees go to defray the cost of providing planning advice and fund selection. Investors pay the fees, hoping that the professional manager's decisions result in better returns than they, as individual investors, could achieve.

Load funds
Fees associated with mutual funds.

A commission fee may be charged that could go as high as 8 1/2 percent of the purchase price for investments under $10,000. Usually the higher the investment, the lower the percentage of commission charged. Low-load funds charge less than 3 percent of the purchase price.

Mutual funds sold to investors directly from fund companies that impose no sales charge are called **no-load funds**. Many people assume that "no-load" means no fee, but it usually means that the fees are imbedded in the management charges.

No-load funds
No charges directly associated with buying or redeeming fund shares.

Index funds fees are very low (0.18% to 0.62% a year) and are based on the amount of money the investor has in the fund. Investors in recent years have been scooping up index funds in record numbers. The biggest mutual funds are index funds.

Another fee is the **12b-1 fee** named for the Securities and Exchange Commission (SEC) rule that authorizes them. Shareholders pay 12b-1 fees to cover a fund's marketing and distribution efforts (promotion expenses), and may also include account maintenance fees. Both load and no-load funds may charge 12b-1 fees.

12b-1 fee
Mutual fund that assesses shareholders for promotion expenses

The last fee to be discussed is the sales charge known as the contingent deferred sales load (also referred to as the back-end load). If you withdrew a fund during the first year of ownership, a fee of 1–6 percent may be charged. Over time the deferred charge lessens, (usually one percentage point per year) so that after five years there is usually not a withdrawal fee. Exhibit 14.2 summarizes fees typically associated with mutual funds.

To summarize, consider the following regarding fees:

1. *All mutual funds have costs that are passed on to the investor.* Some fees are more obvious and more costly than others.

2. The expense ratio of any mutual fund, as explained in a fund's prospectus, can be an indicator of what an investor is paying. While there are other criteria, this is an important one when comparing funds.

3. Cost is meaningful within the context of a fund's total performance. Do not pick a fund solely because it is a no-load fund, a poorly performing fund is not a good investment even if the fees are low.

Open-End Mutual Funds and Closed-End Mutual Funds

Since there are so many funds, this next part of the chapter breaks the funds down into different classifications and types. The most fundamental classification is that they are either open-end or closed-end funds.

Open-end mutual funds
Purchasing and selling shares directly with the fund.

Net asset value (NAV)
Current market value of a fund determined by assets-liabilities, divided by number of shares outstanding.

Open-end mutual funds have shares that are issued and redeemed by the investment company at the request of investors. *About 90 percent of all mutual funds fall under this category.* Investors buy and sell shares at the net asset value plus a small commission, at least for the initial purchase. The **net asset value (NAV)** is defined as the value of the fund's portfolio (assets) minus liabilities, divided by number of shares outstanding as follows:

$$\text{Net asset value} = \frac{\text{Value of fund's portfolio} - \text{Liabilities}}{\text{Number of shares outstanding}}$$

For example, using the formula, assume the portfolio of all investments in the XYZ Mutual Fund has a current market value of $100 million. The fund also has liabilities of $10 million. If this mutual fund has 5 million shares outstanding, the net asset value per share is $18:

$$\text{Net asset value} = \frac{\$100 \text{ million} - \$10 \text{ million}}{5 \text{ million shares}}$$

$$= \$18 \text{ per share}$$

Net asset value is calculated at least once a day and reported over the Internet or other media. Most open-end funds are easily accessible for investors. Details about how to buy open-end funds will be given later in the chapter.

Closed-end mutual funds
Funds with a fixed number of shares listed on a stock exchange.

Closed-end mutual funds represent about 10 percent of the market. They are offered by investment companies. Only a certain number of shares are offered to investors, hence the term "closed-end." After all the shares are sold, then investors purchase shares only from another investor who is willing to sell. Closed-end funds are traded at stock exchanges. Since new closed-end mutual funds do not have a long track record, investors should consider supply and demand, their expectations, and the risk involved.

Many investment companies (Dean Witter, Scudder, Templeton, Vanguard, and Putnam) offer both open-end and closed-end funds to meet the varied objectives of their investors.

EXHIBIT 14.2
Typical Mutual Fund Fees

Type of Fee or Charge	Usual Amount
Load fund	Up to 8 1/2% of purchase price
Low-load fund	1% to 3% of purchase price
No-load fund	No direct sales charge
Index funds' fees	Investors pay annual expenses of 0.18% to 0.62% of the money they have in the fund
Management fee	0.25% to 2% per year of the fund's total assets
12b-1 fee	0.25% to 8.5%
Contingent deferred sales load or back-end lease	1–5% of withdrawals the first and declining thereafter ending at five years of ownership.

Index Funds

Index funds
Track the performance of the broader stock or bond markets.

As has been mentioned, an increasingly popular choice is index funds, which are a type of mutual funds with the difference that they are managed (for the most part) by computer. Often these are called passively managed funds, and the savings are passed on to the investor in the form of lower fees. **Index funds** are designed to track the performance of the broader stock or bond markets. An example would be the Standard & Poor's 500 Index, an index of 500 large U.S. company stocks. Someone who owns a Standard & Poor's 500 Index fund would benefit when the S&P goes up. Another example would be the Vanguard 500 Index fund, which is one of the 29 index funds Vanguard has that forsake stock-picking in favor of matching the market.

An investor in index funds believes in efficient market theory that assumes that trying to beat the market averages over the long run is ineffectual, you would do better just keeping up with the market. Investments in index funds will at least keep pace with the index being tracked. Another advantage to index funds is that they cost less to manage so they are cheaper than an actively managed portfolio. In recent years index funds have outperformed three quarters of professionally managed funds. One of the disadvantages is that they are considered boring or conservative. Other disadvantages depend on which index fund you buy: most are not as balanced as the S&P 500 fund. There are technology stock, small stock, midcapitalization, and international stock indexes. Also their advantage (going up with the market) is also their disadvantage when the part of the market that they represent goes down.

OBJECTIVES

Objective 2:
Classify mutual funds by investment objectives and evaluate for your investment purposes.

Before investing in a mutual fund, besides checking fees and general fund types, you should know the fund's objectives. Investors should classify the mutual fund by the investment objectives of the fund, and evaluate these objectives to see if they match one's own investment objectives.

Here is how matching objectives works. An investor may look for funds that provide:

* *current income.* If current income (high levels of current interest and dividends) is one of your investment objectives, look for funds whose dividends are well protected—probably through investments in blue-chip income

stocks, preferred stocks, or highly rated bonds.

* capital growth. If you would like to participate in the long-term growth of the economy, then seek equity-type securities (stock-based). In the 1990s and early 2000 most of the growth was in the technology sector, so many tech funds experienced high growth.
* conservation of principal. If conserving the principal (the money initially invested) is important to you, then this type of fund would protect the principal amount by careful selection of conservative investments and the constant monitoring of these investments.

Mutual funds exist that combine objectives. For example there are funds that have growth and income objectives. Typically these would have a variety of common stocks. Another type of fund, called a balanced fund—often with the word "balanced" in the name, such as Twentieth Century Balanced—may offer growth, income, and conservation of principal by adding high-quality bonds to stocks.

Prospectus

Prospectus
Mutual fund's written summary of its investment objectives and other information.

As the Twentieth Century Balanced fund shows, a fund's title usually indicates what type it is, but to investigate further, one should obtain the company's prospectus. A **prospectus** is a formal written summary of a commercial venture or offering. The prospectus can be obtained from a company or a broker.

Besides the objectives the prospectus will also provide

* a summary of fees, dividends, distributions, and taxes.
* a statement about risk factors.
* a statement describing the types of investments involved.
* a description of the fund's past performance.
* information about the fund's management.

Annual Report

Annual report
Yearly record of a mutual fund's or corporation's financial condition distributed to shareholders.

For more information a prospective investor can also obtain a fund's **annual report** by contacting the company. Once you buy into the fund, as a shareholder you will automatically receive this yearly report. The annual report reveals the fund's financial condition.

Although most shareholders focus on the bottom line ("How much money did I make?"), it is wise to read over the annual report to notice other aspects of the fund in order to predict future performance. For example, is there a lot of management turnover? A sudden change in objectives or management may warrant a closer look at the fund. The main factor to focus on in evaluating an annual report is the fund's past long-term performance. It is not a guarantee of future success, but it is the best indicator available.

Another way to check annual performance of funds is through a number of media sources that rank mutual funds by performance. For example, *Kiplinger's Personal Finance Magazine* gives an annual mutual fund comparison report. As another example, *The Wall Street Journal* gives quarterly reports on the performance of mutual funds listed by investment objectives over the last five years. Similar lists can be found in *Business Week*, *Forbes*, and *Money*. See Exhibit 14.3 for an example of a ratings list.

EXHIBIT 14.3
A Mutual Funds Performance List

Performance Yardsticks

How Fund Categories Stack Up

Investment Objective		Year-to-Date		Four Weeks		One Year		3 Yrs (annualized)		5 Yrs (annualized)
		On a Total Return Basis								
Capital Appreciation	+	11.26%	+	4.07%	+	33.20%	+	22.94%	+	17.06%
Growth	+	12.93	+	4.33	+	37.36	+	27.72	+	19.51
Small-Cap Stock	+	9.84	+	3.87	+	38.01	+	25.10	+	18.28
Mid-Cap Stock	+	11.80	+	4.28	+	36.71	+	24.51	+	18.07
Growth & Income	+	12.07	+	4.78	+	36.17	+	28.82	+	19.70
Equity Income	+	10.36	+	4.84	+	34.13	+	26.84	+	18.61
Global (inc U.S.)	+	13.91	+	5.73	+	25.07	+	20.33	+	15.67
International (non U.S.)	+	15.65	+	6.79	+	20.02	+	16.04	+	13.89
European Region	+	21.27	+	9.00	+	36.22	+	26.80	+	21.17
Latin America	–	0.18	+	6.91	+	9.59	+	23.50	+	10.58
Pacific Region	+	4.92	+	1.56	–	23.34	–	5.55	+	0.77
Emerging Markets	+	5.44	+	5.38	–	7.91	+	5.83	+	6.02
Science & Technology	+	16.42	+	1.31	+	34.17	+	24.09	+	22.54
Health & Biotech	+	11.32	+	3.11	+	32.89	+	26.44	+	23.67
Natural Resources	+	3.42	+	5.19	+	6.12	+	16.60	+	12.13
Gold	+	2.25	+	2.71	–	39.01	–	12.18	–	2.31
Utility	+	9.53	+	7.28	+	35.66	+	23.07	+	13.83
Balanced	+	7.80	+	3.07	+	25.46	+	20.23	+	14.15
Immediate Corp. Debt	+	1.53	+	0.51	+	10.47	+	8.36	+	6.37
Intermediate Gov't	+	1.43	+	0.43	+	10.04	+	7.65	+	5.75
Long-Term Govt.	+	1.37	+	0.46	+	12.41	+	8.54	+	6.34
High-Yield Taxable	+	3.90	+	1.23	+	15.61	+	14.36	+	10.91
Mortgage Bond	+	1.38	+	0.47	+	9.17	+	7.77	+	5.71
Short-Term US	+	1.30	+	0.39	+	7.43	+	6.56	+	5.05
Long Term	+	1.57	+	0.58	+	11.52	+	9.53	+	7.26
General US Taxable	+	2.31	+	1.04	+	11.35	+	11.97	+	8.47
World Income	+	2.60	+	1.10	+	8.19	+	11.38	+	6.38
Short-Term Muni	+	0.94	+	0.17	+	5.62	+	4.99	+	4.41
Informed-Term Muni	+	0.87	+	0.00	+	8.07	+	6.54	+	5.54
General L-T Muni	+	0.96	+	0.05	+	10.41	+	7.68	+	6.25
High-Yield Muni	+	1.39	+	0.19	+	10.87	+	8.25	+	6.88
Insured Muni	+	0.87	+	0.03	+	9.37	+	7.42	+	6.08

Largest Stock and Bond Funds

Year-to-date performance

Stock		Total Return	Bond		Total Return
Fidelity Invest:Magellan Fund	+	14.29%	PIMCO Funds Instl:Tot Rtn:Instl	+	1.62%
Vanguard Idx Fds:Idx:500 Port	+	13.92	Franklin Class I:Cust:US Govt:I	+	1.56
Amer Funds:Invest Co of Amer	+	11.85	Vanguard Fds:Fxd:GNMA Port	+	1.68
Amer Funds:Wash Mutual Inv	+	12.30	Amer Funds:Bond Fund Amer	+	2.44
Fidelity Invest:Growth & Income	+	12.84	Merrill Lynch B:Corp:Hi Inc	+	3.22
Fidelity Invest:Contrafund	+	12.75	Vanguard Idx Fds:Bd Idx:Total	+	1.65
Vanguard Fds:Windsor II	+	13.17	Dean Witter A & B:US Govt:B	+	1.53
Amer Century 20th:Ultra	+	16.37	Vanguard Fds:Fxd:Hi Yld	+	3.08
Fidelity Invest:Equity-Inc	+	12.05	Vanguard Fds:Fxd:Sht-Tm Corp	+	1.52
Fidelity Advsor T:Growth Opp	+	10.51	AARP Invst:GNMA	+	1.49

Source: Lipper Analytical Services

CONSUMER ALERT

Do not be overly impressed by news about dramatic short-term performance. Last month's best mutual fund may be this month's worst. Be skeptical of top ten lists of mutual funds headlined in financial magazines (XYZ's magazine lists the top 10 funds to invest in next year). Look for a strong track record of several years' running. If you like top ten lists, one way to sort through the choices is to see if a certain fund makes the top ten list in several sources.

Professional Advisory Services

Besides the media rankings, rankings are provided by a number of professional advisory services. The services do not represent one company, but rather look at the whole field of mutual funds and report their progress. Examples of professional advisory service firms are Lipper Analytical Services, Inc., (as shown in Exhibit 14.3) Morningstar, Inc., Standard & Poor's Corporation, and Wiesenberger Investment Companies. Some of their information is free (online and at public libraries), but more in-depth analysis may involve a fee. For example, Morningstar Inc. offers Morningstar Principia Plus, a software program that rates funds. They also have a website at www.morningstar.net that posts bulletin boards for individual funds, lengthy fund-manager interviews each month, and in-depth commentary and analysis.

Size and Type of Company

Since there are more than 11,000 stock, bond, and money market funds, you can not possibly analyze them all.[3] Besides determining the fund's objectives, you could focus on one category of funds, such as stock-mutual funds; but even if you did this there are still over 5,000 stock-mutual funds from which to choose. The next step would be to narrow stock-mutual funds down into categories by size or type, such as domestic versus foreign-stock funds. Stock-mutual funds are divided into large-company stock funds, midsize-company stock funds, small-company stock funds, and foreign-stock funds. A long-term strategy may be to diversify holdings by allocating money to several different types of funds. Many employer-sponsored retirement plans require the employee to choose 5–10 funds from a list of 20–30 preselected mutual funds.

Socially Responsible Funds

Another way to choose a fund is to look for ones that label themselves as socially responsible. What this means is that they invest in only certain types of industries and

CONSUMER ALERT

Investing in foreign markets is a way to diversify holdings, but the prospective investor is warned that foreign markets may be more risky due to variances in markets and local economies. "Investigate before investing" is always a wise dictum, but it is particularly appropriate when investing in foreign markets. It is safest to go domestic first and as you become more experienced, branch out into foreign funds.

ignore some that people find objectionable, such as tobacco. Other objections to a company may be that they are known to be a heavy polluter, or they use child labor in foreign countries. If you are interested in investing in a socially responsible way, then find out the securities that a fund invests in, and especially look for those that define themselves as socially responsible—usually their title indicates this.

Summary of Mutual Fund Information Sources

To summarize, many different kinds of mutual funds exist and the individual investor has many sources of information from which to choose. Fund ratings from professional advisory services are useful, and prospectuses and annual reports provide more information. Other information is available from brokers, fund managers, and Value Line (an investment tracking service report). These can be found in libraries, *The Wall Street Journal*, *Barron's*, *USA Today* and other newspapers, financial magazines, and the Internet (for specific listings see the Electronic Resources section near the end of the chapter).

SUMMARY OF STEPS TO FOLLOW IN SELECTING FUNDS

To find the fund or combination of funds that are best for you follow these steps:

* Decide if a mutual fund fits into your investment plan. For recent college graduates one of the first investment decisions to make is whether to purchase an individual stock or a mutual fund. Consider this advice:

> You can always invest in individual stocks, but it's probably wise for an investor who has been 100% in cash to buy mutual funds to gain exposure to a portfolio of diversified stocks. Purchasing stocks directly can leave an investor vulnerable to the swings of just a few share prices. "Start boring" and start with a stock fund, advises Gary Schatsky, a fee-only financial planner based in New York.[4]

* Find out about different funds' objectives, financial record, and management.
* If you are in a 28 percent or higher income tax bracket, consider tax-free bonds and money funds.[5]
* Consider no-load funds with low annual expenses.[6] If you pay fees, find what they are and what services you get for them. Also inquire about any required minimum investments ($1,000 or $5,000) or required monthly investments ($50 or $100).
* Read any agreement carefully.

How to Read the Mutual Funds Section of the Newspaper

Daily updates on mutual funds are available in newspapers, and quicker updates are available on America Online, CompuServe, and Prodigy. A guide to reading a newspaper listing of a mutual fund is given in Exhibit 14.4. Your local newspaper may use slightly different notations, but explanations of these can be found in the financial pages' footnotes that accompany the mutual fund quotations. For example, newspapers are consistent on basic notations such as "NL" that stands for no-load, meaning that no sales fee is charged.

Mutual Funds for Children

There are mutual funds designed for children. For example, the Stein Roe Young Investor Fund accepts initial investments as low as $100, but you have to agree to

EXHIBIT 14.4
A Guide to Reading Newspaper Listings of Mutual Funds

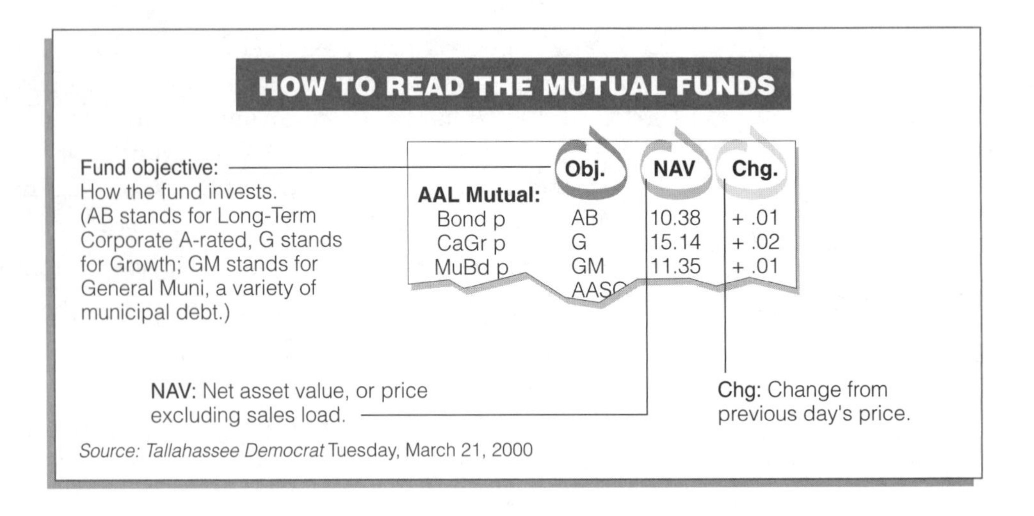

invest at least $50 a month until the account's value grows to $1,000. Another option for parents is to invest in mutual funds for their children through custodial accounts, as explained in Chapter 7. Account executives or financial planners can set up these accounts.

Objective 3:
Describe how and why mutual funds are bought and sold.

PURCHASING MUTUAL FUNDS

As mentioned earlier, there are closed-end and open-end funds. Closed-end funds (representing only 10% of the market) are bought through stock exchanges, so a broker or dealer is needed to purchase them. They are usually initial offerings announced in financial newspapers or other sources. A more likely choice is an open-end fund, so the rest of this section will focus on open-end funds.

No-load open-end funds can be purchased by contacting the investment company that sponsors the fund. After a prospective investor contacts the fund, they send a prospectus and an enrollment/order form. To purchase a fund, the investor fills out the form and sends in a check. Open-end load funds can be purchased through a salesperson authorized to sell them or through account executives. More and more, all of this can be done over the Internet.

Once you invest in a fund, you will probably want to keep buying more shares. The following are the four purchase options available for open-end funds:

Regular Account Transactions. *This is the most common option.* Investors decide how much they want to invest and how many shares they can afford. Commissions, if any, are deducted from the purchase price. Typically money market funds and stock and bond funds require an initial investment of at least $1,000.

Voluntary Savings Plans. In this option you open an account with an investment company and make an initial purchase (say for $1,000) and declare an intent to make regular minimum purchases of $50 to $100 a month. This could be done by arranging an automatic transfer from your bank account. Putting in $50 a month would be an example of dollar cost averaging because you would be investing equal dollar amounts at regular intervals. Examples of diversified no-load funds requiring a $50 to $100 monthly investment are Dreyfus Appreciation Fund, Invesco Industrial Income Fund, Neuberger & Berman Partners Fund and T. Rowe Price Spectrum Growth Fund.[8]

Contractual Plans. These plans (also called front-end loads) require investors to make regular purchases over a number of years, usually 10 to 15. Most of the commissions are paid immediately or within the first few years, and the investor may incur penalties if payments are not kept up. Because of this the Securities and Exchange Commission and several states have imposed rules about how these operate. This should be an indicator that investors should look carefully into this option before purchasing.

Reinvestment Plans. These are a service, provided by the fund, in which shareholder dividends and capital gain distributions are automatically reinvested to purchase more shares of the fund. Usually reinvested funds do not incur more commissions or sales fees, but there are taxes on dividends or capital gain distributions that are reinvested. This is a fairly passive approach to investing, a buy-and-hold strategy.

Selling or Rebalancing Mutual Funds

When do you decide to sell mutual funds? Usually this decision is based on changes in your investment goals or retirement or a need for rebalancing. Other signs that selling may be in order include fund management turnover, lagging performance over several years, and rising fees. For example, you may have invested in a small cap fund that has underperformed the market for several years. After visiting with your retirement advisor, you may decide to move out of the small cap fund and put more into technology funds.

Since closed-end funds are listed on securities exchanges, they can be sold to other investors. A broker or dealer is needed to sell a closed-end fund.

Open-end funds can be sold on any business day to the investment company that sponsors the fund. There are several ways to sell open-end funds:

- The simplest one would be that you notify the fund that you want to sell, and they would send you a check or, in some cases, electronically transfer the money into your account. You could also opt not to sell all your shares but only part of them. Once they are sold, you would receive a check or an electronic transfer of funds.
- Another option would be to withdraw all income that results from interest, dividends, and capital gains earned by the fund while it was held. In this instance the investor's principal would be untouched.
- A final option would be to withdraw a fixed percentage of asset growth. For example, if you arranged to receive 50 percent of the asset growth of your investment and the asset growth amounted to $1,000 during a particular period, for that period you would receive a check for $500 ($1,000 x 50% = $500). If no asset growth occurred, you would receive no payment, but your principal would remain untouched.

Taxation and Mutual Funds

Investors should keep records of all buying and selling transactions for tax-reporting purposes. Before buying or selling mutual funds, an investor should talk with a tax advisor or an account executive about any tax implications arising from transactions. As mentioned earlier, a person in a high tax bracket may want to invest in tax-free bond or money funds.

Funds as a Gateway: The Competition Grows

A final comment to make before leaving this section on mutual funds is that increasingly no-load fund groups are branching out into brokerage operations and home banking. For example, investors holding shares of a no-load fund through a broker, such as Charles Schwab Corporation, are customers of the broker, not the fund group. Funds are being used as a gateway to other financial products, and more and more transactions can be made over personal computers. In this way one company can be an investor's source for all or most financial transactions, and this is the wave of the future. For example, large no-load fund groups—such as Fidelity and Dreyfus—also offer brokerage operations and debit cards. The lines between insurance companies, banks, mutual funds, retail stores, traditional stock brokerage firms, etc. are breaking down. Now the chapter turns to another major investment option—real estate.

Objective 4:
Identify and Evaluate Types of Real Estate Investments.

TYPES OF REAL ESTATE INVESTMENTS

Real estate
Piece of land and all physical property related to it.

Real estate is a piece of land and all physical property related to it, including houses, fences, landscaping, plus all rights to the air above and the earth below the property.[9] Usually when one thinks about real estate, the single-family dwelling comes to mind. This makes sense because about 70 percent of American households own their own homes; but real estate investments encompass far more than this. Real estate investments fall into two main categories:

* *direct investments,* in which investors hold legal title to property, such as single family dwellings, apartments, land, and nonresidential real estate, and
* *indirect investments,* in which investors appoint a trustee to hold legal title on behalf of all investors in a group. This includes real estate investment trusts, limited partnerships and syndicates, mortgages, and mortgage pools.

These will be discussed next.

Direct Real Estate Investments

* *Primary Homes.* Homes differ from other investments, since they provide a place to live and may serve as an income shelter if you have a mortgage and also as a possible hedge against inflation. In recent years home prices have risen gradually although there are regions of the country where home prices have skyrocketed. For the most part homes should be looked at primarily as a place to live and secondarily as an investment since usually higher returns can be gained from other forms of investing.
* *Second Homes or Vacation Homes.* Often there is emotion attached to these, a sense of escape, that attracts investors. However, second homes can be a money drain. The tax advantages of owning a second home have diminished since 1987, when the Internal Revenue Service changed the rules. Since then, property is considered a second home as long as it is not rented out more than 14 days a year. In this case the mortgage interest and property taxes can be written off. If the home is rented out regularly (more than 14 days), then the size of the deduction is altered, depending on whether the owner manages it or someone else does and on the size of the homeowner's income. Since tax codes on second homes change quite frequently, current investors have to keep aware of changes and make appropriate adjustments.
* *Time Shares.* Another potential emotional purchase and a money drain is a time share. A **time share** is an apartment, condominium, or house with several owners. There may be as many as 52 owners, wherein each owner gets a one-week stay per year. Developers (not the purchasers) are usually the ones who get rich from time shares. A $100,000 apartment sold as a time share for 52 weeks at $8,000 a week would bring the developer $316,000 (figured by multiplying 52 x $8000 = $416,000 minus the cost of building the unit at $100,000 leaving $316,000 profit). This does not include marketing and other costs, but you get the idea. Time shares may be sold to an individual investor as a real estate investment, but actually buyers are purchasing a vacation for $5,000 to $20,000, and they will also pay an annual maintenance/management fee.

Time share
Apartment, condominium or house with several owners for vacation purposes.

One of the more versatile types offers exchange arrangements, such as trading a week in Aspen for a week in Hawaii. Generally time shares are not a wise investment because they tie up money, are difficult to resell, and if resold rarely return a profit. Before investing, request evidence of the track record on resales.

Commercial property Real estate that includes income producing property.

* *Commercial Property.* Real estate that includes income-producing property is **commercial property**. This includes apartments, duplexes, office buildings, shopping centers, hotels, factories, restaurants, warehouses, and industrial parks. Commercial property is usually zoned for commercial use. You can invest in commercial property directly or as part of a Real Estate Investment Trust (REIT) or Real Estate Limited Partnership.[10] An example of a direct investment would be if you bought a small apartment building. The upcoming section on Indirect Real Estate Investments will explain REITs and partnerships.
* *Raw land.* Raw land does not produce any cash flow. It usually refers to undeveloped acreage far from established communities with few services. Investors in raw land hope to sell it at a profit, build on it themselves, or subdivide it into lots. There may also be oil (remember the TV show "The Beverly Hillbillies?")or lumber on it to harvest. A basic principle is that the more "raw" the land, the more risk. If subdividing is planned, then utilities should be checked into, as well as roads, taxes, and zoning laws. Less risk, but a higher cost, is involved in buying lots in established neighborhoods. The general advice is not to buy land without seeing it and without investigating zoning laws and taxes.

Indirect Real Estate Investments

This section will cover four types of indirect real estate investments: (1) Real Estate Investment Trusts (REITs), (2) real estate syndicates or real estate limited partnerships, (3) first and second mortgages, and (4) participation certificates. These, with the exception of participation certificates, are not recommended for beginners for a variety of reasons, including complexity and risk. In fact many personal finance textbooks do not cover these, but they may be of interest to you in the future or you may work with people or have family members that have these types of investments, so they are introduced here for these reasons.

Real estate investment trusts (REITs) Companies that own, manage, and develop pools of properties.

Real Estate Investment Trusts. **Real Estate Investment Trusts** (REITs, pronounced *reets)* are companies that own, manage, and develop pools of properties, from apartments and office buildings to self-storage facilities and prisons. Until 1960 real estate was owned primarily by private individuals or family-owned companies, but in 1960 Congress created Real Estate Investment Trusts to let individuals invest in real estate as shareholders. This change in the law stated that REITs must distribute at least 95 percent of taxable income to shareholders, but corporate taxes could be avoided.

In the 1970s the REIT market collapsed due to high interest rates and other financial problems, but by the 1990s their popularity rose. By 1997 the REIT equity market hit $140 billion.[11] REITs are sold through stock markets or over the counter.

CONSUMER ALERT

Do not purchase a vacation or second home or a time share until you have checked out all the angles. Many people regret tying up large amounts of money in these and wish they had rented instead.

REITs offer the following advantages:

1. Economies of scale. Think of them as mega-landowners, hotel-owners etc. For example, with many different hotels under one ownership supplies can be bought at great quantities for less and there can be a centralized reservation service.

2. Public financing. Rather than relying solely on yourself or your family for financing, you join a large pool of investors. Because they are public, REITs must be accountable to shareholders, including the disclosure of information.

3. Diversification and professional management. These characteristics make REITs similar to mutual funds.

Even with their rapid growth, REITs represent only a sliver of the commercial real estate market. Consolidations and mergers are rampant. In other words this is a fast growing type of investment that brings with it great risks and rewards tied to the overall fluctuations of the real estate market. In the mid-1970s many REITs went bust; more guards are in place to prevent this from happening again, but still the potential investor should be wary of overbuilding and overreaching. More information on REITs can be obtained from the National Association of Real Estate Investment Trusts, 1101 17th Street, NW, Washington, DC 20036.

Real estate syndicates
Groups of investors form partnership to invest in real estate.

Real Estate Syndicates also known as real estate limited partnerships. **Real estate syndicates** are made up of groups of investors who form a partnership to invest in real estate. The partnership buys properties—such as office buildings, apartment buildings, or hotels—and passes the income from the rentals to the partners. For example, each partner could buy one unit in the partnership for $1,000, or multiples such as five units for $5,000. When the partnership loses money, partners share in the loss, and they could lose all the money they invested. Conversely, if the investment makes money, then each partner shares in the profits. If the property is sold and has increased in value, then the profit is passed to the partners.

Most investors join real estate limited partnerships to obtain high current income and to profit from the appreciation of real estate. If the property depreciates or if renting fully is unsuccessful, then the partners may suffer a loss.

First and second mortgages
Purchasing mortgages on property as an investment.

First and Second Mortgages. Both **first and second mortgages** may be purchased by someone who wants to take on the risk that other financial institutions do not want to take. Purchasers of first or second mortgages provide a mortgage on a property that has an unclear title or in some other way is deemed undesirable by conventional lending institutions. Since higher risk is involved, higher rates of interest are charged.

Participation certificates
Equity investments in a pool of mortgages that are purchased by government agencies.

Participation certificates. **A participation certificate** is an equity investment in a pool of mortgages that are purchased by one of several federal or state agencies, such as the Government National Mortgage Association (Ginnie Mae), the Federal Home Loan Mortgage Corporation (Freddie Mac), the Federal National Mortgage Association (Fannie Mae), and the State of New York Mortgage Agency (Sonny Mae). For example, Ginnie Mae is a government corporation within the Department of Housing and Urban Development. Ginnie Mae has helped over 18.9 million families afford a home by making mortgage funds available. An investor in a Ginnie Mae is providing capital to facilitate these federally insured or guaranteed mortgages.

Participation certificates involve less risk than the previously mentioned forms of indirect real estate investments because they are sold through state or federal agencies. The certificates are secured by government bonds and notes. They offer safety and relatively high yields and the certificates can be invested in directly, or they may be part of a mutual fund. Each month, when payments are made on mortgages, the investor receives interest and principal by check, or in the case of a mutual fund the proceeds can be reinvested.

Other participation certificates cover student loans. Examples of these are the Student Loan Marketing Association (Sallie Mae), created by Congress in 1972 to provide a secondary market for government-backed student loans, and the New England Education Loan Marketing Corporation (Nellie Mae) which gives federal guarantees to student loans issued in Massachusetts and New Hampshire. Nellie Mae bonds are sold in minimum denominations of $5,000. Other types of participation certificates are available from $1,000 and up.

To find out more about participation certificates contact the government agencies that offer them. Examples of government Web sites are given in the Electronic Resources section.

Advantages of Investing in Real Estate

Because there are so many different kinds of real estate investments, it is difficult to summarize their pros and cons. In addition market conditions and tax laws change; but, in general, real estate has the following advantages:

* *a hedge against inflation through escalating property values.* Especially promising are areas where the demand for real estate is up and prices are rising.
* *professional management* in the cases of indirect real estate investments such as limited partnerships, REITs, and participation certificates. They do not require that you do paperwork or repairs.
* *tax deductions.* Certain investments especially owning a primary residence, have tax advantages; in other cases taxes can be deferred until the property is sold.
* *ease of entry.* Limited partnerships can be purchased for as little as $1,000 meaning that an individual can become part of a multimillion dollar real estate investment that normally would be beyond reach.
* *financial leverage,* which is debt in relation to equity in a firm's capital structure.[12] In real estate this means enhancing the return or value without increasing investment. If you bought a $50,000 lot for cash and sold it for $60,000, the $10,000 gain represents a 20 percent return. If you had invested only $5,000 of your own money and borrowed the other $45,000 (90 percent financing), you made $10,000 on your $5,000 investment or a 200 percent return.

Disadvantages of Investing in Real Estate

The following is a list of possible disadvantages:

* *declining property values and depreciation.* During recessions, depressions, and deflationary periods, real estate may not keep up with inflation. Over time many homes and buildings depreciate in value, meaning they lose value. The buildings may be in poor repair or located in neighborhoods that are declining.
* *risk of loss.* In a general partnership, such as a real estate syndicate or partnership, the partners could lose more than their initial investment. So even if they invested only $10,000, they may be liable for $20,000. In a limited partnership the investor is not liable for losses beyond the investment. What this means is that if an individual invested $10,000 in a limited partnership that fails, he or she could lose only as much as $10,000.
* *taxes.* Property taxes are due yearly on houses and land. Tax assessments can go up, so the value of the house or land has to rise more than the amount of taxes due each year if one is to get ahead. There are also capital gains taxes. The Tax Reform Act of 1986 limits the amount of losses that can be claimed to offset income gained from other sources. In other words the tax-shelter advantages of indirect real estate investments have been lessened.

* *damages, repairs, insurance, and other costs* associated with real estate.
* *illiquidity*. It may be difficult to sell property quickly.
* *management problems*. For example, landlords have to fill out paperwork, collect rent, and take care of the property. If as an owner of property, you hire someone to manage the property for you, then there are still the problems involved in having employees.
* *lack of diversification, opportunity costs*. Tying up too much money in illiquid real estate investments may rule out having money for stocks, mutual funds, and other types of investments.

Objective 5:
Analyze the risks and rewards of investing in precious metals, gems, and collectibles.

ESOTERIC INVESTMENTS

This last category of investments describes precious metals, gems, and collectibles that can be purchased as part of a mutual fund or directly from a store, company, private owner or an auction house. They can be bought over Internet websites such as Ebay. This section provides basic guidelines for assessing the value of precious metals, gems, and collectibles. There are few protections to the potential investor, so the risk is great in this category of investments; but if you are a collector, the rewards in terms of satisfaction are great as well.

Precious Metals

There are a number of different precious metals. The biggest one in terms of volume of sales is gold. In recent years the most favored form by individual investors—gold bullion coins—was showing new life.[13] Bullion coin is minted in gold by governments and traded mostly for their gold content. Examples are the American Eagle, the Canadian Maple Leaf, the Australian Kangaroo, and the South African Krugerrand. The ups and downs in the annual sales of American Eagle gold bullion coins over a 13-year period are shown in Exhibit 14.5. According to the U.S. Mint, an upswing in the sale of gold coins is often triggered by world economic turmoil.[14]

> "It's insurance. It's financial security," says Ray DeMoss, a floral-marketing consultant in Ruston, LA., who has been a buyer of gold in various forms since the 1970s. But he has accelerated his gold buying lately and says the 30 South African Krugerrands he has purchased over the last six weeks are the first gold coins he has bought in a long time. "We're looking at a global money meltdown right now," he explains.[15]

This quote expresses only one investor's opinion about how he strategizes about investing in gold, but it is in line with the U.S. Mint observation that, when there is a perceived shake-up in the world economy, many investors revert back to such basics as gold coins.

Most new gold coins are bought through brokers who require a minimum order and charge a commission. Also there are antique coins that are traded more for their historic or aesthetic value than for their gold content.

Other than coins, gold may be purchased in gold bars, gold certificates, and through purchasing gold-based stocks and bonds.

Gold bars are traded in physical form or through futures and options contracts or mutual funds. Gold certificates are paper certificates providing evidence of ownership of gold bullion. If someone did not want to worry about the safety of having gold in the house, he or she may want a gold certificate instead. The real gold on which the certificate is based is kept in a bank vault. Gold-based stocks and bonds are listed on the U.S. stock exchanges.

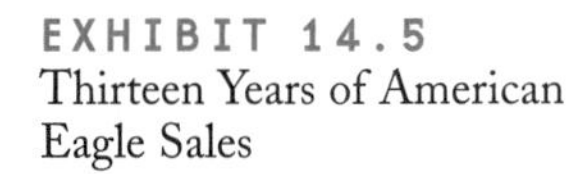

EXHIBIT 14.5
Thirteen Years of American Eagle Sales

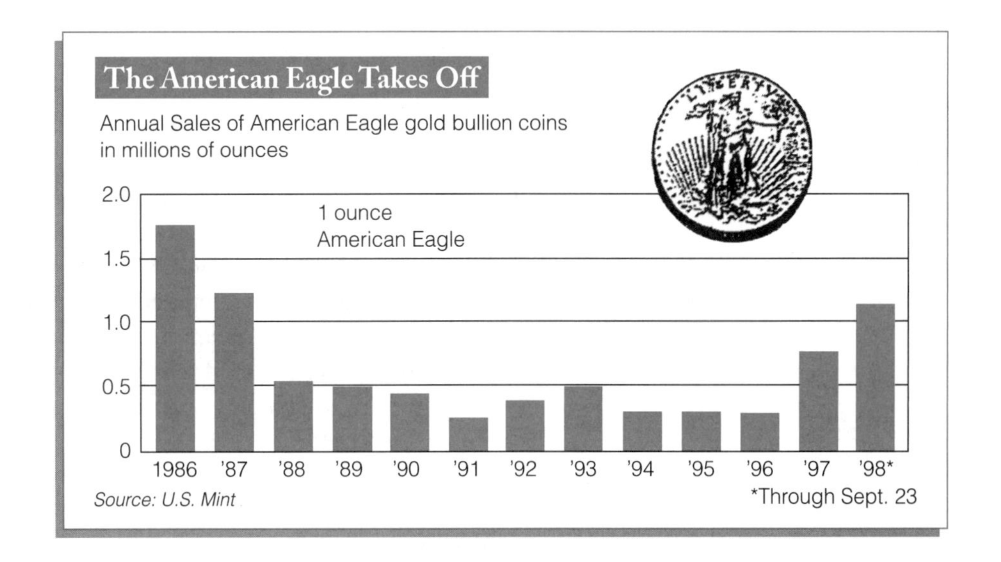

There are also silver-based stocks and bonds. Gold- or silver-based bonds are backed by the gold and silver mining firms who issue the bonds. Silver, platinum, and palladium prices fluctuate wildly. Usually investors buy these as well as gold as hedges against inflation and because, during times of unrest, basic metals tend to hold their value better than troubled currency—but this is not always the case. During the Depression in 1932 an ounce of silver was worth less than 25 cents, but this rebounded to a high of around $50 an ounce in the 1980s. When this book went to press, an ounce of silver was worth about $5.00, gold $285 an ounce, platinum was up to about $400 an ounce, and palladium $190. Thus of the precious metals discussed in this chapter, platinum is the most valuable. Regarding the drop in silver prices:

> It's a classic movie scene: A cat burglar tiptoes into a darkened mansion and, glancing furtively over his shoulder, steals the family silver. Talk about the good ol' days. Today, many owners might help the thief pack up so they could claim the insurance value. "A burglar can probably raise more cash quickly with a compact-disk player than a Victorian tea set," says Phillippa Glanville, the Victoria & Albert Museum's curator of metalwork.[16]

Precious Stones (Gems)

Diamonds, emeralds, rubies, and sapphires are examples of precious stones. One of their chief advantages is portability. When royalty and aristocracy have escaped countries, they have often hidden jewels in the linings of their clothing and shoes or swallowed them. Their main disadvantages are illiquidity and the determination of worth or quality. A standardized grading system has been established for diamonds, but not for emeralds, rubies, or sapphires.

Today diamonds come mostly from Africa, but in the past India was the source for some of the largest gems. Diamonds purchased in the United States can be rated by the American Gem Society or the Gemological Institute of America (GIA). They base their rating systems on a diamond's color, cut, clarity, and carat weight.

Diamonds worth over $1,000 such as those in engagement rings should be certified by an independent gemological laboratory, such as the GIA or the American Gem Society. The GIA was established in 1931 as the nonprofit educational resource for the gem and jewelry industry (see Electronic Resources for their Web site as well as the American Gem Society's). Local jewelry or retail stores display signs indicating that they are affiliated with the GIA or the American Gem Society. An investor in a diamond should obtain a written certificate describing the stone. The certificate would be useful if the stone were lost or stolen.

Regarding gems as an investment, William Boyajian, president of the GIA, says

> GIA does not advocate gems as an investment (nor does the gem and jewelry trade, in general) per se, but rather as a store of value and as portable wealth. We believe people should buy gems to wear and enjoy for emotional reasons, realizing too, that they may be an excellent store of value that can increase over time. However, this depends a lot on the specific gem and its quality. For value determination, I recommend that people go to a jeweler-gemologist in their local community that they know and trust, one that was there yesterday and will be there tomorrow.[17]

Collectibles

Of all the investment categories this one is the most varied and subject to fluctuation and—with the advent of online auctions—one of the fastest moving areas of investing. It includes books, rugs, Barbie dolls, Pokemon, antiques, stamps, sports memorabilia, paintings, and other items that appeal to collectors and investors. Each of these requires knowledge if they are to bring both pleasure and profit.

When buying serious art work, get a condition report, a written statement of the art work that includes flaws and restoration. In art and antiques, forgeries and fakes are common.

> Beginners should carefully weigh the purchase of any prints by these artists: Matisse, Salvador Dali, Marc Chagall and Joan Miro. Fake prints have been sold through telemarketing schemes that were the targets of separate investigations by the Federal Trade Commission and the U.S. Postal Service.[18]

Even legitimate art by the same artist varies greatly in price, depending on the quality of the piece. The highest prices come from work done at the height of the artist's career; often a later work is less valuable.

Usually when one thinks of art, paintings come to mind, but the category of art includes a vast array of decorative objects, including Ming vases and a 1735 silver soup tureen that Sotheby's was auctioning off for more than $8 million.[19]

Nostalgia tends to drive up the price of toys, gadgets, recordings, or memorabilia from celebrities. Investors should carefully weigh the lure of nostalgia versus the likely potential value of a collectible. This is not so easy to do. For example, Beanie Babies that burst onto the toy market in 1994, retailing for $5.95, have had wildly fluctuating prices. Princess, a teddy bear created in memory of Britain's Princess Diana was selling at $75 in May 1998 and went down to $45 in September 1998.[20]

The condition of an item also affects prices. When it comes to toys, mint-in-box (unopened original wrapping) fetches the highest prices. A Xena, the princess warrior doll, mint-in-box was being sold for $59 in 1998.[21]

Another cautionary note is to hesitate before buying art or collectibles while on vacation, when one is vulnerable to making quick decisions and the return of items is difficult. So how does a collector protect himself or herself?

- Use common sense, comparison shop, and keep up with price fluctuations. This may be difficult because—according to Robert Z. Aliber, professor of international economics at the University of Chicago— "There is no New York Stock Exchange for this stuff" offering immediate and widely shared price information.[22]

* Trade with reputable dealers.
* Education is crucial; do not buy blind; discern quality.
* Remember that collections do not pay off until they are sold. Collectibles do not provide regular dividends or income.
* Collectibles in mint condition gain the best prices.
* Figure in the costs of insuring, storing, displaying, and moving collections.

To conclude the investment section of this book, Exhibit 14.6 shows how different types of investments compare over time. As has been mentioned frequently, from a financial viewpoint over the long run stocks perform the best.

ADVICE FOR BEGINNERS

* It is hard to beat mutual funds as a beginning investment. In particular, broad-based index funds, such as the Standard & Poor's 500 or Vanguard 500 Index Fund, are worth investigating.
* If you have an employer-sponsored retirement account, your main option will probably be a selection of mutual funds. Diversify; do not select all one type of fund, such as growth funds or tech stock funds.
* Once purchased, keep buying more shares.
* Try to minimize fees.
* Buy a house primarily as a place to live and secondarily as an investment. Be cautious about real estate investing; check all the angles.
* Gems, collectibles, and precious metals fluctuate widely in value. Knowledge (knowing prices and quality) is the key to making good choices. Buy gems and collectibles primarily because you like them and, secondarily, with the idea of profit.

Tangible/Financial Returns
Period Ending 6/1/95

Asset	Av. Return 1yr.	Av. Return 5 yrs.	Av. Return 10 yrs.	Av. Return 20 yrs.
Stocks	20.1%	11.4%	14.6%	13.8%
Forex	19.8	6.4	9.1	4.4
Bonds	15.8	11.4	11.7	10.5
T-Bills	5.5	4.8	6.2	8.1
Oil	3.5	1.6	3.8	2.7
Stamps	0.8	1.3	0.5	8.2
Gold	0.1	1.1	2.0	4.4
Diamonds	0.0	0.2	5.9	7.7
Sotheby's Stock (Proxy for Art Investment)	0.0	–6.2		
Silver	–0.1	1.1	–1.5	0.9
Housing	–1.4	2.5	3.8	5.8
CPI	2.9	3.3	3.5	5.4

Present source: The Irwin Business and Investment Almanac 1996, edited by Summer Levine and Caroline Levine, Irwin Professional Publishing, Chicago.

EXHIBIT 14.6
Returns on Various Types of Investments

ELECTRONIC RESOURCES

The Internet is crawling with both good and bad investment websites. First, a warning about the bad. Sometimes fraudsters offer stock in fake companies usually based on some late-breaking news such as a new miracle drug discovery. Then they drive up the price on the Web and sell their own shares at a profit before the truth comes out that the drug is in the testing stages or does not work as well as hoped. The fraudsters can do this cheaply because electronic mail costs nothing to send. The advice is to not invest through the Web unless the source or company is legitimate. To report online stock or mutual fund fraud, contact the Securities and Exchange Commission at **enforcement@sec.gov**. Potential investors should also be cautious of real estate investment offerings over the Web or any other type of investment, including collectibles.

Examples of legitimate sources on the Internet for *government securities, participation certificates, named Macs and Maes,* are:
Government National Mortgage Association (GNMA)
(Ginnie Mae) **www.ginniemae.gov/index2.htm**
Federal Home Loan Mortgage Corporation (FHLMC)
(Freddie Mac) **www.freddiemac.com/**
Federal National Mortgage Association (FNMA)
(Fannie Mae) **www.fanniemae./com**
Student Loan Marketing Association
(Sallie Mae) **www.salliemae.com/content.html**
New England Education Loan Marketing Corporation
(Nellie Mae) **www.nelliemae.com/**

There are innumerable real estate and mutual fund websites. To find a specific fund, use the company name. For example, Vanguard (a family of mutual funds) is found at **www.vanguard.com** and Janus Funds is at **www.janus.com**. To access current market values of mutual funds, use one of the Internet search engines, such as Yahoo!. *Information sites about mutual funds and real estate* include:
No-load fund information **www.no-load-funds.com/**
Mutual Fund Education Alliance **networth.galt.com**
Chicago Board of Trade **www.cbot.com**
Chicago Mercantile Exchange **www.cme.com**
National Association of Real
Estate Investment Trusts **www.nareit.com**
National Association of Realtors **www.realtor.com**
(For real estate websites see also the Chapter 6 Electronic Resources section.)

CONSUMER ALERT

According to financial columnist Jane Bryant Quinn: Whenever investors gather, you're going to find fraud—and the Internet is an especially hospitable spot. Some 7 million U.S. households are currently using the Web's investment related services...scam artists can now reach tens of millions of people at once. Think of it as increased productivity. By becoming technologically hip, fraudsters can greatly increase their take.[23]

Examples of on-line ratings and analyses of better-known funds are:

Morningstar, Inc.	**www.morningstar.net**
Lipper Analytical Services Inc.	**www.wechsler.com/index.html**
Standard & Poor's Corp.	**www.standardandpoors.coms**

As of the writing of this book, there were 21,500 sites that start with the word "art," according to registers of domain names. Examples of the bigger sites for *art, antiques, and collectibles auctions* are:

eBay Great Collections (art, antiques & high-end collectibles)	**www.ebaygreatcollections.com**
eBay (general site)	**www.ebay.com**
Amazon.com	**www.amazon.com**
eHammer.com (antiques & fine art)	**www.edhammer.com**
Yahoo!Auctions	**auctions.yahoo.com**
icollector (U.K.-based, broad-based collectibles)	**www.icollector.com**

For information on gems and jewelry, go to Gemological Institute of America at **www.gia.edu** or the American Gem Society website at **www.ags.com**.

FAQs

1. How much money is needed to start a mutual fund?

   ■ Most require an initial investment of $1,000. The exception to this are funds that require regular, monthly investments.

2. What is the most common type of mutual fund?

   ■ The most usual is an open-end fund wherein both existing and new investors may add any amount of money that they want to the fund (beyond the minimum requirement).

3. How do mutual funds increase in value?

   ■ The worth of a mutual fund fluctuates in response to the value of the investments in the fund.

4. What is a no-load fund?

   ■ An open-end fund that charges no fee to purchase shares is called a no-load fund.

5. What is a REIT?

   ■ A REIT is a real estate investment trust. REITs provide a way for many investors to invest in commercial real estate.

6. Who invests in REITs?

   ■ Anyone can invest in REITs. The usual investors are individuals, pension or retirement funds, insurance companies, banks, and mutual funds.

7. How many REITs are there?

   ■ As of this writing there are about 300 REITs in the U.S. that are publicly traded, most over the NYSE.

SUMMARY

Mutual funds are a form of investment operated by companies that raise money from shareholders and invest in stocks, bonds, currencies, futures, options, or money market securities. Mutual funds are groups of investments.

Main advantages of mutual funds are diversification, low initial payment, and professional management.

Mutual funds have costs that are passed down to investors. The price you pay to buy or sell a fund, plus the ongoing fund operating expenses and taxation, affect the return on investment.

Specific kinds of mutual funds are socially responsible funds and index funds.

Mutual funds should be selected based on a number of factors including the company's objectives, the investor's objectives (income, growth, and/or conservation of principal), and historic performance.

There are open-ended (90% of the market) and closed-ended (10% of the market) mutual funds. Most open-ended funds are bought and sold from the investment companies that sponsor the funds.

Real estate is a piece of land and all physical property related to it, including houses, fences, and landscaping, and all rights to the air above and the earth below the property.[24]

Real estate investments are direct (homes, commercial property, and raw land) or indirect (real estate syndicates or partnerships, REITs, first and second mortgages, and participation certificates).

The main advantages of real estate include offering a hedge against inflation through escalating property values, tax deductions, ease of entry, and financial leverage.

The main disadvantages of real estate include declining property values and depreciation; risk of loss; property taxes; damages, repair, insurance, and other costs; illiquidity; lack of diversification; opportunity costs; and management problems (if a landlord).

When there is turmoil in global financial markets, many investors turn to gold.

Diamonds are rated, but other gems are not. Most diamonds in the U.S. now come from Africa. Gems are portable wealth.

Collectibles have the advantages of attraction and satisfaction, but these may be offset by the disadvantages in fluctuations in value, lack of current income provision, and lack of liquidity.

Potential investors should always be careful when purchasing over the Web, but especially cautious when it comes to art, antiques, and collectibles.

CONSUMER ALERT

The potential for misrepresentation exists whenever buying over the Web, but it is extremely likely when it comes to art, antiques, and collectibles. Some auction services screen sellers and merchandise; others do not. Shoddy authentication is another problem. Plus there is a practice called shilling, in which sellers use a separate email account to make offers on their own items, initiating a bidding war between legitimate buyers and a phantom. State consumer affairs departments and other governmental units are stepping in to protect consumers, but this is such a vast, changing market, the best advice is caveat emptor, "may the buyer beware." For further information check AuctionWatch.com, a Web site for consumers.

REVIEW QUESTIONS

(See Appendix A for answers to questions 1–10).

Identify the following statements as true or false

1. An advantage of mutual funds is that they are diversified.

2. No-load mutual funds involve no charges of any kind (hidden or otherwise).

3. A prospectus describes how a company did last year. It is sent to all shareholders.

4. Commercial property includes homes and time shares.

5. In the long run, collectibles outperform stocks.

Pick the best answers to the following five questions.

6. ______________ is a practice in which sellers bid on the items they are selling to get the bidding up from legitimate buyers.

a. Churning
b. Shilling
c. Narrowing
d. Double kicking

7. ______________ are a form of investing in a group of stocks, bonds, currencies, futures, options, or money market securities.

a. Limited partnerships
b. Ginnie Maes
c. Certificates of deposit
d. Mutual funds

8. About ______________ of mutual funds are open-ended.

a. 10 percent
b. 30 percent
c. 70 percent
d. 90 percent

9. Which of the following is not an indirect real estate investment?

a. Real Estate Investment Trusts (REITs)
b. Real estate syndicates or real estate limited partnerships
c. First and second mortgages
d. Single-family homes

10. Today most diamonds come from

a. South America.
b. Europe.
c. Africa.
d. India.

DISCUSSION QUESTIONS

1. In a movie song sung by Marilyn Monroe there is the line "diamonds are a girl's best friend." Based on what you have learned in this chapter and the previous chapters on investments, is this true? If not, what is (financially speaking)?
2. What fees and charges are involved in investing in a mutual fund? How does an investor buy or sell a mutual fund?
3. What is the difference between a REIT and a real estate syndicate (or limited partnership)?
4. Is it possible to lose not only your investment in real estate but also more than your investment? If so, explain under what circumstances this could occur.
5. Why should someone be suspicious of a magazine's list of the top ten best mutual funds? How can one determine the potential worth of a fund?

DECISION-MAKING CASES

1. Sean, age 30, has the opportunity to buy the duplex he is living in and rent out the other half. Right now he is paying $600 a month for his half of the duplex and the other side rents out for the same amount. So the current cash flow for the duplex, is $1200. He has the downpayment, and his monthly mortgage payment would be $950, so it seems to be a poten-

tial money-maker, but what other expenses should he find out about before buying the duplex?

2. Devon, age 27, went to a two-day stay at a time share resort for $99. Included in the deal was a sales pitch to buy a unit. A one bedroom furnished unit that would cost him $10,000 upfront would allow him a one- to-two-week stay (depending on the time of year) for $350 a year in ownership/maintenance fees. He liked the resort, but $10,000 would take all his savings and he is wondering if it is a good investment. What would you tell him to look into before signing up?
3. Montana has never invested before. In her first job as a store buyer she is making $30,000 a year. The store chain she works for does not offer stock options so she is on her own to make investments. She has set aside $1,000 to invest in a mutual fund. What should she consider in making her selection? In addition her roommate, Rene, has not saved up $1000, but she is in a position with her new sales job to put $50 a month into a mutual fund. Would any mutual fund investment companies sign Rene up as an investor? If so, under what conditions?
4. Josh and Marie own their own home. Now they are thinking about buying a second home at a lakefront resort that they would use part of the time and rent out during the high season for three months in the summer. What are the tax advantages and disadvantages to their plan?
5. Stefan saw an advertisement on television for a newly minted limited edition coin whose face value is $1.00, yet it is being offered to the public at $30. Why is the price so high? Can he be sure that he can sell it for more than $30 in five years?

GET ON LINE

Two activities to try:

- Go to one of the auction websites and see what they have to offer. Select an object to buy and find the price and condition of the object. See if you can find the same object on two or more web sites and compare prices and quality.
- Go to Morningstar at www.morningstar.net and click on the research tab. Next find the fund selector option. Find the three top-performing stock funds and research each. Which one would you select and why?

INFOTRAC

To access InfoTrac follow the instructions at the end of Chapter 1.

Exercise 1: This chapter covered a number of subjects including investing in mutual funds, real estate, precious metals, gems, and collectibles. Type in the subject word "real estate." Click on the topic of "real estate investment." There you will find articles and subdivisions. Select a subject within real estate investment and analyze if this is a good time to buy real estate. If so, what kind of real estate is recommended?

Exercise 2: One area of investing that is fun and varied is collectibles. Type in the subject word "collectibles" and you will see subdivisions such as film collectibles, political collectibles, sports collectibles and so on. Select a collectible category that interests you and find an article that reports on this type of collectible. Report on what you find.

part four

planning your financial future

chapter fifteen
retirement planning

Making Money Last for Life

DID YOU KNOW?

* The average American spends over 60,000 hours working, but only 10 hours planning for retirement.
* Traditional pension plans cover less than half of the people working today.

OBJECTIVES

After reading Chapter 15 you should be able to do the following:

1. State the reasons why retirement planning is important and explain common pitfalls to sound retirement planning.
2. Estimate retirement needs and develop a plan.

CHAPTER OVERVIEW

A book on personal finance would be incomplete without retirement planning because achieving a comfortable retirement is a common financial goal. **Retirement** is defined as the permanent withdrawal of an employee from gainful employment. It should be noted, however, that some people never fully retire, by choosing to work part-time or by owning their own business.

The principles that guide retirement planning are based on many of the key strategies already discussed in the book, such as being forward looking, considering the impact of inflation, career planning, and benefits. The investment chapters are particularly pertinent because the investment stakes go up when you are not earning money anymore. If you are just starting out, you will need to save a substantial amount of money in order to retire comfortably. So this chapter views retirement planning as a lifelong process that benefits from accumulated wealth begun in early working years.

Retirement
Permanent withdrawal from gainful employment.

Usually, during the first week of employment, you will visit your employer's human resource department to discuss various benefits, including retirement plan options. This chapter will give you the steps to follow to make those initial decisions and to initiate an investment plan that will supplement employer-sponsored programs. As in previous chapters, there is an Electronic Resources section before the Summary.

Objective 1:
State the reasons why retirement planning is important and explain common pitfalls to sound retirement planning.

IMPORTANCE OF RETIREMENT PLANNING

Even though the average American worker spends over 60,000 hours working, the same person spends only 10 hours planning for retirement.[1] People are so busy trying to make a living that they rarely slow down enough to make plans for the future, but it is particularly important to plan for retirement because of:

1. *greater life expectancy.* Retirement savings have to stretch further, possibly 30 years or more. Today's 65-year-old male has a 25 percent chance of reaching 90, and a 65-year-old female has a 40 percent chance of reaching 90.

2. *the gender gap.* Even though women live longer, they are less likely to have pensions than men. According to the U.S. Department of Labor, only 39 percent of female workers are covered by private pension plans, compared with 46 percent of men; and 32 percent of female retirees over 55 get pension benefits, compared with 55 percent of male retirees.[2]

3. *early retirement.* People retire early for a variety of reasons, including being tired of working or being fed up with their jobs, poor health (their own or their spouse's), family problems, wanting to travel and enjoy life more, or being the beneficiary of an inheritance that frees them. Or an employer may offer early retirement incentives. *Although there is no longer a mandatory retirement age, many corporations consider retirement age around age 65, and the federal government has a retirement age of 70.* You can begin withdrawing money from most retirement plans at age 59 1/2 or 60.

4. *Social Security and pensions not stretching far enough.* Most people supplement their Social Security and company-sponsored retirement programs or pensions with their own savings and Individual Retirement Accounts (IRAs). According to the Social Security Administration, retired persons who during their full time employment years earned over $31,000 a year typically in retirement receive income from several sources such as investments (see Exhibit 15.1). In the past, partial Social Security benefits started at age 62 and full benefits at age 65. Those turning 65 before 2000 received full benefits at this age, but the age moves to 67 by 2027 on a sliding scale. In the past Social Security recipients who worked between the ages of 65 and 69 were penalized for earning extra income. New legislation allows employees to continue working past age 65 and receive their full Social Security benefit.

5. *uncertainty.* No one knows for sure how long he or she will spend in retirement, although the *average number of years is 18.*[3]

Important factors include health status and how soon a person retires. Uncertainties include changes in Medicare or Social Security and fluctuations in interest and inflation rates. Currently, Medicare begins the first day of the month a person turns 65. Retirement planning should be flexible enough to adjust to uncertainties. The rising costs of prescriptions and health care are other concerns. One year in assisted living or in a nursing home can cost over $40,000.

EXHIBIT 15.1
Retirement Income Sources

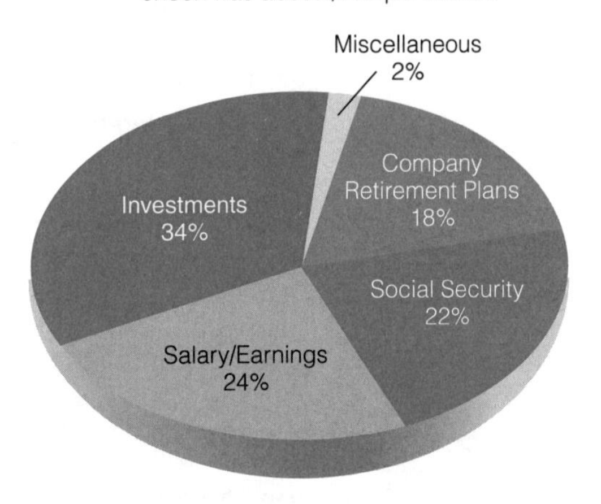
SOCIAL SECURITY PROVIDES A FRACTION OF
THE ASSETS YOU WILL NEED IN RETIREMENT
Social Security accounts for just 22% of the total income for
retired people with incomes of $31,000 or more.
In 1999, the average Social Security retirement
check was about $780 per month.
Miscellaneous
2%
Company
Retirement Plans
18%
Investments
34%
Social Security
22%
Salary/Earnings
24%

Source: Social Security Administration

*Note: Lower income people may mostly rely on Social Security

6. *anticipated retirement lifestyle needs.* The overriding goal already stated is that most people would like a comfortable retirement, but individuals have to define for themselves what that means. Retirement planning requires a vision, an idea of what retirement should include— perhaps travel; golf; fishing; a cabin in the mountains; a boat or airplane; or the time and money to pursue hobbies, gardening, home and family activities, or a small business.

Common Pitfalls to Sound Retirement Planning

Arriving at retirement without a plan can be disastrous, but many people do because:

- they start too late.
- they save or invest too little (perhaps by being too conservative in their investment choices).
- they switch jobs too often and do not build up a pension or employee-sponsored retirement account. Or, if they have worked part-time or in temp jobs, they may not have had a pension offered to them.
- they underestimate how much they will need.
- they do not figure in such factors as inflation or rising property taxes. Retirement portfolios can be "weatherproofed" by using the principles of diversification and asset allocation (increasing potential return and decreasing risk).

Individuals can feel positive about their world in retirement if they create a sound plan, implement it, and stick to it. This is easier said than done because many people find it hard to put money away for retirement. If they are in their twenties or thirties, retirement may seem too far away, and more immediate needs take precedence.

Another explanation for the lack of retirement planning is that it is linked to personality traits. Some people do not want to think about aging or retirement. Margaret Mead, the famous anthropologist, said, "Sooner or later I'm going to die, but I'm not going to retire." A study by Public Agenda, a nonprofit research foundation, found that the general population fits into the following types:

* *Planners,* who are twice as likely as others to have learned about the importance of saving in childhood.
* *Strugglers,* who believe every time they get ahead, something happens to set them back.
* *Deniers,* who live for today and figure the future will take care of itself.
* *Impulsives,* who like to shop and spend; long-distance planning is difficult for them.[4]

Which of these four types or what combination of types describes you best or members of your family?

Retirement Savings and Level of Education

Studies show a link between level of education and retirement savings. Since the higher the education level, the greater the income, it follows that savings would be greater too. For example, it has been found that 31 percent of college graduates have saved more than $50,000 toward retirement compared to only 11 percent of high school graduates.[5] Further, 37 percent of college graduates save at least 10 percent of their incomes for retirement, compared to 22 percent of high school graduates.[6] In order to put these percentages into perspective, nearly half of all Americans set aside less than $10,000 for retirement.[7]

Objective 2:
Estimate retirement needs and develop a plan.

ESTIMATING RETIREMENT NEEDS AND DEVELOPING A PLAN

Estimating how much is needed for retirement is part finance and part educated guessing. The logical way to begin is to figure out your current financial status and how much will be needed later. This can be done by following three steps.

Step 1: Estimating Net Worth (Assets - Liabilities)

To estimate current financial status, a net worth statement is needed. This is done by subtracting liabilities (what is owed) from assets (what is owned), as covered earlier in the book. An estimate of net worth provides a starting base.

Step 2: Estimate How Much Money Will Be Needed at Retirement

In this step individuals estimate how much money they will need for retirement, given their goals, inflation, and other factors. Two ways to make beginning estimates are as follows.

1. The simplest way is provided by the U.S. Department of Labor's estimate that individuals will need 70–90 percent of preretirement income to maintain their standard of living at retirement. A range is given because the lower percentage is suggested as workable for high income earners, and the higher percentage is suggested for low income earners. For example, if a person makes $50,000 a year, that person will need at least $35,000 a year (based on 70 percent of preretirement

income) during retirement. To estimate how much of a nest egg (assets put aside for retirement) will be needed, multiply $35,000 by the number of years until retirement. If a person has 18 years to go until retirement, then the nest egg would be figured as $35,000 x 18 = $630,000. Rather than providing a range of 70–90 percent, some financial planners have set 80 percent as a guide.

2. The second, a method of estimating how much to save each year to reach a nest egg, is shown in Exhibit 15.2. In this exhibit a case example of Barbara, age 50, is given.

Retirement Planning Advice When people consult their insurance agents, financial planners, accountants, and account executives, they will find there are a number of ways to estimate retirement needs that include many more factors, such as possible inheritances and a person's investment in mutual funds, stocks, bonds, and property. They will also find that these professionals have their own software that they use to

Retirement calculators (estimating formulas) can be simple or complex. The more complicated ones will ask for an estimate of the value of 401(k) plans and Social Security. Another approach is to estimate how much should be saved beyond these. For example, Barbara's goal is to have $100,000 for retirement. You can find the calculator that she used and insert your age, expected retirement age, goal, and suggested yearly contribution at Bloomberg.com.

Retirement Calculator

With a current fund amount of $15,000
and a desired total fund amount of $100,000
earning 6% interest, you will need to contribute $2,751.83
each year, in order to reach your goal by a retirement age of 65.

Use the calculator below to help you plan for your retirement.
Enter the fields below and select the type of calculation to perform.
(Below the calculator is a table of accumulated fund balances.)

Retirement Fund Calculator

Your Current Age: 50

Expected Retirement Age: 65

Current Amount in Fund: $ 15000

Expected Annual Rate of Return: 6 %

○ Enter annual contribution to calculate fund balance at retirement
● Enter desired fund amount to calculate required annual contribution

100000

Calculate | Reset

Fund Values for the next 15 years

Year	Total Annual Contribution	Total Fund Value
1	2,751.83	18,651.83
2	5,503.66	22,522.77
3	8,255.49	26,625.97
4	11,007.32	30,975.35
5	13,759.15	35,585.71
6	16,510.98	40,472.68
7	19,262.81	45,652.87
8	22,014.64	51,143.87
9	24,766.47	56,964.33
10	27,518.30	63,134.02
11	30,270.13	69,673.89
12	33,021.96	76,606.16
13	35,773.79	83,954.36
14	38,525.62	91,743.45
15	41,277.45	99,999.89

EXHIBIT 15.2
Estimating a Retirement Nest Egg
Used with permission of Bloomberg L.P.

calculate retirement planning needs. Because they use different formulas and approaches, estimates by financial advisors may vary, but if they are all given the same data and are provided a sense of the individual's goals and comfort level with risk, one would assume that the final estimates would be near each other. Naturally the closer one is to retirement, the easier it is to provide an accurate estimate of the nest egg needed. Regardless, however, of the age or income of the client, financial advisors should be able to provide answers about how much more needs to be saved and at what age a person might feasibly retire.

Two examples of the results of consulting financial advisors follow:

- Joe and Bonnie Blackstone, 41 and 38, who own two children's clothing stores, make $100,000 a year, and their current assets are $542,500. A financial advisor suggested that they need to save $12,100 a year, and should keep working for at least ten years, to build a secure retirement and a college fund.
- Leah Everly, a single 29-year-old family and consumer sciences teacher, has a current income of $27,900, and her current assets are $3,800. Her advisor suggested that she needed to save $2,500 a year (assuming a 10 percent return on assets) so she could retire at age 55 if she desired. To take full advantage of her early start at savings, it was suggested that she should put the maximum possible into her employer-sponsored tax-deferred savings plan (a 403(b) plan to be covered later in the chapter).

Besides turning to advisors or filling out worksheets, such as the one given in Exhibit 15.2, retirement planning software is available from a number of sources such as Intuit's Quicken Financial Planner and MS Money. There are also websites to help in estimating retirement needs. For interactive websites see the listings in the Electronic Resources section including the "Get Online" activity at the end of the chapter. As one example, a recent search on Yahoo.com alone revealed 933 sites with retirement calculators.

Step 3: Building and Maintaining Retirement Income

After estimating how much money is needed to support retirement years, then the person is ready to decide how to build income for and during retirement. Using the example in the first method described earlier, if a person estimated he or she needed a nest egg of $630,000 to retire on, how would he or she go about accumulating that much money? According to Exhibit 15.1, the main sources of retirement income are earned income, Social Security, savings and investments, and pensions (employer-sponsored retirement plans). Building savings and income for retirement can start as early as the 20s and go on for 40 years or more. Exhibit 15.3 shows that saving an amount as low as $100 a month can make a tremendous difference at retirement. Also, the earlier started, the better as shown in Exhibit 15.4.

Since it makes so much financial sense to set money aside for retirement during the early working years, why do so few people do it? As mentioned earlier, more immediate needs come first, such as buying a house or meeting the costs involved in going to school or raising a family. If a secure retirement is a goal, then individuals would be wise to meet immediate needs while saving and investing as much as they can for the future.

Individual Retirement Accounts and Homes as Assets

So far the chapter has emphasized the need to build a nest egg. It used to be that for many people their single largest asset was their house, and it still is for some. But more

EXHIBIT 15.3
The Power of Regular Saving

SAVING $100 PER MONTH

Rate of Return	YEARS 5	10	15	20	25	30
5%	6,809	15,499	26,590	40,746	58,812	81,870
6%	6,982	16,325	28,831	45,565	67,958	97,926
7%	7,160	17,202	31,286	51,041	78,747	117,606
8%	7,341	18,128	33,976	57,266	91,484	141,761
9%	7,527	19,109	36,928	64,346	105,531	171,438
10%	7,717	20,146	40,162	72,399	124,316	207,929

This example assumes no withdrawals

recently the trend is toward individual retirement accounts (IRAs) being the single largest asset.

The problem with homes is that even though they are a valuable asset, they also tie up wealth. One way to gain money at retirement is to downsize by selling a larger home and moving to a less expensive, smaller home. The money from the sale of the larger home could be put into a safe income-producing investment. Another strategy is selling the larger home and using the money gained to pay cash for a smaller home, thus eliminating monthly mortgage payments, and investing the rest. However, during retirement people usually spend more time in their homes, and so the function and appearance of the home is important. Some retirees have downsized too much and found that they are not happy in a smaller home.

Another option is a reverse annuity mortgage (RAM) or equity conversion introduced in Chapter 6. In RAMs an elderly person lives off the equity in a fully paid-for house. Such a homeowner enters into a reverse annuity mortgage agreement with a financial institution, such as a bank (the lender), which guarantees a lifelong fixed monthly income in return for gradually giving up ownership of the house. The longer the payments continue, the less equity the elderly person would retain. At the owner's death the lender gains title to the property.[8] Sons or daughters can also enter into a RAM with their parents. This provides a home for the elderly parents and some cash

CONSUMER ALERT

Although reverse annuity mortgages (RAMs) seem straightforward enough, there have been lawsuits against them. In one lawsuit it was reported that in less than four years, some homeowners wound up owing two to six times the amount of money they received. The consumers had contracts that spelled out the terms of agreement, but lawyers contended that many of the homeowners were in their 80s and 90s and did not fully understand the contracts they signed. To guard against this situation, contracts should be carefully read and questions about reverse annuity mortgages should be addressed to lawyers, the American Association of Retired Persons (AARP), and the Department of Housing and Urban Development (HUD).

EXHIBIT 15.4
The Benefit of Starting Early

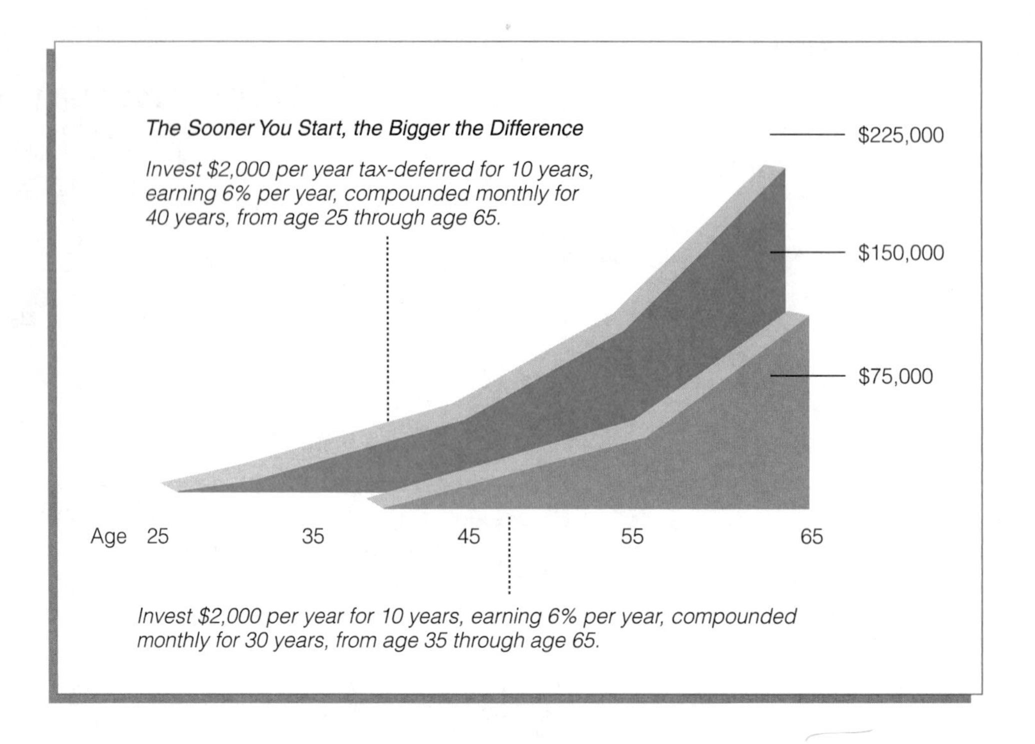

to invest or live off of, and the son or daughter receives tax benefits, such as depreciation from real estate ownership.

To summarize, reverse annuity mortgages work as follows:

* A lender makes monthly payments to the homeowner. The home's equity serves as collateral.
* The equity is used to repay the loan, including interest, when the homeowner sells the house, moves out, or dies.
* The homeowner purchases an annuity to receive payments after the home's equity has been depleted.

Whole Life Insurance as an Asset

A retired person with a whole life insurance policy may consider converting some of this asset into cash or an income-producing annuity. The face value of the policy could also be decreased so payments would be lower. Some financial advisors would recommend that at retirement, life insurance (primarily a protection to loved ones) is no longer needed if children are grown and independent. If life insurance is eliminated, this would be one less yearly payment to be made. At retirement a lot of people want to simplify their lives, and reducing bills is one way to do it.

Net Worth/Assets Summary

Selling a large home, taking IRA distributions (money from accumulated individual retirement accounts), or reducing life insurance are options to consider within a general strategy of reducing expenses and increasing income during the retirement years. All other investments should be reevaluated as one nears retirement. Some financial advisors recommend thorough retirement "check-ups" at age 55, 60, and 65 with yearly reviews as needed. Life situations should also be considered such as changes in marital or family status (the birth of grandchildren) and health.

CONSUMER ALERT

Traditional retirement plans are based on the assumption of a long conventional marriage with a husband, wife, and children. The reality is that nearly half of American marriages end in divorce; couples live together without marriage; there are single parents, blended families, lifelong singles without children, strained relations with children and other relatives, and other life/relationship situations. Because of this retirees should make sure that their plans fit their needs.

Reduced Expenses During Retirement

Expenses change during retirement, but generally the following expenses will be lower:

- *Work-related expenses,* such as transportation to and from work, work clothes, lunches out, and membership in associations or unions.
- *Housing.* Housing may be paid off or the cost of housing may decrease or even return income with a RAM. Different forms of retirement housing other than single-family homes and condominiums—such as congregate, nursing home, and assisted living—were described in Chapter 7.
- *Federal and state income taxes.* Since federal and state income taxes are progressive taxes, it follows that as income declines taxes should drop. Retirees often have the pleasant surprise of getting a tax refund from the government after many years of having had to pay extra on April 15. Also some retirees move to states with no state income tax to reduce their taxes.
- *Transportation.* Besides saving money from less commuting, savings may also be gained from reducing from two cars to one or relying on public transportation.

Increased Expenses During Retirement

Whenever there is a lifestyle change, some expenses go up while others go down. Retirement is no exception. The following may go up.

- *Leisure-related expenses* for travel or fees for health, golf, or tennis club memberships, or for buying a recreational vehicle.
- *Health care, prescriptions, and medical expenses.*
- *Insurance.* This is an "it depends" category because spending can go up, especially if a person takes early retirement and his or her company does not provide health insurance to fill in the gap before Medicare sets in, or it could go down by eliminating life insurance. A potential added expense may be the purchase of long-term care insurance.

CONSUMER ALERT

Retirees should be careful before signing up for retirement housing or moving to a new location. It may be wise to rent before making a long-term commitment. One suggestion is for individuals not to do anything drastic the first six months after retirement. This time period will allow them to adjust to their new financial state and figure out what they want to do next.

Sources of Retirement Income

Social Security One of the ways to accumulate funds for retirement is to work and pay into Social Security. *People born after 1928 have to work 40 quarters (10 years) to qualify for benefits (or be married to someone who worked 10 years or more). The most widely used source of retirement income is Social Security which also provides survivors' and disability income.* Out of every dollar a worker pays into Social Security, nearly 70 cents goes to a trust fund that pays monthly benefits to retirees and to the children and spouses of workers who died; 19 cents goes to Medicare, 11 cents goes to a trust fund that pays benefits for people with disabilities; and less than one cent goes to administrative costs. Further details about Social Security were given in Chapter 7, but the following provides information pertinent to this chapter:

- Three months before retirement a person should apply for Social Security benefits by calling the Social Security Administration (SSA) at 800-772-1213 or by visiting an SSA office (there are 1300 across the nation). The SSA will want proof of age, a Social Security number, W-2 withholding forms for the past two years, a marriage license if applying for a spouse's benefits, and birth certificates of children if their benefits are involved.

- As discussed earlier, most people qualify for a reduced benefit from Social Security starting at age 62. The full benefit traditionally began at age 65, but the age at which full benefits are paid will increase in graduated steps from 65 to 67 (by 2027). For example, a person born in 1950 will receive full retirement at age 66. Someone born in 1960 or after will be eligible for full retirement benefits at 67.

- Up to 85 percent of Social Security benefits may be taxed by the federal government. To check on taxation and Social Security benefits, call the Internal Revenue Service at 800-829-3676 and ask for publications *Tax Benefits for Older Americans (#554) and Tax Information on Social Security* (#915).

- Each January benefits are increased based on the rise in the cost of living during the previous year.

- Spouses who do not work are eligible for benefits based on what their working spouses earned under certain conditions (length of marriage, how much was paid in, age at which working spouse retired, etc.). The SSA should be contacted for specifics. If you are eligible both as a spouse and as a working person who paid into the system, then the SSA pays your personally earned benefit first. But if your benefit as a spouse is higher, then you will get a combination of benefits equal to the higher-paid spouse.

- Confidence about the future of the Social Security system has been dropping because of the rise in the number of potential recipients and the decline in workers paying into the system.[10] According to the Social Security Administration (SSA), the system is financially sound, and they are closely monitoring what is paid in and what flows out. Moving the full retirement benefit age up is one way they are coping with the rising number of baby boomers getting ready to access funds. Even with reassurances from the SSA the public, politicians, the Executive branch of the federal government, and Congress continue to be concerned about the future of the system. New legislation, regulations, and options (such as taking federal government revenue and using it to bolster Social Security rather than spending it on other government services) are being considered including letting workers have more say about how the money they paid in is invested. Will Social Security be there for you? It depends on the changes that will be made.

Public Pension Plans Besides Social Security the federal government offers other types of retirement plans. These include pensions for federal employees and from the Veterans Administration and the Railroad Retirement System. Other pension plans are available for state, county, and city employees.

Employer-Sponsored Retirement Plans Including 401(k)s Employer-sponsored retirement plans—also known as pension plans—vary widely. Usually an employer contributes to an employee's retirement benefits, and the employee may contribute also. These contributions are tax-free until retirement. One of the best ways to build a retirement nest egg is to participate in these employer-sponsored plans through automatic contributions.

Employer plans are classified as defined-contribution plans or defined-benefit plans. A **defined-contribution pension plan** (also called an **individual account plan**) is a plan in which the amount of contributions is fixed at a certain level, while benefits vary, depending on the return from investments.[11] These are popular because they limit a company's pension outlay (such as 2 percent to 10 percent of an employee's annual salary) and shift the responsibility of choosing investment performance from the company to the employees. In other words defined-contribution plans give employees options of where to invest their accounts, such as in stock, bond, and money market accounts. Examples are 401(k), **403(b) plans** (offered by tax-exempt institutions such as universities, schools, social service agencies, or hospitals), and **457 plans** in which employees make voluntary contributions into a tax-deferred account, which may or may not be matched by an employer. The numbers 401(k), 403(b), and 457 come from the sections of the federal tax law governing the plans.

Defined-contribution pension plan or individual account plan
Pension plan in which the amount of contributions is fixed at a certain level, while benefits vary depending on return on investments.

403(b) plan
Retirement plan offered by non-profit organization employers.

457 plan
Retirement plan in which employees make voluntary contributions into a tax-deferred account which may or may not be matched by employers.

401(k) plan
Retirement plan offered by for-profit company employers.

The best known plan is the **401(k) plan** in which an employer makes nontaxable contributions to the plan and reduces the employee's salary for the same amount. Sometimes these are referred to as salary-reduction plans or 401(k)TSA plans. TSA stands for tax-sheltered annuity. Obviously the higher the employer match, the better for the employee, so this is something to consider when comparing company benefits.

Financial planners suggest that employees should make the maximum contribution possible so they can take advantage of matching funds to build retirement income and to reduce tax burden. Employees with 401(k) plans can tuck away a percentage of their pretax salary, commonly as much as 15 percent, to an annual maximum of $10,000. Many companies match a portion of employee contributions, often to 5 or 6 percent. Employees should direct questions about how much can be contributed and how much is matched to accountants and the benefits expert at the workplace.

Subgroups under defined-contribution plans are:

1. *Stock bonus plans,* in which the employer's contribution is used to buy stock in the company of employment. The stock is held in trust until retirement, at which time the employee can receive the shares accumulated or sell them.

2. *Profit-sharing plans,* in which an employer's contribution depends on the company's profits during a certain time period.

The Tax Reform Act of 1986 put a cap on how much can be put into these plans yearly. Since then updates on the Act have altered these amounts and options. Since tax laws change, specific questions regarding taxes should be addressed to accountants and benefits experts.

Defined benefit pension plans promise to pay a specified amount to each person who retires after a set number of years of service.[12] Employers can contribute totally or partially to these plans. These are becoming less common, as the defined-contribution pension plans, such as 401(k) plans, have grown.

Defined benefit pension plan
Pension plan that pays a specified amount to each employee who retires after a set number of years of service.

Popularity of 401(k)s

According to the U.S. Department of Labor, over 32 million Americans are now eligible for 401(k)s. That figure represents nearly three times as many as ten years ago. At the same time the number of people covered by traditional pensions, where the employer is in charge of investing and payout decisions, remains static at about 40 million.[13]

Fees and 401(k) Plans The trend toward 401(k)s has brought with it more responsibility for individual decision making. Besides deciding how retirement funds are to be invested, people also have to realize that the costs of 401(k) plans are being shifted to workers, reducing their retirement accounts. Knowing fees is important because

> For instance, say you salt away $3,000 a year at a 10 percent annual return. If you pay 1 percent annually in fees, your nest egg would grow to $409,000 over 30 years. Pay 2 percent in fees, and the balance shrinks by $69,000. "It's like having an account with a hole in it," said Ted Benna, president of the 401(k) Association in Pennsylvania.[14]

A 10 percent annual return cannot be counted on, so Benna is being optimistic in that regard, but his point about the substantial effect of fees on the eventual worth of a nest egg is important. There are three types of fees:

1. *Investment management.* The bulk of the fees go to pay a professional to manage the money in the plan. Fees in this category can range from 0.05 to 3 percent.
2. *Trustee.* These fees are small and often picked up by the employer. They cover the duties of a trustee, usually a bank, who holds the plan's securities and pays out benefits and expenses.
3. *Administrative costs.* Fees for this cover record keeping, rule compliance, and communication. Most companies pay for administrative costs, but there is a trend toward shifting these fees to employees.

Changing Jobs

When employees change jobs, there are several options regarding retirement funds.

* They may *cash out, meaning they take the money they have built up in the fund. This is the most popular option.* In 1996 *six in ten job-changers cashed out their retirement savings* instead of rolling it over into an Individual Retirement Account (IRA) or their new employer's plan.[15] Although it is the most popular option, financial experts would say, "This is your retirement money; do not spend it." If there is any way you can afford to, choose one of the following options.
* *Roll it over into an IRA or over to the new employer's plan.*
* *Leave some or all of it behind* with your old employer and later move it to a new employer's plan or roll it into an IRA.

Vesting

Vesting
Right an employee acquires after being employed for a set number of years.

One of the most important considerations is a right called vesting. Not all employers offer vesting, and those that do may offer it as an option rather than a requirement.

Under **vesting** an employee acquires, by length of service at a place of employment, the right to receive employer-contributed benefits, such as a pension or profit-sharing plan. Accrued funds under an employer pension plan remain ready for employees to access at retirement even if they leave the company before retirement. For example, if a teacher in a vested school system left after five years of employment, he or she would have five years vested in the school system's retirement fund. Before changing jobs an employee should know what the options are and plan accordingly.

Personal Retirement Plans

Individuals who do not want to rely solely on Social Security and employer-sponsored plans—such as 401(k) plans—for their retirement funds, or who are not employed by companies with sponsored plans, may set up individual retirement accounts (IRAs, including Roth IRAs, SIMPLE IRAs, and Education IRAs), simplified employee pension (SEP) plans, or Keogh accounts.

Individual retirement accounts (IRAs)
Tax-deferred retirement accounts, not necessarily tied to an employer.

Individual Retirement Accounts (IRAs). **Individual Retirement Accounts (IRAs)** are tax-deferred retirement accounts not necessarily tied to an employer. Employed people can set these up with a deposit up to a maximum of $2,000 a year, or $4,000 for a couple who both work, or $2250 for a couple when one works and the other's income is $250 or less. IRAs can be set up at banks, brokerage firms, and other institutions that offer them. In many ways IRAs offer the most straightforward approach to retirement savings. As stated earlier, IRAs are the largest single asset owned by many Americans. This is due to a booming stock market and rollovers from 401(k) retirement plans.

Individuals are eligible to have IRAs if they are under age 70 1/2 and have compensation (work for pay) even if they already participate in other government plans, tax-sheltered annuities, or simplified employee pension (SEP) plans (to be discussed later in the chapter). If an employee works part-time and does not earn $2,000 a year, he or she can contribute up to 100 percent of the compensation.

Earnings on IRA contributions are tax deferred. Withdrawals from IRAs prior to age 59 1/2 are subject to 10 percent (of principal) penalty tax.

A holder of an IRA may be eligible for a maximum tax deduction, a partial, or no deduction. The basic rules for determining IRA deductibility are as follows:

- Singles who are not active participants in employer-maintained retirement plans are eligible for a full $2,000 deduction, no matter how large their income.
- If employees and spouses are active participants or if singles are active participants they may be eligible for either a full or partial deduction, depending on modified adjusted gross income (MAGI).
- If a person is not an active participant, but the spouse is, and the couple files a joint federal income tax return, then they are eligible for a full $2,000 deduction if their joint MAGI is less than $150,000. A partial deduction is possible if the joint MAGI is between $150,000 and $160,000.
- Heirs can be named as beneficiaries of individual retirement accounts. It is critical that the bank or whoever holds the IRA has a beneficiary form. The person whose IRA it is should have a copy, and family members should know where the form can be found.

Roth IRAs Traditional IRAs used to be the only IRA choice until the Roth IRAs, the Education IRAs, and the SIMPLE IRAs came along. The best- known of these are the traditional IRAs and new Roth IRAs. Exhibit 15.5 gives a comparison of these two.

Roth IRA
Type of individual retirement account started in 1998 that differs from a traditional IRA.

The **Roth IRA**, a type of individual retirement account started in 1998, was named after Senate Finance Committee Chairman William V. Roth, Jr. In Roth

IRAs contributions are made only with aftertax dollars. In addition withdrawals are tax free if

* made five taxable years from the later of (a) establishing the Roth IRA or (b) the most recent rollover/conversion that relates to a non-Roth IRA, or
* the withdrawal is after age 59 1/2, on account of death or disability, or used for a first-time home purchase up to $10,000.

For example, if a 50-year-old set up a Roth IRA in 1998 and contributed $2,000 a year until age 60, that individual could make a tax-free withdrawal because the IRA had been held for five years and the person is over 59 1/2.

There are two basic requirements for participation: (1) you must have earned income (or your spouse must have earned income) and (2) your combined Adjusted Gross Income can not exceed $160,000. **Adjusted Gross Income** is the income a taxpayer earns from salary, self-employment, interest and dividends, capital gains, pensions and other sources—less any adjustments. A usual adjustment could be an IRA contribution.

Adjusted gross income
Income a taxpayer earns from salary and other forms of income less any adjustments.

An individual can contribute to both a traditional IRA and a Roth IRA in the same year, but the total contribution to both can not exceed $2,000. Contributions can continue even after age 70 1/2.

Roth IRA Conversion When the Roth IRA came into existence financial experts did not foresee that investors would convert back and forth between a Traditional IRA and a Roth IRA as a way to lower taxes. Although this was saving investors some money, it was an administrative nightmare for IRA custodians, such as mutual fund companies and banks.[17] Effective November 1, 1998, the number of times investors could reconvert to a Roth IRA was restricted to one time in each of 1998 and 1999 for the maneuver to be counted for tax purposes.

Education IRA
Nondeductible account that has tax-free withdrawals for children's higher education expenses.

Education IRAs Another new IRA begun in 1998 is the **Education IRA,** a nondeductible account that features tax-free withdrawals for a child's higher-education expenses. In an Education IRA a child is named as the beneficiary. Contributions are limited to $500 per year. There are two eligibility requirements:

1. The child (defined as under the age of 18) may not have had any contributions made on his or her behalf to a state prepaid-tuition program in that year, and
2. Adjusted Gross Income cannot exceed certain limits. Since this limit is subject to change, contact a financial planner, tax advisor, university financial aid office, or bank for the latest limits.

An individual can contribute up to $500 in an Education IRA and also contribute up to $2,000 in a Roth or a traditional IRA per year. If there are several children in a family, each child can have an Education IRA set up in his or her behalf. There are

CONSUMER ALERT

Bum advice and basic mistakes by banks, brokers and mutual funds are wreaking havoc with many people's individual retirement accounts. The result: Their heirs may end up losing a bundle. Tama McAleese, a Cleveland-area financial adviser, tells of a client who stands to lose, "hundreds and hundreds of thousands of dollars" on an IRA she inherited from her mother. The bank, it seems, can't find the beneficiary form. And she is not alone.[16]

	Traditional IRAs	New Roth IRAs
Who is eligible to contribute?	Anyone with earned income (including earnings from self-employment)	Couples filing jointly whose adjusted gross income is less than $160,000 and singles whose adjusted gross income is less than $110,000[1]
How much can be contributed and for how long?	Up to $2,000[2] per person annually up to age 70 1/2	Up to $2,000[2] per person annually for life
Are contributions tax deductible?	Yes. For those who are not actively participating in an employer-sponsored retirement plan, contributions are fully deductible. However, contributions are only partially deductible for active participants with adjusted gross income between $30,000 and $40,000 (single) or $50,000 and $60,000 (married) (1998 figures)	No
Does it offer tax-free distribution?	No	Yes, if conditions are met
Does it provide tax-deferred growth?	Yes	Yes
Lifetime minimum distribution requirements?	Yes, at age 70 1/2	No

1. Singles with AGI between $95,000 and $110,000 and married couples (filing jointly) with AGI between $150,000 and $160,000 may contribute up to $2,000 on a sliding scale.
2. Cumulative for both traditional and Roth IRA contributions.

EXHIBIT 15.5 Comparing Two IRAs' Features and Advantages

no taxes on the earnings to be paid by the parent or the child. When the child is ready to take his or her withdrawal for school, there are no taxes due on any interest that the money has earned.

An eligible higher-education institution is a vocational school, college, or university. If a child decides not to go on to higher education, the account can be transferred to another member of the same family.

SIMPLE IRA
Individual retirement account that allows workers—mostly of small businesses—to make contributions to with an employer match.

SIMPLE IRAs A program begun in 1997 allows for the Savings Incentive Match Plan for Employees nicknamed the **SIMPLE IRA**. As its name implies, the SIMPLE IRA has fewer reporting and paperwork rules than traditional 401(k) plans making it attractive for small businesses. This plan lets workers contribute up to $6,000 a year through salary deductions. A business has to put in a match of about three percent of participants' eligible pay.

Simplified employee pension plans (SEPs)
Pension plans in which the employee and the employer contribute to an IRA, can be used by self-employed individuals.

Simplified Employee Pension Plans (SEPs). **Simplified Employee Pension Plans (SEPs)** are pension plans in which both the employee and the employer contribute to an IRA. *SEPS are ideal for self-employed individuals and owners of companies with fewer than 25 employees.* An account executive (stockbroker) or financial planner can help someone set up a SEP. Unlike some employer-sponsored plans, SEPs have no annual contribution requirements. This allows for varied annual funding based on how well the business or other outside employment is doing. For example, an artist, writer, or craftsperson working out of the home may earn over $20,000 one year and less than $5,000 the next. The SEP allows for those individuals to contribute the amount that they want up to a maximum limit. Special provisions allow the coverage of self-employed persons, so many people who work from their home as a main business or as a side business have SEPs. SEPs are the simplest form of retirement plan available for those who are fully or partially self-employed.

If you are a self-employed individual with a SEP, your money can be invested in savings accounts, certificates of deposit, mutual funds, stocks, bonds, annuities, coins, real estate and others. Certain types of investments are not allowed such as precious metals (other than U.S gold and silver coins), securities bought on margin, and life insurance.

Keogh plan
Tax-deferred pension account for employees of unincorporated businesses and self-employed individuals.

Keogh Plans. **Keogh plans** are tax-deferred pension accounts for employees of unincorporated businesses or for persons who are self-employed (part-time or full-time). The Keogh plan was established in 1962 by Congress and was expanded in 1976 and 1981. In many ways it acts like an IRA, but it is slightly more complex. The Keogh plan allows all investment earnings to grow, tax-deferred, until money is withdrawn between the ages of 59 1/2 and 70 1/2. If you wait until after 70 1/2, there is a penalty.

The acceptable forms of investments for Keoghs are much like those in SEPs. Eligible people can contribute up to 25 percent of earned income, up to *a maximum of $30,000.*

How are Keoghs different from SEPs? SEPs are newer (established in 1986) and cover more people. Jobs that are eligible for SEPs and Keoghs are lawyers, doctors, dentists, plumbers, actors, taxi drivers, waiters, freelance writers, carpenters, and other self-employed people or small-business owners.

Selecting a Plan

Given the vast array of retirement account options, how does a person choose which one or ones to pick? To aid in the selection, individuals or families should educate themselves. This includes keeping up with the latest options and rules. They should also consult experts, such as financial advisors including the benefits people at places of employment.

Unfortunately *there is no universal formula.* What might be right for you may not be right for the person in the cubicle next to you. In addition since the rules change and since your life circumstances change, what might have been the best choice for you a year ago may no longer be the best choice. That is why annual financial check-ups are recommended.

At Retirement

Once a retirement plan or plans are selected and contributed to, what happens at retirement? Here are the options for individuals with IRAs: they could withdraw the IRA in a lump sum, withdraw it in installments, or invest it in an annuity that guarantees payments over the rest of the lifetime.

There are tax consequences involved. For example, in the first option the entire amount would be taxable as ordinary income. Since each retirement plan has its own rules and tax implications, educating oneself and consulting advisors is wise before choosing or accessing a retirement plan.

Annuities

The third option mentions an annuity. Annuities were covered in Chapter 10 on Life Insurance because they are *purchased directly from life insurance companies,* although they can also be purchased indirectly through stock brokerage firms. An annuity was defined as a contract in which the insurer promises the insured a series of periodic payments. An annuity can be bought in a lump sum (such as $50,000 for one, perhaps from money from the sale of a house) or in periodic payments.

Straight life annuity
Contract sold by life insurance companies that provides payments as long as a person lives.

Joint and survivorship annuity
Contract sold by life insurance companies that provides payments until the death of the last designated survivor.

Straight life annuities pay as long as the person lives. **Joint and survivor annuities** provide payments until the death of the last designated survivor (usually a spouse or it could be someone else, such as one's own child or a niece or nephew). Because of the elongated benefit, a joint or survivor annuity is going to cost more than a straight life annuity. Opinions differ about which a person should choose. Two questions to ask are "Can the person afford the higher monthly payment or initial lump sum for a joint and survivor annuity?" and is the spouse or child financially independent?"

There are different types of annuities:

* *Immediate annuities* are purchased by people at retirement age usually in a lump sum payment. They provide income payments that begin immediately.
* *Deferred annuities* provide income at a later date and are often purchased by younger-than-retirement-age people. They are a savings device.
* *Cash value from life insurance.* A person over age 65 with grown independent children may choose to convert the cash value of a life insurance policy to an annuity.
* *Fixed versus variable annuities.* When you pick an annuity, you will be asked how you want the annuity premiums to be invested. Fixed-dollar annuities are invested in bonds and mortgages with guaranteed returns. In variable annuities money may be invested in stocks, and thus there is more variability in outcome. Options should be discussed with insurance agents along with costs, fees, and other features such as penalties for early withdrawal.

Distribution

So far Social Security, retirement plans, and annuities have been covered, along with changing expenses. During retirement your goal is to receive income from your assets to meet expenses and pursue dreams, such as travel or a comfortable home. Another goal is to preserve assets for later years and/or your heirs. Concerns include protecting income and assets from inflation and unnecessary taxation, making income last as long as you do, providing for health care, protecting a surviving spouse, and planning your estate (to be covered in the next chapter).

Income from Investments During Retirement

Retirement income gap
Shortfall between the amounts of retirement income generated and retirement living needs.

Even with the best pension or retirement plans there still may remain a **retirement income gap**, which is the shortfall between the amounts of retirement income generated and the retirement living needs. In other words you do not have enough money. There are a number of ways to cope with this gap including working part-time or reaping the benefits of investments accumulated during the main working years.

A general investment principle is that younger people can take on more risk than older people. As one nears retirement, a more conservative (although not stagnant) approach to investing is usually recommended, such as investing in more bonds. The idea is to reduce risk since lost income cannot be regained when one is not working full-time. The investment ladder given in Chapter 11 may be used as a guide because it shows the progression of investments from the most secure at the bottom to the most risky at the top.

Suggestions for accumulating funds through investing have been given throughout the investment chapters in this book. As with all age groups, retirees should keep abreast of interest rates and inflation rates, watch volatility and liquidity, diversify investments, and procure professional management and advice when they need it. So it is a myth that retirement planning ends at retirement. Nothing could be farther from the truth since retirement may last 30 years or longer. Investments and insurance policies should be reassessed as situations dictate, as needs change, or even if things are progressing uneventfully.

The following should be addressed on a yearly basis and more often if there is a crisis:

* A review of retirement goals.
* A recalculation of retirement income needs.
* A recalculation of retirement income projections.
* A recalculation of retirement income gaps.
* A review of the performances of investments chosen.

Susan's Parents: A Case Study

As an illustration of changes that occur during the retirement years, consider the progression of the lives of Susan's parents. They traveled abroad during the first three years of their retirement. Then her mother became ill, and the couple could no longer travel outside the country. They switched their travel plans and bought a recreational vehicle. For several years they crisscrossed the country, visiting children and grandchildren, until her mother could no longer travel. Within a year her mother passed on, and her father was in ill health.

He stayed alone on the family farm. Eventually he could no longer take care of himself or the farm. Susan and her brothers, who lived in other states and had jobs and families of their own, took turns visiting him and hired a part-time nurse to take care of him. Eventually he died in a nursing home. During their retirement years Susan's parents' goals and income needs changed dramatically, and when her father became too ill to take care of himself, the children had to step in and make arrangements.

This case study shows how complex the retirement years can be and that at some point children or other relatives may have to handle financial matters. For example, during the transition time in Susan's father's life from his need for care until his death, many subjects came up including estate planning, wills, and inheritances. These topics will be discussed in the next chapter.

ADVICE FOR BEGINNERS

* Start saving at a young age and stick with a saving's program.
* Take advantage of all tax-deferred retirement savings plans. Retirement plans at work are a great tax shelter as, are SEP and IRA accounts. Pour in all you can into retirement funds.
* If you change jobs, rollover your retirement account or reinvest the money.
* The younger you are, the more important it is to invest in growth stocks and mutual funds, don't be too conservative.

ELECTRONIC RESOURCES

The government provides websites on subjects for the later years. The Pension and Welfare Benefits Administration provides information on pension plans and health benefit plans at **www.dol.gov/dol/pwba**. The Social Security Administration can be reached at **www.ssa.gov**; the U.S. Dept. of Labor at **www.dol.gov**; and Medicaid at **www.hcfa.gov/medicaid.html**. The Consumer Information Center at **www.pueblo.gsa.gov** has a number of publications on retirement, including a free booklet entitled *About Social Security*.

The National Institute on Aging Information Center Web site is **www.aoa.dhhs.gov**. Information about the Service Corps of Retired Executives Association (SCORE) is available at www.sba.gov. Information about RAMs (reverse mortgages) can be obtained from **www.hud.gov/rvsmort.html**. Other official government sites can be reached at **www.ipcress.com/writer/gov.html**.

For comparing annuity prices go to **www.annuityshopper.com** and **www.annuity. com**. The most secure annuities come from top-rated insurance companies. For more about savings for retirement go to the American Saving Education Web site at www.asec.org. The Employee Benefits Research Institute for more on benefits can be reached at **www.ebri.org**.

The American Association of Retired Persons (AARP) provides information on Social Security, Medicare, and retirement at **www.aarp.org**, as does the National Council on Aging at **www.ncoa.org**. For Roth IRA information try **www.rothira.com**. CareGuide (**www.careguide.net**) provides an interactive search for support services for elders by state or city. Yahoo! Seniors' Guide (**seniors.yahoo.com**) has chat rooms for elders to discuss topics of mutual interest, such as financial planning and the best places to retire and travel.

Examples of websites offering interactive worksheets to estimate retirement nest eggs are

SmartMoney	**www.smartmoney.com**
T. Rowe Price Inc.	**www.troweprice.com**
Vanguard Group	**www.vanguard.com**
Quicken	**www.quicken.com/retirement**
Investor Guide	**www.investorguide.com/Retirement.htm**
Valic	**www.valic.com**

1. How solid is Social Security?

   ■ It depends on whom you ask, but the Social Security trustees estimate that the program will continue to pay full benefits until 2034 and about 75 percent thereafter unless there are significant changes.

2. What is the 3-legged stool of retirement security?

   ■ This refers to having the following three: pensions (or other employer-sponsored retirement plans), savings, and Social Security benefits.

3. If I have a pension plan at work, can I also have an IRA?

   ■ The answer is "yes" and it is a good idea. In so doing, you would have two retirement accounts—one you individually contribute to and manage and another through your employer. Another advantage of IRAs is that they are tax-deferred investments.

4. What can a person do to retire early?

   ■ An individual should save money, invest well, keep living expenses down, and consider working part-time during retirement.

5. Should someone borrow from his or her 401 (k) program?

   ■ The answer to this is "no" or only under extreme circumstances of need. An individual should consider other loan sources first.

6. When are IRAs not a good idea?

   ■ If money is tight and a person needs money in the next few years for a house or college tuition, he or she should wait on an IRA.

SUMMARY

Retirement planning is important because of greater life expectancy, the gender gap in pension benefits, the chance of early retirement, the inadequacy of Social Security and pensions, uncertainty, and retirement goals and needs.

College graduates save more for retirement than do high school graduates.

Less than half of Americans have traditional pensions.

The U.S. Department of Labor estimates that a person needs 70–90 percent of preretirement income to support a similar lifestyle in retirement.

One of the best ways to build a retirement nest egg is to pay yourself first with automatic contributions to your employer-sponsored retirement plan.

The three steps to follow in the retirement planning process are to estimate net worth, to estimate how much money will be needed, and to build and maintain retirement income.

Increasingly, many people's largest single asset is their individual retirement account (IRA).

Sources of retirement income include Social Security, public pension plans, employer pension plans—such as 401(k)s—and personal retirement plans—such as IRAs, Roth IRAs, SEPs, SIMPLE IRAs, and Education IRAs. Also some people work part-time during retirement to supplement assets accumulated during their main working years.

Social Security legislation revisions are being considered. If it keeps on its present course, Social Security's projected assets will no longer meet promised full benefits in the year 2034.

Some expenses go up during retirement (for example, more may be spent on leisure activities and prescriptions) and some go down (for example, work-related expenses).

A common myth is that retirement planning ends at retirement. Since the retirement years may last thirty years or more (although the average number of years is 18) and laws, regulations, and lifestyles change, retirement plans need to be revised at least on a yearly basis.

Another common myth is that retirement planning starts around age 50. In reality, the earlier one starts saving and planning the better.

REVIEW QUESTIONS

(See Appendix A for answers to questions 1–10).

Identify the following statements as true or false

1. College graduates save more for retirement than high school graduates.
2. The most widely used source of retirement income is Social Security.
3. Full retirement benefits begin at age 62.
4. Simplified Employee Pension Plans (SEPs) are made for large companies with over 100 employees.

5. There is a universal formula for retirement that everyone should follow.

Pick the best answers to the following five questions.

6. The U.S. Department of Labor estimates that retirees need ________ of preretirement income to support a similar lifestyle in retirement
   a. 40–50 percent
   b. 60–65 percent
   c. 70–90 percent
   d. nearly 100 percent
7. The average number of years spent in retirement is______.
   a. 5
   b. 10
   c. 12
   c. 18
8. A study by Public Agenda, a nonprofit research foundation, reported that ____________ affects how a person plans for retirement.
   a. Personality
   b. Friends
   c. Number of children
   d. Place of residence
9. In ______________________, an elderly person lives off the equity in a fully paid for house.
   a. retained earnings mortgages
   b. reverse annuity mortgages (RAMs)
   c. oligopsony
   d. dividend mortgages
10. Across all Americans the most widely used source of retirement income is __________.
   a. pensions
   b. IRAs
   c. Social Security
   d. Bonds

DISCUSSION QUESTIONS

1. In retirement usually some costs go up and some go down. Give two examples of each.
2. A 401(k) is one of the better-known employer retirement plans. What are the others, and for whom are they appropriate?
3. If a person wanted to supplement the employer retirement plan and Social Security with his or her own personal retirement plan what are examples of options? What retirement plans can a self-employed person set up?
4. What are the reasons why someone needs a retirement plan?
5. What are the three steps in the retirement planning process? Explain the purpose of each step.

DECISION-MAKING CASES

1. Estimate the retirement nest egg that Courtney Lewis needs. She wants to retire at age 60 and she is 30 years old now. Her current income is $30,000 a year, and she is setting aside $50 a month. This is her only retirement savings plan. Given what you know from reading the chapter, is she setting enough aside?
2. Kevin, a single man who has a salaried job and also is a freelance writer, has heard about the Roth IRA and wonders if he would be better off contributing to one of those instead of to a traditional IRA. He makes $100,000 a year. Is he allowed to have a Roth IRA at his income level? What other retirement plan options could he consider?
3. Jim owns a small business with 20 employees. He knows it is easier to keep and recruit good employees if he offers a retirement plan. He is considering offering an IRA, an SEP, a SIMPLE IRA, or a Keogh plan. He wants to keep his paperwork to a minimum. Which one(s) should Jim choose?
4. Ryan is changing jobs after eight years with the same company. At this point he has accumulated $15,000 in his retirement account. He can take it out and spend it or roll it over into the new company's plan. Why would it be wise for Ryan to choose the rollover option?

5. Wiley is in a similar situation as Leah was in the chapter. He is single, 26 years old, and working full-time. His current income is $26,000, and his assets amount to $4,000. If he puts $2,000 a year into a tax-deferred savings plan, earning 10 percent a year, Wiley will have $803,000 by age 65. This sounds like a lot of money, but given what you know about retirement expenses (and how long retirement can last), is this a suitable amount? Why or why not? Wiley says he can not possibly save $2,000 a year for retirement at this point in his life because he needs money for other things. Is this a typical or an atypical response, given his age?

GET ON LINE

Here are two activities:

1. Try www.vanguard.com or www.smartmoney.com to get a rough idea of how much you will need for retirement. Be prepared to answer questions about your retirement goals (high income, leaving money to children). You might put in your own answers or make up a hypothetical situation. For example, if you figure you can live on $35,000, and Social Security pays $15,000, your nest egg will have to provide the missing $20,000 (from investments, figuring in an upward adjustment every year to cover the inflation rate). You will be asked to pick an investment return. If you pick 7 percent, the calculators assume you will get this every year and in the real world sometimes you will and sometimes you won't.
2. Vanguard, a large mutual fund company, has a Web site at www.vanguard.com/cgi-bin/Roth which provides an interactive calculator to help you determine whether you should select a traditional IRA or a Roth IRA. Fill out the worksheet and select the best option for you.

INFOTRAC

To access InfoTrac follow the instructions at the end of Chapter 1.

Exercise 1: This chapter covered the topic of retirement planning. Using InfoTrac, when you run a subject search on "retirement planning" you will find articles and subdivisions. Under the subdivisions, select finance and review one article. What does the article say regarding financing retirement?

Exercise 2: As in Exercise 1, go to the subject "retirement planning," but this time look at the subdivisions entitled demographic aspects, economic aspects, or management. Select one of these and look at the articles listed. Select one article of interest and report on what the article discusses or recommends.

chapter sixteen

estate planning

Establishing a Plan for Accumulated Wealth

DID YOU KNOW?

* About 70 percent of adults do not have wills.
* Estate taxes can be as high as 55 percent.

OBJECTIVES

After reading. Chapter 16, you should be able to do the following:

1. Explain the importance of estate planning.
2. Describe the three steps in the estate planning process.
3. Discuss the various types of wills and trusts and the taxation and legal issues involved in estate planning.

CHAPTER OVERVIEW

This chapter builds on the previous one because if individuals do little or no planning, their retirement plan assets (401(k)s, IRAs, etc.) will be diminished by estate and income taxes. As much as 40–80 percent of retirement plan assets may not benefit families, friends, institutions or causes that individuals intend. The good news is there are solutions to this predicament. For example, trusts can be set up to provide income for donors or heirs and substantially reduce taxes.

Trusts and the avoidance or minimization of estate taxes are part of the discussion in this chapter on **estate planning,** the planning for the orderly handling, disposition, and administration of an estate during one's lifetime and at one's death. It includes drawing up a will, setting up trusts, minimizing estate taxes, and consulting advisors. All of these will be discussed as well as related topics. The chapter concludes with an Electronic Resources section.

Objective 1:
Explain the importance of estate planning.

IMPORTANCE OF ESTATE PLANNING

Estate planning
The orderly handling, disposition, and administration of an estate during a lifetime and at death.

Estate planning is an integral part of personal finance. Every adult with substantial financial assets should have an estate plan. Many people put off estate planning because they think it is morbid or it is only for the elderly. But if it is ignored, *the lack of a plan can cost family and friends more than time or anxiety; it can cost them thousands or even millions of dollars.* Estate planning takes far less time than most people realize, and it provides peace of mind.

Certain documents to be discussed in this chapter are key to the estate planning process and, regardless of age or circumstances, they are a good idea to have. Financial advisors at banks (trust officers) and brokerage firms (trust specialists), as well as others—including tax advisors and attorneys—are also key to the estate planning process. In addition, software and Web sites are available that provide information and interactive experiences to get started on formulating a plan.

Estate Defined

Estate
What is owned at the time of death less what is owed.

Most simply stated, your estate is everything that you own, your total property. An **estate** is all the assets a person possesses at the time of death, including securities, real estate, interests in business, physical possessions, and cash less liabilities (what is owed).

In the previous chapter the case study of Susan's parents discussed what happened during their retirement years. To pick up where the case left off, during his final year of life Susan's father did not want to talk about wills or estate planning or leave his house or give away his things, leaving his adult children unsure what to do. The children decided their only option was to wait until he passed on.

After his death in 1998 they gathered at the farm to sort out what they could in a few days' time before returning to their jobs and families. They did not find a will, but what they found was money, important papers, and jewelry hidden in "secret" places, such as in the bottom of house plants and coffee tins. This was only the beginning of a long process of sorting out his possessions and eventually settling his estate. This is an example of what often happens when there is no estate planning or discussion with heirs, attorneys, or anybody else about where wealth is hidden. The children doubt that they will find all the secret places before they sell the farm.

Probate
Judicial process whereby the will of a deceased person is presented at court and an executor is appointed.

Nonprobate property
Property that does not go through probate.

An estate is distributed to heirs according to the dictates of the person's will or, if there is no will, a court ruling.[1] The court referred to is probate court. **Probate** is a judicial process whereby the will of a deceased person is presented to a court, and an executor is appointed to carry out the will's instructions.[2] **Nonprobate property** does not go through probate. Examples of this would be property transferred directly to survivors by contract, such as pension plans or annuities. Asset transfer can be arranged in such a way that an estate is settled out of court. More on probate and executors will be covered later in the chapter.

Estate planning varies in complexity due to the number of people involved (the spouse, children, and other heirs), taxes, legal issues, state and federal laws, and the size of the estate. A common misconception is that estate planning is only for the wealthy. If an estate is worth more than the excess of the applicable exemption amount (see Exhibit 16.1), then the estate may be subject to estate tax.

Estate taxes
Federal and state taxes assessed on an estate's value and paid for by the estate.

Estate taxes are federal and state taxes assessed on the value of an estate and paid by the estate. As the exhibit shows federal and state estate taxes are based on a sliding scale so that the higher the estate value the higher the potential taxes. Estate taxes can be as high as 55 percent. Although $700,000 sounds like a lot of money, once a person's home, cars, investments, life insurance policies, and retirement savings are added up, it is not unusual for their worth to be over this amount. Even in estates worth less than this, estate planning would be wise in order to ensure that the disposition of assets is carried out in the way an individual desires.

Inheritance taxes
Taxes levied in sixteen states paid for by the heirs.

Besides the federal estate taxes, there are two kinds of state death taxes: estate (levied in Mississippi, New York, Ohio, and Oklahoma) and inheritance. **Inheritance taxes** are levied in sixteen states (Connecticut, Delaware, Indiana, Iowa, Kansas, Kentucky, Louisiana, Maryland, Montana, Nebraska, New Hampshire, New Jersey, North Carolina, Pennsylvania, South Dakota, and Tennessee) and are paid by the heirs based on the value of the assets they receive.

Do not confuse inheritance taxes with estate taxes that are paid out of the estate's assets. In inheritance taxes the recipient pays. So an Iowa estate worth $2,000,000 with heirs would be subject to federal estate taxes and inheritance taxes. Ways to lessen taxes will be discussed throughout the chapter, but next the chapter turns to the overall process of estate planning, of which minimizing taxes is a part.

Objective 2:
Describe the three steps in the estate planning process.

PROCESS OF ESTATE PLANNING

Estate planning is considered a process because it involves three steps and plans need to be updated. For example, an estate plan established at age 45 will need to be altered

Taxable Estate Size	2000	2001	2002	2003	2004	2005	2006
$700,000	$9,250	$9,250	$0	$0	$0	$0	$0
$850,000	$66,750	$66,750	$57,500	$57,500	$0	$0	$0
$950,000	$105,750	$105,750	$96,500	$96,500	$39,000	$0	$0
$1,000,000	$125,250	$125,250	$116,000	$116,000	$58,500	$19,500	$0
$2,000,000	$560,250	$560,250	$551,000	$551,000	$493,500	$454,500	$435,000
$5,000,000	$2,170,250	$2,170,250	$2,161,000	$2,161,000	$2,103,500	$2,064,500	$2,045,500
$10,000,000	$4,920,500	$4,920,500	$4,911,000	$4,911,000	$4,853,500	$4,814,500	$4,795,000
$20,000,000	$10,929,250	$10,929,250	$10,911,000	$10,911,000	$10,853,500	$10,814,500	$10,795,000
$50,000,000	$27,500,000	$27,500,000	$27,500,000	$27,500,000	$27,500,000	$27,500,000	$27,500,000

These estimated figures include both federal and state death taxes. Actual figures will vary by state and may change as legislation changes.

EXHIBIT 16.1 Taxable Estates: Minimum size needed to be taxed

several times if the person lives until 90. There are many reasons (death of a spouse or a child, divorce or remarriage, changes in health or financial status, winning the lottery) that would suggest that a revision is necessary.

The process can be divided into three basic steps: (1) the setting of goals and objectives, (2) the making of a will, and (3) the establishing of a trust. This section of the chapter provides an overview of these aspects of estate planning. Tax advisors, bankers, financial planners, and attorneys should be consulted for specifics regarding individual cases. Also estate laws vary by states so how an estate may be settled varies around the country, not to mention around the world. For example, having dual citizenship can complicate estate planning; an attorney should be consulted if there is dual citizenship. He or she may suggest that there is an advantage to declaring citizenship in one country over another.

Goals and Objectives

As with other forms of financial planning, estate planning should be based on goals and objectives. What does the individual want to achieve? Possible objectives may be

- to establish the optimal disposition and management of assets.
- to minimize inconvenience to family members.
- to provide for the guardianship of minor children.
- to appoint an agent through a durable power of attorney to administer a person's affairs in the event of disability.
- to avoid or minimize taxes, expenses, and other costs associated with the transfer of assets or settling of an estate.

Assessment of Current Situation: Taking Stock

Once goals and objectives are determined, then estate planning proceeds to an analysis of the current situation including locating assets and debts and, on the basis of these, estimating the value of an estate. Individuals should inventory what they own and owe and assign a value to each asset (residence, savings, investments, motor vehicles, life insurance policies, annuities, etc.) and to each debt. Then decisions should be made about the possible disposition of assets. Exhibit 16.2 shows an example of how John Adkins, a retired executive, estimated the value of his estate and provides a space for you to estimate your estate. When making estimates, individuals should use **fair market value**, which is not what was paid for items, but what they are currently worth.

Fair market value
Current worth.

Providing for a Spouse and Children

An objective many people have when making an estate plan is providing for a spouse and children. Most married couples own their property jointly, so if one of the spouses dies, the other usually gets the entire value of the property. **Joint tenants with rights of survivorship** is an arrangement whereby usually two people own property together, and if one of them dies, the property automatically goes to the other owner. Many individuals plan to pass property directly to a spouse and/or to children. **Heirs** (children, a spouse, sisters and brothers) are persons who are entitled to receive property.

Joint tenants with rights of survivorship
Arrangement whereby usually two people own property together, and if one dies, the property automatically goes to the other.

Heirs
Persons entitled to receive property from an estate.

If there are children, parents usually want their assets to be divided equally among their children, although in some cases they may want to give more to one who has potential health or education needs or less to one who has not reached a certain level of financial maturity. Someone with several children with a wide span of ages may want to provide differently for each.

EXHIBIT 16.2
Estimating the value of an estate.

Estate contents	John Adkins	You
Primary residence	$300,000	______
Other homes/real estate	0	______
Mutual funds, stocks, bonds	200,000	______
Certificates of deposit	100,000	______
Bank accounts	5,000	______
Interest and dividend payments	500	______
Automobiles and other tangible personal property	75,000	______
Life insurance policy	100,000	______
No-fault insurance payments owned	0	______
Value of IRAs and company pension	400,000	______
Pension death benefits	0	______
Income-tax refunds	0	______
Other assets/items (miscellaneous)	0	
Total value of items listed above	$1,180,500	______
Minus liabilities (funeral expenses, debts such as credit card bills, cost of settling estate)	22,500	______
Estate value	$1,158,000	______

Never Married, No Children

Even individuals who have never married nor have children need an estate plan. The information given in this chapter applies to a variety of lifestyles and lifestages. Everyone should take steps especially five to ten years before retirement, to assess where they are financially and where they want their assets to go (perhaps to nieces or nephews, friends, or charities). In addition setting up probable health-care decisions such as designating a preferred caregiver or institution, is important.

Records and Estate Plans

As explained in Chapter 2, for many reasons it is essential for everyone to have important documents and personal information pulled together in an accessible form. It is especially important as one nears the later years of life. Most people do not realize how complex their financial lives are and how much leaving organized records will help others.

In the event of death, the important papers that will be needed by those settling the estate include:

- birth certificates.
- marriage certificates.
- legal name change documentation.
- military service records.
- wills.

Other important documents requiring proof include:

- veteran documents.
- insurance policies.
- safe deposit box records.
- Social Security documents.
- bank account documents.
- stock, bond, and mutual fund certificates or evidence.
- automobile registrations.
- home ownership documents.
- pensions or 401(k)s or other retirement plans.

Objective 3:
Discuss the various types of wills and trusts and the taxation and legal issues involved in estate planning.

WILLS

A **will** is a legal document designating the transfer of property and assets in the event of death—in other words how property should be divided. Wills have been around for centuries. The oldest known will was written in 2448 B.C. by the Egyptian pharaoh Uah.

Will
Legal document designating transfer of property and assets in the event of death.

Intestate
Dying without a will.

Every adult should have a will, although *about 70 percent of adult Americans die without one.* Wills can be written by anyone over the age of 18 or 21 (depending on the state) who is mentally capable ("being of sound mind and memory"). **Intestate** means dying without a valid will.

Types of Wills

There are three main types of wills:

Formal wills
Properly drafted, signed, and witnessed wills usually prepared by an attorney.

Formal Wills The most common are **formal wills**—properly drafted, signed, and witnessed (two or three witnesses, depending on the state) and usually prepared by an attorney. Within this category there may be individual wills or joint wills written by a couple, which leave the bulk or all of the estate to one spouse if the other dies first, or to beneficiaries if they should die together.

Holographic wills
Handwritten wills without the advice of an attorney.

Holographic Wills A **holographic will** is handwritten by the person signing it, dated, and signed in the same handwriting without the advice of an attorney. Most states do not recognize holographic wills even if there are witnesses.

Nuncupative wills
Oral wills.

Nuncupative Wills Oral wills spoken by a dying individual to another person or persons are called **nuncupative wills**. Most states do not recognize these either.

Forms for wills are available in bookstores or office stores, in software such as Quicken Family Lawyer by Intuit, or over the Internet. These have their drawbacks in terms of acceptability in certain states.

The most reliable choice is the first, the formal will, prepared by an attorney.

Other choices involve risk, which varies by the type chosen, by how it is prepared and witnessed, and by state laws.

Legal Terms

The legal terms in a will include:

* **testator**: the person who makes a will.
* **beneficiary**: person receiving property identified in a will.
* **codicil**: a legal document that amends a will. Rather than write a new will, a person may choose to add a codicil that would change a bequest or other provision.
* **residuary clause**: catchall clause in a will that usually states, "I give the remainder of my estate to ..." If this clause is not there, a judge may distribute the remainder of an estate according to state law.
* **executor** (male) or **executrix** (female) also known as **personal representative**: the person responsible for carrying out the directions and requests in a will. For simplicity the word "executor" will be used for the remainder of this discussion. An executor is legally obligated to act in the interests of the deceased. An executor may hire an attorney to help with the probate process. The attorney's fees are charged to the estate. Anyone over the age of 18 who has not been convicted of a felony can be an executor.

Testator
Person who makes a will.

Beneficiary
Person identified in a will to receive property.

Codicil
Legal document that amends a will.

Residuary clause
Catchall clause in a will that distributes the remainder of an estate.

Executor (Executrix) or personal representative
Person responsible for carrying out the directions and requests in a will.

Can Husbands and Wives Have the Same Executor and Wills?

They may or may not have the same executors. Either can name any legally competent individual or corporate entity as executor that he or she wants. In addition any individual can be named as an agent in a power of attorney.

Usually joint wills are not recommended. Some people try this to save money (making one will instead of two), but married couples need separate wills even if their wills are very similar. Among the reasons for this are the possibilities of divorce and of one spouse predeceasing the other. Separate wills actually make it easier to settle estates rather than more complicated.

With a married couple wills can be similar or differ greatly. For example, if she was a multimillionaire entertainer and he was a house painter, their assets would be very different. Further, if she had been married five times and had children from three of the marriages, her will could be extremely complicated.

Regardless of circumstances, as much as possible wills and executors should be coordinated to create an organized estate plan of most benefit to the entire family or other beneficiaries. For example, in the event that the couple dies in a joint catastrophe (a car accident or plane crash), then totally different or conflicting wills may lead to legal difficulties in settling the estate. To avoid these problems, most states follow the Uniform Simultaneous Death Act, which provides for this type of situation.

As a final note, if a person is legally married at the time of death, it is very difficult to leave nothing to the spouse, nor is it easy to totally leave out children. In nearly all states something has to be left to survivors.

Executor Duties

The first duty of an executor is to initiate probate, which is the formal process of proving the authenticity of a person's will and confirming the assignment as executor. This involves filing an application to appear before a probate court. The application form is obtained from the clerk of the probate court (listed in the government section of a telephone directory).

Next, all beneficiaries are notified that the will is being processed, and a court hearing is set up. At the court hearing the executor will need the copies of the papers previously listed. The court will determine the validity of the will. If it is valid, the executor can begin to pay the bills and taxes of the estate and distribute the assets to the beneficiaries. The court costs will be charged to the estate. The estate will also pay the executor a fee, which may be specified in the will. This fee may go toward paying the executor's expenses, such as traveling to another state to go through the probate process.

Exhibit 16.3 provides a checklist for executors.

Visiting an Attorney to Make a Will

Before meeting with an attorney, you should gather the following information: a summary of data about yourself and beneficiaries—including names, addresses, social security numbers, and dates of birth—and financial documents or a summary or description of assets, liabilities, and ownership (a net worth statement).

In addition you should have an idea of what your will should accomplish. Who should receive your assets? Other important considerations, such as the naming of guardians, are covered in the next section.

The attorney will begin by reviewing the information that you have brought to the meeting, and he or she will discuss with you your wishes and goals. Questions about beneficiaries and executors will be asked. In the case of a simple will the usual procedure would be to have an initial meeting with an attorney, who will then draft

Settling an estate, even a simple one, involves lots of details. The first step is to initiate probate by locating and authenticating the will. Other steps include:

Check when accomplished

_____ Obtain an attorney, if necessary

_____ Apply to appear before probate court, if estate amount is high enough

_____ Notify beneficiaries named in the will

_____ Arrange for publication of notice to creditors

_____ Send notices of the person's death to businesses

_____ Inventory assets, have appraisals made

_____ Collect debts owed to the estate

_____ Check with deceased's employer for unpaid salary, benefits, insurance

_____ File for Social Security, civil service, veteran's benefits, life insurance

_____ Fill out tax returns, pay estate taxes if necessary

EXHIBIT 16.3
Executor's Checklist for Settling an Estate

the will. The client then reviews the draft and makes any changes necessary along with the attorney; then a finalized will is prepared for signature.

Elements of a Will Including Naming a Guardian

A very important part of a will is naming a guardian for minor children. Guardians should be over the age of 18 and willing to assume the responsibility of caring for children. Most people pick a close relative— usually a brother or a sister. The named guardian should be informed about your selection of him or her and should agree and accept the role. A first and second choice of guardians or coguardians can be named. If a person does not name a guardian or an executor in a will, a judge will appoint one.

Family Heirlooms. It is not recommended to make lengthy lists of personal assets in a will since items and wishes may change over time. Instead a person should consider directing the executor to follow the instructions outlined in a separate letter about specific bequests of personal effects such as antiques or jewelry. Such a letter should be dated. It is not binding, does not cost anything to make, and can be changed as often as wished. Previous letters or lists should be destroyed to avoid confusion.

Witnesses

Once the will is drawn up, witnesses will watch as the person signs and the witnesses will sign also. As a further step, experts suggest that an affidavit (a written statement made under oath before an authorized person, such as a notary public) be attached to the will. If you were making a will, then you and the witnesses would sign the affidavit. The reason this is suggested is that an affidavit would speed up the probate process because it eliminates the need for witnesses to appear before probate court. Often what happens in attorney's offices is that one of the paralegals or office assistants is also a notary public, and he or she can take care of the affidavit at the same time that you and the witnesses are signing the will.

Witnesses should not be beneficiaries of the will. If a will is updated, the old will should be destroyed.

Storage of Wills

Wills should be stored in a safe place, yet accessible to significant others. Most people keep them in a safe-deposit box or at home. Laws in many states restrict access to a safe-deposit box after death, so that is why a copy should be kept with other important personal documents at home and should also be given to named representatives. If kept at home, a friend or relative should know where the will is kept. If an attorney prepares the will, he or she should retain a copy with a note attached about the location of the original will. If a trust company is named as an executor, it should have a copy of the will also. One of the reasons it is wise not to have too many copies of wills is that if there is a change, all previous wills have to be destroyed; there can be confusion if too many versions exist when a person dies.

Two other types of documents, living wills and letters of last instruction, are not wills in the sense of the ones previously described. They are, however, related to the estate planning process and need to be considered.

Living will
Document that lets family members know what type of care a person wants or does not want.

Living Will. A **living will** is a separate document that lets family members know what type of care you do or do not want to receive if you become terminally ill or permanently unconscious. Living wills should be discussed with an attorney and family members. Family members should have a copy. The reason an attorney should be involved is that states vary in their acceptance of living wills, and an attorney will be knowledgeable about these variances. In essence a living will specifies what you want done should you become physically or medically incapacitated, and in so doing, the document relieves loved ones of making difficult decisions. Occasionally doctors, hospitals, and family members go against the wishes stated in the living will.

Health care proxy or **health care power of attorney**
Document that permits the signer to name individuals to make health care decisions on his or her behalf.

Health-Care Proxy. A number of states have a **health care proxy** also called a **health-care power of attorney**, a document that permits the signer to name individuals to *make health-care decisions* on his or her behalf if the signer is incapable of making decisions. A designated health- care proxy can make life-support decisions but cannot address financial affairs.

Durable power of attorney
Document that addresses financial affairs on behalf of a severely incapacitated individual.

Durable Power of Attorney. A **durable power of attorney** is designated in a legal document and can *address financial affairs* on behalf of a person who is severely incapacitated. To be more specific:

> a person with assets (the principal) appoints another person (the agent) to act on the principal's behalf, even if the principal becomes incompetent. If the power of attorney is not "durable," the agent's authority to act ends if the principal becomes incompetent. The agent's power to act for the principal may be broadly stated, allowing the agent to buy and sell securities, or narrowly stated to limit activity, for example, to selling a car.[3]

Letter of last instruction or **letter precatory**
Signed letter directing the executor to documents and funeral and burial instructions.

Letter of Last Instruction. A **letter of last instruction** (also called a **letter precatory**) is a signed letter directing the executor to documents and describing funeral and burial instructions. The letter guides family members about how you want things to be in the event of your death, but it is not part of a will unless it is specifically referenced in, or made part of, the will. It is usually given to a family member or an executor for safe keeping, or the letter's location is told to them.

Laws on letters of last instruction vary by state. Although the letters are not legally binding in most states, executors and family members try to honor the wishes of the writer.

PROBATE

Probate, as defined earlier, is the judicial process whereby the will of a deceased person is presented to a court and an executor is appointed. Probate exists because at the end of life, all wealth must be given away, as recognized in the expression "You can't take it with you." If there is not a will, the court will decide how property will be divided up.

State laws vary on how assets will be divided; thus it is important to have a will so that the testator's wishes are followed. Since going through probate will incur costs for administrative actions, and the outcomes may not be what the testator wanted, many people try to distribute assets as they get older in order to avoid probate court and to distribute wealth as they see fit. The following section describes options in this regard.

Asset Distribution Outside of Probate

As explained earlier, there is nonprobate property that does not go through probate. There are two ways that assets may be distributed using prearranged actions outside of probate court: transfers by operation of law and transfers by contract. Examples in this latter category include annuities, life insurance proceeds, IRAs, Keoghs, government bonds, profit-sharing plans, bank accounts when there is a payable-on-death beneficiary named, and certain types of trusts that are established before death.

EXHIBIT 16.4
Your Guide to Trust Types

- *Living Trusts:* established while you are alive. At death, the trust assets are distributed to beneficiaries.
- *Testamentary Trusts:* created by a will at the time of death.
- *Revocable Trusts:* can be modified at any time until death or incapacity. Assets can be transferred in and out. You control the trust and assets and you pay taxes on the trust.
- *Irrevocable Trusts:* can be changed in only a limited number of ways (i.e., assets can be added to certain trusts, the distribution of income can be flexible) so this type is more "fixed" than a revocable trust.
- *Generation-Skipping Trusts:* created to hold assets tax-free for grandchildren.
- *Charitable Remainder Annuity Trusts:* a type of irrevocable trust that transfers assets from individuals to charities or other organizations from which you (and/or others) receive income for life or a fixed term of years. When the trust terminates, the assets are transferred to the charity designated.

Trusts

Trust
Legal structure or contract typically used for asset management.

Trustee
Person holding title to property for the benefit of another.

Trust fund
Property itself.

Testamentary trust fund
Trust created by a will.

Living trust
Trust established between living persons, most typically between a parent and a child.

A **trust**, a legal structure or contract, can be used for a variety of purposes ranging from competent asset management, to avoidance of probate, to flexibility in planning for others. Trusts can also provide estate and income tax savings opportunities particularly as asset values increase. In trusts a person called a **trustee** holds title to property for the benefit of another person called the beneficiary. The property itself is called a **trust fund.**

Types of Trusts. Exhibit 16.4 shows the six different kinds of trusts that will be explained further in this section. One way to categorize trusts is by the way in which they are created. For example, a trust created by a will is called a **testamentary trust fund.**[4]

If the trust is created while the donor is living it is called a **living trust.** For example, a father could establish a trust for his daughter during his lifetime, designating himself as a trustee and the daughter as beneficiary. As the beneficiary, the daughter does not own the property, but she can receive profits from it in accordance with the

CONSUMER ALERT

If one spouse predeceases the other, the surviving spouse can change or alter the distribution of assets in the living trust, provided the deceased spouse gave the survivor either a general or limited power of appointment over the trust assets. Giving anyone the power to appoint property is an important decision that should be discussed with an attorney.

terms of the trust fund. In a living trust a person—such as the father in this case—can be a trustee, managing the investments, or a person could arrange for professional management of the trust.

Another way to categorize trusts is by how they can be modified: irrevocable or revocable. **Irrevocable living trusts** can be changed only in limited ways; because of this they should not be entered into lightly. They may be created for tax benefits or because it provides a way for managing assets of a person the creator believes cannot or should not be managing his or her own property. For example, a **generation-skipping trust** can be created to hold assets tax-free for grandchildren.

Irrevocable living trusts
Trusts that cannot be changed without the agreement of the beneficiary.

Generation-skipping trusts
Trusts that hold assets tax-free for grandchildren.

The most popular kind of living trust, because it is a more flexible choice, is a **revocable living trust.** This allows the person creating the trust to change the terms of the trust or cancel it at any time. A revocable living trust operates much like a will and is preferred over a will in some states with especially high probate costs.

Revocable living trusts
Trusts that allow the person creating the trust to change terms within it or cancel it.

In revocable trusts you control the trust and its assets, and you pay the taxes on the trust. Because of all the financial and legal aspects involved in trusts, it is wise to consult attorneys, tax advisors, and other financial personnel before the selection of a trust is made.

Reasons for Trusts. To summarize, there are a number of different kinds of trusts and they are set up for a variety of reasons, including:

- to manage and protect assets.
- to safeguard inheritances.
- to reduce estate taxes.
- to provide for spouses, children, or grandchildren.
- to maintain privacy for heirs (their names and inheritances are kept private).
- to simplify finances and increase income.
- to avoid costly probate or estate taxes.
- to donate to charities or institutions, such as universities, religious organizations, or hospitals.

Here is an example of how Sara Clawson, a widow age 70, found a charitable solution for California coastal property that she had inherited. When she decided to sell, she wanted her church to ultimately benefit from the property, but she also felt it was important to augment her income. With help from her church headquarters' Gift Services Officer, she set up a charitable trust in which the trust was able to sell the property without incurring capital gains taxes, leaving a larger sum to invest to produce an income for her. The church staff negotiated with potential purchasers and oversaw the closing. Everyone was pleased with this outcome: Mrs. Clawson, the church, and the buyer of the property. So this is a story with a happy ending, but a gift of money to any organization requires planning and an understanding of the tax consequences.

A person considering a charitable trust should feel confident about the organization's purpose and their foundation or gift and trust department, should understand the alternatives, and should have his or her personal tax advisor or attorney look over any agreements before signing. Signing over assets is serious because of its permanency. Also, it should be noted, to receive the sort of services that Mrs. Clawson did, her particular church required a minimum of $150,000 for gifts in trust (in other words her property had to be worth more than $150,000 for the church to get involved in establishing a trust).

What Mrs. Clawson set up with the help of her church was a **charitable remainder annuity trust** (a type of irrevocable trust). In this type of trust you transfer assets

Charitable remainder annuity trust
A type of irrevocable trust that transfers assets from individuals to charities or other organizations.

to a trust from which you (and/or others) receive income for your choice of life or a fixed term of years. When the trust terminates, its assets are transferred to the church, university and/or other charitable organizations of your choice. This type of trust is selected by people who desire a fixed-income stream regardless of inflation. Mrs. Clawson gave property, but she could have also given stocks, bonds, mutual funds, life insurance, and cash that could have been used to set up a charitable remainder annuity trust.

Reducing Estate and Gift Taxes

It cannot be emphasized enough that one of the reasons people make estate plans is to avoid or minimize estate taxes, which are taxes imposed by a state or the federal government on assets left to heirs in a will.[5] The main exceptions are estates that are too small to be taxed, and transfers of property between spouses.[6] Under the Economic Recovery Tax Act of 1981, estate taxes are not paid on spousal transfers. You can pass along an unlimited amount of property to your spouse without paying federal estate taxes.

Other than these exceptions, individuals seek to minimize estate taxes by

- spending all their assets—purposely dying penniless ("you can't take it with you" philosophy).
- giving away assets to others during their lifetime.
- using trusts.
- giving charitable gifts. These are not taxed as long as the organization operates for educational, religious, or charitable purposes.

One overarching strategy is to give to charity the assets that will cost your family the most to inherit. When left to a university, for example, retirement plan assets can escape both estate and income taxation while entitling the estate to a charitable estate tax deduction.

Gift taxes
Graduated taxes levied on the donor of a gift by the federal government and most state governments.

Gift taxes are graduated taxes levied on the donor of a gift by the federal government and most state governments when assets are passed from one person to another. The more money given in a gift, the higher the tax rate.[7] Taxes are not levied on gifts between spouses, but there may be a limit—such as $10,000 per year—that can be given to a child or grandchild. The gift tax is computed on the amount over $10,000 per person per year.

Single people and unmarried couples do not receive the benefit of the unlimited gifts exchange between spouses. If they have sizable estates they should consider setting up trusts to transfer assets and to reduce estate taxes. An unmarried couple living together for a long time should have an attorney check into state laws about community property and the transfer of assets. If one of the individuals in the relationship is not willing to go to an attorney about this, the other one should go to look out for himself or herself and their dependents.

As a final note, in many cases fees are charged to set up and manage trusts, so these should be considered when estate planning. The bottom line in estate planning as in all areas of financial planning is that a person should take the time to look before leaping.

TAX ADVISORS

Since estate planning is very complex and tax rates and exemptions vary by state (for example, Nevada does not have death or inheritance taxes), a person with a sizable estate should consult a tax attorney or an estate planner about state as well as federal estate taxes. One of the drawbacks of relying on the Internet for estate planning advice is that it tends to offer generic information, and websites often do not take into account state laws.

The tax advisor's main role is to review your overall situation and to recommend specific action to help you achieve your goals and objectives. The motivation to do this is that money spent upfront for the best advice means more money for beneficiaries. The advisor's secondary role is to make sure that you follow through by implementing the plan of action, which may include meeting with another type of financial advisor.

Trust Departments in Banks

Commercial banks often operate trust companies. These trust companies, usually designated as trust departments, engage in estate planning and are regulated by state law. Trust departments offer investment management services, usually by charging a fee based on the assets managed. Initial consultations are free. If the amount invested with them is small (for example, under $500,000, note that the amount considered small varies by bank), it will often be invested in funds managed by the bank (in consumer loans rather than individual stocks and bonds). You may visit a trust department to set up a trust or as a beneficiary of a trust or to inquire about estate planning.

If the assets managed are of a substantial amount (over $500,000), then many times a banker or trust officer will take care of the investments just as a broker or financial planner would. The banker or trust officer, in this case, may be compensated by commission or by a percentage of the assets managed.

ADVICE FOR BEGINNERS

General advice is difficult to give on this subject because there is so much variance, depending on an individual's age, health, values, and goals; the number and age of dependents; size of the estate; and so on. As a minimum, most adults (even young ones) should have, or work toward having:

- an overall estimate of net worth (the value of an estate) and organized financial records.
- most importantly, a will, especially if there is a spouse or children.
- a living will regarding life support and other directives such as a health care proxy, a person designated to make health care decisions if necessary.
- a statement of durable power of attorney, someone with the legal right to handle finances in case of incapacitation.

ELECTRONIC RESOURCES

Many of the government websites provided in the previous chapter are useful for estate planning as well. In addition a government website that provides federal estate tax information is **www.irs.ustreas.gov**.

Estate planning information can be found at **www.webtrust.com** and estate tax information at **www.lifenet.com/estate**. The National Association of Financial & Estate Planning has a Web site at **www.nafep.com**.

The National Network of Estate Planning Attorneys at **www.netplanning.com** will help you search for attorneys in your area. Wills are available at many legal association or firm sites, as well as at **www.ca-probate.com/wills.htm**.

Most insurance companies have information about estate planning on their websites. An example is the Prudential Insurance Company of America at **prudential.com/estateplan**.

Deloitte & Touche LLP, an accounting firm, has extensive estate planning coverage at **www.dtonline.com**. Quicken's Web site (**www.quicken.com**) has a database of articles on estate planning.

FAQs

1. What is an estate?
   - The total property (assets-liabilities) owned by an individual.
2. What is property?
   - Property can include real estate, automobiles, household goods, investments, cash, and insurance.
3. Why does someone need estate planning?
   - It is needed to avoid dying "intestate" and paying too much in taxes, to set up trusts, and to preserve an estate for heirs.
4. Why do people try to avoid probate?
   - Avoiding probate is desirable because it is expensive and time consuming.
5. What are estate and inheritance taxes?
   - Taxes placed on an estate and the transfer of that estate to heirs.
6. What is a trust?
   - A trust is a legal entity or device that takes care of property distribution.
7. What are the six kinds of trusts?
   - They are living, revocable, irrevocable, testamentary, generation-skipping, and charitable remainder annuity trusts .
8. Who makes decisions regarding health care if someone becomes incapacitated?
   - A durable power of attorney can be set up for health-care decisions prior to a person becoming incapacitated.
9. Where should estate planning documents be kept?
   - In a safe, secure place that is accessible when needed.
10. Can a person designate who will get what heirlooms in his or her will?
    - Yes, a person can designate who gets what in a will, but it is not recommended that wills have extensive lists of specific possessions. This is because items change and relationships change, and wills should not be rewritten often.

SUMMARY

Estate planning is a process—based on goals and objectives—that involves the orderly handling, distribution, and administration of an estate. Your estate consists of everything you own (assets) minus your debts.

A common misconception is that estate planning is for the wealthy only. Another misconception is that it is only for the elderly.

The estate planning process includes three steps: the setting of goals and objectives; the making of a will; and the establishing of a trust.

About 70 percent of adult Americans do not have wills. If a person dies without a will, the court decides how property will be divided, who will be the executor of the estate, and who will be the guardian of the children.

Probate is the judicial process whereby the will of a deceased person is presented to a court and an executor is appointed.

The main types of wills are formal, holographic, and nuncupative.

Living wills are separate documents that let family members know what type of care individuals want if they become terminally ill or permanently unconscious.

Trusts are legal documents in which a person called a trustee holds title to a property for the benefit of another person called the beneficiary.

Trusts are set up for a variety of reasons, including the reduction or avoidance of probate and estate taxes (federal and state).

There are federal estate taxes and four states have state estate taxes, paid by the estate. Also sixteen states have inheritance taxes that beneficiaries have to pay.

REVIEW QUESTIONS

(See Appendix A for the answers to questions 1–10).

Identify the following statements as true or false.

1. Executors pay attorney and probate court fees out of their own pockets.
2. A health-care proxy can make financial decisions on behalf of an incapacitated person.
3. A trust created after a person's death is called a living trust.
4. There are fees involved in setting up and maintaining a trust.
5. Nearly 90 percent of Americans have wills.

Pick the best answers to the following five questions.

6. _______________ is the planning for the orderly handling, disposition, and administration of an estate during one's lifetime and at one's death.
   a. Estate planning
   b. Fiduciary planning
   c. Inventory planning
   d. Declination planning
7. ___________ is a judicial process whereby the will of a deceased person is presented to a court and an executor is appointed to carry out the will's instructions.
   a. Claimation
   b. A jury of executive opinion
   c. Probate
   d. Scripophily
8. Which of the following is not one of the three main parts in the estate planning process?
   a. The setting of goals and objectives
   b. The making of a will
   c. The establishing of a trust
   d. The selection of life insurance
9. A __________ is designated in a legal document and can address financial affairs on behalf of a person who is severely incapacitated.
   a. proxy
   b. durable power of attorney
   c. judge
   d. will specialist

10. The first duty of an executor is to
    a. collect debts owed to the estate.
    b. initiate probate.
    c. inventory all assets.
    d. file for Social Security, civil service, or veteran benefits.

DISCUSSION QUESTIONS

1. Under what circumstances is it especially wise to have a will? What might be the reasons why about 70 percent of Americans die intestate?
2. What are the main kinds of wills, and how do they differ?
3. Why would someone set up a trust? What are the six basic types of trusts? How do they differ?
4. Darla's father says he is setting up a living trust and naming her as a beneficiary. What does this mean exactly?
5. What is a charitable remainder annuity trust? What types of organizations are considered to be charities? What types of assets can go into funding a charitable remainder annuity trust?

DECISION-MAKING CASES

1. Jerome, age 81, with health problems and his wife, Frances, age 78, have not done any estate planning or written wills even though their net worth is high and they have children and grandchildren for whom they want to provide. Their stockbroker and their children are aware of this neglect of estate planning and do not know what to do. When the subject comes up, Jerome says there is no hurry, there is plenty of time. Given what you have learned in this chapter, what do you think of his answer?
2. Hannah's grandfather wants to avoid going through probate and wants to reduce estate taxes. His net worth is about $850,000. He is no longer married, but has children and Hannah is his only grandchild. He would like to help Hannah with her college expenses. What could he do over the next several years to meet his goals of avoiding probate, reducing estate taxes, and helping Hannah financially?
3. Marie Treschman, an elderly widow, is in very poor health and would like to get her financial affairs in order. A relative suggested that she sign a durable power of attorney. Is this an option she should consider? Why or why not? Whom should she consult? In addition Marie's maid wants her to sign an irrevocable living trust. Why is this a bad idea?
4. Ray Borderland is 56 and in good health. He has a wife and grown children for whom he would like to provide. He has done the first step in the estate planning process by figuring that his net worth is about $150,000. What should he do next?
5. John James McMillan, a former football player, has always been a big fan of college football. He goes to every game that he can at the college he attended. His wife has passed away and he is alone. He has decided to leave all his money to his alma mater to fund scholarships. John is considering a living trust so he can be around to see some of the money dispersed. He would also like to make a will designating his college for the remainder of his estate. Whom should he consult? Are their tax or estate benefits or fees associated with his plan of action?

GET ON LINE

Go to either the Quicken Web site or the Deloitte and Touche Web site and find their estate planning information. Find a topic of interest to you (wills, trusts, discussing estate planning with parents) and read the Web site's articles or advice. Summarize the key points on the subject you have selected.

INFOTRAC

To access InfoTrac follow the instructions at the end of Chapter 1.

Exercise 1: Type in the subject "estate planning" and you will find about 60 periodical references, 18 subdivisions, and 11 related subjects. Under subdivisions, click on management and select one article to review that is about the personal financial aspects of estate planning. Summarize the main points in the article.

Exercise 2: Type in the subject "estate planning" and click on subdivisions. This time click on technique or analysis and select two articles to compare. What advice do they give? Describe the similarities and differences between the articles.

Appendix A

Answers to the End of Chapter Review Questions

Chapter 1

1. F
2. T
3. F
4. F
5. T
6. d
7. b
8. c
9. a
10. a

Chapter 2

1. F
2. F
3. T
4. T
5. F
6. d
7. b
8. d
9. a
10. b

Chapter 3

1. F
2. T
3. F
4. F
5. T
6. a
7. c
8. c
9. b
10. d

Chapter 4

1. T
2. F
3. T
4. F
5. T
6. c
7. a
8. b
9. a
10. c

Chapter 5

1. F
2. F
3. F
4. T
5. T
6. d
7. c
8. a
9. d
10. b

Chapter 6

1. F
2. T
3. T
4. T
5. T
6. b
7. c
8. d
9. a
10. c

Chapter 7

1. F
2. T
3. F
4. F
5. F
6. b
7. c
8. c
9. a
10. d

Chapter 8

1. T
2. T
3. T
4. T
5. F
6. a
7. c
8. b
9. d
10. a

Chapter 9

1. T
2. F
3. F
4. F
5. T
6. c
7. b
8. a
9. b
10. d

Chapter 10

1. T
2. F
3. F

4. F
5. T
6. c
7. b
8. a
9. d
10. c

Chapter 11

1. F
2. T
3. T
4. F
5. F
6. a
7. d
8. b
9. b
10. c

Chapter 12

1. F
2. F
3. F
4. F
5. F
6. b
7. d
8. c
9. a
10. d

Chapter 13

1. T
2. F
3. F
4. F
5. T
6. c
7. a
8. d
9. b
10. a

Chapter 14

1. T
2. F
3. F
4. F
5. F
6. b
7. d
8. d
9. d
10. c

Chapter 15

1. T
2. T
3. F
4. F
5. F
6. c
7. c
8. a
9. b
10. c

Chapter 16

1. F
2. F
3. F
4. T
5. F
6. a
7. c
8. d
9. b
10. b

References

Chapter 1 References

[1] "Two-Tier Marketing." *Business Week*, March 17, 1997,82–87.
[2] *Merrill Lynch Insights and Strategies*, 1998, 2 (2), 2.
[3] Crispell, D. "Workers in 2000." *American Demographics*, March 1990, 12, 36–40.
[4] R. Rubin & B. Riney, *Working Wives and Dual-Career Families.*,(Westport, CT: Praeger, 1994).
[5] D. Hawkins, R. Best, and K. Coney, *Consumer Behavior*, Sixth Edition (Chicago: Irwin, 1995).
[6] R. Lathrop, *Who's Hiring Who, How to Find that Job Fast*, (Berkeley, CA: Ten Speed Press, 1989).
[7] P. Marion, *Crisis-Proof Your Career* , (New York: Carol Publishing Group, 1993).
[8] L. Asinof "Click & Shift: Workers Control Their Benefits OnLine, *The Wall Street Journal*, Nov. 21 1997, C1.
[9] R. Quick, "Your Cyber Career: Using the Internet to Find a Job." *The Wall Street Journal*, March 5, 1998, B8.

Chapter 2 References

[1] "Two Family Budgets: Different Means, Similar Ends," *The Wall Street Journal*, April 14, 1997, B1.
[2] Marilyn Harris. "Personal Business." *Business Week*, May 11, 1998, 106E4.
[3] Eric Weiner, "Independent Financial Advisers are Major Subjects for Regulators," *The Wall Street Journal*, May 22, 1998, C26.
[4] Ibid.
[5] E. Schultz. "Tips on Finding a Financial Advisor." *The Wall Street Journal*, May 23, 1998, C1 and C18.
[6] Harris, op. cit.

Chapter 3 References

[1] *The Wall Street Journal*, September 23, 1998, A1.
[2] M. McNamee. "Audits: Can Your Life Pass the Muster?" *Business Week*, February 3, 1997, 143.
[3] M. Loeb, *Marshall Loeb's Lifetime Financial Strategies* (NY: Little, Brown, and Company, 1996), 290–295.
[4] E. Book. "Red Flag Deductions: When Less Isn't More" *Working Woman*, February 1997, 34.
[5] L. Sloane, *The New York Times Personal Finance Handbook* (NY: Random House), 45.

Chapter 4 References

[1] "Business Bulletin." *The Wall Street Journal*, April 10, 1997, A1.
[2]"Tax Report." *The Wall Street Journal*, May 14, 1997, A1.
[3]"Supermarket Banking Grows Popular, But Analysts See Limits on Potential," *The Wall Street Journal*, December 16, 1996, B7a.
[4]Ibid.
[5]M.Mannix. "Money & Markets: Debit Card Dangers," *U.S. News & World Report*, April 7, 1997. 73.
[6]"Paper Checks Persist." *The Wall Street Journal*, January 23, 1997. A1.
[7], Board of Governors of the Federal Reserve System, *Making Sense of Savings* (Washington, DC).
[8] D. Skidmore. "Better Returns Could Boost Sales of Savings Bonds," *Tallahassee Democrat*, May 1, 1997. 8D.
[9]Ibid.
[10] J. Siegel, *Stocks for the Long Run* (New York: McGraw-Hill, 1999).
[11] "No Lunch-Hour Lines and No Tellers," *U.S. News & World Report*, April 7, 1997, 76.

Chapter 5 References

[1] J. Wuorio. "The World of Online Finance." *Business Week*, December 6, 1999.
[2] F. Lalli. "Consumer Alert: Avoid the New Credit Penalties," *Money*, November 1996, 84–87.
[3] Consolidated Credit Counseling Services, Inc., *Budgeting 101*, (Ft. Lauderdale, FL, 1998).

[4] J. Shenk. "In Debt All the Way Up to Their Nose Rings," *U.S. News & World Report*, June 9, 1997, 38–39.
[5] D. Copponi, "Fast and Easy Credit Results in Hard Times for University Student," *FSUVIEW*, December 6, 1996, 22.
[6] J. Shenk, op. cit.
[7] J. B. Quinn, "Student Loans, "*Tallahassee Democrat*, November 16, 1999.
[8] K. Davis, *Financing College*, 1997 Kiplinger Washington, Inc., 1997
[9] F. Lalli, op. cit.
[10] "Money Track." CNBC, November 27, 1998.
[11] A. Fisher. "Will My Bad Credit Sink Me?...When Not to Whine," *Fortune*, October 26, 1998, 302.
[12] F. Bleakley. "After Bankruptcy, Shopping Like Scrooge." *The Wall Street Journal*, December 12, 1996, B1–B2.

Chapter 6 References

[1] C. Tejada. "House Calls," *The Wall Street Journal*, December 1, 1997, R19.

[2] J. B. Quinn, *Making the Most of Your Money*, (NY: Simon & Schuster, 1997) 474.
[3] *MGIC's Field Guide to Buying, Selling & Owning Your Home*, (Englewood Cliffs, NJ: ECV).
[4] "Take a Step in the Right Direction." *U.S. News & World Report*, Special Advertising Section, October 6, 1997, 2.
[5] A. Guerlin. "Mortgage Maze: How to Cash in on Ever-Lower Rates." *The Wall Street Journal*, November 22, 1996, C2.
[6] J. Fletcher. "Every House Is a Money Pit." *The Wall Street Journal*, Sept. 4, 1998, W10.
[7] MGIC, op. cit.
[8] K. Morris & A. Siegel, *The Wall Street Journal Guide to Understanding Personal Finance* (NY: Lightbulb Press), 76.
[9] A. Wellner. "Keeping Old Cars on the Road." *American Demographics*, July 1997, 53.
[10] Ibid.
[11] T. O'Loughlin & J. Norris, "Ten Compelling Reasons Why You May Not Want to Lease a Car." *At Home with Consumers, 17* (3), (Washington, DC; Direct Selling Education Foundation) 5.
[12] A. Wellner, 54.
[13] T. O'Loughlin & J. Norris.
[14] Ibid.
[15] D. Bank. "Microsoft Moves to Rule On-line Sales." *The Wall Street Journal*, B1.

Chapter 7 References

[1] S. Orman, *The 9 Steps to Financial Freedom*, (NY: Crown, 1997), 114 & 115.
[2] M. Loeb, *Marshall Loeb's Lifetime Financial Strategies*. NY: Little, Brown and Company, 1996).
[3] Ibid.
[4] T. Pugh. "White House Takes on Child Care." *Tallahassee Democrat*, October 24, 1997, 3A.
[5] Ibid.
[6] L. Asinof, "No Free Lunch." *The Wall Street Journal*, December 12, 1996, R26.
[7] M. Baker. "Now or Later." *The Wall Street Journal*, December 12, 1996, R8.
[8] Asinof, op.cit., R10.
[9] J. Clements. "Dumb Moves: Ten Financial Errors That Parents Often Make." *The Wall Street Journal*, December 12, 1996, R6.
[10] Ibid.
[11] Ibid.
[12] K. Koppe. "Now Is Time to Fill Out Aid Form." *Tallahassee Democrat*, November 24, 1996, C1.
[13] R. Dogar. "Here Comes the Billion Dollar Bride." *Working Woman*, May 1997, 32–36, 69–70.
[14] Ibid.
[15] Loeb, *Financial Strategies*.
[16] Dogar, "Billion Dollar Bride."
[17] Loeb, op.cit.
[18] Ibid.
[19] Ibid.
[20] K. Hannon. "Suddenly Single." *Working Woman*, September 1996, 32–37, 75.
[21] D. Harris. "I Don't Need Your Money, Honey." *Money*, November 1996, 162.
[22] Hannon, op.cit.
[23] Ibid., 34.
[24] Loeb, op.cit.
[25] S. Dallas. "Making Room for Grandaddy." *Business Week*, February 17, 1997, 102E4.
[26] Ibid.
[27] Loeb, op.cit.
[28] Ibid., 410.
[29] E. Stark. "Which Charities Merit Your Money." *Money*, November 1996, 100–102.

Chapter 8 References

[1] E. Tyson, *Personal Finance for Dummies*, (Foster City, CA: IDG Books Worldwide, Inc., 1997).
[2] L. Asinof. "Homeowner's Insurance: Don't Overlook the Details," *The Wall Street Journal*, Feb. 13, 1998, C1.
[3] Tyson, op.cit.
[4] J. Clements. "Keeping Finances Simple Isn't Easy, But These 22 Strategies Should Help." *The Wall Street Journal*, June 3, 1997, B1.
[5] Asinof, op.cit.
[6] Ibid.
[7] E. Kinsella. "A Matter of Policy." *The Wall Street Journal*, December 1, 1997, R11.
[8] Ibid.
[9] J. Trieschmann & S. Gustavson, *Risk Management & Insurance, 9th edition*, (Cincinnati, OH: South-Western, 1995).
[10] Asinof, op.cit
[11] D. Kennedy, "Cut Your Homeowners Insurance Bill," *Family Money*, Summer 1998, 41–42.
[12] Kinsella, op.cit.
[13] Ibid.
[14] Ibid.

Chapter 9 References

[1] "Healthcare Overview," *U.S. News & World Report*, November 18, 1996, 57.
[2] R. Winslow. "Health-Care Inflation Kept in Check Last Year." *The Wall Street Journal*, January 20, 1998, B1.
[3] Ibid.
[4] G. Jaffe. "Corporate Carrots, Sticks Cut Health Bill." *The Wall Street Journal*, February 3, 1998, B10.
[5] Ibid.
[6] "Healthcare," op.cit.
[7] Ibid.
8 K. Jacobs. "On Your Own." *The Wall Street Journal*, October 23, 1997, R10.
[9] "Healthcare," op,cit.
[10] R. Zaldivar. "Medicaid Offers Coverage That Millions Don't Take." *Tallahassee Democrat*, February 16, 1997, 3A.
[11] Ibid.
[12] Ibid.
[13] K. Cheney. "Why Middle Class Kids Are Losing Out." *Money*, January 1998, 96–104.
[14] Ibid.
[15] N. Jefrey. "Easing the Financial Pains of Childbirth Requires

Planning by Expectant Parents." *The Wall Street Journal*, January 30, 1998, C1.

[16] Ibid.

[17] Ibid.

[18] N. Jefrey. "Say Aah," *The Wall Street Journal*, December 1, 1997, R10.

[19] B. Cottrell. "Chiles' Budget Repeals State Pay Raises." *Tallahassee Democrat*, January 13, 1998, 8A–9A.

[21] J. Downes & J. Goodman, *Dictionary of Finance and Investment Terms, Fourth Edition*, (New York: Barron's, 1995), 140.

[20] Jacobs, op.cit.

[22] *About Long-Term Care*, (NY: MetLife, 1998), 1–8.

[23] *About Health Insurance*, (NY: MetLife, 1998), 1–11.

[24] F. Whitsitt. "Who Will Change Your Diaper?" *The Wall Street Journal*, May 29, 1997.

[25] J. Quinn. "Think Hard Before Buying Insurance for Long-Term Care." *Tallahassee Democrat*, Feb. 1, 1998, 1C.

[26] C. Adams. "Long-Term Protection." *Family Money*, Spring 1998, 50.

[27] Quinn, op.cit.

[28] Ibid.

[29] "What to Expect from Medicaid." *Consumer Reports*, October 1997, 36–38.

[30] "A Special Background Report on Trends in Industry and Finance." *The Wall Street Journal*, December 3, 1998, C1.

[31] Ibid.

[32] "Self-Employed Workers," *The Wall Street Journal*, October 28, 1998, A1.

Chapter 10 References

[1] Florida Life, Health and Variable Annuity Study Manual, 12th edition, (Tallahassee: Florida Department of Insurance).

[2] T. Lauricella. "Just in Case." The Wall Street Journal, Dec. 1, 1997, R8.

[3] K. Morris & A. Siegel, Guide to Understanding Personal Finance, (New York: Lightbulb Press, 1997).

[4] L. Scism. "Despite Strides Made Against Churning, Caution Is Key in Purchasing Decisions," The Wall Street Journal, January 20, 1998, C1 & C25.

[5] C. Dickens. "A Christmas Carol." A Charles Dickens Christmas, (New York: Oxford University Press, 1976), 3–4.

[6] E. Tyson, Personal Finance for Dummies, 2nd Edition, (Foster City, CA: IDG Books 1999), 346.

[7] M. Dorfman, Introduction to Risk Management and Insurance, 5th Edition, (Englewood Cliffs, NJ: Prentice Hall 1994), 293.

[8] Tyson, op.cit.

[9] Dorfman, op.cit.

[10] Morris & Siegel, op.cit.

[11] Lauricella, op.cit, R21.

[12] Ibid.

[13] Tyson, op.cit, 346.

[14] Scism, op.cit., C1.

[15] A. Landers. "Advice." Tallahassee Democrat, January 27, 1998, 6B.

[16] D. Folsom, Assets Unknown: How to Find Money You Didn't Know You Had, 2nd edition (Billings, Montana: 2 Dot Press, 1996).

Chapter 11 References

[1] J. Clements, "Sticking to an Investment Plan Isn't Easy." *The Wall Street Journal*, November 3, 1998, C1.

[2] "Freshmen Have Investment Plans." *Tallahassee Democrat*, March 18, 1998, 18D.

[3] J. Downes & J. Goodman, *Dictionary of Finance and Investment Terms, 4th edition*, (NY: Barron's, 1995).

[4] Ibid.

[5] Clements, "Sticking....."

[6] C. Jones, *Investments: Analysis and Management, 4th edition*, (New York: Wiley, 1994).

[7] J. Clements. "How Many Ways Are Thee Invested? Let Us Count(and Appraise) the Ways." *The Wall Street Journal*, February 17, 1998, C1.

[8] R. Zaldivar. "Social Security Changes Could Be a Tough Remedy to Swallow." *Tallahassee Democrat*, April 20, 1998, A1.

[9] Ibid.

[10] Clements, February 17, 1998, Ibid.

[11] Ibid.

[12] J. Quinn. "Avoiding All the Trouble in Beardstown." *Tallahassee Democrat*, April 19, 1998, C1.

[13] Downes & Goodman, op. cit., 538.

[14] J. Clements, "When to Trim the Risk in Your Portfolio," *The Wall Street Journal*, Febuary 11, 1997, C1.

[15] R. Goldsmith, E. Goldsmith, and J. Heaney, "Sex Differences in Financial Knowledge: A Replication and Extension," *Psychological Reports, 81*, 1169–1170.

[16] E. Goldsmith & R. Goldsmith. "Gender Differences in Perceived and Real Knowledge of Financial Investments." *Psychological Reports, 80*, 1997, 1–3.

[17] C. Markovich and S. DeVaney. "College Seniors' Personal Finance Knowledge and Practices," *Journal of Family and Consumer Sciences*, Fall 1997, 61–65.

[18] A. Baldwin. "Why Women Need to Learn to Manage Their Money." *Tallahassee Democrat*, April 19, 1998, C1.

[19] Are Women Investors Different? (Boston, MA: Putnam Investments, 1997).

[20] Ibid.

[21] Baldwin, op. cit.

[22] Ibid.

[23] Ibid.

[24] Ibid.

[25] H. McClane, *The Investors Manual*, (Madison Heights, MI: National Association of Investors Corporation).

[26] T. Stanley and W. Danko, *The Millionaire Next Door*, (Atlanta, GA: Long Street Press 1998), 131.

Chapter 12 References

[1] K. Swisher. "Yahoo! Finds Way to Top Spot in '97." *The Wall Street Jo*urnal, Feb. 26, 1998, R1.

[2] L. Lee. "Booby Prize Goes to Boston Chicken." *The Wall Street Journal*, Feb. 26, 1998, R1.

[3] C. Wiegold. "Don't Forget Dividends When Judging Returns." *The Wall Street Journal*, Feb. 26, 1998, R15.

[4] J. Downes and J. Goodman, *Dictionary of Finance and Investment Terms*, (Woodbury, NY: Barron's, 1995).
[5]Ibid.
[6]R. Wurman, A. Siegal, and K. Morris, *The Wall Street Journal Guide to Understanding Money & Markets*, (NY: ACCESS-PRESS Ltd., 1990).
[7]S. McGee. "Continued Boom in U.S. Economy Upsets Timing on Cyclical Stocks." *The Wall Street Journal*, March 30, 1998, C1.
[8]R. Lowenstein. "Charles Dow's Vision Was Well Above Average." *The Wall Street Journal*, January 8, 1996, C1.
[9] "Industry Rankings of the S&P 500." *Business Week*, March 30, 1998, 123.
[10]R. O'Brien. "Blue Chips Post 2nd-Straight Loss; Texaco, Chevron, Exxon Advance." *The Wall Street Journal*, March 27, 1998, C2.
[11] Wurman, Siegal, and Morris, op.cit.
[12] Downes and Goodman, op.cit.
[13] Ibid.
[14] D. Lohse "NASDAQ and Amex Boards Approve a Merger Agreement." *The Wall Street Journal*, March 19, 1998, C1.
[15] Wurman, Siegal, and Morris, op.cit.
[16] G. Rosenberg "The Stock Trap" *Working Woman*, April 1998, 61.
[17] "Mousing Around the Market." *U.S. News & World Report*, December 9, 1996, 112–114.
[18] Ibid.
[19] "The Perils of Buying on the Buzz." *U.S. News & World Report*, December 22, 1997, 114.

Chapter 13 REFERENCES

[1]K. Hays. "Are you overstocked?" *Working Woman*, June 1998, 78 & 80.
[2]C. Farrell. "Why Bonds Should Get More Respect." *Business Week*, November 13, 1995, 126.
[3]J. Clements "A Tip for Longer-Term Investors: Choosing Bonds Is Shortsighted." *The Wall Street Journal*, May 30, 1995.
[4]Ibid.
[5]G. Zuckerman. "Electronic Trading in Bond Market Is Slow to Catch On." *The Wall Street Journal*, June 3, 1998, C1.
[6]J. Downes and J. Goodman, *Dictionary of Finance and Investment Terms*, 4th Ed., (NY: Barrons, 1995), 116.
[7]K. Morris and A. Siegel, *The Wall Street Journal Guide to Understanding Personal Finance*, (NY: Prentice Hall, 1992).
[8]Ibid.
[9]K. Kristof. "Municipal Bonds." *Family Money*, Summer 1998, 24–29.
[10]J. Downes and J. Goodman, op.cit., 284.
[11]T. Gutner. "Junk Bonds Have Grown Up." *Business Week*, May 18, 1998, 202.
[12]Ibid.
[13]Ibid.
[14]T. Gutner. "Bonds." *Business Week*, Feb. 23, 1998, 129.

Chapter 14 References

[1]R. McGough. "Bull Milestone: Mutual Funds Cross the $5 Trillion Mark." *The Wall Street Journal*, May 29, 1998, C1.
[2]J. Clements. "A Failure to Diversify Stock Investments Could Prove Disastrous in the Long Run." *The Wall Street Journal*, C1.
[3]J. Clements. "It Pays for You to Take Care of Asset Allocation Needs Before Latching Onto Funds." *The Wall Street Journal*, April 6, 1998, R27.
[4]Ibid.
[5]Ibid.
[6]Ibid.
[7]P. Tam. "Getting Back After Missing the Bull Market." *The Wall Street Journal*, May 29, 1998, C1.
[8]J. Clements. "For Graduates Joining the Work Force: Some Ideas for Building Your Nest Egg." *The Wall Street Journal*, May 26, 1998, C1.
[9]J. Downes and J. E. Goodman, *Dictionary of Finance and Investment Terms*, 4th ed., (NY: Barrons, 1995), 460.
[10]*Business Week*, Sept. 22, 1997, 86.
[11]D. Kirkpatrick and E. S. Browning. "REITs May Be About to Cool Off Now, After Luring Investors via Big Returns." *The Wall Street Journal*, Feb. 18, 1998, C1.
[12]Downes and Goodman, 297.
[13]T. Ewing. "Gold Coins Haven't Lost Their Luster." *The Wall Street Journal*, September 25, 1998, C1.
[14]Ibid.
[15]Ibid.
[16]C. Brown. "Tarnished Silver." *The Wall Street Journal*, May 8, 1998, W12.
[17]B. Boyajian, E-mail communication, October 7, 1998.
[18]"Eight Collecting Mistakes." *The Wall Street Journal*, April 17, 1998, W14.
[19]C. Brown, op.cit.
[20]R. Gibson. "Bear Market? Some Worry That Beanies Are Ripe for a Fall. *The Wall Street Journal*, September 25, 1998, A1.
[21]K. Bensigner. "Racing for Mint-Condition Toys." *The Wall Street Journal*, Sept. 25, 1998, C1.
[22]R. Gibson, op.cit.
[23]J. Quinn. "Web's Crawling with Unsavory Financial Scams." *Tallahassee Democrat*, July 26, 1998, E1.
[24]Downes and Goodman, 460.

Chapter 15 References

[1]*Florida Retirement System Bulletin*, April 1997, 1.
[2]S. Warner. "Women Are Far Behind Men in Investing for Retirement." *Tallahassee Democrat*, October 25, 1998, 4E.
[3] U.S. Department of Labor, Washington: DC.. Top 10 Ways to Beat the Clock and Prepare for Retirement.
[4]"Retirement Planning." *Business Week*, Summer 1998, Special Advertising Section.
[5]*Business Week*, July 21, 1997.
[6]Ibid.
[7]U.S. Department of Labor, op.cit., and Retirement Planning, a publication of MetLife.
[8]J. Downes and J.E. Goodman, *Dictionary of Finance and Investment Terms*, 4th ed. (NY: Barrons, 1995), 486–487.
[9]R. King. "Insurers Face 'Elder Abuse' Charge in Mortgage Suit." *The Wall Street Journal*, October 15, 1998, B1.
[10]Employment Benefit Research Institute Issue Brief, January 1997.
[11]Downes and Goodman, op.cit., 131.

[12]Ibid.
[13]L. Pullman. "Get the Most From Your 401(k)." *Family Money*, Fall 1998, 64.
[14]E. Ambrose. "Be Aware of Fees in Your 401(k) Plan." *Tallahassee Democrat*, Feb. 4, 1996, 10D.
[15]Viewpoints, Vol. 14, No. 1, (Houston, TX: VALIC), 1.
[16] L. Asinof. "Oops...How a Variety of Basic Foul-Ups Are Bedeviling the Beneficiaries of IRAs." *The Wall Street Journal*, March 29, 1999, C1.
[17]K. Hube. "IRS Rules to Close Loophole in Roth IRA. "*The Wall Street Journal*, October 21, 1998, C1.

Chapter 16 References

[1] J. Downes and J.E. Goodman, *Dictionary of Finance and Investment Terms*, 4th ed., (NY: Barrons, 1995), 166–167.
[2] Ibid., 441.
[3] Ibid., 153.
[4] Ibid., 621.
[5] Ibid., 167.
[6] Ibid., 167.
[7] Ibid., 219.

Glossary

125 loans Home equity loans that allow the homeowner up to 125 percent of the home's value.

12b-1 fee Mutual fund charge to shareholders for the fund's promotion expenses and maintenance.

401(k) plan Retirement plan offered by for-profit company employers.

403(b) plan Retirement plan offered by non-profit organization employers.

457 plan Retirement plan in which employees make voluntary contributions into a tax-deferred account which may or may not be matched by employers.

Ability-to-pay philosophy Those with higher incomes pay proportionally more taxes than those with lower incomes.

Abstract Written history of property's ownership.

Accidental death and dismemberment (AD&D) Policies that provide benefits for accidental death, paralysis, loss of limb, speech, hearing, or eyesight.

Account executives (stockbrokers) Employees of brokerage firms who advise and handle orders for clients and perform other financial services.

Actuarial risk Risk an insurance underwriter takes in exchange for premiums.

Adjusted gross income Income a taxpayer earns from salary and other forms of income less any adjustments.

Affinity cards Credit cards offered by institutions and organizations that lend their names for a percentage return.

Age of majority Age when young adults can access custodial accounts in their name.

Agent Represents the insurer and acts on the insurer's behalf.

Agreements Description of extent of coverage.

Alimony Court ordered support to a former spouse.

Amortization Process of paying off a loan through a series of payments.

Annual percentage rate Interest rate paid per dollar per year for credit.

Annual percentage yield Amount of interest earned on a yearly basis expressed as a percentage.

Annual report Written yearly report of a company's financial condition that is distributed to stockholders or mutual fund holders.

Annuity Contract in which the insurer promises the insured a series of periodic payments.

Appraisal Estimate of home's current value.

Appreciation Increase in value.

Assessed value A percentage of fair market value.

Asset allocation Target mix of different types of investments.

Asset management account Multi-purpose account for handling cash, checking, credit, and investments.

Assets What is owned.

Attitudes Likes and dislikes.

Au pairs Young Europeans who live-in and provide child care.

Automated teller machines Computerized, automated banking.

Average daily balance method A way finance charges are determined, there are four variations.

Average propensity to consume Percentage of each dollar of income that an individual spends, on average, for current consumption.

Average tax rate Amount of taxes paid divided by income.

Balance sheets or **net worth statements** List the value of assets and liabilities.

Bank credit cards Versatile cards that can be used almost anywhere.

Basic liquidity ratio Length of time an individual or a household can meet expenses based on liquid assets.

Basic medical insurance Covers hospitalization and outpatient treatment.

Bear market The market is down.

Beneficiary Person identified in a will to receive property.

Beneficiary Recipient of policy proceeds if insured dies.

Benefits-received philosophy Those who receive benefits should pay taxes for them.

Beta coefficient Numerical measure of a stock's volatility.

Bill presentment Bills sent and paid for electronically.

Binder or **earnest money** Security deposit by the buyer as evidence of a serious offer.

Blue chip stocks Stocks of nationally known companies with quality reputations, long records of profit growth, and which usually pay dividends.

Bondholders Owners of bonds.

Bonds Investments involving lending money to organizations.

Book value Numerical measure of a company derived by deducting liabilities from a corporation's assets and dividing the remainder by the number of outstanding shares of common stock.

Broad named perils Extended perils coverage.

Budgets Trace the flow of income and expenses over a period of time.

Bull market The market is up.

Buyer's agent Represents the buyer.

Cafeteria-style benefits Allow for employee choice of benefits.

Call Gives the holder the right to purchase.

Callable bonds Type that allows corporations to call in or buy back outstanding bonds.

Callable preferred stock Clause that allows for the corporation to recall the stock.

Capital gains Income that results from an increase in the value of an investment

Capital gains Taxes on the profit from the sale of an asset.

Capital losses Lost income resulting from a decrease in an investment's value.

Cash value Accumulated savings that provide a refund to the owner of the policy if the contract is terminated prior to death.

Cashier's checks A guaranteed check purchased by a customer drawn on a bank's general funds.

Certificate of deposit Fixed-time deposit account.

Certified checks Personal checks on which the bank has guaranteed payment.

Charitable remainder annuity trust A type of irrevocable trust that transfers assets from individuals to charities or other organizations.

Child support gap Difference between what the parent receives and how much it costs to raise a child.

Churning Practice of encouraging insureds to switch policies often to generate commissions.

Closed-end mutual funds Less common type, limited number of shares offered by an investment company.

Closing Meeting in which real estate is transferred from seller to buyer.

Codicil Legal document that amends a will.

Coinsurance Sharing of an insurance risk between two or more insurers.

Collateral Property acceptable to secure a loan

Collision Violent striking of an auto by another object.

Commercial property Real estate that is income producing.

Common stock Shares or units of ownership in a public corporation.

Compound interest Calculation of interest on interest.

Conditions Obligations imposed on policy holder and insurer.

Condominiums The homeowner has title to his or her unit and jointly owns common areas.

Consumer price index Widely used measure of average changes in prices.

Consumption Using up of goods and services.

Contingent deferred sales load or **back-end load** Fee charged to fund holders if they sell prior to five years of ownership.

Continuous compounding Compounding of interest continuously during the day.

Contractual plans or **front-end loads** Plans in which sales charges are paid up front and the mutual funds are regularly bought over ten to fifteen years.

Convertibility A provision that allows the conversion of a term policy to whole life.

Convertible bonds Bonds that can be exchanged for stock.

Cooperative housing Units owned by the co-op, not the home-owner who owns shares in the building.

Copayment The portion of the bill that the patient pays.

Cosigner Person who agrees to repay a loan if the borrower defaults.

Coupon bonds Bonds issued with detachable coupons that are presented to a paying agent or issuer for payments.

Coupon rate Interest rate expressed as an annual percentage of face value.

Covered options Contract backed by shares of underlying option.

Credit Receiving goods, money, and services based on an agreement between the lender and the borrower.

Credit bureau Reporting agency that collects, stores and sells credit information to potential lenders.

Credit cards Cards used to purchase something or to get cash now with the promise of future payment.

Credit life insurance Policy used to repay debts.

Credit statement Monthly summary of credit transactions.

Current income Money received from an investment.

Current yield Rate of return determined by dividing a bond's annual income amount by its current market value.

Custodial accounts Accounts created for minors usually at brokerage firms, banks, or mutual funds.

Cyclical stocks React quickly to changes in the economy.

Debenture Debt obligation backed by the borrower and documented by an indenture, a formal agreement.

Debit cards Cards that deduct purchase amounts from checking or savings accounts.

Debt safety ratio Proportion of monthly consumer credit obligations to monthly take-home pay.

Debt to income ratio Measures financial health by dividing loan and debt payments by income.

Debt to total assets ratio Liabilities divided by assets.

Declarations Description of what is insured.

Decreasing term Policy has decreasing face value but premium remains the same.

Deductible Amount policy holder pays toward a loss before insurance coverage begins.

Deed Document that transfers property ownership.

Default Failure of debtor to make payments of interest and principal when due.

Defaults Unpaid loans.

Defensive stocks Provide consistent dividends and remain stable regardless of changes in the economy.

Defined benefit pension plan Pension plan that pays a specified amount to each employee who retires after a set number of years of service.

Defined-contribution pension plan or individual account plan Pension plan in which the amount of contributions is fixed at a certain level, while benefits vary depending on return on investments.

Demand Amount of money borrowers are willing to pay.

Dependent-care assistance plan Employer sponsored plan to reimburse parents for child care costs.

Depreciation Loss of the vehicle's value due to time and use.

Depression Downward trend, high unemployment, growth at standstill.

Deregulation Fewer controls, opening up of competition.

Direct deposit Paycheck directly deposited into accounts.

Direct investment plans Investors buy stock directly from a corporation.

Disability income insurance Pays benefits to policyholders when they become incapable of working.

Discretionary order Lets the account executive decide when to buy a stock or bond.

Diversify Putting money into different types of investments.

Dividend reinvestment plans (DRIPs) Automatic reinvestment of shareholder dividends to buy additional stock.

Dividends Distributions of money from a corporation or government to investors.

Dollar-cost averaging Systematic investing of equal sums of money at regular intervals regardless of price fluctuations in an investment.

Double or triple-indemnity clause A clause that doubles or triples the amount of insurance proceeds if the insured dies in certain types of accidents.

Dow Jones Industrial Average (DJIA) Market index of the movement of 30 leading stocks.

Drawee Financial institution with the account.

Drawer or payer Person who writes a check.

Durable power of attorney Document that addresses the financial affairs on behalf of a severely incapacitated individual.

E-banking Electronic access to cash and accounts.

Earnings per share Numerical measure of the portion of a company's profit allocated to each outstanding share of common stock.

Economics The study of the economy, how wealth is created and distributed.

Economy The economic system of a country or region.

Education IRA Nondeductible account that has tax-free withdrawals for children's higher education expenses.

Effective rate Actual rate of interest including compounding.

Efficient market theory or random walk theory Assumes that it is impossible to beat the market, that it is futile to forecast market movements.

Electronic funds transfer Making withdrawals, paying bills, depositing money or in some other way moving money electronically.

Emergency fund Fund kept in safe, readily available assets in the event of a financial crisis.

Employee stock ownership programs or stock options Program in which employees can buy stock in the company they work for at a reduced price or with matching funds from employers.

Endorsements Amend or add to a policy.

Endowment insurance Policy offers protection for a limited time with a savings feature.

Escrow Earnest money held by the mortgage lender until closing.

Estate What is owned at the time of death less what is owed.

Estate planning The orderly handling, disposition, and administration of an estate during a lifetime and at death.

Estate tax Tax imposed on the value of a deceased person's net worth.

Estate taxes Federal and state taxes assessed on an estate's value and paid for by the estate.

Excise taxes Taxes on specific goods and services.

Exclusions Items insurer will not pay for.

Executor (executrix) or personal representative Person responsible for carrying out the directions and requests in a will.

Expansion Growing economic activity, low unemployment rate.

Expected returns Anticipated future returns.

Exposures Sources of risks.

Face amount Stated sum of money on a policy.

Fair market Current worth.

Fair market value What a willing buyer would pay a willing seller for property.

Family household Two or more related persons, one of whom owns or rents the living quarters.

Family income policies Combine whole life with decreasing term insurance typically for the main breadwinner.

Family policies Combine whole life with decreasing terms for all family members.

Federal Deposit Insurance Corporation Government insurance of bank and s&l accounts.

Federal Reserve System Regulates U.S. monetary system including maintaining an adequate money supply.

Filing status Marital or household status that affects tax rates.

Financial assets Intangibles or paper assets such as savings and securities.

Financial institutions Multi-purpose institutions offering banking and other financial services.

Financial statement Assessment of current status of finances.

First and second mortgages Purchasers of these provide a mortgage on a property.

Fixed expenses The same amount each time period.

Flat tax Charges the same percentage of tax to everyone regardless of income.

Formal wills Properly drafted, signed, and witnessed wills usually prepared by an attorney.

Fundamental theory Assumes that a stock's real value is determined by a company's future earnings.

Future value The value in the future of a current amount of money or asset.

Gift tax Federal tax a giver of a sizable gift may have to pay.

Gift taxes Graduated taxes levied on the donor of a gift by the federal government and most state governments.

Goals End results.

Grace period Amount of time to pay for purchases without incurring finance charges.

Grace period Time allowed for overdue payment without penalty or the policy lapsing.

Gross domestic product Market value of goods and services produced in one year.

Growth stocks Earn profits above average for all firms and expected to prosper.

Health maintenance organizations (HMOs) Type of managed care that provides all one's health care for a fixed monthly fee.

Health-care proxy or health-care power of attorney Document that permits the signer to name individuals to make health-care decisions on his or her behalf.

Heirs Persons entitled to receive property from an estate.

Holographic wills Handwritten wills without the advice of an attorney.

Home equity loans Borrowing against the equity in homes.

Home warranty Provides additional protection to the buyer.

Homeowner's fees Monthly fees paid by home owners.

Homestead exemptions Reductions in property taxes.

Income stocks Pay higher than average dividends.

Income taxes Taxes on income, earned and unearned.

Index funds Mutual fund that has holdings matching that of a broad-based portfolio such as the S & P 500 Index.

Index of Leading Economic Indicators Composite index averaging 11 components of growth.

Individual retirement accounts (IRAs) Tax-deferred retirement accounts, not necessarily tied to an employer.

Inflation Rise in prices.

Inheritance tax Taxes paid by beneficiaries of an estate.

Inheritance taxes Taxes levied in sixteen states paid for by the heirs.

Insolvent Person owes more than they own.

Installment credit An agreement that requires a fixed number of regular payments of principal and interest.

Insurance Financial arrangement between individuals and insurers to provide protection from loss or injury.

Interest The cost of using money.

Interest rates The cost of borrowing money.

Inventory Location list of financial and legal records.

Investment Commitment of funds to achievement of long-term goals.

Investment clubs Groups that form partnerships to invest in stocks and to learn about investing.

Investment grade Bonds rated BBB or above.

Investment risk Uncertainty over an investment's actual return.

Investment strategy Plan to allocate assets among diverse investments.

Invoice price Car cost to dealer.

Irrevocable living trusts Trusts that cannot be changed without the agreement of the beneficiary.

Joint and survivorship annuity Contract sold by life insurance companies that provides payments until the death of the last designated survivor.

Joint tenants with rights of survivorship Arrangement whereby usually two people own property together, and if one dies, the property automatically goes to the other.

Junk bonds or high yield bonds Bonds with ratings of BB or lower, volatile.

Keogh plan Tax-deferred pension account for employees of unincorporated businesses and self-employed individuals.

Kiddie-tax Tax filed by parents for the investment income of children under age 14.

Laddering Buying issues of varying maturities so that bonds expire at different intervals.

Lapsed policy Termination of a policy because of non-payment.

Large cap stocks Companies with market capitalization of $1 billion or more.

Lease Legal document between renter and landlord.

Leasing Contractual agreement outlining monthly payments, security deposit, and terms of lease.

Letter of last instruction or letter precatory Signed letter directing the executor to documents and funeral and burial instructions.

Level of living Current state of living.

Level-premium term or guaranteed-premium Policy with constant face values but premiums increase with each renewal.

Liabilities What is owed.

Liability Claim on the assets of a business, organization, or individual.

Liability insurance Pays for losses from negligence resulting in bodily injury or property damage to others for which the policy holder is responsible.

Liens Legal rights to take and hold property or sell it for repayment of a loan.

Life insurance Contract between an insurer and policyholder specifying a sum to be paid to a beneficiary on the insured's death.

Limit order Request to buy or sell a stock at a specified price.

Limited named perils Usual basic perils covered by insurance.

Liquidity How readily something can be turned into cash.

Living trust Trust established between living persons, most typically between a parent and a child.

Living will Document that lets family members know what type of care a person wants or does not want.

Load funds Mutual funds sold that charge fees.

Loans Sums of money lent at interest.

Long term care insurance (LTC) Policies that provide for nursing home or in-home care not covered by Medicare or Medigap.

Low-load funds Mutual funds that charge less than 3 percent of the purchase price.

Major medical insurance Covers large medical expenses.

Managed care A prepaid plan that combines financing and delivery of health care.

Manufactured housing Units assembled at a factory and moved to a building site.

Margin Amount an investor borrows when buying a stock.

Marginal tax rate Tax rate applied to the last dollar of earnings.

Market order Request to buy or sell a stock at the best price available currently.

Market value Price determined by net asset value, yield, and investor's expectations of future earnings.

Marriage penalty Effects of a tax code that makes married people pay more than two single people.

Maturity A designated future date.

Maturity or redemption date When the borrower must repay the maturity value of a bond.

Medicaid Federal health care program for low income people.

Medicare Federal health insurance program for elderly and many people with disabilities.

Medicare supplement insurance or medigap Covers medical expenses, services, and supplies not paid for by Medicare.

Mid cap stocks Companies with market capitalization

Money market deposit account Type of account that offers a higher rate of interest than most checking, savings, or NOW accounts.

Money market mutual fund A money market account that is in a mutual fund.

Money orders Types of checks purchased for use instead of personal checks.

Mortgages Loans to purchase real estate.

Municipal bonds (munis) Bonds issued by cities, state, or local governments for improvements.

Mutual funds Groups of stocks, bonds, or other securities managed by an investment company.

NASDAQ Index Market index of fast moving technology and financial services stocks.

National Foundation for Consumer Credit Non-profit organization that provides credit and debt counseling.

Needs What you must have.

Net worth Assets minus liabilities.

No-fault insurance After an auto accident, each party collects from his or her insurer.

No-load funds Mutual funds that have no sales charge.

No-load funds Mutual funds that impose no sales charge on shareholders.

Nominal rate of interest Stated annual rate of interest.

Nonfamily household Householders who either live alone or with others to whom they are not related.

Noninstallment credit Rare type of agreement in which a single payment is due at the end of the loan period.

Nonprobate property Property that does not go through probate.

NOW account Type of checking account that pays interest or dividends.

Nuncupative wills Oral wills.

Odd lots Securities traded for less than 100 shares.

Open-end mutual funds Most typical kind of mutual fund that are issued and redeemed by an investment company.

Open-ended credit Extended in advance of transactions, borrowers are pre-approved up to a limit.

Open-enrollment periods Times when insurance companies or employers offer the opportunity to add on to existing policies.

Opportunity costs The cost of what is given up.

Option premium Amount of money the seller receives from the buyer.

Options Securities transactions agreements to have the option to buy or sell stocks or other types of investments.

Overdrafts Checks written against insufficient funds.

Par value Face value of a security.

Participation certificates Equity investments in a pool of mortgages.

Passbook savings account A savings book record of deposits and withdrawals in an account.

Payee Person or company for whom a check is written.

Perils Events that cause financial loss.

Perquisites Job extras beyond usual benefits.

Personal auto policy Legal liability arising out of the ownership or operation of a covered auto.

Personal bankruptcy Legal recourse open to insolvent debtors.

Personal finance The study of how people spend, save, invest, and protect their financial resources.

Personal Financial Planning The process of managing finances to reach goals and to provide satisfaction.

Personal Loans The borrowing of cash that is repaid.

Personal risk management Process of identifying and evaluating risk faced by individuals and families.

Personalized identification number Number assigned to customers for use at ATMs and for other forms of cash management.

PITI Principal, interest, taxes, and insurance.

Point of service organization (POSs) Type of managed care that combines traditional insurance with HMOs.

Points Fees paid to the lender at the closing.

Policy Contract between individual or group and insurer.

Pooling Spreading of loss of a few over an entire group.

Portfolio Combined holdings to increase diversification and to reduce risk.

Preferred provider organizations (PPOs) Type of managed care that provides health-care services to employees at a discount.

Preferred risks People whom insurance companies prefer to insure.

Preferred stock Shares that pay dividends and have precedence over common stock.

Premiums Payments to an insurer.

Prenuptial agreement Legal contract between a future husband and wife about the splitting of assets in the event of a divorce.

Prepaid tuition plans or college savings plans State-offered tuition savings plans.

Present value An asset's current value.

Price-earnings ratio Numerical measure of the relationship between the price of one share of stock and the annual earnings of the company.

Primary market The market for purchasing new issues of securities.

Prime rate Rate that banks use to determine loans.

Principal Initial deposit or investment.

Probate Judicial process whereby the will of a deceased person is presented at court and an executor is appointed.

Progressive taxes Those earning more, pay more.

Property insurance Pays for losses to homes and personal property due to a number of causes.

Prospectus Written summary of a commercial venture or offering.

Proxy Written statement assigning shareholder votes to another person who votes for them.

Purchasing power Amount of goods and services an individual's money can buy.

Pure risk Risk that has a threat of loss without possibility of gain.

Put Gives the holder the right to sell.

Real estate Piece of land and all physical property related to it.

Real estate agent or broker Offers services to buyers and sellers of property.

Real estate investment trusts (REITs) Companies with many owners that manage and develop properties.

Real estate syndicates or real estate limited partnerships Groups of investors form partnerships to invest in real estate.

Realized return Actual return on investments.

Rebalancing Adjusting asset allocation.

Recession Temporary slowing of the economy.

Recordkeeping Process of recording sources and amounts of money, earned, saved, spent, and invested over a period of time.

Recovery Hopeful stage, unemployment lessening, economy moving upward.

Refinance Revising a payment schedule to reduce monthly payments and to modify interest charges.

Registered bonds Bonds recorded in name of bondholder.

Regressive taxes Approach that takes a smaller percentage of income as income rises.

Renewable A clause that allows for renewing the policy.

Rent Payment for use of property.

Residuary clause Catchall clause in a will that distributes the remainder of an estate.

Resources What individuals use to get what they want and to reach goals.

Retirement Permanent withdrawal from gainful employment.

Retirement income gap Shortfall between the amounts of retirement income generated and retirement living needs.

Return Total income from an investment.

Reverse mortgages Arrangement wherein owner borrows against home equity.

Revocable living trusts Trusts that allow the person creating the trust to change terms within it or cancel it.

Right of rescission The right to cancel a contract or agreement after it is signed.

Risk The possibility of experiencing harm, suffering, danger, or loss.

Risk aversion Avoidance of risk.

Roll over When a CD or other form of investment moves rom one investment to another.

Roth IRA Type of individual retirement account started in 1998 that differs from a traditional IRA.

Round lots Transactions of 100 shares or multiples of 100 shares or $1,000 or $5,000 par value on bonds.

Russell 2000 small stock index Measure of the stock price performance of small companies.

Secondary market Previously owned shares of stock that are resold on this market.

Secured bonds Bonds backed by collateral, mortgages, or other liens.

Securities exchange The marketplace where member brokers represent investors who want to buy and sell.

Security deposit Payment required in advance to cover wear on unit.

Selling short A technique of selling a securities or commodities contract that was borrowed.

Shilling Practice of sellers driving up bids by making their own bids.

Simple interest Interest on original sum of money.

Simple IRA Individual retirement account that allows workers mostly of small businesses to contribute to with an employer match.

Simplified employee pension plans (SEPs) Pension plans in which the employee and the employer contribute to an IRA, can be used by self-employed individuals.

Single limit Auto insurance pays up to a stated amount.

Small cap stocks Companies with market capitalization of $500 million or less.

Smartcards Cards embedded with computer chips for a prepaid amount of money.

Solvency Ability to pay debts.

Solvency ratio Extent to which an individual is exposed to insolvency.

Specified disease policies Provide medical benefits for a specific disease such as cancer.

Speculative risk Risk that has the possibility of gain and loss.

Speculative stocks Risky, usually newer stocks invested in with the anticipation of gain and the realization of possible loss.

Spenddown Process by which medical and nursing home care reduces a person's assets.

Split limit Auto insurance coverage is split between three categories.

Standard & Poor's Index (S & P 500) Market index of the movement of 500 stocks.

Standard deduction Deduction set by the government based on filing status.

Standard of living Quality of life one seeks, implies prosperity.

Statement savings account A statement of withdrawals and deposits in an account.

Sticker price Price listed on the car.

Stock split When a corporation declares that stocks will split.

Stocks Ownership in a corporation represented by shares.

Stop order A type of limit order in which a request is made to sell a stock if it falls below a specified price.

Straight life annuity Contract sold by life insurance companies that provides payments as long as a person lives.

Strike price Agreed upon price.

Sublease Property can be leased to another.

Subordinated debentures Unsecured bonds that give bondholders a claim secondary to that of designated bondholders.

Supply Amount of money lenders are willing to lend.

Survivorship or 2nd-to-die Policy that pays a death benefit at the later death of two insured individuals.

Tangible assets Physical assets such as homes or cars.

Tax audit Detailed examination by the IRS of a tax return.

Tax avoidance Legitimate methods of reducing taxes.

Tax credits Items that reduce tax liability.

Tax deduction Items that reduce taxes.

Tax evasion Illegal methods of reducing taxes.

Tax-deferred investments Investments that are not taxed until later.

Tax-exempt investments Investments that are tax-free.

Tax-free No taxes.

Tax-sheltered No taxes are charged on profits until a later date.

Taxable income Everything earned that can be taxed.

Taxes Required payments of money to governments.

Teaser rates Lower rates on signing that jump to a higher rates later.

Technical theory Assumes that trends and patterns can be found in the overall market.

Tender form Application form to purchase U.S. Treasury securities.

Term life insurance Policies that offer pure protection, no savings feature, over a specified period of time.

Testamentary trust fund Trust created by a will.

Testator Person who makes a will.

Tiered rate Interest-rate structure tied to a balance level.

Time deposits Accounts with time limits, early withdrawals are penalized.

Time horizon factor Number of years before money is needed.

Time share Sharing ownership in a unit such as an apartment, condominium or house usually for vacation purposes.

Time value of money Theory that the value derived from the use of money over time increases its total by investment and reinvestment.

Title Legal right of property ownership.

Total return Annual return on an investment, including appreciation and dividends or interest.

Transaction costs Charges for buying and selling securities.

Traveler's checks Types of checks pre-purchased for set amounts for travel.

Treasury bills (T-bills) U.S. securities sold in minimum units of $1,000 that mature in one year or less.

Treasury bonds U.S. securities that are long-term debt instruments ranging from 10 to 30 years issued in minimum denominations of $1,000.

Treasury notes U.S. securities that mature in one to 10 years and range in denomination from $1,000 to $1 million or more.

Trust Legal structure or contract typically used for asset management.

Trust fund Property itself.

Trustee Person holding title to property for the benefit of another.

Truth in savings act Federal law requiring lenders to tell customers APY and other information.

Underwriter Person who conducts a policy application review.

Underwriting Insurance company process of evaluating a policy application.

Uniform Transfer to Minors Act or Uniform Gifts to Minors Act Provide simple ways to transfer wealth to minors without a formal trust.

Uninsured motorist coverage Pays for bodily injury caused by an uninsured motorist or hit-and-run driver, or negligent driver with an insolvent insurer.

Universal life insurance Flexible-type of policy that combines term life insurance with a tax-deferred savings account.

Usury laws State laws that limit interest rate.

Values Principles that guide behavior.

Variable annuities Investments that fluctuate with those of underlying securities.

Variable expenses Expenditures over which individual has more control.

Variable life insurance Policy that allows for flexibility in types of investments.

Vertical equity People in different income groups pay different amounts of taxes.

Vesting Right an employee acquires after being employed for a set number of years.

Volunteer Income Tax Assistance Organization that helps individuals with filling out tax forms.

Wants What you would like to have.

Wealth Total value of all items owned.

Whole life insurance Policy that combines insurance with a savings feature.

Will Legal document designating transfer of property and assets in the event of death.

Wilshire 5000 equity index Broadest of all market indexes, the market value of all U.S. equities on the main stock exchanges.

Withholding Amount employers deduct from employees pay and send to the federal government for income tax purposes.

Workers' compensation Money paid if a disability is due to illness or injury from a job.

Yield Return on a capital investment, typically bonds.

Yield-to-maturity Calculation that determines the rate of return on a long-term investment such as a bond.

Zero-coupon bonds Bonds sold at a deep discount.

Index